I0887865

HANDBOOK OF HEAVY-TAILED DISTRIBUTIONS IN ASSET MANAGEMENT AND RISK MANAGEMENT

World Scientific Handbook in Financial Economics Series

(ISSN: 2010-1732)

Series Editor: William T. Ziemba
(University of British Columbia, Canada) (Emeritus) and London School of Economics, UK

Advisory Editors:

Kenneth J. Arrow
Stanford University, USA
George C. Constantinides
University of Chicago, USA
Espen Eckbo
Dartmouth College, USA
Harry M. Markowitz
University of California, USA
Robert C. Merton
Harvard University, USA
Stewart C. Myers
Massachusetts Institute of Technology, USA

The Handbooks in Financial Economics (HIFE) are intended to be a definitive source for comprehensive and accessible information in the field of finance. Each individual volume in the series presents an accurate self-contained survey of a sub-field of finance, suitable for use by finance, economics and financial engineering professors and lecturers, professional researchers, investments, pension fund and insurance portfolio mangers, risk managers, graduate students and as a teaching supplement.

The HIFE series will broadly cover various areas of finance in a multi-handbook series. The HIFE series has its own web page that include detailed information such as the introductory chapter to each volume, an abstract of each chapter and biographies of editors. The series will be promoted by the publisher at major academic meetings and through other sources. There will be links with research articles in major journals.

The goal is to have a broad group of outstanding volumes in various areas of financial economics. The evidence is that acceptance of all the books is strengthened over time and by the presence of other strong volumes. Sales, citations, royalties and recognition tend to grow over time faster than the number of volumes published.

Published

Vol. 7 *Handbook of Heavy-Tailed Distributions in Asset Management and Risk Management*
by Michele Leonardo Bianchi, Stoyan V. Stoyanov, Gian Luca Tassinari, Frank J. Fabozzi & Sergio M Focardi

Vol. 6 *Handbook of the Economics of Wine (In 2 Volumes)*
Volume 1: Prices, Finance, and Expert Opinion
Volume 2: Reputation, Regulation, and Market Organization
edited by Orley Ashenfelter, Olivier Gergaud, Karl Storchmann & William T. Ziemba

Vol. 5 *The World Scientific Handbook of Futures Markets*
edited by Anastasios G. Malliaris & William T. Ziemba

More information on this series can also be found at
http://www.worldscientific.com/series/wshfes

World Scientific Handbook in Financial Economics Series — Vol. 7

HANDBOOK OF HEAVY-TAILED DISTRIBUTIONS IN ASSET MANAGEMENT AND RISK MANAGEMENT

Michele Leonardo Bianchi
Banca d'Italia, Italy

Stoyan V Stoyanov
Stony Brook University, USA

Gian Luca Tassinari
University of Bologna, Italy

Frank J Fabozzi
EDHEC Business School, France

Sergio M Focardi
Léonard De Vinci University, France

World Scientific

NEW JERSEY • LONDON • SINGAPORE • BEIJING • SHANGHAI • HONG KONG • TAIPEI • CHENNAI • TOKYO

Published by

World Scientific Publishing Co. Pte. Ltd.
5 Toh Tuck Link, Singapore 596224
USA office: 27 Warren Street, Suite 401-402, Hackensack, NJ 07601
UK office: 57 Shelton Street, Covent Garden, London WC2H 9HE

Library of Congress Cataloging-in-Publication Data
Names: Bianchi, Michele Leonardo, author. | Stoyanov, Stoyan V., author. | Tassinari, Gian Luca, author.
Title: Handbook of heavy-tailed distributions in asset management and risk management / Michele Leonardo Bianchi (Banca d'Italia, Italy), Stoyan V Stoyanov (Stony Brook University, USA), Gian Luca Tassinari (University of Bologna, Italy), Frank J Fabozzi (EDHEC Business School, France) and Sergio M Focardi (Léonard De Vinci University, France).
Description: Mew Jersey : World Scientific, [2019] | Series: World scientific handbook in financial economics series ; Volume 7
Identifiers: LCCN 2018050278 | ISBN 9789813274914 (hc : alk. paper)
Subjects: LCSH: Asset management accounts. | Distribution (Probability theory) | Risk management.
Classification: LCC HG1660 .B53 2019 | DDC 332.01/51924--dc23
LC record available at https://lccn.loc.gov/2018050278

British Library Cataloguing-in-Publication Data
A catalogue record for this book is available from the British Library.

Copyright © 2019 by World Scientific Publishing Co. Pte. Ltd.

All rights reserved. This book, or parts thereof, may not be reproduced in any form or by any means, electronic or mechanical, including photocopying, recording or any information storage and retrieval system now known or to be invented, without written permission from the publisher.

For photocopying of material in this volume, please pay a copying fee through the Copyright Clearance Center, Inc., 222 Rosewood Drive, Danvers, MA 01923, USA. In this case permission to photocopy is not required from the publisher.

For any available supplementary material, please visit
https://www.worldscientific.com/worldscibooks/10.1142/11118#t=suppl

Desk Editors: V. Vishnu Mohan/Lum Pui Yee

Typeset by Stallion Press
Email: enquiries@stallionpress.com

MLB

To my wife Giorgia and my sons Adriano Maria and Elia

SVS

To my wife Petya and my children Ava and Nia

GLT

To my mother Anna and to the memory of my father Pietro Paolo

FJF

To my wife Donna and my children Karly, Patricia, and Francesco

SMF

To the memory of my parents Teresa and Umberto

Preface

In 1987, Per Bak, Chao Tang, and Kurt Wiesenfeld published a now-famous paper where they introduced the sandpile as a model of what was called self-organized criticalities.[1] Randomly adding grains of sands to a sandpile might produce a collapse of the sandpile when the slope reaches a certain value. The sandpile model was the first dynamic model of a system that displays self-organized criticalities. In other words, the sandpile self-organizes into a system that collapses at a random time and displays cascades whose size has a heavy-tailed distribution.

The sandpile is a much studied model. Many other models that produce heavy-tailed distributions have been theoretically studied and implemented in simulations. Models range from geometrical models such as random graphs, percolation, and network models to dynamical models such as interacting particle systems and even familiar econometric models such as ARCH and GARCH models. It was found that heavy-tailed distributions are not rare and arcane phenomena, but are almost ubiquitous in finance and economics as well as the physical sciences.

Heavy-tailed distributions form a bridge between equilibrium and non-equilibrium systems. In a dynamical system which is not in equilibrium there is typically some tension which accumulates until the

[1] P. Bak, C. Tang, and K. Wiesenfeld. Self-organized criticality: An explanation of 1/f noise. *Physical Review Letters*, 59(4):381–384, 1987.

system reaches a critical state and a discharge happens. In finance, non-equilibrium phenomena include large market swings and even market crashes.

The study of heavy-tailed distributions allows one to represent phenomena that occasionally exhibit very large deviations from the mean. The dynamics underlying these phenomena are an interesting theoretical subject but the study of their statistical properties is in itself a very useful endeavor from the point of view of managing assets and controlling risk. In this book we are primarily concerned with the statistical properties of heavy-tailed distributions and with the processes that exhibit jumps.

In this book, we offer a detailed overview with a Matlab implementation[2] of heavy-tailed models applied to asset management and risk management. It is not intended as a theoretical treatise on probability or statistics, but as a tool to understand the main concepts regarding heavy-tailed random variables and processes as applied to real-world applications in finance. Accordingly, we review approaches and methodologies whose realization will be useful for developing new methods for forecasting of financial variables where extreme events are not treated as anomalies, but as intrinsic parts of the economic processes.

In addition to these models enhancing our understanding of the behavior of financial markets, we seek to fill the gap between theoretical models and the numerical analysis needed to turn them into a useful computer code. Depending on the data at hand and on the phenomena one wants to explain, many of the proposed algorithms can be also used for applications in other fields.

The book is divided into three parts. Each chapter begins with what the reader can expect to learn from the chapter. Simple definitions and examples are then provided. We then move on to more complex examples in a systematic way. Each chapter concludes with the key points covered in the chapter.

---

[2]For more details, see the *How to run the Matlab code* section at the end of the book.

The three chapters in Part I cover the preliminaries. Specifically, after the introductory chapter, Chapters 2 and 3 review some basic concepts pertaining to random variables and stochastic processes with a focus on heavy-tailed random variables and processes with jumps.

In Part II, we review the theory and methodologies related to some heavy-tailed models. In Chapter 4, we review the main properties of generalized hyperbolic distributions. Chapter 5 focuses on the class of stable distributions as a model for investigating the properties of asset returns. The history, definitions, formulas, simulation, and estimation methods related to univariate tempered stable distributions are covered in Chapter 6. The last two chapters in Part II, Chapters 7 and 8, cover the multivariate time-changed Brownian motion, focusing on extending the multivariate framework of the univariate generalized hyperbolic and tempered stable model. Extreme value theory is the subject of Chapter 9.

In Part III, we provide applications to finance. In the first of the three chapters in this part of the book, Chapter 10, we apply Gaussian, multivariate generalized hyperbolic and multivariate normal tempered stable models to the portfolio selection problem. Chapter 11 provides an application of non-Gaussian processes to option pricing; and in Chapter 12 we apply extreme value theory to estimate tail thickness for asset return distributions.

The authors are grateful to William T. Ziemba and Svetlozar T. Rachev who encouraged our writing of this book. This book does not represent the views of the employers of any of the authors. In particular, although Michele Leonardo Bianchi coauthored this book while an employee of the Banca d'Italia, the views expressed in this book do not necessarily reflect those of the Banca d'Italia.

Michele Leonardo Bianchi
Stoyan V. Stoyanov
Gian Luca Tassinari
Frank J. Fabozzi
Sergio M. Focardi

About the Authors

Michele Leonardo Bianchi is an Advisor in the Macroprudential Analysis Division at the Regulation and Macroprudential Analysis Directorate of the Banca d'Italia. From 2008 to 2014 he worked in the Division of Risk and Financial Innovation Analysis at the Specialized Intermediaries Supervision Department of the Banca d'Italia. He earned an Italian *Laurea* in Mathematics in 2005 from the University of Pisa and completed his PhD in Computational Methods for Economic and Financial Decisions and Forecasting in 2009 from the University of Bergamo. In 2017 he received the Italian national scientific qualification (associate professorship). He has extensive experience in the analysis and statistical modeling of financial data. He coauthored a book published by Wiley in quantitative finance and authored articles on quantitative finance, financial econometrics, probability theory, and nonlinear optimization published in international scientific journals, including *Journal of Banking and Finance*, *International Journal of Theoretical and Applied Finance*, *Theory of Probability and Its Applications*, *Probability and Mathematical Statistics*, *The Econometrics Journal*, *Computational Economics*, and *The Journal of Derivatives.*

Stoyan V. Stoyanov is a Research Professor of Finance in the Center for Finance and the College of Business, Stony Brook University. Prior to joining Stony Brook University, he was a professor of finance at EDHEC Business School and head of research at EDHEC Risk

Institute-Asia. He worked for more than six years as head of quantitative research for FinAnalytica, a financial technology firm. His research focuses on probability theory, extreme risk modeling, and optimal portfolio theory. He has published more than 40 articles in leading academic and practitioner-oriented scientific journals such as *Annals of Operations Research*, *Journal of Banking and Finance*, *Quantitative Finance*, *Studies in Nonlinear Dynamics and Econometrics*, *Applied Mathematical Finance*, *Journal of Derivatives*, *International Journal of Theoretical and Applied Finance*, and *Journal of Portfolio Management*. He has contributed to many professional handbooks and co-authored several books on probability and stochastics, financial risk assessment, and portfolio optimization. He holds a PhD in Mathematical Finance from Karlsruhe Institute of Technology.

Gian Luca Tassinari is an Adjunct Professor of Mathematical Methods of Economics, Finance and Actuarial Sciences at the School of Economics, Management and Statistics of the University of Bologna. He earned a four-year Italian *Laurea* in Economics with honors in 2005 from the University of Bologna and a PhD in Computational Methods for Economics and Financial Decisions and Forecasting in 2009 from the University of Bergamo. Since then he has taught courses on mathematical finance and financial econometrics at the University of Bologna. His scientific research interests mainly concern the applications of multivariate stochastic models to finance. He has authored articles on quantitative finance published in international scientific journals, including *Quantitative Finance* and *International Journal of Theoretical and Applied Finance*.

Frank J. Fabozzi is Professor of Finance at EDHEC Business School and a Senior Scientific Adviser at EDHEC-Risk Institute. He has published numerous books in financial econometrics and operations research, including *Basics of Financial Econometrics*;

Financial Econometrics: From Basics to Advanced Modeling Techniques; *Simulation and Optimization Modeling in Finance*; *Advanced Stochastic Models, Risk Assessment, and Portfolio Optimization*; *Robust and Non-Robust Models in Statistics*; *Robust Portfolio Optimization and Management*; *The Methods of Distances in the Theory of Probability and Statistics.* His papers have appeared in the *Journal of Finance, Journal of Financial and Quantitative Analysis, Econometric Theory, The Econometrics Journal, Journal of Economic Dynamics and Control, Probability and Mathematical Statistics, Operations Research, European Journal of Operational Research,* and *Annals of Operations Research.* He earned a doctorate in economics from the Graduate Center of the City University of New York.

Sergio M. Focardi is Professor of Finance and Researcher in the Finance Group, ESILV EMLV, of the Léonard De Vinci University (Paris). He is a member of the editorial board of the *Journal of Portfolio Management* and has authored numerous articles on mathematical finance and financial econometrics. He has authored several books, including *Basics of Financial Econometrics*; *Robust Portfolio Optimization and Management*; *Quantitative Equity Investing, and Financial Econometrics: From Basics to Advanced Modeling Techniques*, and a series of monographs for the Research Foundation of The CFA Institute. His research interests include the econometrics of large equity portfolios, the theory of economic crises, and the theory of economies as complex systems. He holds a degree in electronic engineering from the University of Genoa, Italy, and a PhD in Mathematical Finance from Karlsruhe Institute of Technology.

Contents

Part I

Preliminaries

Chapter 1

Introduction

After explaining the need for modeling the tails of a distribution, we describe the univariate and multivariate framework for analyzing the tails of a distribution. We then provide the definition of extreme events. The main topics covered in this chapter are as follows:

- brief history of empirical applications of heavy tails to finance and the role of heavy tails in other fields;
- the reasons for the failure of models built on traditional economic theories and the avenues for reformulating economic theory to build better models to explain a financial crisis;
- stylized facts about the behavior of financial markets.

The first empirical application of heavy tails to finance is usually attributed to Mandelbrot (1963), who approximated the returns of cotton prices with an α-stable distribution. In the last 55 years, a large body of research dealing with the modeling of phenomena through heavy tails has appeared. Interest in these models is not limited to finance, insurance, risk management, and economic applications.[1] Heavy-tailed distributions play a major role in the analysis of many stochastic systems. For example, they are largely used to model inputs to computer and communications networks and there is statistical evidence for their appropriateness in physics, geoscience, engineering, and epidemiology. The applications range from the identification of a physically and mathematically consistent distribution type for modeling the probability of intense daily precipitation (Cavanaugh *et al.* (2015)) and natural disasters (Pisarenko and Rodkin (2010)) to the commercially uncertain success rates of motion pictures (Walls (2005)) and to the *retweets* on Twitter (Mathews *et al.* (2017)).

Some applications of heavy-tailed distribution appeared also in *Nature*. Barabasi (2005) observed that the timing of many human activities, ranging from communication to entertainment and work patterns, follow non-Poisson statistics, characterized by bursts of rapidly occurring events separated by long periods of inactivity. For example, Internet traffic is driven by two separate heavy-tailed processes: the size distribution of the files sent by users, and the human decision-driven timing of various Internet-mediated activities in which individuals engage. Heavy tails have been observed in the foraging patterns of birds as well, raising the intriguing possibility that animals also use some evolutionarily encoded priority-based queuing mechanisms to decide between competing tasks, such as caring for offspring, gathering food, or fighting off predators (see Viswanathan *et al.* (1999) and Reynolds (2018)).

As noted above, there is a large number of research papers on this topic. It is enough to try to search on Google Scholar the word *finance* combined with the following words: *heavy-tailed distributions*, *fat tails*, *non-Gaussian distributions*, *Lévy processes*, *Lévy flight*, *extreme events*, or something like these. These words are more or less synonyms. Heavy-tailed distributions and Lévy processes are different sides of the same coin. In some case (i.e., for infinitely divisible

[1] See Adler *et al.* (1998).

distributions), heavy-tailed random variables are the law of the increments of Lévy processes.

In applications to finance, researchers have studied both heavy and semi-heavy tails distributions. The tails of a semi-heavy tail distribution are lighter than those of α-stable laws, but much heavier than the Gaussian law (see Misiorek and Weron (2012)). Just looking at the main books on this topic and without considering the hundreds of articles published in dozens of scientific journals, we mention that some monographs are focused on extreme events and heavy-tailed distributions (e.g., Embrechts *et al.* (1997) and Kemp (2011)). Other works describe the applications of particular distributions, like the α-stable (e.g., Rachev and Mittnik (2000), Rachev (2003), and Rachev *et al.* (2005)) or the semi-heavy tailed distributions for explaining the behavior of stock returns and to define an option pricing framework (e.g., Cont and Tankov (2003), Schoutens (2003), McNeil *et al.* (2005), and Rachev *et al.* (2011)). Chernobai *et al.* (2008) and Peters and Shevchenko (2015) cover the modeling of operational risk. The books of Boyarchenko and Levendorskii (2002), Jondeau *et al.* (2007), and Habart-Corlosquet *et al.* (2013) put the attention on the word *non-Gaussian* distributions, even if they discuss the topic of heavy tails modeling applied to finance. Finally, Ibragimov *et al.* (2015) provide a fairly detailed introduction to the analysis and applications of heavy-tailed distributions in a number of important problems in economics, finance, risk management, and insurance.

1.1. The Need for Modeling Tails

Stock market crashes such as those in October 1987 and October 1997, the burst of the "dotcom bubble", the highly volatile period after the subprime mortgage crisis that began in the summer of 2007, during the recent European sovereign-debt crisis that began in 2009 and after the Italian political elections in 2018 have presented challenges to investors, financial institutions, regulators, and central governments. Traditional theories in financial economic theory have failed to give recognition to extreme events or, equivalently, the tails of the distribution of the financial variables of interest. It has to be pointed out that some researchers distinguish between crashes and fat-tailed price variations (or tail events). Although events are the result of endogenous self-organizing processes, they are not the same phenomenon. The largest market crashes are outliers (see Johansen and Sornette (1998)) and may differ from fat-tailed price variations. In this book, we do not disentangle crashes from tail events.

Financial economics draws upon economic theory for the formulation of models about the behavior of financial markets and econometrics to empirically examine the actual behavior of financial markets. Despite the preponderance of empirical evidence as to how real-world financial markets behave, the economic models that are the bedrock of traditional financial economic theory and risk measures have changed little over the last five decades since they were first proposed. As a result, many financial crises cannot be explained by prevailing statistical models of markets based on prevailing financial economic theories.

With the failure of such models, two avenues for reformulating financial economic theories have emerged. The first has been the recasting of the underlying assumptions about how investors behave and what is popularly referred to as "behavioral finance". The second avenue is the examination of the microstructure of financial markets to incorporate the observed behavior of such markets into models of portfolio selection and asset pricing. In early 2013, the *Journal of Econometrics* published a special issue on heavy-tailed distributions stating: "the study and use of heavy-tailed distributions has grown from a niche field to one of great relevance" (see Paolella *et al.* (2013)). In 2010, the *Journal of Empirical Finance* dedicated a special issue on heavy tails and Paretian distributions in empirical finance (see Dufour *et al.* (2010)). This "empirical", or for lack of a better term "financial econometric", approach to financial economic theory is the focus of this book.

The financial econometric approach we pursue in this book builds on the stylized facts that are accepted about the behavior of financial markets: non-normal and asymmetric distribution of asset returns and a dependence structure of multivariate return distributions that goes beyond simple linear correlation. Formulating and estimating alternative models for pricing, portfolio selection, and risk measurement that take into account these stylized facts is far from simple. It requires drawing on and extending statistical models that may not be as well known in finance and economics as they are to statisticians and mathematicians. The model proposed in this book can be further extended to capture the volatility clustering in time series (i.e., calm periods followed by highly volatile periods and vice versa), but we do not analyze these enhanced models in this book.[2]

These observations about the behavior of financial markets have led to increased efforts to improve the flexibility and statistical reliability of existing models that seek to capture the dynamics of economic variables. The common practice before the recent subprime mortgage crisis was to assume that the economy was developing without much turbulence, where extreme events were infrequent enough to be regarded as rare shocks. The essence of the methods which are going to be discussed in the chapters to follow is the application of non-normal theoretical distributions for explaining financial uncertainties, where tails and extreme events are treated as intrinsic characteristics of the model.

A considerable number of empirical studies performed by academic researchers and the financial industry confirm that in modeling volatile markets, the following phenomena should be taken into account:

- heavy-tailed distributions;
- asymmetry from both univariate and multivariate viewpoints;
- non-Gaussian dependence.

First, as discussed above, financial return series are asymmetric and heavy-tailed, where the tails are important because bad news are tail events. The normal distribution is symmetric and has tails

[2]See McNeil *et al.* (2005) and Rachev *et al.* (2011).

that are too light to match real-world market data, but semi-heavy-tailed distributions introduce heavier tails and skewness (Schoutens (2003), Cont and Tankov (2003), and Rachev *et al.* (2011)). Second, a dependence model has to be flexible enough to account for several empirical phenomena observed in the data, in particular asymmetry of dependence during severe market crashes and dependence of the tail events. The dependence model implied by the multivariate normal distribution fails to describe both phenomena — the covariance is a symmetric dependence concept and tail events are asymptotically independent. A dependence structure of multivariate distribution is needed beyond simple linear correlation (McNeil *et al.* (2005), Kring *et al.* (2009), Bianchi *et al.* (2016), Bianchi and Tassinari (2018), and Alcock and Satchell (2018)).

These well-documented findings reported for asset returns are not mere academic conclusions that hold little interest for practitioners. Rather, they have important implications for asset managers, risk managers, and regulators. Not properly accounting for these stylized facts can give rise to models that result in inferior investment performance by asset managers and disastrous financial consequences for financial institutions that rely upon them for risk management. Discussions with practitioners as to why models that can deal with these stylized facts are not used in practice suggest that it is due to the complexity of estimating the non-normal multivariate models proposed in the literature. The computational complexity is a very important aspect in the designing of front office and risk management systems, particularly for large banks having to deal with the revaluation of hundreds of positions on a daily basis, for example, for computing profit-and-loss or value-at-risk.

For the above reasons, in this book, we discuss models which take into account the stylized facts that have been observed for real-world financial markets and explain how to deal with these models in real-world applications. The purpose of this book is to provide a practical framework to model and forecast the behavior of financial variables (e.g., asset returns, volatility surfaces, derivatives prices) by paying attention to the modeling of tails. In particular, we want to present a general theory and a unifying framework with the following aims:

(1) review efficient algorithms to calibrate and simulate non-normal models;
(2) defining non-Gaussian dependency structures;
(3) analyzing applications to finance and risk management;
(4) testing the proposed models on real financial market data.

Of course, there are many possible alternative models to try to explain the stylized facts observed in financial market returns. For example, the smile or smirk observed in the graph of the Black and Scholes option pricing model, implying volatility against the strike price for a series of equity call options with the same maturity, can be explained in different ways: it is possible to obtain a smile with local volatility, heavy tails, stochastic volatility, or some mixture of all three (see Derman (2004)). Furthermore, in derivative pricing, there is not a single model that can be used in all cases. It usually depends on the asset class as well as on the current market behavior. Large banks have for the same asset class different possible models that can be used for different pay-offs. Additionally, a model implemented for derivative pricing may not be well suited for an asset management firm looking for possible arbitrage opportunities (i.e., a statistical arbitrage desk). This means that the same instruments may be priced differently by market players. The same derivative can be priced differently by different banks, and even within a given bank, the same portfolio can be priced within a model by the front office that differs from another model for risk management purposes. More in the past than today, it could be the case that different desks at the same financial institution priced an instrument with different models. After the advent of model validation, within a single institution, such cases can occur only in very particular circumstances.[3]

However, even if there is not a perfect model at hand in seeking to explain the behavior of the prices of financial assets, one has

[3]Nash (2017) detailed the process of internal control functions in investment banking.

to select a given model and eventually compare it with its possible alternatives. Modeling reality is a very difficult task: a model is just a model, and unfortunately reality is usually much more complex than the model itself.

1.2. A Univariate Framework

Asset management strategies and asset pricing require the proper modeling of the return distribution of financial assets. While the return distribution used in the traditional theories of asset pricing and portfolio selection is the normal distribution, numerous studies that have investigated the empirical behavior of asset returns in financial markets throughout the world reject the hypothesis that asset return distributions are normally distributed. Returns from financial assets show well-defined patterns of leptokurtosis and skewness that cannot be captured by the normality assumption.

Several researchers have proposed alternative distributions that they believe would better describe returns on financial assets. For a univariate model, desirable properties are as follows:

- the density function can be written in closed form or in quasi-closed form (i.e., through a simple numerical method);
- the characteristic function has a closed form;
- the model is able to explain some stylized facts about financial time series — heavy tails, negative skew, and volatility clustering;
- the model can be extended to price derivatives;
- the number of parameters with respect to the normal model is limited (i.e., the model should be flexible enough, but it should not be overparameterized).

More than 50 years ago, Mandelbrot (1963) and Fama (1963) proposed the α-stable distribution. Despite being superior to the normal distribution and its use being advocated by some researchers, its popularity did not find its way into mainstream financial theory until the turn of the century (see Rachev and Mittnik (2000)). Although the α-stable distribution was superior to the normal distribution as a description of return behavior, the tails are so fat that even its variance is infinite. To overcome this drawback and partially maintain the heavy-tailed and asymmetric shape of the α-stable distribution, an alternative model has been proposed. The tails of an α-stable random variable can be appropriately tempered or truncated in order to obtain "tempered stable distributions" (see Mantegna and Stanley (1995)). By construction, this random variable is semi-heavy-tailed. This family of distributions allows for flexible modeling of various types of extreme scenarios, asymmetries, and rare events.

The formal and elegant definition of tempered stable distributions and processes has been proposed in the work of Rosinski (2007). A tempered stable distribution combines both α-stable and normal behavior (see Grabchak and Samorodnitsky (2010)). By construction, a tempered stable distribution is infinitely divisible (Sato (1999)) and therefore we can define a tempered stable process and apply it to constructing an asset pricing model. Tempered stable random variables may have all moments finite and exponential moments of some order. This last property is essential in the construction of tempered stable option pricing models. Furthermore, the sum of independent tempered stable random variables with tail index α is tempered stable with tail index α; indeed, the class of tempered stable laws with a fixed tail index α is closed under convolution.

This class of distributions has an infinite-dimensional parametrization by a family of measures (Terdik and Woyczyński (2006)), which makes its use in practice difficult. For this reason, some parametric examples have been proposed and, in some cases, an analytical form for the characteristic function has been derived. The few applications of univariate tempered stable distributions to finance studied in academia (see Carr *et al.* (2002), Kim *et al.* (2010), Bianchi *et al.* (2017), and references therein) have been only partially extended to a multivariate setting due to the difficulty of calibrating and simulating these models (see Kawai and Masuda (2011), Baeumer and Kovács (2011), Kim *et al.* (2012), Bianchi *et al.* (2016), and Fallahgoul *et al.* (2019)).

Another semi-heavy-tailed model proposed in the literature is based on the generalized hyperbolic (GH) random variable. The GH distribution has received a lot of attention in the financial-modeling literature (see, for example, Eberlein and Keller (1995), Prause (1999), and Eberlein *et al.* (2002)). The GH parametric family includes many familiar distributions such as the Student's t, the skew-t, the hyperbolic, the variance gamma, and the normal inverse Gaussian. Furthermore, it can be easily extended to a multivariate framework as described by McNeil *et al.* (2005).

1.3. A Multivariate Framework

Many important problems in finance are multidimensional. These problems are usually solved by considering a multivariate normal distribution, mainly because the complexity of a model increases as soon as one moves from a normal to a non-normal distribution framework. The dependence structure implied by the multivariate normal distribution is symmetric and it is not capable of capturing asymmetry of dependence during extreme market movements and the dependence of tail events. Bedendo *et al.* (2010) find that, in several cases and during stable market conditions, the choice of the dependence structure (i.e., the copula function) does not significantly affect option prices. The specification of the dependence structure, however, becomes more relevant when the market scenarios for future share prices are very volatile, and both linear correlation and tail dependence strengthen. A more complex dependence structure of multivariate distributions is needed rather than simple linear correlation matrix. Additionally, the margins of a multivariate normal distribution are normally distributed and they do not assign enough weight to events in the tails and are not able to explain negative skew.

For a multivariate model, the desirable properties are as follows:

- the density function can be written in closed form or in quasi-closed form (i.e., through a simple numerical method);

- the characteristic function has a closed form;
- the univariate distribution of the portfolio defined as the weighted sum of the margins can be easily computed and can be applied to evaluate portfolio risk measures and solve portfolio optimization problems;
- the model is able to explain four stylized facts about financial time series — heavy tails, negative skew, asymmetric dependence, and volatility clustering (see Allen and Satchell (2014) and Bianchi *et al.* (2016));
- the model can be extended to price derivatives;
- the number of parameters with respect to the multivariate normal does not increase linearly by the number of margins (i.e., the model should be flexible enough but it should not be overparameterized).

Note that in this book we review only continuous-time models that usually are not able to capture neither the volatility clustering effect nor the leverage effect. The volatility clustering effect is the tendency of large changes in asset prices (either positive or negative) to be followed by large changes and small changes to be followed by small changes. The leverage effect is the empirically observed fact that negative shocks have a stronger impact on the variance than positive shocks of the same magnitude (i.e., bad news raises the future volatility more than good news). However, if considered in combination with a discrete-time multivariate GARCH model or with a continuous-time stochastic volatility structure, the models reviewed in this book can be enhanced to explain both the volatility clustering and the leverage effect, but at the cost of losing analytical tractability and/or increasing the computational complexity. Some of the multivariate distributions reviewed in this book are elliptical: the class of elliptical distributions is consistent with mean–variance portfolio optimization (see Gupta *et al.* (2013, Chapter 10)).

There are different approaches to build a multivariate non-Gaussian model. It is possible (1) to consider linear mixture of independent non-Gaussian processes, as done by Kawai (2009) and Ballotta and Bonfiglioli (2016), and references therein, (2) to subordinate a multivariate Brownian motion with a univariate or a multivariate stochastic time (e.g., Barndorff-Nielsen *et al.* (2001)), (3) to build the multivariate process by defining its Lévy measure, as done by Rosinski (2007) and Bianchi *et al.* (2010), (4) to specify the models for the marginal distributions (processes) separately from the copula (i.e., the dependence structure) that links these distributions (processes) to form a joint distribution (multivariate processes) (see Genest *et al.* (2009), Patton (2012), and Tankov (2016) for a review). In this book, we consider only the second and fourth approaches. Even if the third approach is elegant from a theoretical perspective, it is not simple to use in practical applications. Then, copula methods provide a simple framework for thinking about dependence, but it is not always clear how to find a change of measure needed to price derivatives. Mikosch (2006) observed that there is no particular advantage of using copulas when dealing with multivariate distributions (processes). However, in some cases, copulas are very useful in practical applications.

1.4. Extreme Events

Extremes are events which are rare in frequency and huge in magnitude with respect to the majority of observations (i.e., tail events). Extremes in the realm of finance manifest themselves most clearly through stock market crashes (see Embrechts *et al.* (1999)). As observed by Diebold *et al.* (2000),

> Financial risk management is intimately concerned with tail quantiles (e.g., the value of x such that $P(X > x) = 0.05$) and tail probabilities (e.g., $P(X > x)$, for a large value x). Extreme quantiles and probabilities are of particular interest because the ability to assess them accurately translates into the ability to manage extreme financial risks effectively. (p. 30)

These extreme quantiles and probabilities can be estimated using only the extreme data rather than all the data, thereby fitting the tail and only the tail. By following this approach, the functional form of return distributions can be left unknown, and one can focus only on the functional form of the tail decay.

Extreme value theory is concerned with the study of the asymptotic distribution of extreme events, or, in simple terms, it deals with the quantification of extreme events and their consequences in a statistically optimal way. Statistical methods derived from it can be employed in finance. The tail decay has important implications for portfolio risk measurement and stress-testing, as well as option pricing where the so-called smile effect can be explained by the presence of fat tails. For example, a risk manager may be interested in estimating the downside of a financial institution's trading book, which typically can be reformulated in terms of a quantile of a profit-and-loss distribution.

The rationale behind the extreme value theory approach is essentially contained in two theorems: (1) the Fisher–Tippett–Gnedenko theorem on the distribution of rescaled and recentered block maxima,[4] and (2) the Pickands–Balkema–de Haan theorem on the distribution of exceedances over a threshold. Both theorems consider a sequence of independent and identically distributed random variables X_i with distribution F.

The first theorem studies the distribution of block maxima of this sequence of random variables. The limit distribution of rescaled and recentered block maxima depends on the behavior of the upper tail of F, whether it is power-law decaying (the fat-tailed case, the most important for financial applications), exponentially decaying (e.g., if F is the normal distribution), or with upper bounded support (the least relevant case for finance). The three asymptotic distributions of block maxima can be written in a unified manner by means of the generalized extreme value (GEV) distribution.

The second theorem studies the distribution of exceedances over some fixed high threshold u, that is, the distribution of the random variables $Y_i = X_i - u$, for those observations exceeding u (i.e., $X_i > u$). The theorem shows that as the threshold tends to infinity (or to the right endpoint of the support of F, if X_i has a bounded

[4] Given a sample of N observations divided into subsamples of a given length, a block maximum is defined by picking the maximum of each subsample.

support), then Y_i is generalized Pareto distributed (GPD). Furthermore, if the distribution of block maxima of the sequence X_i follows a GEV distribution with shape parameter ξ then for a sufficiently high threshold u, the distribution of exceedances over u follows a GPD with shape parameter ξ.

As we will show in Chapters 9 and 12, both the results stated above can be used to define estimation methods focused on the tails of the distribution, compute well-known portfolio risk measures, and can be considered in combination with a GARCH model. The extension to a multivariate framework is possible but is not as straightforward as its univariate counterpart and can still encounter some computational limitations (see Rocco (2014)).

Key Points

- Heavy-tailed distributions play a major role in the analysis of many stochastic systems.
- In finance, the terms heavy-tailed distributions, fat-tailed distributions, non-Gaussian distributions, Lévy processes, Lévy flight, and extreme events are more or less synonyms.
- Heavy-tailed distributions and Lévy processes are different sides of the same coin. In some cases (i.e., for infinitely divisible distributions), the increments of Lévy processes are heavy-tailed random variables.
- Traditional theories in financial economics have failed to give recognition to extreme events or, equivalently, the tails of the distribution of financial variables of interest.
- Some researchers distinguish between crashes and fat-tailed price variations (or tail events).
- Many financial crises cannot be explained by prevailing statistical models of markets based on traditional financial economic theories, resulting in the failure of models built on those theories.
- There are two avenues for reformulating financial economic theory: (1) modifying the underlying assumptions about investor behavior (i.e., applying behavioral finance theory), and (2) examination of

the microstructure of financial markets to incorporate the observed behavior of such markets into models of portfolio selection and asset pricing.

- There are stylized facts about the behavior of asset returns in financial markets: non-normal and asymmetric distribution of asset returns and a dependence structure of multivariate return distributions that goes beyond simple linear correlation.
- The stylized factors about the behavior of financial markets has led to increased efforts to improve the flexibility and statistical reliability of existing models that seek to capture the dynamics of economic variables which involves the application of non-normal theoretical distributions for explaining financial uncertainties, where tails and extreme events are treated as intrinsic characteristics of the model.

References

R.J. Adler, R. Feldman, and M. Taqqu. *A Practical Guide to Heavy Tails: Statistical Techniques and Applications.* Birkhäuser, Boston, 1998.

J. Alcock and S. Satchell, editors. *Asymmetric Dependence in Finance: Diversification, Correlation and Portfolio Management in Market Downturns.* Wiley, Hoboken, NJ, 2018.

D. Allen and S. Satchell. The four horsemen: Heavy-tails, negative skew, volatility clustering, asymmetric dependence. Discussion paper 4, University of Sydney, 2014.

B. Baeumer and M. Kovács. Approximating multivariate tempered stable processes. *Journal of Applied Probability*, 49(1):167–183, 2011.

L. Ballotta and E. Bonfiglioli. Multivariate asset models using Lévy processes and applications. *The European Journal of Finance*, 22(13): 1320–1350, 2016.

A.L. Barabasi. The origin of bursts and heavy tails in human dynamics. *Nature*, 435:207–211, 2005.

O.E. Barndorff-Nielsen, J. Pedersen, and K.I. Sato. Multivariate subordination, self-decomposability and stability. *Advances in Applied Probability*, 33(1):160–187, 2001.

M. Bedendo, F. Campolongo, E. Joossens, and F. Saita. Pricing multiasset equity options: How relevant is the dependence function? *Journal of Banking and Finance*, 34(4):788–801, 2010.

M.L. Bianchi and G.L. Tassinari. Forward-looking portfolio selection with multivariate non-Gaussian models and the Esscher transform. Preprint, 2018. https://arxiv.org/abs/1805.05584.

M.L. Bianchi, S.T. Rachev, Y.S. Kim, and F.J. Fabozzi. Tempered infinitely divisible distributions and processes. *SIAM Theory of Probability and Its Applications*, 55(1):59–86, 2010.

M.L. Bianchi, S.T. Rachev, and F.J. Fabozzi. Tempered stable Ornstein-Uhlenbeck processes: A practical view. *Communications in Statistics — Simulation and Computation*, 46(1): 423–445, 2017.

M.L. Bianchi, G.L. Tassinari, and F.J. Fabozzi. Riding with the four horsemen and the multivariate normal tempered stable model. *International Journal of Theoretical and Applied Finance*, 19(4):1650027, 2016.

S.I. Boyarchenko and S.Z. Levendorskii. *Non-Gaussian Merton–Black–Scholes theory.* World Scientific Publishing, River Edge, NJ, 2002.

P. Carr, H. Geman, D.B. Madan, and M. Yor. The fine structure of asset returns: An empirical investigation. *Journal of Business*, 75(2): 05–332, 2002.

N.R. Cavanaugh, A. Gershunov, A.K. Panorska, and T.J. Kozubowski. The probability distribution of intense daily precipitation. *Geophysical Research Letters*, 42(5):1560–1567, 2015.

A.S. Chernobai, S.T. Rachev, and F.J. Fabozzi. *Operational Risk: A Guide to Basel II Capital Requirements, Models, and Analysis.* Wiley, Hoboken, NJ, 2008.

R. Cont and P. Tankov. *Financial Modelling with Jump Processes.* Chapman & Hall/CRC, Boca Raton, FL, 2003.

E. Derman. *My Life as a Quant: Reflections on Physics and Finance.* Wiley, Hoboken, NJ, 2004.

F.X. Diebold, T. Schuermann, and J.D. Stroughair. Pitfalls and opportunities in the use of extreme value theory in risk management. *The Journal of Risk Finance*, 1(2):30–35, 2000.

J.M. Dufour, J.R. Kurz-Kim, and F.C. Palm. Editorial introduction: Heavy tails and stable Paretian distributions in empirical finance: A volume honoring Benoît B. Mandelbrot. *Journal of Empirical Finance*, 17(2): 177–179, 2010.

E. Eberlein and U. Keller. Hyperbolic distributions in finance. *Bernoulli*, 1:281–299, 1995.

E. Eberlein and K. Prause. The generalized hyperbolic model: Financial derivatives and risk measures. In H. Geman, D. Madan, S.R. Pliska, and T. Vorst, editors, *Mathematical Finance, Bachelier Congress.* Springer, Berlin, 2002, pp. 245–267.

P. Embrechts, C. Klüppelberg, and T. Mikosch. *Modelling Extremal Events for Insurance and Finance.* Springer, Berlin, 1997.

P. Embrechts, S.I. Resnick, and G. Samorodnitsky. Extreme value theory as a risk management tool. *North American Actuarial Journal*, 3(2): 30–41, 1999.

H. Fallahgoul, Y.S. Kim, F.J. Fabozzi, and J. Park. Quanto option pricing with Lévy models. *Computational Economics*, forthcoming, 2019.

E.F. Fama. Mandelbrot and the stable Paretian hypothesis. *Journal of Business*, 36(4):420–429, 1963.

C. Genest, M. Gendron, and M. Bourdeau-Brien. The advent of copulas in finance. *The European Journal of Finance*, 15(7–8):609–618, 2009.

M. Grabchak and G. Samorodnitsky. Do financial returns have finite or infinite variance? A paradox and an explanation. *Quantitative Finance*, 10(8):883–893, 2010.

A.K. Gupta, T. Varga, and T. Bodnar. *Elliptically Contoured Models in Statistics and Portfolio Theory.* Springer, 2013.

M. Habart-Corlosquet, J. Janssen, and R. Manca. *VaR Methodology for Non-Gaussian Finance.* Wiley, Hoboken, NJ, 2013.

M. Ibragimov, R. Ibragimov, and J. Walden. *Heavy-tailed Distributions and Robustness in Economics and Finance.* Springer, 2015.

A. Johansen and D. Sornette. Stock market crashes are outliers. *The European Physical Journal B — Condensed Matter and Complex Systems*, 1(2):141–143, 1998.

E. Jondeau, S.H. Poon, and M. Rockinger. *Financial Modeling Under Non-Gaussian Distributions.* Springer, Berlin, 2007.

R. Kawai. A multivariate Lévy process model with linear correlation. *Quantitative Finance*, 9(5):597–606, 2009.

R. Kawai and H. Masuda. On simulation of tempered stable random variates. *Journal of Computational and Applied Mathematics*, 235: 2873–2887, 2011.

M. Kemp. *Extreme Events — Robust Portfolio Construction in the Presence of Fat Tails.* Wiley, Hoboken, NJ, 2011.

Y.S. Kim, S.T. Rachev, M.L. Bianchi, and F.J. Fabozzi. Tempered stable and tempered infinitely divisible GARCH models. *Journal of Banking and Finance*, 34(9):2096–2109, 2010.

Y.S. Kim, R. Giacometti, S.T. Rachev, F.J. Fabozzi, and D. Mignacca. Measuring financial risk and portfolio optimization with a non-Gaussian multivariate model. *Annals of Operations Research*, 201(1):325–343, 2012.

S. Kring, S.T. Rachev, M. Höchstötter, F.J. Fabozzi, and M.L. Bianchi. Multi-tail generalized elliptical distributions for asset returns. *The Econometrics Journal*, 12(2):272–291, 2009.

B.B. Mandelbrot. The variation of certain speculative prices. *Journal of Business*, 36:394–419, 1963.

R.N. Mantegna and H.E. Stanley. Scaling behaviour in the dynamics of an economic index. *Nature*, 376:46–49, 1995.

P. Mathews, L. Mitchell, G. Nguyen, and N. Bean. The nature and origin of heavy tails in retweet activity. In *Proceedings of the 26th International Conference on World Wide Web Companion*, 2017, pp. 1493–1498.

A. McNeil, R. Frey, and P. Embrechts. *Quantitative Risk Management: Concepts, Techniques, and Tools.* Princeton University Press, Princeton, NJ, 2005.

T. Mikosch. Copulas: Tales and facts. *Extremes*, 9(1):3–20, 2006.

A. Misiorek and R. Weron. Heavy-tailed distributions in VaR calculations. In J.E. Gentle and W.K. Härdle, editors, *Handbook of Computational Statistics: Concepts and Methods.* Springer, Berlin, 2012, pp. 1025–1059.

P. Nash. *Effective Product Control: Controlling for Trading Desks.* Wiley, Chichester, 2017.

M. Paolella, E. Renault, G. Samorodnitsky, and D. Veredas. Latest developments on heavy-tailed distributions. *Journal of Econometrics*, 172(2):183–185, 2013.

A.J. Patton. A review of copula models for economic time series. *Journal of Multivariate Analysis*, 110:4–18, 2012.

G.W. Peters and P.V. Shevchenko. *Advances in Heavy Tailed Risk Modeling: A Handbook of Operational Risk.* Wiley, Hoboken, NJ, 2015.

V. Pisarenko and M. Rodkin. *Heavy-tailed Distributions in Disaster Analysis.* Springer, 2010.

K. Prause. The Generalized Hyperbolic Model: Estimation, Financial Derivatives, and Risk Measures. PhD thesis, Universität Freiburg i. Br., Freiburg i. Br, 1999.

S.T. Rachev, editor. *Handbook of Heavy Tailed Distributions in Finance*, Handbooks in Finance, Vol. 1. Elsevier, 2003.

S.T. Rachev and S. Mittnik. *Stable Paretian Models in Finance.* Wiley, Chichester, 2000.

S.T. Rachev, C. Menn, and F.J. Fabozzi. *Fat-tailed and Skewed Asset Return Distributions: Implications for Risk Management, Portfolio Selection, and Option Pricing.* Wiley, Hoboken, NJ, 2005.

S.T. Rachev, Y.S. Kim, M.L. Bianchi, and F.J. Fabozzi. *Financial Models with Lévy Processes and Volatility Clustering.* Wiley, Hoboken, NJ, 2011.

A.M. Reynolds. Current status and future directions of Lévy walk research. *Biology Open*, 7(1):bio030106, 2018.

M. Rocco. Extreme value theory in finance: A survey. *Journal of Economic Surveys*, 28(1):82–108, 2014.

J. Rosinski. Tempering stable processes. *Stochastic Processes and their Applications*, 117(6):677–707, 2007.

K.I. Sato. *Lévy Processes and Infinitely Divisible Distributions*. Cambridge University Press, Cambridge, 1999.

W. Schoutens. *Lévy Processes in Finance*. Wiley, Chichester, 2003.

P. Tankov. Lévy copulas: Review of recent results. In M. Podolskij, R. Stelzer, S. Thorbjørnsen, and A.E.D. Veraart, editors, *The Fascination of Probability, Statistics and their Applications: In Honour of Ole E. Barndorff-Nielsen*. Springer International Publishing, Cham, 2016, pp. 127–151.

G. Terdik and W.A. Woyczyński. Rosiński measures for tempered stable and related Ornstein–Uhlenbeck processes. *Probability and Mathematical Statistics*, 26:213–243, 2006.

G.M. Viswanathan, S.V. Buldyrev, S. Havlin, M.G.E. Da Luz, E.P. Raposo, and H.E. Stanley. Optimizing the success of random searches. *Nature*, 401:911–914, 1999.

W.D. Walls. Modelling heavy tails and skewness in film returns. *Applied Financial Economics*, 15(17):1181–1188, 2005.

Chapter 2

Random Variables

After formally introducing the notion of random variable and some related concepts, in this chapter we look at discrete and absolutely continuous random variables. We focus our attention on definitions and properties that will be needed in the chapters that follow. The main topics covered in this chapter are:

- basic definitions of probability theory;
- the definition of the characteristic function and other generating functions, and their main properties for real-valued random variables;
- a review of some univariate discrete random variables (Bernoulli, binomial and Poisson);
- a review of some univariate continuous random variables (gamma, exponential, inverse Gaussian, normal, log-normal, variance gamma, normal inverse Gaussian);
- a review of the multivariate normal distribution.

2.1. Preliminaries

In simple terms, a *random variable* is a variable whose possible values are outcomes of a random phenomenon. As we will see in the following, the definition of a random variable in mathematical terms is a little more challenging.[1]

An *event* ω is mathematically defined as a subset of a given sample space Ω. If we have an event A, then the set of outcomes which are not included in A is also an event. For example, if we consider an event that the return of a stock tomorrow will be positive, then the set of outcomes that the return tomorrow will be negative is also an event. If we have two events A and B, then a set of outcomes included in both A and B is also an event. For instance, consider the following two events: the return on the stock of Apple Inc tomorrow will be positive and the return on the stock of Alphabet Inc tomorrow will be positive. Then a set of outcomes that both returns will be positive tomorrow is an event.

The mathematical description of a collection of events is referred to as a *σ-algebra*. A σ-algebra, denoted by $\mathcal{F}$, is a collection of the subsets of Ω satisfying the following properties:

Property 1. $\emptyset \in \mathcal{F}$ and $\Omega \in \mathcal{F}$.
Property 2. If $A \in \mathcal{F}$, then $A^c = \{x \in \Omega | x \notin A\} \in \mathcal{F}$.
Property 3. If $A_1, A_2, A_3, \cdots \in \mathcal{F}$, then $\bigcup_{n=1}^{\infty} A_n \in \mathcal{F}$.

The collection of all Borel sets on $\mathbb{R}(\mathbb{R}^d)$, denoted by $\mathcal{B}(\mathbb{R})(\mathcal{B}(\mathbb{R}^d))$, is called the Borel σ-algebra. It can be generated by the open sets in $\mathbb{R}(\mathbb{R}^d)$; that is, it is the smallest σ-algebra that contains all open sets in $\mathbb{R}(\mathbb{R}^d)$.

Let $\mathcal{G}$ denote a collection of subsets contained in Ω. Then the smallest σ-algebra containing $\mathcal{G}$ is referred to as the *σ-algebra generated by* $\mathcal{G}$, and is denoted by $\sigma(\mathcal{G})$. For a given function X, consider

[1]See Feller (1971), Letta (1993), Landenna *et al.* (1994), Billingsley (1995), Jacod and Protter (2004), Paolella (2007) for an in-depth discussion on this topic.

the class

$$\mathcal{G} = \{A \subseteq \Omega : A = X^{-1}(I)\},$$

for all open interval I in $\mathbb{R}$, where X^{-1} is the inverse image of X. Then the σ-algebra generated by $\mathcal{G}$ is referred to as the *σ-algebra generated by X*, and denoted by $\sigma(X)$. If there is a σ-algebra $\mathcal{F}$ such that $\sigma(X) \subseteq \mathcal{F}$, then we say that X is $\mathcal{F}$-measurable.

A *probability space* is a triple $(\Omega, \mathcal{F}, P)$, where Ω is a non-empty set, $\mathcal{F}$ is a σ-algebra of subsets of Ω, and P is a finite measure on $(\Omega, \mathcal{F})$ with $P(\Omega) = 1$ (P is a map from $\mathcal{F}$ to the unit interval $[0, 1]$).

In the following, we will consider maps X on $(\Omega, \mathcal{F})$ which take values in a measurable space $(\mathbb{R}, \mathcal{B}(\mathbb{R}))$, that is

$$X : (\Omega, \mathcal{F}) \longrightarrow (\mathbb{R}, \mathcal{B}(\mathbb{R}))$$

and we will refer to X as a *real-valued random variable.* In this case, we have that the event

$$\{X \leq x\} = X^{-1}(] - \infty, x]) \in \mathcal{F}.$$

The function F_X (or simply F) defined by

$$F_X(x) = P(X \leq x)$$

is called the distribution (or law) of X.

Similarly, we will consider maps X which take values in a measurable space $(\mathbb{R}^d, \mathcal{B}(\mathbb{R}^d))$ and we will refer to X as a *real-valued random vector.*

2.1.1. Discrete random variables

A random variable X is *discrete* if it assumes a finite number or a countable infinity of values $x_1, \ldots, x_n$ with probabilities $f_X(x_1), \ldots, f_X(x_n)$. The function f_X is called a *probability function* and possesses the following properties:

- f_X is discrete;
- $f_X(x_j) = P(X = x_j) > 0$, $j \in \mathbb{N}$;
- $\sum_j f_X(x_j) = 1$.

Let $x_1, \ldots, x_n$ be the sequence of the possible values of X arranged in increasing order, then the function

$$F_X(x) = P(X \leq x) = \sum_{x_j \leq x} f_X(x_j) \tag{2.1}$$

is referred to as *cumulative distribution function.*

The cumulative distribution function of a discrete random variable X has the following properties:

- $F_X(-\infty) = 0$;
- $F_X(+\infty) = 1$;
- $0 \leq F(x) \leq 1, \forall x \in \mathbb{R}$;
- $F_X(x)$ is right continuous;
- $F_X(x)$ is an increasing step function;
- $F_X(x_j) - F_X(x_{j-1}) = f_X(x_j)$.

Some common examples of discrete random variables are considered in Section 2.4. It is possible to extend the notion of a random variable to a random vector. A d-dimensional random vector has a discrete distribution if its components are one-dimensional discrete random variables. Let X be a vector whose components $X_1, \ldots, X_d$ are discrete random variables. Then, its *probability function* $f_X(x_1, \ldots, x_d)$ is a function on the real variables $x_1, \ldots, x_d$ that are null in $\mathbb{R}^d$ except for a finite number or a countable infinity of points $x_{1j_1}, \ldots, x_{dj_d}$, $j_r = 1, 2, \ldots$ and $r = 1, \ldots, d$ such that

$$f_X(x_{1j_1}, \ldots, x_{dj_d}) > 0,$$

and

$$\sum_{j_1} \cdots \sum_{j_d} f_X(x_{1j_1}, \ldots, x_{dj_d}) = 1.$$

The cumulative distribution function of the discrete random vector X

$$F_X(x_1, \ldots, x_d) = \sum_{x_{1j_1} \leq x_1} \cdots \sum_{x_{dj_d} \leq x_d} f_X(x_{1j_1}, \ldots, x_{dj_d}) \tag{2.2}$$

allows one to compute the probability

$$P(X_1 \le x_1, \ldots, X_d \le x_d), \quad \forall (x_1, \ldots, x_d) \in \mathbb{R}^d. \tag{2.3}$$

From (2.2), it is possible to recover the *marginal cumulative distributions*:

$$\begin{aligned} F_{X_r}(x_r) &= F_X(\infty, \ldots, x_r, \ldots, \infty) \\ &= \sum_{x_{1j_1}} \cdots \sum_{x_{rj_r} \le x_r} \cdots \sum_{x_{dj_d}} f_X(x_{1j_1}, \ldots, x_{rj_r}, \ldots, x_{dj_d}), \end{aligned}$$

for $r = 1, \ldots, d$.

We conclude this discussion with an important property characterizing discrete independent random variables. The discrete variables $X_1, \ldots, X_d$ are *stochastically independent* if and only if

$$F_X(x_1, \ldots, x_d) = \prod_{j=1}^{d} F_{X_j}(x_j), \quad \forall (x_1, \ldots, x_d) \in \mathbb{R}^d.$$

2.1.2. Absolutely continuous random variables

A random variable X is *absolutely continuous*[2] if there exists a function $f_X(x)$, called a *probability density function* or simply *density* of X, such that

$$P(a \le X \le b) = \int_a^b f_X(x)dx,$$

where $a, b \in \mathbb{R}$ with $a < b$.

For an absolutely continuous random variable $f_X(x) \ne P(X = x)$ since the probability that X assumes that a particular value x is null for all $x \in \mathbb{R}$. Thus, for a continuous random variable

$$P(a \le X \le b) = P(a < X \le b) = P(a \le X < b) = P(a < X < b).$$

[2]In the following, for simplicity we will use the term *continuous* to refer to an absolutely continuous random variable.

The density function satisfies the properties:

- $P(x \leq X \leq x + dx) = f_X(x)dx$, where $dx \to 0$;
- $f_X(x) \geq 0$, $\forall x \in \mathbb{R}$;
- $\int_{-\infty}^{\infty} f_X(x) = 1$.

The cumulative distribution function (or simply the distribution function) of an absolutely continuous random variable X is

$$F_X(x) = P(X \leq x) = \int_{-\infty}^{x} f_X(x)dx.$$

The cumulative distribution function $F_X(x)$ is a continuous function with the following properties:

- $0 \leq F(x) \leq 1$, $\forall x \in \mathbb{R}$;
- $F_X(x)$ is increasing;
- $F_X(x)$ may have discontinuities but is right continuous;
- $\lim_{x \to -\infty} F_X(x) = 0$;
- $\lim_{x \to +\infty} F_X(x) = 1$;
- $P(a < X < b) = F(b) - F(a)$, $a < b$;
- $\frac{dF_X(x)}{dx} = f_X(x)$, $\forall x$, where $f_X(x)$ is continuous.

The normal random variable and other relevant examples of continuous random variable are illustrated in Section 2.5.

We will consider

$$L^p = L^p(\Omega, \mathcal{F}, P)$$

for $p \in [1, \infty)$, the space of all real-valued random variables X such that $|X|^p$ is integrable.

A d-dimensional random variable X is *absolutely continuous* if there exists a function $f_X(x_1, \ldots, x_d)$ such that

$$\begin{aligned} F_X(x_1, \ldots, x_d) &= P(X_1 \leq x_1, \ldots, X_d \leq x_d) \\ &= \int_{-\infty}^{x_1} \cdots \int_{-\infty}^{x_d} f_X(x_1, \ldots, x_d)dx_1 \ldots dx_d, \end{aligned} \tag{2.4}$$

where $F_X(x_1, \ldots, x_d)$ is the cumulative distribution function of the random vector X.

For all the points in $\mathbb{R}^d$ except at most for a set of points with null probability, the following relation holds:

$$f_X(x_1, \ldots, x_d) = \frac{dF_X(x_1, \ldots, x_d)}{dx_1 \ldots dx_d}. \tag{2.5}$$

The probability density function $f_X(x_1, \ldots, x_d)$ satisfies the properties:

- $f_X(x_1, \ldots, x_d) \geq 0, \ \forall (x_1, \ldots, x_d)$ satisfying (2.5);
- $\int_{-\infty}^{+\infty} \cdots \int_{-\infty}^{+\infty} f_X(x_1, \ldots, x_d) dx_1 \ldots dx_d = 1$.

In Section 2.5 as a first example of a continuous random vector, the multivariate normal random variable is discussed.

From (2.4) the margins can be derived:

$$\begin{aligned} F_{X_r}(x_r) =& F_X(\infty, \ldots, x_r, \ldots, \infty) \\ =& \int_{-\infty}^{\infty} \cdots \int_{-\infty}^{x_r} \cdots \int_{-\infty}^{\infty} f_X(x_1, \ldots, x_d) dx_1 \ldots dx_d, \end{aligned}$$

for $r = 1, 2, \ldots, d$.

The components of the random vectors X are stochastically independent if and only if the joint density is equal to the product of the marginal densities

$$f_X(x_1, \ldots, x_d) = \prod_{j=1}^{d} f_{X_j}(x_j).$$

2.2. Parameters of a Random Variable

The form of the distribution and of the density of a random variable are characterized by a set of constants, known as *parameters*. In this section, we focus on different classes of parameters:

- moments;
- quantiles or order parameters;
- location, scale, and shape parameters.

2.2.1. Expected value

Let X be a discrete random variable with probability function $f_X(x_j) = P(X = x_j)$ for $j = 1, \ldots, n$. Consider the random variable $g(X)$ function of X. Then, the quantity

$$E[g(X)] = \sum_{j=1}^{\infty} g(x_j) f_X(x_j) \tag{2.6}$$

is called the *expectation* or *expected value* or *mean value* or simply *mean* of $g(X)$. If n is not finite, (2.6) is the expectation of $g(X)$ if and only if the series $\sum_{j=1}^{\infty} g(x_j) f_X(x_j)$ is absolutely convergent, i.e.,

$$\sum_{j=1}^{\infty} |g(x_j)| f_X(x_j) < \infty. \tag{2.7}$$

Let X be a continuous random variable with density function $f_X(x)$ and $g(X)$ a function of X. Then, the integral

$$E[g(X)] = \int_{\mathbb{R}} g(x) f_X(x) dx \tag{2.8}$$

is said to be the *expected value* of $g(X)$ if and only if

$$\int_{\mathbb{R}} |g(x)| f_X(x) dx < \infty. \tag{2.9}$$

If $g(X) = X$, then (2.6) with (2.7) ((2.8) with (2.9)) defines the mean of a discrete (continuous) random variable. If it exists, the expectation of a random variable X is denoted as $E[X]$.

The expectation is a linear operator. Let a and b be real constants, then

$$E[aX + b] = aE[X] + b. \tag{2.10}$$

From (2.10), it follows

$$E[X - E[X]] = 0. \tag{2.11}$$

Let $X_1, \ldots, X_n$ be a sequence of random variables with expectations $E[X_1], \ldots, E[X_n]$. Then, the following property holds:

$$E\left[\sum_{j=1}^{n} X_j\right] = \sum_{j=1}^{n} E[X_j]. \tag{2.12}$$

By considering (2.10) and (2.12), the expectation of a linear combination of n random variables is computed as

$$E\left[\sum_{j=1}^{n} a_j X_j + b\right] = \sum_{j=1}^{n} a_j E[X_j] + b.$$

This property is particularly useful in some financial applications because it allows one to derive the expected return of a portfolio of assets from the portfolio weights and the expected returns of the assets in the portfolio.

Let $X_1, \ldots, X_n$ be a sequence of independent random variables with expectations $E[X_1], \ldots, E[X_n]$. Then,

$$E\left[\prod_{j=1}^{n} X_j\right] = \prod_{j=1}^{n} E[X_j].$$

Here we provide an alternative way to express the expected value of a continuous random variable. Let X be a non-negative continuous random variable with mean $E[X]$ and cumulative distribution function F_X, then

$$E[X] = \int_0^{\infty} [1 - F_X(x)]\, dx.$$

Similarly, if X is a non-positive continuous random variable, then

$$E[X] = -\int_{-\infty}^{0} F_X(x) dx.$$

More generally, if X is a continuous random variable with mean value $E[X]$, then

$$E[X] = \int_0^{\infty} [1 - F_X(x)]\, dx - \int_{-\infty}^{0} F_X(x) dx.$$

2.2.2. Moments

All the characteristics of a random variable are represented by a family of parameters known as *moments.*

Let X be a random variable and define the random variable $g(X) = X^k$, where k is a positive integer. Then, the quantity $E[X^k]$ is referred to as *raw moment of order* k of the random variable X if and only if $E[|X|^k] < \infty$. Thus, the raw moment of order one is the expectation of X.

Let X be a random variable and define the random variable $g(X) = (X - c)^k$, where c is an arbitrary constant and k a positive integer. Then, the quantity $E[(X - c)^k]$ is called *moment of order* k *from the origin* c of the random variable X if and only if $E[|X - c|^k] < \infty$. In particular, setting $c = E[X]$, we get the *central moment of order* k: $E[(X - E[X])^k]$. The second central moment is referred to as the *variance*

$$\text{var}[X] = E[(X - E[X])^2]. \tag{2.13}$$

The variance of a random variable is a measure of dispersion around its mean value. A measure of dispersion expressed in the same unit of measure as X is the *standard deviation* which is defined as the (positive) square root of the variance

$$\text{std}[X] = \sqrt{\text{var}[X]}.$$

The third central moment contains information about the asymmetry of a distribution. In particular, if $E[(X - E[X])^3] = 0$, then the distribution of X is symmetric around the mean. The fourth central moment contains information about the fatness of the tails of the distribution. By normalizing the third and the fourth central moments, it is possible to get two extremely important quantities characterizing the distribution of the random variable X: *skewness* and *kurtosis.* The skewness is defined as

$$\text{skew}[X] = \frac{E[(X - E[X])^3]}{(\text{var}[X])^{3/2}},$$

and kurtosis as

$$\text{kurt}[X] = \frac{E[(X - E[X])^4]}{(\text{var}[X])^2}.$$

Skewness is a measure of the asymmetry of a distribution. If $\text{skew}[X] > 0$ (< 0), then the right (left) tail is fatter than the left (right) one. If $\text{skew}[X] = 0$, then X is symmetric.

Kurtosis contains information about the heaviness of the tails of a distribution. The quantity

$$\text{ex.kurt}[X] = \frac{E[(X - E[X])^4]}{(\text{var}[X])^2} - 3$$

is referred to as *excess kurtosis* and is a measure of the fatness of the tails of a distribution relative to the normal distribution. If $\text{ex.kurt}[X] > 0$, then X has fatter tails than the normal distribution (i.e., $\text{kurt}[X] > 3$). In contrast, if $\text{ex.kurt}[X] < 0$, then X has lighter tails than the normal distribution (i.e., $\text{kurt}[X] < 3$). In the first case, the distribution of X is called *leptokurtic*, in the second one *platykurtic*. In this book, we will work with leptokurtic distribution, that is with distribution whose tails decay more slowly than the tails of the normal distribution, and therefore tail events occur with greater frequency. We refer to expectation, variance, skewness, and kurtosis as the *main moments* or simply *moments* of a distribution.

Now, we review some properties of the variance. By rewriting equation (2.13) in a different form, it is possible to express the variance as a function of the first and second raw moments, that is

$$\text{var}[X] = E[X^2] - E[X]^2.$$

Let X be a random variable and a and b be two real constants. Then,

$$\text{var}[aX + b] = a^2 \ \text{var}[X].$$

Let $X_1, \ldots, X_n$ be a sequence of mutually independent random variables and $a_1, \ldots, a_n$ a sequence of constants. Then,

$$\text{var}\left[\sum_{j=1}^{n} a_j X_j\right] = \sum_{j=1}^{n} a_j^2 \text{ var}[X_j]. \tag{2.14}$$

Dropping the hypothesis of independent random variables, (2.14) becomes

$$\text{var}\left[\sum_{j=1}^{n} a_j X_j\right] = \sum_{j=1}^{n} a_j^2 \text{ var}[X_j] + \sum_{j=1}^{n} \sum_{k \neq j}^{n} a_j a_k \text{ cov}[X_j, X_k], \tag{2.15}$$

where

$$\text{cov}[X_j, X_k] = E[(X_j - E[X_j])(X_k - E[X_k])] \tag{2.16}$$

is the *mixed central moment of order* 2 of the bivariate random variable (X_j, X_k), better known as the *covariance*. The covariance is a measure of the *linear dependence* between the random variables X_j and X_k. If it exists,

$$E[(X_j - E[X_j])^g (X_k - E[X_k])^h]$$

is called the *mixed central moment of order* $g+h$ of the *bi*-dimensional random variable (X_j, X_k).

Developing equation (2.16), the covariance can be expressed as a function of the *raw mixed moment of order* 2 and the expectations of X_j and X_k:

$$\text{cov}[X_j, X_k] = E[X_j X_k] - E[X_j]E[X_k].$$

If it exists,

$$E[X_j^g X_k^h]$$

is called the *mixed raw moment of order* $g + h$ of the *bi*-dimensional random variable (X_j, X_k).

Since

$$\mathrm{cov}[X_j, X_k] = \mathrm{cov}[X_k, X_j],$$

equation (2.15) can be written as

$$\mathrm{var}\left[\sum_{j=1}^{n} a_j X_j\right] = \sum_{j=1}^{n} a_j^2 \ \mathrm{var}[X_j] + 2\sum_{j=1}^{n-1} \sum_{k=j+1}^{n} a_j a_k \ \mathrm{cov}[X_j, X_k]. \tag{2.17}$$

Taking into account that

$$\mathrm{cov}[X_j, X_j] = \mathrm{var}[X_j], \tag{2.18}$$

equation (2.15) can be further simplified as

$$\mathrm{var}\left[\sum_{j=1}^{n} a_j X_j\right] = \sum_{j=1}^{n}\sum_{k=1}^{n} a_j a_k \ \mathrm{cov}[X_j, X_k]. \tag{2.19}$$

We emphasize that in (2.15), (2.17), and (2.19) we used implicitly the property

$$\mathrm{cov}[a_j X_j, a_k X_k] = a_j a_k \ \mathrm{cov}[X_j, X_k].$$

Since the Cauchy–Schwartz inequality given by

$$\mathrm{cov}[X_j, X_k]^2 \leq \mathrm{var}[X_j]\mathrm{var}[X_k]$$

holds, it is possible to define the normalized measure of dependence as follows:

$$\mathrm{corr}[X_j, X_k] = \frac{\mathrm{cov}[X_j, X_k]}{\sqrt{\mathrm{var}[X_j]\mathrm{var}[X_k]}},$$

and is referred to as the *linear correlation coefficient* with

$$|\mathrm{corr}[X_j, X_k]| \leq 1.$$

When we deal with a set of random variables, it is convenient to represent all their variances and covariances in matrix form. Specifically, if X is an n-dimensional column random vector with expected value $\mu = [\mu_1, \ldots, \mu_n]'$, then

$$\Sigma = E[(X - \mu)(X - \mu)']$$

is an $n \times n$ symmetric positive semidefinite matrix called the *variance–covariance matrix.* If $\mathrm{Det}(\Sigma) \neq 0$, the variance–covariance matrix is positive definite. In probability, the determinant of Σ, typically denoted as $|\Sigma|$, is called the *generalized variance.* Using matrix notation, (2.15) can be written as

$$\mathrm{var}[a'X] = a'\Sigma a,$$

where $a = [a_1, a_2, \ldots, a_n]$.

2.2.3. Quantiles

The *quantiles* or *order parameters* are always defined even if a random variable does not admit any moments. Let X be a random variable with distribution function F_X. Then, the value x_p satisfying the condition

$$p \leq F_X(x_p) \leq p + P(X = x_p)$$

is called the *quantile of order* p of the random variable X.

In particular, if X is a continuous random variable, its quantile of order p satisfies

$$F_X(x_p) = p,$$

and, therefore, is

$$x_p = F_X^{-1}(p).$$

The value x_m satisfying

$$\frac{1}{2} \leq F_X(x_m) \leq \frac{1}{2} + P(X = x_m)$$

is called the *median* of the distribution. For a continuous random variable, the median is

$$x_m = F_X^{-1}\left(\frac{1}{2}\right).$$

The quantiles are very important in risk management: the value-at-risk measure is a quantile at a given tail level (see Chapters 5, 6, 8–10 and 12).

2.2.4. Location, scale, and shape parameters

The distribution of a random variable can be a function of one or more parameters. These parameters can coincide with some moments or order parameters or they can be expressed as a function of them. Here, we provide the definition of three classes of parameters:

- location parameters;
- scale parameters;
- shape parameters.

A *location parameter* shifts the distribution of a random variable of a quantity equal to the value of the parameter itself. More formally, a constant μ characterizing the distribution of the random variable X is a location parameter if the distribution of the random variable $Y = X - \mu$ does not depend on μ.

In contrast, a *scale parameter* does not affect the position (location) of a distribution, but shrinks or stretches the distribution. To be precise, a constant σ is a scale parameter of X if the distribution of the random variable $Y = X/\sigma$ does not depend on σ.

A *shape parameter* is a parameter of a probability distribution that is neither a location parameter nor a scale parameter and must not be a function of them. Such a parameter must affect only the shape of a distribution.

2.3. Some Useful Functions

In this section, we provide the definition and discuss the main properties of the characteristic function and other important generating functions. These functions will be useful both from a theoretical and a practical perspective in the chapters that follow. Thanks to its two-way relationship with the probability (density) function, it allows one to completely characterize the distribution of a random variable. In Chapter 3, we will make use of the characteristic function to describe infinitely divisible distributions and Lévy processes. In the discussion of the distributions of Chapters 5–7, a special focus is given to their characteristic functions. From an application perspective, in Chapters 8 and 11, the characteristic function will be used in both parameter estimation and option pricing.

2.3.1. Characteristic function

Let X be a real-valued random variable with probability distribution F_X. The function $\phi_X : \mathbb{R} \to \mathbb{C}$, defined as

$$\phi_X(u) = E[\exp(iuX)] = E[\cos(uX) + i\sin(uX)], \quad u \in \mathbb{R},$$

is called the *characteristic function* of the random variable X or, equivalently, characteristic function of the distribution F_X.

In particular, if X is a discrete random variable with probability function $f_X(x_j) = P(X = x_j)$, its characteristic function is given by

$$\phi_X(u) = \sum_j \exp(iux_j) f_X(x_j). \tag{2.20}$$

The characteristic function of a discrete random variable always exists because the series in (2.20) is absolutely and uniformly convergent. Thus, it is also a continuous function of u.

If X is a continuous random variable with density function f_X, its characteristic function is given by

$$\phi_X(u) = \int_{-\infty}^{\infty} \exp(iux) f_X(x) dx. \tag{2.21}$$

This characteristic function always exists because the integral in (2.21) is absolutely and uniformly convergent. Therefore, $\phi_X(u)$ is a continuous function of u.

Thus, if a random variable admits a probability density function, then its characteristic function is the Fourier transform of its probability density function.

The characteristic function of the random variable X has the following properties:

- $\phi_X(0) = 1$;
- $|\phi_X(u)| \leq 1$ is bounded since $|E[\exp(iuX)]| \leq E[|\exp(iuX)|] = 1$;
- $\phi_X(u)$ is a uniformly continuous function;
- $\phi_X(-u) = \overline{\phi_X(u)}$, where the bar indicates complex conjugation.

If a random variable X admits the raw moment of order k, then this moment is given by

$$E[X^k] = i^{-k} \frac{d}{du^k} \phi_X(u)|_{u=0}.$$

The opposite does not hold. Indeed, a random variable admitting a characteristic function with finite first-order derivative for $u = 0$ does not necessarily have a mean. In general, if $\frac{d}{du^k} \phi_X(u)|_{u=0}$ is finite and k is a positive integer, then there exist all moments up to order $\lfloor k/2 \rfloor$, where $\lfloor k/2 \rfloor$ is the largest integer number not bigger than $k/2$. Clearly, if $\phi_X(u)$ is not differentiable for $u = 0$, then X does not possess moments. The previous theorem can be generalized to random vectors. In particular, let (X_1, X_2) be a vector of two real-valued random variables admitting the raw mixed moment of order $l + k$, then

$$E[X_1^l X_2^k] = i^{-(l+k)} \frac{d^{(l+k)}}{du^l dv^k} \phi_{X_1,X_2}(u, v)|_{(u=0,v=0)},$$

where

$$\phi_{X_1,X_2}(u, v) = E[\exp(i(uX_1 + vX_2))], \quad (u, v) \in \mathbb{R}^2.$$

If X admits moments of every order k, using the power series expansion of $\exp(iux)$ and the linearity property of the expectation operator, the characteristic function can be written as

$$\phi_X(u) = 1 + \sum_{k=1}^{\infty} \frac{(iu)^k}{k!} E[X^k],$$

where the coefficient of the term $E[X^k]$ is the raw moment of order k.

Here we present some properties of the characteristic function useful in applications.

Recall that a function is analytic if it can be locally represented by a convergent power series and a complex function is said to be entire if it is analytic at all finite points of the complex plane $\mathbb{C}$. If X is a random variable on $\mathbb{R}$ and $\phi_X(u)$ is an entire function on $\mathbb{C}$, then X has finite exponential moments: that is, it has finite $e^{c|x|}$-moment for every $c > 0$.

Let Y be a random variable defined as $Y = aX + b$, where $a \in \mathbb{R}^*$ ($\mathbb{R}^* = \mathbb{R}\backslash\{0\}$) and $b \in \mathbb{R}$ are constants. The characteristic function of Y can be derived from the one of X:

$$\phi_Y(u) = E[\exp(iu(aX + b))] = \exp(iub)\phi_X(ua). \tag{2.22}$$

In particular, the characteristic function of the standard random variable

$$Z = \frac{X - E[X]}{\text{std}[X]}$$

can be obtained by setting $a = 1/\sigma$ and $b = -\mu/\sigma$ in (2.22), i.e.,

$$\phi_Z(u) = E\left[\exp\left(\frac{iu(X-\mu)}{\sigma}\right)\right] = \exp\left(-i\frac{u\mu}{\sigma}\right)\phi_X\left(\frac{u}{\sigma}\right),$$

where $\mu = E[X]$ and $\sigma = \sqrt{\text{var}[X]}$.

Let $X_1, \ldots, X_n$ be a sequence of mutually independent random variables and define

$$W = \sum_{j=1}^{n} X_j.$$

Then, the characteristic function of W is

$$\phi_W(u) = E\left[\exp\left(iu\sum_{j=1}^{n} X_j\right)\right] = \prod_{j=1}^{n} E\left[\exp(iuX_j)\right] = \prod_{j=1}^{n} \phi_{X_j}(u). \tag{2.23}$$

If we additionally assume that the X_js are identically distributed to X, we get

$$\phi_W(u) = E\left[\exp\left(iu\sum_{j=1}^{n} X_j\right)\right] = [\phi_X(u)]^n.$$

Equation (2.23) can be extended to get the characteristic function of any linear combination of n independent random variables with coefficients a_j, for $j = 1, \ldots, n$,

$$\begin{aligned}\phi_W(u) &= E\left[\exp\left(iu\sum_{j=1}^{n} a_j X_j\right)\right] \\ &= \prod_{j=1}^{n} E\left[\exp(iua_j X_j)\right] = \prod_{j=1}^{n} \phi_{X_j}(ua_j).\end{aligned}$$

Here, we provide two fundamental theorems showing the two-way link between characteristic function and the distribution function. The first one, due to Paul Lévy and known as *inversion theorem*, allows one to get the distribution function of a random variable from the knowledge of its characteristic function.

Theorem 2.1 (Inversion theorem). *Let $F_X(x)$ and $\phi_X(u)$ be the cumulative distribution and the characteristic function of a real-valued random variable X, respectively. If $(c+h)$ and $(c-h)$ are continuity points of $F_X(x)$, then*

$$F_X(c+h) - F_X(c-h) = \lim_{U\to\infty} \frac{1}{\pi}\int_{-U}^{U} \frac{\sin(hu)}{u}\exp(-iuc)\phi_X(u)du. \tag{2.24}$$

Since both c and h are arbitrary, the knowledge of $\phi_X(u)$ allows one to compute $P(x_1 < X \leq x_2) = F_X(x_2) - F_X(x_1)$, where x_1 and x_2 are continuity points of $F_X(x)$.

Theorem 2.2. *Two probability distributions are identical if and only if the corresponding characteristic functions are identical.*

Theorem 2.2 does not imply the existence of such a correspondence between characteristic function and random variables because two different random variables can have the same distribution and the same characteristic function. As an intuitive example, consider the random variable X, symmetric around zero, and $Y = -X$. Even if they are the opposite of each other, they have the same distribution and characteristic function.

As a consequence of Theorem 2.1, we want to show how to derive the density function of a continuous random variable X from the knowledge of its characteristic function. If $\int_{-\infty}^{+\infty} |\phi_X(u)|du < \infty$, then $F_X(x)$ is the distribution function of a continuous random variable X and, setting $x = c$, (2.24) becomes

$$F_X(x+h) - F_X(x-h) = \frac{1}{\pi}\int_{-\infty}^{+\infty} \frac{\sin(hu)}{u} \exp(-iux)\phi_X(u)du.$$

Now, dividing both sides by $2h$ and sending h to 0, we can write

$$\lim_{h\to 0} \frac{F_X(x+h) - F_X(x-h)}{2h} = \frac{1}{2\pi}\int_{-\infty}^{+\infty} \exp(-iux)\phi_X(u)du.$$

Since

$$\lim_{h\to 0} \frac{F_X(x+h) - F_X(x-h)}{2h} = f_X(x),$$

the density function of the random variable X can be obtained from its characteristic function as

$$f_X(x) = \frac{1}{2\pi}\int_{-\infty}^{+\infty} \exp(-iux)\phi_X(u)du. \tag{2.25}$$

Equation (2.25) is the so-called *inverse Fourier transform*. This equality will turn out to be particularly useful in Chapters 5, 6, 8, and 11 to evaluate numerically the densities of stable, tempered stable, and normal tempered stable random variables which are not known in closed form.

Some properties of random variables

We conclude this discussion on the characteristic function with the definition of three important properties that may characterize the distribution of a random variable:

- infinite divisibility;
- reproductivity;
- stability.

A random variable X is said to be *infinitely divisible* if there exists a random variable X_n such that

$$\phi_X(u) = [\phi_{X_n}(u)]^n, \quad \forall n \in \mathbb{N}.$$

In Chapter 3, a more formal definition of infinitely divisible distributions is given and their link with Lévy processes is analyzed in detail. A systematic use of infinitely distributions is made throughout this book.

Let $X_1, \ldots, X_n$ be a sequence of independent random variables with characteristic functions $\phi_X(u, \alpha_1), \ldots, \phi_X(u, \alpha_n)$. Then, the random variable $X = \sum_{j=1}^n X_j$ is said to be *reproductive* with respect to the parameter α if its characteristic function is

$$\phi_X(u, \alpha) = \prod_{j=1}^n \phi_X(u, \alpha_j),$$

where $\alpha = \sum_{j=1}^n \alpha_j$. Under appropriate conditions, the infinitely divisibility implies the reproductive property. In Sections 2.4 and 2.5 some examples of distributions, both infinitely divisible and reproductive, are provided.

A random variable X is referred to as *stable* if there exist two numbers $\alpha_n > 0$, and β_n such that the characteristic function of the random variable $Y = \alpha_n X + \beta_n$ is

$$\phi_Y(u) = \exp(iu\beta_n)\phi_X(u\alpha_n) = [\phi_X(u)]^n, \quad \forall n \in \mathbb{N}. \tag{2.26}$$

Stable and tempered stable distributions are discussed in depth in Chapters 5 and 6. Furthermore, a multivariate normal tempered stable model is developed in Chapter 7 and its estimation methodology is discussed in Chapter 8.

2.3.2. Moment-generating function

The *moment-generating function* of a real-valued random variable X is defined as

$$\vartheta_X(u) = E[\exp(uX)], \quad u \in \mathbb{R}. \tag{2.27}$$

While the characteristic function always exists because it is the integral of a bounded function on a space of finite measure, the moment-generating function may not exist. In particular, the moment-generating function of a random variable X is well defined if and only if $E[\exp(uX)]$ is finite for values of u in a interval around 0.

If X admits a moment-generating function, then its distribution function is uniquely determined and vice versa. When it does exist, the moment-generating function can be obtained from the expression of the characteristic function as follows:

$$\vartheta_X(u) = E[\exp(uX)] = \phi_X(-iu).$$

Furthermore, if X has moments of every order, (2.27) can be written using the MacLaurin series expansion of the function $\exp(ux)$

$$\vartheta_X(u) = 1 + \sum_{k=1}^{\infty} \frac{u^k}{k!} E[X^k],$$

where the coefficient of the term $E[X^k]$ is the raw moment of order k, which can be computed as

$$E[X^k] = \frac{d}{du^k} \vartheta_X(u)|_{u=0}.$$

As we did for the characteristic function, we provide some properties of moment-generating function that can be helpful in applications.

The moment-generating function of the random variable $Y = aX + b$, where $a \in \mathbb{R}^*$ are constants, is given by

$$\vartheta_Y(u) = E[\exp(u(aX + b)] = \exp(ub)\vartheta_X(ua).$$

The moment-generating function of the random variable $W = \sum_{j=1}^{n} a_j X_j$ can be computed as

$$\begin{aligned}\vartheta_W(u) &= E\left[\exp\left(u\sum_{j=1}^{n} a_j X_j\right)\right]\\ &= \prod_{j=1}^{n} E\left[\exp(ua_j X_j)\right] = \prod_{j=1}^{n} \vartheta_{X_j}(ua_j),\end{aligned} \tag{2.28}$$

where $X_1, X_2, \ldots, X_n$ are independent random variables and $a_1, a_2, \ldots, a_n$ are constants.

Setting $a_j = 1$ for all j in (2.28), we get the moment-generating function of the sum of n independent random variables

$$\vartheta_W(u) = E\left[\exp\left(u\sum_{j=1}^{n} X_j\right)\right] = \prod_{j=1}^{n} E[\exp(uX_j)] = \prod_{j=1}^{n} \vartheta_{X_j}(u).$$

If the X_js are also identically distributed to a random variable X, the previous expression becomes

$$\vartheta_W(u) = E\left[\exp\left(u\sum_{j=1}^{n} X_j\right)\right] = [\vartheta_X(u)]^n.$$

2.3.3. Cumulant-generating functions

The *cumulant-generating function*, also known as *Laplace exponent* in the theory of Lévy processes, is defined as the natural logarithm of the moment-generating function

$$l_X(u) = \log \vartheta_X(u) = \log E[\exp(uX)] = \log \phi_X(-iu). \tag{2.29}$$

If the moment-generating function exists, (2.29) can be written using the MacLaurin series expansion

$$l_X(u) = \sum_{k=1}^{\infty} \frac{u^k}{k!} c_k(X),$$

where the coefficient of the term $c_k(X)$ is the *cumulant of order k*, which can be computed as

$$c_k(X) = \frac{d}{du^k} l_X(u)|_{u=0}. \tag{2.30}$$

The *cumulant-characteristic function*, sometimes referred to as the *second characteristic function* and known in the literature of Lévy processes as the *characteristic exponent*, is defined as the natural logarithm of the characteristic function, that is

$$\psi_X(u) = \log E[\exp(iuX)] = \log \phi_X(u).$$

In probability and statistics, both the cumulant-generating function and the cumulant-characteristic function are usually called the cumulant-generating function. However, since the second one is obtained from the characteristic function, we prefer to call it as the cumulant-characteristic function to distinguish it from the first one.

While the cumulant-generating function does not always exist, the cumulant-characteristic function is defined for all $u \in \mathbb{R}$ and can be expanded in the following way:

$$\psi_X(u) = \sum_{k=1}^{\infty} \frac{(iu)^k}{k!} c_k(X),$$

where the kth cumulant can be computed as

$$c_k(X) = i^{-k} \frac{d}{du^k} \psi_X(u)|_{u=0}.$$

The cumulant-characteristic function of the random variable $Y = aX + b$, where $a \in \mathbb{R}^*$ and $b \in \mathbb{R}$ are constants, is given by

$$\psi_Y(u) = iub + \psi_X(ua). \tag{2.31}$$

Consider now the random variable W, defined as $W = \sum_{j=1}^{n} X_j$, where the X_js are mutually independent random variables. The second characteristic function of W is

$$\psi_W(u) = \sum_{j=1}^{n} \psi_{X_j}(u).$$

Thus, the cumulant-characteristic function of the sum of n independent random variables is simply given by the sum of the cumulant-characteristic functions of the margins. Furthermore, the kth cumulant of the sum of n independent random variables is obtained by summing the kth cumulants of the n random variables.

Since

$$\psi_X(u) = \log \phi_X(u),$$

the kth cumulant can be expressed in terms of raw moments by evaluating

$$c_k(X) = i^{-k} \frac{d}{du^k} \log \phi_X(u)|_{u=0}.$$

In particular, the first four cumulants are

$$\begin{aligned}
c_1(X) &= E[X], \\
c_2(X) &= E[X^2] - (E[X])^2 = E[(X - E[X])^2] = \text{var}[X], \\
c_3(X) &= E[X^3] - 3E[X^2]\ E[X] + 2(E[X])^3 = E[(X - E[X])^3], \\
c_4(X) &= E[X^4] - 4E[X]\ E[X^3] + 12(E[X])^2\ E[X^2] \\
&\quad - 3(E[X^2])^2 - 6(E[X])^4 \\
&= E[(X - E[X])^4] - 3(\text{var}[X])^2.
\end{aligned} \tag{2.32}$$

Thus, expectation, variance, skewness, and kurtosis can be obtained as follows:

$$\begin{aligned}
E[X] &= c_1(X), \\
\text{var}[X] &= c_2(X),
\end{aligned}$$

$$\text{skew}[X] = \frac{c_3(X)}{c_2(X)^{3/2}}, \tag{2.33}$$

$$\text{kurt}[X] = 3 + \frac{c_4(X)}{c_2(X)^2}.$$

The formulas in equations (2.32) and (2.33) will be considered in Chapters 6 and 7.

2.4. Particular Discrete Random Variables

In the following, we review some well-known discrete random variables.

2.4.1. Bernoulli

Consider a random experiment Υ whose realization can produce one of the two incompatible events ω_s and ω_f, called, respectively, *success* and *failure*, or *true* and *false*, or *yes* and *no*. It is possible to associate with the result of this random experiment a random variable X such that $X(\omega_s) = 1$, with $P(\omega_s) = P(X = 1)$, and $X(\omega_f) = 0$, with $P(\omega_f) = P(X = 0)$. The random variable X follows a *Bernoulli distribution* and has probability mass function

$$f_X(x) = p^x(1-p)^{1-x}, \quad x = 0, 1,$$

where $p = P(X = 1)$ is the probability of success. To denote that X is Bernoulli-distributed with parameter p, we write $X \sim \text{Ber}(p)$. The distribution of heads and tails in coin tossing is an example of a Bernoulli distribution.

The characteristic function of the Bernoulli distribution is

$$\phi_X(u) = 1 + p[\exp(iu) - 1],$$

and its moment-generating function is given by

$$\vartheta_X(u) = 1 + p[\exp(u) - 1].$$

Observing that $E[X^k] = p$ for $k = 1, 2, \ldots$ and using the relationship between raw and central moments, it is possible to get

$$\begin{aligned} E[X] &= p, \\ \mathrm{var}[X] &= p(1-p), \\ \mathrm{skew}[X] &= \frac{1-2p}{\sqrt{p(1-p)}}, \\ \mathrm{kurt}[X] &= \frac{3p^2 - 3p + 1}{p(1-p)}. \end{aligned}$$

The Bernoulli distribution will be considered in value-at-risk backtesting in Chapter 8.

2.4.2. Binomial

Let $X_1, \ldots, X_n$ be a sequence independent Bernoulli-distributed random variables with parameter p. Then, the random variable $X = \sum_{j=1}^{n} X_i$, describing the number of successes in n independent and identical Bernoulli experiments, follows a *binomial distribution* with parameters n and p, i.e., $X \sim \mathrm{Bin}(n, p)$, and its probability function is

$$f_X(x) = \frac{n!}{x!(n-x)!} p^x (1-p)^{1-x}, \quad x = 0, 1, \ldots, n. \tag{2.34}$$

From (2.34) it is immediate to get the Bernoulli cumulative distribution function

$$F_X(k) = \sum_{x=0}^{\lfloor k \rfloor} \frac{n!}{x!(n-x)!} p^x (1-p)^{1-x},$$

where $\lfloor k \rfloor$ denotes the greatest integer less than or equal to k.

The characteristic and moment-generating functions are given by

$$\phi_X(u) = [1 + p(\exp(iu) - 1)]^n, \tag{2.35}$$

$$\vartheta_X(u) = [1 + p(\exp(u) - 1)]^n. \tag{2.36}$$

Using (2.35) or (2.36) we can get the following characteristics of the binomial distribution:

$$\begin{aligned} E[X] &= np, \\ \text{var}[X] &= np(1-p), \\ \text{skew}[X] &= \frac{1-2p}{\sqrt{np(1-p)}}, \\ \text{kurt}[X] &= 3 + \frac{1-6p(1-p)}{np(1-p)}. \end{aligned}$$

The binomial distribution is reproductive with respect to the parameter n. Let $X_1, \ldots, X_d$ be independent binomial random variables such that $X_j \sim \text{Bin}(n_j, p)$, $j = 1, 2, \ldots, d$. Then,

$$\sum_{j=1}^{d} X_j \sim \text{Bin}\left(\sum_{j=1}^{d} n_j, p\right). \tag{2.37}$$

The binomial distribution will be applied in Chapter 6 to demonstrate that the normal distribution can be adopted as an approximate model for the binomial distribution and in Chapter 9 to define the order statistics.

2.4.3. Poisson

A random variable X with values in $\mathbb{N}$ and probability function

$$f_X(x) = \frac{1}{x!}\lambda^x \exp(-\lambda), \quad x = 0, 1, 2, \ldots, \quad \lambda > 0, \tag{2.38}$$

is *Poisson* distributed with shape parameter λ, and we write $X \sim \text{Poiss}(\lambda)$.

From (2.38), it is immediate to get the Poisson cumulative distribution function

$$F_X(k) = \sum_{x=0}^{\lfloor k \rfloor} \frac{1}{x!}\lambda^x \exp(-\lambda),$$

where again $\lfloor k \rfloor$ indicates the greatest integer less than or equal to k.

The characteristic and moment-generating functions are given by

$$\phi_X(u) = \exp[\lambda(\exp(iu) - 1)], \tag{2.39}$$

$$\vartheta_X(u) = \exp[\lambda(\exp(u) - 1)]. \tag{2.40}$$

Using (2.39) or (2.40), the following properties of the Poisson distribution can be demonstrated:

$$\begin{aligned} E[X] &= \lambda, \\ \text{var}[X] &= \lambda, \\ \text{skew}[X] &= \lambda^{-1/2}, \\ \text{kurt}[X] &= 3 + \lambda^{-1}. \end{aligned} \tag{2.41}$$

The Poisson distribution is infinitely divisible since

$$\phi_X(u) = \exp[\lambda\,(\exp(iu) - 1)] = \exp\left[\frac{\lambda}{n}\,(\exp(iu) - 1)\right]^n, \quad \forall n \in \mathbb{N},$$

and is reproductive since

$$\sum_{j=1}^{n} X_j \sim \text{Poiss}\left(\sum_{j=1}^{n} \lambda_j\right),$$

where $X_1, \ldots, X_n$ is a vector of independent Poisson random variables with parameter λ_j, for $j = 1, \ldots, n$. The Poisson distribution will be applied in Chapter 3 to define the Poisson process and in Chapter 9.

2.5. Particular Continuous Random Variables

In the following, we review some well-known continuous random variables and compare the shape of their density function to the density of the normal random variable with the same mean and variance.

2.5.1. Gamma

A *gamma random variable* in its most general form has the following density function:

$$f_X(x) = \frac{b^a}{\Gamma(a)}(x-c)^{a-1}\exp(-b(x-c)), \quad x > c,\ a, b > 0,\ c \geq 0, \tag{2.42}$$

where

$$\Gamma(a) = \int_0^{\infty} x^{a-1}\exp(-x)dx \tag{2.43}$$

is the gamma function.

Setting $c = 0$, equation (2.42) provides the density function of the most used gamma random variable

$$f_X(x) = \frac{b^a}{\Gamma(a)}x^{a-1}\exp(-bx). \tag{2.44}$$

We denote the gamma random variable with density (2.44) as $X \sim \Gamma(a,b)$.

The characteristic function of the gamma distribution with shape parameter a and scale parameter b is

$$\phi_X(u) = \left(\frac{b}{b-iu}\right)^a,$$

and the moment-generating function, defined for $u < b$, is

$$\vartheta_X(u) = \left(\frac{b}{b-u}\right)^a.$$

Since

$$\frac{d^k\vartheta_X(u)}{dk^r} = b^a\frac{\Gamma(a+k)}{\Gamma(a)}(b-u)^{-a-k},$$

the raw moment of order k is

$$E[X^k] = \frac{\Gamma(a+k)}{b^k\Gamma(a)}. \tag{2.45}$$

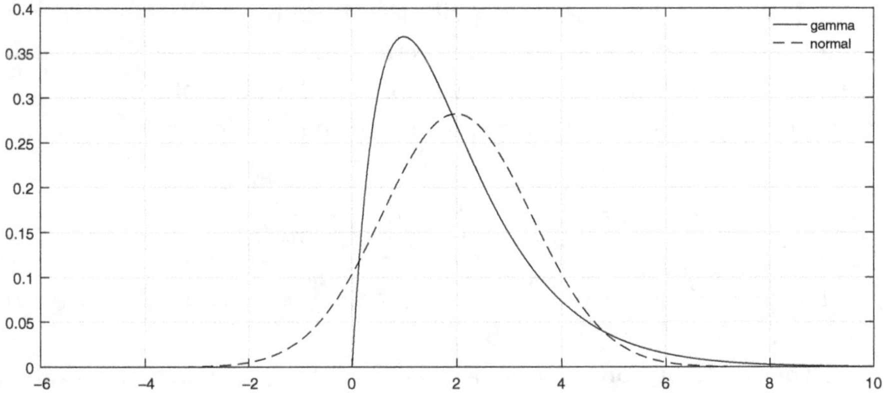

Figure 2.1. Shape of the probability density function of a gamma law with parameters $a = 2$ and $b = 1$.

Using (2.45), the following characteristics can be derived:

$$\begin{aligned} E[X] &= \frac{a}{b}, \\ \text{var}[X] &= \frac{a}{b^2}, \\ \text{skew}[X] &= \frac{2}{\sqrt{a}}, \\ \text{kurt}[X] &= 3\left(1 + \frac{2}{a}\right). \end{aligned}$$

In Figure 2.1, we show the shape of the density function of a gamma law and compare it to the normal density with the same mean and variance. The figure is obtained by means of the Matlab function *gampdf*. Note that this Matlab function has a slightly different parameterization.

The gamma distribution is reproductive with respect to the parameter a. Thus, if $X_1, \ldots, X_n$ are independent gamma random variables with parameters $a_1, \ldots, a_n$ and with the same parameter b, then

$$\sum_{j=1}^{n} X_j \sim \Gamma\left(\sum_{j=1}^{n} a_j, b\right).$$

Additionally, the gamma distribution is infinitely divisible. In particular, for every $n \in \mathbb{R}$, the characteristic function of $X \sim \Gamma(a,b)$ can be expressed as the nth power of the characteristic function of gamma distributed random variables with parameters a/n and b, i.e.,

$$\phi_X(u) = \left(\frac{b}{b-iu}\right)^a = \left[\left(\frac{b}{b-iu}\right)^{a/n}\right]^n.$$

The gamma has also the following scaling property: if $X \sim \Gamma(a,b)$, then $kX \sim \Gamma(a, b/k)$ for $k > 0$.

Other distributions can be derived from the gamma. For example, setting $b = 1$ in (2.44), we get the *reduced gamma*. Instead, if we set $a = d/2$ and $b = 1/2$, where $d \in \mathbb{N}$, equation (2.44) provides the density function of *chi-square* random variable with d degrees of freedom. While if we set $a = 1$ and $b = \lambda$, we get the density of the exponential random variable with parameter λ.

The gamma distribution will be considered in Chapters 4 and 7 to define the univariate and multivariate variance gamma distribution.

2.5.2. Exponential

A random variable X follows an exponential distribution with scale parameter λ, and we denote it as $X \sim \mathrm{Exp}(\lambda)$, if its density is given by

$$f_X(x) = \lambda \exp(-\lambda x), \quad x \geq 0, \quad \lambda > 0.$$

The characteristic function and the moment-generating function are given by

$$\phi_X(u) = a\frac{\lambda}{\lambda - iu}, \tag{2.46}$$

$$\vartheta_X(u) = \frac{\lambda}{\lambda - u}, \quad u < \lambda. \tag{2.47}$$

The kth raw moment of $X \sim \mathrm{Exp}(\lambda)$ is

$$E[X^k] = \frac{k!}{\lambda^k},$$

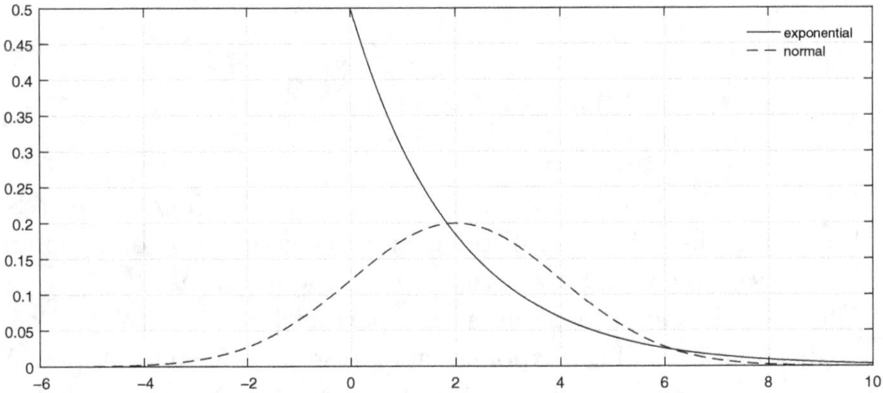

Figure 2.2. Shape of the probability density function of an exponential law with parameter $\lambda = 0.5$.

and the four main moments are

$$\begin{aligned} E[X] &= \lambda^{-1}, \\ \text{var}[X] &= \lambda^{-2}, \\ \text{skew}[X] &= 2, \\ \text{kurt}[X] &= 9. \end{aligned}$$

Thus, both skewness and kurtosis of exponential random variables do not depend on the parameter λ.

In Figure 2.2, we show the shape of the density function of an exponential law and compare it to the normal density with same mean and variance. The figure is obtained by means of the Matlab function *exppdf*. Note that this Matlab function has a slightly different parameterization.

The exponential is a *memoryless* random variable, that is

$$\text{Prob}\,(X > h + k | X > h) = \text{Prob}\,(X > k) = \exp(-\lambda k),$$

where h and k are two positive numbers.

If $X_1, \ldots, X_n$ are independent exponential random variables with the same parameter λ, then

$$\sum_{j=1}^{n} X_j \sim \Gamma\,(n, \lambda)\,.$$

Since

$$\vartheta_X(u) = \frac{\lambda}{\lambda - iu} = \left[\left(\frac{\lambda}{\lambda - iu}\right)^{1/n}\right]^n,$$

X is infinitely divisible because its characteristic function can be obtained as the nth power of the characteristic function of a gamma random variable with parameters $1/n$ and λ, for all $n \in \mathbb{N}$.

There exists an interesting and important link between the Poisson distribution and the exponential one (see also Chapter 3). Let X be an exponential random variable measuring the time to wait before a given event will occur. Let also Y be the random variable counting the number of times the event has occurred up to time t. Then, the probability that the event has not occurred by time t is

$$P(Y = 0) = P(X > t) = \exp(-\lambda t),$$

which is the probability that a Poisson random variable with intensity λt assumes a zero value.

2.5.3. Inverse Gaussian

An *inverse Gaussian* random variable X, with parameters $a > 0$ and $b > 0$, is a positive continuous random variable with density function

$$f_X(x) = \frac{a}{\sqrt{2\pi}} \exp\left[-\frac{1}{2}\frac{(bx-a)^2}{x}\right] x^{-3/2}, \quad x > 0,$$

and characteristic function

$$\phi_X(u) = \exp[a(b - \sqrt{b^2 - 2iu})].$$

The moment-generating function, defined for $u \leq b^2/2$, is

$$\vartheta_X(u) = \exp[a(b - \sqrt{b^2 - 2u})].$$

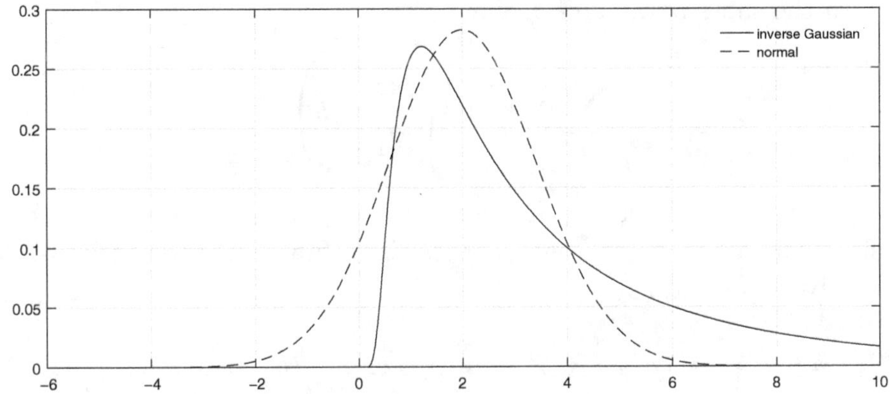

Figure 2.3. Shape of the probability density function of an inverse Gaussian law with parameters $a = 2$ and $b = 1$.

and the first four moments are

$$E[X] = \frac{a}{b},$$
$$\text{var}[X] = \frac{a}{b^3},$$
$$\text{skew}[X] = \frac{3}{\sqrt{ab}},$$
$$\text{kurt}[X] = 3\left(1 + \frac{5}{ab}\right).$$

In Figure 2.3, we show the shape of the density function of an inverse Gaussian law and compare it to the normal density with same mean and variance. The figure is obtained by means of the Matlab function *pdf* with *name* equal to *InverseGaussian*. Note that this Matlab function has a slightly different parameterization.

The inverse Gaussian random variable possesses the following scaling property: if k is a positive constant and $X \sim \text{IG}(a, b)$, then $kX \sim \text{IG}(a\sqrt{k}, b/\sqrt{k})$.

The inverse Gaussian distribution is also reproductive with respect to the parameter a. Indeed, if $X_1, \ldots, X_n$, are independent inverse Gaussian random variables with parameters $a_1, \ldots, a_n$ and with the same parameter b, then

$$\sum_{j=1}^{n} X_j \sim \mathrm{IG}\left(\sum_{j=1}^{n} a_j, b\right).$$

Lastly, the inverse Gaussian distribution is infinitely divisible since for every $n \in \mathbb{R}$, the characteristic function of $X \sim \mathrm{IG}(a, b)$ can be expressed as

$$\phi_X(u) = \exp[a(b - \sqrt{b^2 - 2iu})] = \exp\left[\frac{a}{n}(b - \sqrt{b^2 - 2iu})\right]^n.$$

The inverse Gaussian distribution will be used in Chapters 4 and 7 to define the univariate and multivariate normal inverse Gaussian distribution.

2.5.4. Normal

A *normal*, or *Gaussian*, random variable X with location parameter μ and scale parameter σ has density function

$$f_X(x) = \frac{1}{\sigma\sqrt{2\pi}} \exp\left[-\frac{1}{2}\left(\frac{x-\mu}{\sigma}\right)^2\right], \quad x \in \mathbb{R}, \quad \mu \in \mathbb{R}, \quad \sigma > 0. \tag{2.48}$$

The characteristic function and the moment-generating function are

$$\phi_X(u) = \exp\left(i\mu u - \frac{\sigma^2 u^2}{2}\right), \tag{2.49}$$

$$\vartheta_X(u) = \exp\left(\mu u + \frac{\sigma^2 u^2}{2}\right).$$

All the central moments of odd order are null, while the even are given by

$$E[(X-\mu)^k] = 2^{1-\frac{k}{2}} \sigma^k \frac{\Gamma(k)}{\Gamma(k/2)}, \quad k = 2n, \quad n \in \mathbb{N}.$$

Furthermore, the first four moments are

$$\begin{aligned} E[X] &= \mu, \\ \text{var}[X] &= \sigma^2, \\ \text{skew}[X] &= 0, \\ \text{kurt}[X] &= 3. \end{aligned}$$

The normal random distribution is infinitely divisible since

$$\begin{aligned} \phi_X(u) &= \exp\left(i\mu u - \frac{\sigma^2 u^2}{2}\right) \\ &= \left[\exp\left(i(\mu/n)u - \frac{(\sigma^2/n)u^2}{2}\right)\right]^n, \quad \forall n \in \mathbb{N}, \end{aligned}$$

and is reproductive with respect to both μ and σ^2. Specifically, given a sequence $X_1, \ldots, X_n$ of independent $N(\mu_j, \sigma_j^2)$ random variables, $j = 1, \ldots, n$, we have

$$\sum_{j=1}^n X_j \sim N\left(\sum_{j=1}^n \mu_j, \sum_{j=1}^n \sigma_j^2\right).$$

Furthermore, the Gaussian distribution is stable with respect to both linear transformations and linear combinations, i.e.,

- if $X \sim N(\mu, \sigma^2)$, then

  $$aX + b \sim N(a\mu + b, a^2\sigma^2),$$

  where a, b are any real constants;
- if $X_j \sim N(\mu_j, \sigma_j^2)$ and $\text{corr}(X_j, X_k) = \rho_{jk}$ for $j, k = 1, 2, \ldots, n$, then

  $$\sum_{j=1}^n a_j X_j \sim N\left(\sum_{j=1}^n a_j \mu_j, \sum_{j=1}^n \sum_{k=1}^n a_j a_k \sigma_j \sigma_k \rho_{jk}\right),$$

  where the a_js are real constants.

Let X be a Gaussian random variable with expectation μ and standard deviation σ. Then,

$$Z = \frac{X - \mu}{\sigma} \sim N(0, 1)$$

and we refer to it as a *standard normal random variable.* Setting $\mu = 0$ and $\sigma = 1$ in (2.48) and in (2.49), it is possible to get the density

$$f_Z(z) = \frac{1}{\sqrt{2\pi}} \exp\left(-\frac{1}{2}z^2\right), \quad z \in \mathbb{R},$$

and the characteristic function

$$\phi_Z(u) = \exp\left(-\frac{u^2}{2}\right).$$

The Gaussian distribution was the first distribution used to model asset returns. The normality assumption for the distribution of the underlying stock is also one of the main assumption of the Black–Scholes option pricing model (see Chapter 11).

2.5.5. Log-normal

Let Y be a $N(\mu, \sigma^2)$ and let X be a positive random variable such that $Y = \ln X$. Then X is called a *log-normal random variable* and its density is

$$f_X(x) = \frac{1}{\sigma\sqrt{2\pi}x} \exp\left[-\frac{1}{2}\left(\frac{\ln x - \mu}{\sigma}\right)^2\right], \quad x > 0.$$

We refer to the distribution of X as Lognormal (μ, σ^2).

The moments of a log-normal can be derived making use of the moment-generating function of the normal random variable. Thus, the raw moment of order k can be computed as

$$E[X^k] = E[\exp(kY)] = \exp\left(k\mu + \frac{k^2\sigma^2}{2}\right). \tag{2.50}$$

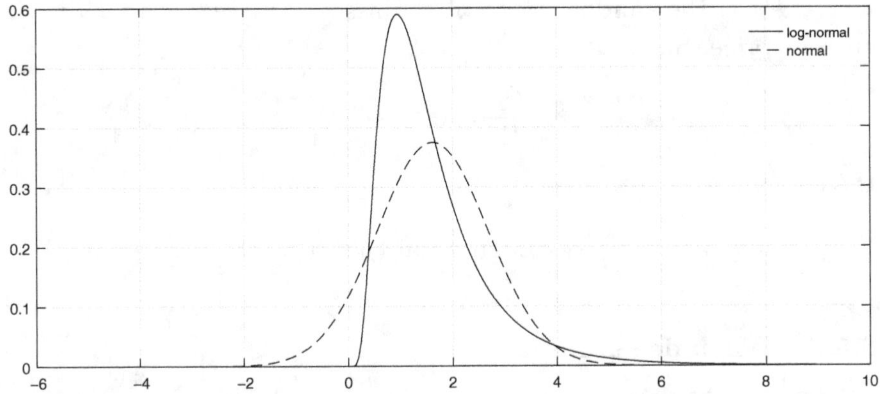

Figure 2.4. Shape of the probability density function of a log-normal law with parameters $\mu = 0.3$ and $\sigma = 0.6$.

For $k = 1$ and $k = 2$, equation (2.50) allows to compute

$$\begin{aligned} E[X] &= \exp(\mu + \sigma^2/2), \\ \mathrm{var}[X] &= \exp(2(\mu + \sigma^2))(1 - \exp(-\sigma^2)). \end{aligned}$$

In Figure 2.4, we show the shape of the density function of a log-normal law and compare it to the normal density with the same mean and variance. The figure is obtained by means of the Matlab function *lognpdf.*

Let $X_1, \ldots, X_n$ be independent log-normal random variables with parameters μ_j and σ_j^2, for $j = 1, \ldots, n$. Then,

$$\prod_{j=1}^{n} X_j \sim \text{Lognormal}\left(\sum_{j=1}^{n} \mu_j, \sum_{j=1}^{n} \sigma_j^2\right).$$

The log-normal distribution has been used in finance to model asset prices. Specifically, let r_T be a $N(\mu T, \sigma^2 T)$ random variable representing the log-return of an asset on the time interval $[0, T]$, i.e.,

$$r_T = \log \frac{P_T}{P_0},$$

where P_T is the random price of the asset at time T and P_0 is the current price. Since

$$\frac{P_T}{P_0} = \exp(r_T),$$

then

$$\frac{P_T}{P_0} \sim \text{Lognormal } (\mu T, \sigma^2 T).$$

2.5.6. Variance gamma

Let Z be a $N(0,1)$ and G an independent $\Gamma(a,b)$ random variable defined as

$$X = \mu + \theta G + \sigma\sqrt{G}Z,$$

where μ and $\theta \in \mathbb{R}$ and $\sigma > 0$. Then, the random variable X follows a *variance gamma distribution* (VG) with parameters μ, θ, a, b, σ, and we write $X \sim \text{VG}(\mu, \theta, a, b, \sigma)$.

By definition, the VG is a normal variance–mean mixture distribution with a gamma mixing distribution. Since $X|(G = g) \sim N(\mu + \theta g, \sigma^2 g)$, the density of X can be written as

$$\begin{aligned} f_X(x) = \int_0^\infty \frac{1}{\sqrt{2\pi\sigma^2 g}} \exp\left[-\frac{1}{2}\left(\frac{x-\mu-\theta g}{\sigma\sqrt{g}}\right)^2\right] \\ \times \frac{b^a}{\Gamma(a)} g^{a-1} \exp\left(-gb\right) dg. \end{aligned} \tag{2.51}$$

This representation emphasizes the nature of the VG random distribution as a normal variance–mean mixture.

The density in (2.51) can be developed to obtain

$$\begin{aligned} f_X(x) = \frac{\sqrt{2}b^a e^{\frac{(x-\mu)\theta}{\sigma^2}}}{\sigma\sqrt{\pi}\Gamma(a)} \left[\frac{(x-\mu)^2}{\theta^2 + 2b\sigma^2}\right]^{\frac{h}{2}} \\ \times K_h\left(\frac{1}{\sigma^2}\sqrt{(x-\mu)^2(\theta^2 + 2b\sigma^2)}\right), \end{aligned}$$

where $h = a - 1/2$ and K_h is the modified Bessel function of the second kind of order h.

The characteristic function of X is

$$\phi_X(u) = \exp(iu\mu)\left[1 - \frac{1}{b}\left(iu\theta - \frac{1}{2}u^2\sigma^2\right)\right]^{-a},$$

and its first four moments are

$$\begin{aligned} E[X] &= \mu + \frac{a}{b}\theta, \\ \mathrm{var}[X] &= \left(b\sigma^2 + \theta^2\right)\frac{a}{b^2}, \\ \mathrm{skew}[X] &= \frac{\theta\left(3b\sigma^2 + 2\theta^2\right)}{\sqrt{a}\left(b\sigma^2 + \theta^2\right)^{3/2}}, \\ \mathrm{kurt}[X] &= 3\left[1 + \frac{2}{a} - \frac{b^2\sigma^4}{a(b\sigma^2 + \theta^2)^2}\right]. \end{aligned}$$

In Figure 2.5, we show the shape of the density function of a variance gamma law with zero mean and compare it to the normal density with the same mean and variance. The figure is obtained by means of the Matlab function *densityMVG_matrix* explained in Chapter 7.

The parameter θ determines the sign of the skewness. If $\theta > 0$ ($\theta < 0$), then the right tail is longer (shorter) than the left one. If $\theta = 0$, the distribution is symmetric around the mean. The VG distribution is leptokurtic because it has a positive excess of kurtosis.

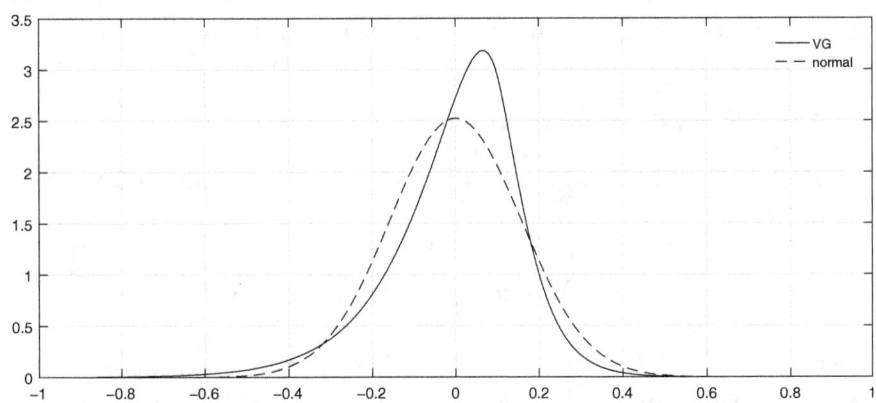

Figure 2.5. Shape of the probability density function of a Variance Gamma (VG) law with parameters $a = 2$, $b = 1$, $\theta = -0.05$, $\sigma = 0.1$, and $\mu = -a/b\theta$.

The VG random variable is infinitely divisible because

$$\begin{aligned}\phi_X(u) &= \exp(iu\mu)\left[1-\frac{1}{b}\left(iu\theta-\frac{1}{2}u^2\sigma^2\right)\right]^{-a}\\ &= \left[\exp\left(iu\frac{\mu}{n}\right)\left[1-\frac{1}{b}\left(iu\theta-\frac{1}{2}u^2\sigma^2\right)\right]^{-\frac{a}{n}}\right]^n, \quad \forall n\in\mathbb{N},\end{aligned}$$

and is reproductive with respect to μ and a. In particular, if $X_1, X_2, \ldots, X_n$ are independent VG $(\mu_j, \theta, a_j, b, \sigma^2)$ random variables, $j = 1, 2, \ldots, n$, then

$$\sum_{j=1}^n X_j \sim \text{VG}\left(\sum_{j=1}^n \mu_j, \theta, \sum_{j=1}^n a_j, b, \sigma^2\right).$$

The symmetric VG distribution was introduced into finance literature in Madan and Seneta (1990) and subsequently extended in Madan *et al.* (1998) as a model for the increments of the log-price process of stocks. In Chapter 4, the VG distribution will be derived as a special case of the generalized hyperbolic distribution. In Chapter 7, the link between multivariate normal variance–mean mixtures and time-changed Brownian motion will be analyzed and, in particular, the multivariate VG distribution and process will be studied. In Chapter 8, an expectation–maximization algorithm to estimate the parameters of the VG gamma distribution will be presented.

2.5.7. Normal inverse Gaussian

Let Z be an $N(0,1)$ and S an independent IG(a,b) random variable. Then, define

$$X = \mu + \theta S + \sigma\sqrt{S}Z,$$

where μ, $\theta \in \mathbb{R}$ and $\sigma > 0$. The random variable X follows a *normal inverse Gaussian distribution* (NIG) with parameters μ, θ, a, b, σ, and we write $X \sim$ NIG$(\mu, \theta, a, b, \sigma)$. Since

$X|(S=s) \sim N(\mu+\theta s, \sigma^2 s)$, the density of X can be written as

$$f_X(x) = \int_0^\infty \frac{1}{\sqrt{2\pi\sigma^2 g}} \exp\left[-\frac{1}{2}\left(\frac{x-\mu-\theta g}{\sigma\sqrt{g}}\right)^2\right] \times \frac{a}{\sqrt{2\pi}} \exp\left[-\frac{1}{2}\frac{(bx-a)^2}{x}\right] x^{-3/2}. \quad (2.52)$$

Equation (2.52) shows the NIG density as the density of a normal variance–mean mixture with an inverse gamma mixing variable. This density can also be written as

$$f_X(x) = \frac{ae^{\frac{(x-\mu)\theta+ab\sigma^2}{\sigma^2}}}{\sigma\pi}\left[\frac{(x-\mu)^2+a^2\sigma^2}{\theta^2+b^2\sigma^2}\right]^{-\frac{1}{2}} \times K_1\left(\frac{1}{\sigma^2}\sqrt{[(x-\mu)^2+a^2\sigma^2]\left(\theta^2+b^2\sigma^2\right)}\right),$$

where K_1 is the modified Bessel function of the second kind of order 1.

The characteristic function of the normal inverse Gaussian random variable is

$$\phi_X(u) = \exp\left[iu\mu + a\left(b - \sqrt{b^2 - 2\left(iu\theta - \frac{1}{2}u^2\sigma^2\right)}\right)\right]. \quad (2.53)$$

Using (2.53), the main moments are

$$\begin{aligned}
E[X] &= \mu + \frac{a}{b}\theta, \\
\text{var}[X] &= (\sigma^2 b^2 + \theta^2)\frac{a}{b^3}, \\
\text{skew}[X] &= \frac{3\theta}{\sqrt{ab\left(\sigma^2 b^2 + \theta^2\right)}}, \\
\text{kurt}[X] &= 3\left[1 + \frac{5\theta^2 + \sigma^2 b^2}{ab\left(\sigma^2 b^2 + \theta^2\right)}\right].
\end{aligned}$$

The parameter θ determines the sign of the skewness. If $\theta > 0$ ($\theta < 0$), then the NIG distribution has a positive (negative) asymmetry. If $\theta = 0$ the distribution is symmetric. Since the excess of kurtosis

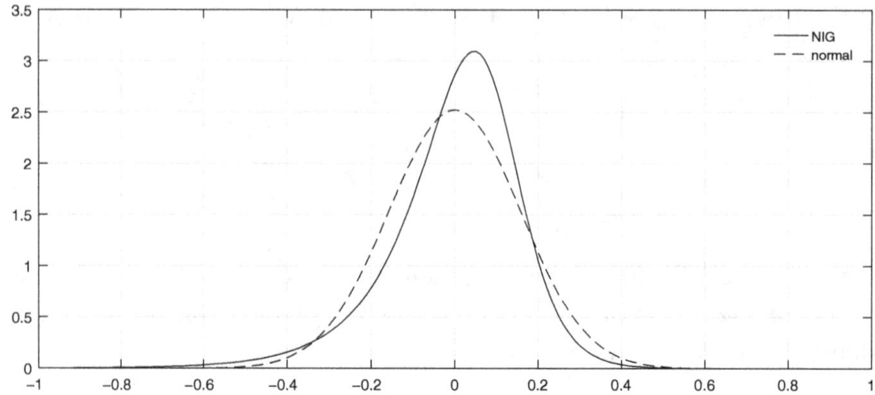

Figure 2.6. Shape of the probability density function of an NIG law with parameters $a = 2$, $b = 1$, $\theta = -0.05$, $\sigma = 0.1$, and $\mu = -a/b\theta$.

is positive, the NIG distribution has fatter tails than the Gaussian distribution.

In Figure 2.6, we show the shape of the density function of an NIG law with zero mean and compare it to the normal density with same mean and variance. The figure is obtained by means of the Matlab function *densityMNIG_matrix* reported in Chapter 7.

Since

$$\begin{aligned}\phi_X(u) &= \exp\left[iu\mu + a\left(b - \sqrt{b^2 - 2\left(iu\theta - \frac{1}{2}u^2\sigma^2\right)}\right)\right] \\ &= \exp\left[iu\frac{\mu}{n} + \frac{a}{n}\left(b - \sqrt{b^2 - 2\left(iu\theta - \frac{1}{2}u^2\sigma^2\right)}\right)\right]^n, \quad \forall n \in \mathbb{N},\end{aligned}$$

the NIG is infinitely divisible.

Furthermore, it is reproductive with respect to μ and a. In particular, if $X_1, \ldots, X_n$ are independent NIG $\left(\mu_j, \theta, a_j, b, \sigma^2\right)$ random variables, $j = 1, \ldots, n$, then

$$\sum_{j=1}^{n} X_j \sim \text{NIG}\left(\sum_{j=1}^{n}\mu_j, \theta, \sum_{j=1}^{n} a_j, b, \sigma^2\right).$$

This distribution was introduced into the finance literature in Barndorff-Nielsen (1995) as a model for the stock return distribution. In Chapter 4, the NIG distribution will be presented as a special case of the generalized hyperbolic. In Chapter 7, the multivariate NIG distribution and process will be constructed. In Chapter 8, an expectation–maximization algorithm to estimate the NIG model will be discussed.

2.5.8. Multivariate normal

A d-dimensional random variable $X = (X_1, \ldots, X_d)'$ follows a multivariate normal (or Gaussian) distribution if, for any constant vector $a \in \mathbb{R}^d$, the random variable $Y = a'X$ has a Gaussian distribution.

Alternatively, we can say that a d-dimensional random vector X has a multivariate normal distribution, with mean vector μ and covariance matrix Σ, if there exist an n-dimensional vector of standard normal random variables $Z = (Z_1, Z_2, \ldots, Z_n)'$, a $d \times n$ matrix A with $AA' = \Sigma$, and a vector $\mu \in \mathbb{R}^d$ such that $X = \mu + AZ$.

Note that these equivalent definitions do not require the covariance matrix to have full rank. In general, the covariance matrix is positive semidefinite and if $\mathrm{Det}(\Sigma) = 0$, X does not have a density.

A d-dimensional random vector X normally distributed with expectation $\mu \in \mathbb{R}^d$ and a positive definite covariance matrix Σ has density function

$$f_X(x) = \frac{1}{(2\pi)^{d/2}\sqrt{Det\,(\Sigma)}} \exp\left[-\frac{1}{2}(x-\mu)'\,\Sigma^{-1}\,(x-\mu)\right], \quad x \in \mathbb{R}^d.$$

As observed, if $\mathrm{Rank}(\Sigma_d) < d$ (i.e., Σ is only positive semidefinite), then X is degenerate and does not have a density.

The characteristic function can be written as

$$\phi_X(u) = \exp\left(i\mu'u - \frac{u'\Sigma u^2}{2}\right), \quad u \in \mathbb{R}^d, \tag{2.54}$$

and the moment-generating function can be written as

$$\vartheta_X(u) = \exp\left(\mu' u + \frac{u'\Sigma u^2}{2}\right), \quad u \in \mathbb{R}^d. \tag{2.55}$$

In scalar notation, (2.54) and (2.55) become

$$\phi_X(u) = \exp\left(i \sum_{j=1}^{n} \mu_j u_j - \frac{1}{2}\sum_{j=1}^{n}\sum_{k=1}^{n} u_j u_k \sigma_j \sigma_k \rho_{jk}\right), \tag{2.56}$$

$$\vartheta_X(u) = \exp\left(\sum_{j=1}^{n} \mu_j u_j + \frac{1}{2}\sum_{j=1}^{n}\sum_{j=1}^{n} u_j u_k \sigma_j \sigma_k \rho_{jk}\right), \tag{2.57}$$

where $\rho_{jk} = \text{corr}(X_j, X_k)$, $j, k = 1, 2, \ldots, n$.

Setting $u_k = 0$ for $k \neq j$ in (2.56) and in (2.57), the characteristic function and the moment-generating function of the jth margin are obtained as follows:

$$\phi_{X_j}(u) = \exp\left(i\mu_j u_j - \frac{\sigma_j^2 u^2}{2}\right), \quad u_j \in \mathbb{R},$$

and

$$\vartheta_{X_j}(u) = \exp\left(\mu_j u_j + \frac{\sigma_j^2 u^2}{2}\right), \quad u_j \in \mathbb{R}.$$

Multidimensional financial applications for decades have been, and in part are still, dominated by the multivariate normality assumption. The mean–variance approach to portfolio selection introduced by Markowitz is based on the assumption that asset returns follow a multivariate Gaussian distribution. The use of the multivariate distribution is particularly convenient because if the assets returns are jointly normal, then the portfolio return is normally distributed. Specifically, let $r \sim N(\mu, \Sigma_n)$ be an n-dimensional random variable that models the return distribution of n stocks. Let r_p be the return of a portfolio invested in these assets with weights

$\omega = (\omega_1, \omega_2, \ldots, \omega_n)'$, then

$$r_p \sim N(\omega'\mu, \omega'\Sigma_n\omega).$$

Applications of multivariate normal distribution to asset pricing are discussed in Chapters 8 and 11.

Key Points

- A random variable is a variable whose possible values are outcomes of a random phenomenon.
- A discrete random variable assumes a finite number or a countable infinity of values.
- For a continuous random variable, the probability of assuming a particular value is null.
- A random variable is characterized by different classes of parameters: moments, order parameters, location, scale, and shape parameters.
- The characteristic function of a random variable takes values in the complex plane $\mathbb{C}$, although the density function (if it exists) is real.
- Contrary to the characteristic function, which is always well-defined, the moment and cumulant-generating functions are not always defined.
- If the characteristic function of a random variable is entire on $\mathbb{C}$, then the random variable has moments and exponential moments of any order.
- The distribution of the random variable describing the result in a coin tossing is an example of a Bernoulli distribution.
- The sum of a sequence of independent Bernoulli-distributed random variables is referred to as binomial random variable.
- The Poisson random variable is infinitely divisible.
- The gamma and the inverse Gaussian random variables are infinitely divisible. The exponential random variable is a special case of the gamma one.

- The normality assumption for the distribution of the underlying stock is one of the main assumptions of the Black–Scholes option pricing model.
- The log-normal random variable is used in finance to model asset prices.
- The variance gamma and normal inverse Gaussian random variables are normal variance–mean mixture random variables.
- Multidimensional financial applications for decades have been, and in part are still, dominated by the multivariate normality assumption.

References

O.E. Barndorff-Nielsen. Normal inverse Gaussian distributions and the modeling of stock returns. *Department of Theoretical Statistics, Aarhus University, Research report*, 300, 1995.

P. Billingsley. *Probability and Measure.* Wiley, New York, 3rd edition, 1995.

W. Feller. *An Introduction to Probability Theory and Its Applications*, Volume 2. Wiley, New York, 3rd edition, 1971.

J. Jacod and P. Protter. *Probability Essentials.* Springer, Berlin, 2004.

G. Landenna, D. Marasini, and P.A. Ferrari. *Probabilità e variabili casuali.* Il Mulino, Bologna, 1994.

G. Letta. *Probabilità elementare: compendio di teoria-problemi risolti.* Zanichelli, Bologna, 1993.

D.B. Madan and E. Seneta. The variance gamma (VG) model for share market returns. *Journal of Business*, 63(4):511–524, 1990.

D.B. Madan, P.P. Carr, and E.C. Chang. The variance gamma process and option pricing. *Review of Finance*, 2(1):79–105, 1998.

M. Paolella. *Intermediate Probability: A Computational Approach.* Wiley, Chichester, 2007.

Chapter 3

Stochastic Processes with Jumps

In this chapter, we summarize the main results for Lévy processes. We focus our attention on definitions and properties that we will need in the chapters that follow in this book. The main topics covered in this chapter are:

- basic definitions of stochastic processes;
- the definition of Lévy processes;
- some basic results for Lévy processes;
- the modeling of asset log-prices with Lévy processes;
- a discussion about the change of measure for Lévy processes.

In simple terms a *stochastic process* is a mathematical model of a random phenomenon in time evolution. Even in this case, the definition in mathematical terms is not trivial. For an introduction to stochastic and Lévy processes see Samorodnitsky and Taqqu (1994), Sato (1999), Boyarchenko and Levendorskii (2002), Schoutens (2003), Cont and Tankov (2003), Jacod and Protter (2004), Koralov and Sinai (2007), and Klebaner (2012).

Let $(\mathcal{F}_t)_{t\geq 0}$ be a sequence of σ-algebras with continuous index $t \geq 0$ (or discrete index $t = 0, 1, \ldots$). If $\mathcal{F}_s \subseteq \mathcal{F}_t$ for all $0 \leq s \leq t$, then $(\mathcal{F}_t)_{t\geq 0}$ is referred to as a *filtration*, where $\mathcal{F}_t$ can be interpreted as the "information" available to all market agents at time t. The filtration describes increasing information for time t.

Let $(\Omega, \mathcal{F})$ be a measurable space, on which a probability measure can be placed. We can consider $(\Omega, \mathcal{F})$ as the source of randomness and refer to it as a *sample space*. A stochastic process is a collection of random variables $X = \{X_t\}_{0\leq t<\infty}$ on $(\Omega, \mathcal{F})$, which can take values in a measurable space $(E, \mathcal{E})$. A process may be considered as a mapping from $(\Omega \times \mathbb{R}_+, \mathcal{B}(\mathbb{R}_+) \otimes \mathcal{F})$ into $(E, \mathcal{E})$ (usually $(\mathbb{R}, \mathcal{B}(\mathbb{R}))$), via

$$(\omega, t) \longrightarrow X(\omega, t) = X_t(\omega).$$

Each mapping $t \to X_t(\omega)$, for a fixed $\omega \in \Omega$, is called a *path*, or a *trajectory*, of the process X.

Consider a stochastic process $X = (X_t)_{t\geq 0}$. If X_t is $\mathcal{F}_t$-measurable for all $t \geq 0$, then X is referred to as a $(\mathcal{F}_t)_{t\geq 0}$*-adapted process*. If X_t is $\mathcal{F}_{t-1}$-measurable for all discrete index $t = 0, 1, 2, \ldots$, then X is referred to as an $(\mathcal{F}_t)_{t\geq 0}$*-predictable process*.

For a given process $X = (X_t)_{t\geq 0}$, the simplest choice of a filtration is that generated by the process itself: that is, we can generate a filtration $(\mathcal{F}_t)_{t\geq 0}$ by

$$\mathcal{F}_t = \sigma(X_s; 0 \leq s \leq t),$$

where $\sigma(X_s; 0 \leq s \leq t)$ is the smallest σ-algebra containing all $\sigma(X_s)$ with $0 \leq s \leq t$. Then the process X is $(\mathcal{F}_t)_{t\geq 0}$-adapted and this filtration is referred to as a *filtration generated by* X.

A real-valued, adapted process $X = (X_t)_{t\geq 0}$ is called a *martingale with respect to the filtration* $(\mathcal{F}_t)_{t\geq 0}$ if

(1) $X_t \in L^1$;
(2) $E[X_t|\mathcal{F}_s] = X_s$, a.s. for all $s \leq t$

3.1. Lévy Processes and Infinitely Divisible Distributions

In some of the chapters that follow, we will make an extensive use of Lévy processes based on heavy tail distributions. For this reason, after justifying the need for Lévy processes in modeling financial returns, we introduce them and discuss some of their general properties.

The historical distribution of asset returns exhibits skewness, heavy tails, and asset price processes (or dynamics) characterized by jumps and spikes. These features provide strong empirical evidence against the normal distribution assumption for asset returns and are inconsistent with the geometric Brownian motion hypothesis for asset price dynamics. Nonlinear diffusion processes, like stochastic and local volatility models built on Brownian motion, can be used to generate models with heavy tails. However they are not able to produce sudden and large price movements because their trajectories are continuous. Furthermore, in nonlinear diffusion models fat tails are the result of the aggregation of small movements and, consequently, short-term returns tend to be normal and predictable in size. Additional evidence against normality is provided by the smile phenomenon observed in option markets: implied volatilities are not constant either across strikes or across maturities as postulated by the Black–Scholes option pricing model. Even if the smile can be explained by nonlinear diffusion models, the fact that it increases for short-term options is a signal of the presence of jumps in the underlying asset price process. By contrast, Lévy processes with jumps generate trajectories that are consistent with the discontinuous sudden movements observed for asset prices, and, consequently, infinitely divisible return distributions with heavy tails and asymmetries. Furthermore, exponential Lévy option pricing models, more general than geometric Brownian motion, can lead to different smile patterns and to high skews for short maturity options.

In this section, we introduce Lévy processes and discuss some of their general properties. In particular, we mainly follow Sato (1999) and Cont and Tankov (2003). Here we assume a filtered probability space $(\Omega, \mathcal{F}, P, (\mathcal{F}_t)_{t\geq 0})$ satisfying the usual hypotheses.

An adapted stochastic process $X = (X_t)_{t\geq 0}$ with values on $\mathbb{R}^d$ such that $X_0 = 0$ a.s. is a *Lévy process* if the following conditions are satisfied:

(a) X has independent increments, that is, $X_t - X_s$ is independent of $\mathcal{F}_s$, $0 \leq s < t < \infty$;
(b) X has stationary increments, that is, $X_t - X_s$ has the same distribution as X_{t-s}, $0 \leq s < t < \infty$;
(c) X is stochastically continuous, that is, $\forall \varepsilon > 0 \lim_{s\to t} P(|X_t - X_s| \geq \varepsilon) = 0$.

Broadly speaking, the last condition does not imply the continuity of the sample paths but only allows one to exclude processes with discontinuities at non-random times.

Now we introduce some basic examples of Lévy processes:

- Brownian motion;
- Poisson process;
- compound Poisson process.

Brownian motion is not only a special Lévy process and a basic ingredient for the theory of stochastic processes, but it is also the core of many financial models.

An adapted stochastic process $W = (W_t)_{t\geq 0}$ with values on $\mathbb{R}^d$ such that $W_0 = 0$ a.s. is a *standard Brownian motion*, or a *Wiener process*, if it is a Lévy process and it satisfies the following additional conditions:

(d) $W_t - W_s \sim N(0, (t-s)I)$, $0 \leq s < t$, where I is the $d \times d$ identity matrix;
(e) the map $s \to W_s(\omega)$ is continuous P-almost surely.

The stochastic process $B = (B_t)_{t\geq 0}$, defined as $B_t = \mu t + AW_t$, where $\mu \in \mathbb{R}^n$ and A is an $n \times d$ matrix, is a Lévy process with values on $\mathbb{R}^n$, called a *generalized Brownian motion* (or *arithmetic Brownian motion*, or *scaled Brownian motion with drift*). Its increments follow a Gaussian distribution

$$B_t - B_s \sim N\left((t-s)\mu, (t-s)\Sigma_B\right), \quad 0 \leq s < t,$$

where $\Sigma_B = AA'$ is the covariance matrix for $t - s = 1$. Brownian motions are the only Lévy processes with continuous trajectories and with normally distributed increments.

Let $T_1, T_2, \ldots$ be a sequence of independent and exponentially distributed random variables with parameter λ and let $\tau_n = \sum_{j=1}^{n} T_j$. Then the process $N = (N_t)_{t\geq 0}$ defined by

$$N_t = \sum_{n\geq 1} 1_{t\geq \tau_n}$$

is a *Poisson process* with intensity λ. The Poisson process is an example of a *counting process*. Specifically, a Poisson process counts the number of times an event with exponentially distributed interarrival times occurs in the interval $[0, t]$. Its sample paths are right continuous with left limit (or *cadlag*) piecewise constants and they increase by jumps of size one. The increments of a Poisson process are independent, stationary, and Poisson-distributed

$$N_t - N_s \sim \text{Poiss}\left(\lambda(t-s)\right), \quad 0 \leq s < t.$$

Taking into account (2.39), the characteristic function of N_t can be computed

$$\phi_{N_t}(u) = \exp\left[\lambda t(\exp(iu) - 1)\right]. \tag{3.1}$$

Since $E\left[N_t | N_s\right] = N_s + \lambda(t-s)$, the Poisson process is not a martingale. It is possible to build a centered version of N_t, referred to as *compensated Poisson process*, as follows:

$$N_t^c = N_t - \lambda t.$$

The process $N^c = (N_t^c)_{t\geq 0}$ is a martingale since $E\left[N_t^c | N_s^c\right] = N_s^c$ for all $t > s$. Multiplying equation (3.1) by $\exp(-iu\lambda t)$, the characteristic function of the compensated Poisson process can be derived

$$\phi_{N_t^c}(u) = \exp\left[\lambda t(\exp(iu) - 1 - iu)\right].$$

Let Y_j, with $j \in \mathbb{N}$, be a sequence of independent and identically distributed (iid) random variables with distribution f_Y and let $N = (N_t)_{t\geq 0}$ be a Poisson process with intensity λ independent from the Y_js. Then the process $X = (X_t)_{t\geq 0}$ defined by

$$X_t = \sum_{j=1}^{N_t} Y_j \tag{3.2}$$

is a *compound Poisson process* with jump intensity λ and jump size distribution f_Y. The compound Poisson process is characterized by cadlag piecewise constant sample paths and by independent and exponentially distributed interarrival times of jumps. These features also characterize the Poisson process. However, in a compound Poisson process the jump sizes are iid with distribution f_Y. Thus, if we set $Y_j = 1$ for all j, then we get the Poisson process as a special case of the compound Poisson one. Let $\phi_Y(u)$ be the characteristic function of the jump size distribution, then we can write

$$\phi_{X_t}(u) = \sum_{j=0}^{\infty} \frac{(\lambda t)^j \exp(-\lambda t)}{j!} (\phi_Y(u))^j, \tag{3.3}$$

where

$$\frac{(\lambda t)^j \exp(-\lambda)}{j!}$$

is the probability of observing j jumps in the time interval $[0, t]$ and $(\psi_Y(u))^j$ is the characteristic function of the cumulated jump size, given that j jumps have occurred. Then, computing the sum of the

Taylor series in (3.3), the characteristic function of X_t becomes

$$\begin{aligned}\phi_{X_t}(u) &= \exp\left[\lambda t(\phi_Y(u) - 1)\right] \\ &= \exp\left[t\lambda \int_{\mathbb{R}} (\exp(iux) - 1) f_Y(dx)\right] \\ &= \exp\left[t \int_{\mathbb{R}} (\exp(iux) - 1)\nu_X(dx)\right], \quad u \in \mathbb{R},\end{aligned} \tag{3.4}$$

where ν_X is called *Lévy measure* of the process. Since $\nu_X(dx) = \lambda f_Y(dx)$, then $\nu_X(\mathbb{R}) = \lambda$, that is $\nu_X(\mathbb{R})$ for a compound Poisson process is equal to the expected number of jumps per unit of time. The characteristic function of a d-dimensional compound Poisson process can be easily obtained generalizing (3.4)

$$\phi_{X_t}(u) = \exp\left[t \int_{\mathbb{R}^d} (\exp(iu'x) - 1)\nu_X(dx)\right], \quad u \in \mathbb{R}^d. \tag{3.5}$$

The Gaussian, the Poisson, and the compound Poisson laws are examples of infinitely divisible distributions.

A probability distribution F_X on $\mathbb{R}^d$ is *infinitely divisible* if and only if for each $n \in \mathbb{N}$ it can be represented as the distribution of the sum of n independent random variables with a common distribution F_n, that is,

$$X = X_{1,n} + X_{2,n} + \cdots + X_{n,n}, \tag{3.6}$$

where $X_{j,n} \sim F_n$, $j = 1, 2, \ldots, n$. It is possible to provide an intuitive graphical representation of an infinitely divisible distribution through an infinite triangular array whose rows contain iid random variables

$$\begin{array}{l} X_{1,1} \\ X_{1,2}, X_{2,2} \\ \vdots \\ X_{1,n}, X_{2,n}, X_{3,n}, \ldots, X_{n,n} \\ \vdots \end{array}$$

where $X_{j,k} \sim F_k$ for $j, k \in \mathbb{N}$.

Equivalently, a probability measure F_X is infinitely divisible if, for any positive integer n, there is a probability measure F_n, such that $F_X = F_n^{n*}$, where F_n^{n*} is the n-fold convolution of a probability measure F_n with itself, that is,

$$F_n^{n*} = \underbrace{F_n * \cdots * F_n}_{n\ times}.$$

As already observed in Chapter 2, an infinitely divisible distribution can also be characterized using its characteristic function. A random variable X has an *infinitely divisible* distribution if, for each $n \in \mathbb{N}$, there exists a random variable X_n, such that

$$\phi_X(u) = [\phi_{X_n}(u)]^n. \tag{3.7}$$

The Poisson, the gamma, the inverse Gaussian, the exponential, the compound Poisson, the variance gamma, the normal inverse Gaussian, and the normal are some examples of infinitely divisible distributions. In contrast, the binomial is not infinitely divisible. Note that the infinitely divisibility property, in general, does not require that F_n^{n*} and F_n are distributions of the same type. In Chapters 4 and 7, we will see some examples of infinitely divisible distributions which are not closed under convolution (i.e., the generalized inverse Gaussian, and the generalized hyperbolic).

There exists a two-way correspondence between Lévy processes and infinitely divisible distributions, as stated by the following proposition proven in Sato (1999).

Proposition 3.1. *If $X = (X_t)_{t\geq 0}$ is a Lévy process on $\mathbb{R}^d$, then X_t has an infinitely divisible distribution for every $t > 0$. Conversely, if F_X is an infinitely divisible distribution, then there exists a Lévy process $X = (X_t)_{t\geq 0}$ such that F_X is the distribution of X_1.*

Now, we provide the so-called *Lévy–Khintchine representation* of the characteristic function for both infinitely divisible distributions and Lévy processes.

Theorem 3.2. (Lévy–Khintchine formula for infinitely divisible distributions). *A probability law F_X of a random variable X*

with values on $\mathbb{R}^d$ *is infinitely divisible if and only if there exists a triplet* (a_X, Σ_X, ν_X) *such that*

$$\int_{\mathbb{R}^d} \exp(iu'x) F_X(dx) = \exp(\psi_X(u)), \quad u \in \mathbb{R}^d,$$

and

$$\begin{aligned}\psi_X(u) &= ia_X' u - \frac{1}{2} u' \Sigma_X u \\ &\quad + \int_{\mathbb{R}^d} (\exp(iu'x) - 1 - iu'x 1_{|x| \leq 1}) \nu_X(dx), \end{aligned} \tag{3.8}$$

where $a_X \in \mathbb{R}^d$, Σ_X *is a symmetric positive definite* $d \times d$ *matrix, and* ν_X *is a positive measure on* $\mathbb{R}^d \setminus \{0\}$, *called the Lévy measure of* F_X, *satisfying*

$$\int_{|x| \leq 1} |x|^2 \nu_X(dx) < \infty$$

and

$$\int_{|x| \geq 1} \nu_X(dx) < \infty.$$

Thus, an infinitely divisible distribution is completely characterized by the triplet (a_X, Σ_X, ν_X), called *Lévy characteristic triplet*, or *Lévy generating triplet* or, simply, *Lévy triplet.*

Theorem 3.3. (Lévy–Khintchine formula for Lévy processes). *If* $X = (X_t)_{t \geq 0}$ *is a Lévy process with values on* $\mathbb{R}^d$ *with generating triplet* (a_X, Σ_X, ν_X), *then its characteristic function is*

$$E\left[\exp\left(iu'X_t\right)\right] = \exp\left(t\psi_X(u)\right), \quad u \in \mathbb{R}^d,$$

with characteristic exponent

$$\begin{aligned}\psi_X(u) &= ia_X' u - \frac{1}{2} u' \Sigma_X u \\ &\quad + \int_{\mathbb{R}^d} (\exp(iu'x) - 1 - iu'x 1_{|x| \leq 1}) \nu_X(dx). \end{aligned} \tag{3.9}$$

Thus, the characteristic exponent of a one-dimensional Lévy process with characteristic triplet (a_X, σ_X, ν_X) is given by

$$\psi_X(u) = ia_X u - \frac{1}{2}\sigma_X^2 u^2 + \int_{\mathbb{R}} (e^{iux} - 1 - iux1_{|x|\le 1})\nu_X(dx), \quad u \in \mathbb{R}.$$

By (3.9), the law of a Lévy process is completely identified by its characteristic triplet (a_X, Σ_X, ν_X). The Lévy measure determines the intensity of jumps of different sizes. In particular, given a set $A \in \mathcal{B}(\mathbb{R}^d)$, $\nu_X(A)$ is the expected number of jumps on the time interval $[0,1]$, whose sizes fall in A. Additionally, the cumulant characteristic function of every Lévy process is a linear function of time, that is,

$$\psi_{X_t}(u) = t\psi_{X_1}(u) = t\psi_X(u),$$

where $\psi_{X_1}(u)$ is the cumulant characteristic function of the process for $t = 1$. Therefore, if we choose an infinitely divisible distribution for X_1, then the distribution of X_t is automatically fixed for all t. In other words, to characterize completely a Lévy process the only thing to do is to specify the distribution of X_t for $t = 1$.

Equivalent representations for the characteristic exponent in (3.9) can be provided. Specifically, given a bounded measurable function $h : \mathbb{R}^d \to \mathbb{R}$ such that, as $x \to 0$, $h(x) = 1 + o(|x|)$, and, as $x \to \infty$, $h(x) = 1 + O(1/|x|)$, the characteristic exponent may be written as

$$\psi_X(u) = ib'_X u - \frac{1}{2}u'\Sigma_X u \int_{\mathbb{R}^d} \left(\exp\left(iu'x\right) - 1 - iu'xh(x)\right) \nu_X(dx),$$

where

$$b_X = a_X + \int_{\mathbb{R}^d} x(h(x) - 1_{|x|\le 1})\nu_X(dx).$$

The triplet (b_X, Σ_X, ν_X) is the characteristic triplet of the Lévy process with respect to the *truncation function* $h(x)$. Thus, the Lévy triplet (a_X, Σ_X, ν_X), introduced earlier, is implicitly the generating triplet of the same process with respect to the truncation function $h(x) = 1_{|x|\le 1}$. Both Σ_X and ν_X are not affected by the choice of the truncation function. On the contrary, this choice has an impact on the first component of the characteristic triplet.

If ν_X satisfies

$$\int_{|x|\geq 1} |x|\nu_X(dx) < \infty,$$

we can consider $h(x) = 1$ because it is not necessary to truncate large jumps, and we obtain

$$\psi_X(u) = d'_X u - \frac{1}{2}u'\Sigma_X u \int_{\mathbb{R}^d} (\exp(iu'x) - 1 - iu'x)\,\nu_X(dx),$$

where

$$d_X = a_X + \int_{\mathbb{R}^d} x(1 - 1_{|x|\leq 1})\nu_X(dx) = a_X + \int_{|x|>1} x\nu_X(dx).$$

Since

$$E[X_t] = t\left(a_X + \int_{|x|\geq 1} x\nu_X(dx)\right),$$

the expectation of the process is given by

$$E[X_t] = d_X t.$$

3.1.1. Properties of Lévy processes

In the following, we point out some properties of Lévy processes. Before doing that we need some definitions we are going to use to provide a complete characterization of the path structure of Lévy processes.

Let $(\Omega, \mathcal{F}, P)$ be a probability space, $E \in \mathbb{R}^d$, and $(E, \mathcal{E})$ a measurable space. Then $M : \Omega \times \mathcal{E} \to \mathbb{R}$ is a *random measure* if

- $M(\omega, \cdot)$ is a measure on $\mathcal{E}$ for all $\omega \in \Omega$;
- $M(\cdot, A)$ is measurable for all $A \in \mathcal{E}$.

Let $(\Omega, \mathcal{F}, P)$ be a probability space, $E \in \mathbb{R}^d$, and μ be a measure on $(E, \mathcal{E})$. Then $M : \Omega \times \mathcal{E} \to \mathbb{R}$ is a *Poisson random measure* with intensity μ if

- for all $A \in \mathcal{E}$ with $\mu(A) < \infty$, $M(A)$ follows the Poisson law with parameter $E[M(A)] = \mu(A)$;

- for any disjoint sets $A_1, A_2, \dots, A_n$, $M(A_1), M(A_2), \dots, M(A_n)$ are independent.

In particular, the Poisson random measure is a positive integer-valued random measure.

Let $X = (X_t)_{t \geq 0}$ be a cadlag process with values on $\mathbb{R}^d$. The *jump measure* of X is a random measure on $\mathcal{B}([0,\infty) \times \mathbb{R}^d)$ defined by

$$J_X(B) = N\{(t : \Delta X_t) \in B\}.$$

For every measurable set $A \in \mathbb{R}^d$, $J_X([s,t] \times A)$ counts the number of jumps of X in the time interval $[s,t]$ such that their sizes fall into A.

For example, the jump measure J_X of a compound Poisson process with intensity λ and jump size distribution f is a Poisson random measure $\mathcal{B}([0,\infty) \times \mathbb{R}^d)$ with intensity measure $\mu_X(dx \times dt) = \nu_X(dx)dt = \lambda f(dx)dt$ (see Proposition 3.5 in Cont and Tankov (2003)).

This result implies that every compound Poisson process can be represented as follows:

$$X_t = \sum_{s \in [0,t]} \Delta X_s = \int_{[0,t] \times \mathbb{R}^d} x J_X(dx \times ds). \tag{3.10}$$

The integral in equation (3.10) is a finite sum because the compound Poisson process has a finite number of jumps in the interval $[0,t]$. The representation (3.10) can be seen as a special case of the Lévy–Itô decomposition for Lévy processes.

Theorem 3.4 (Lévy–Itô decomposition). *Let $X = (X_t)_{t \geq 0}$ be a Lévy process with values on $\mathbb{R}^d$. Then X can be decomposed into the sum of three independent Lévy processes as follows:*

$$X_t = B_t + N_t + M_t$$

with

$$B_t = a_X t + A W_t,$$

$$N_t = \int_{|x|>1, s \in [0,t]} x J_X(ds \times dx),$$

and

$$M_t = \int_{0<|x|\leq 1, s\in[0,t]} x(J_X(ds \times dx) - \nu_X(dx)ds), \tag{3.11}$$

where ν_X *satisfies the conditions*

$$\int_{|x|\leq 1} |x|^2 \nu_X(dx) < \infty,$$
$$\int_{|x|\geq 1} \nu_X(dx) < \infty,$$

and where the jump measure J_X *is a Poisson random measure on* $[0, \infty)$ *with intensity measure* $\nu_X(dx)$.

Proof. See Proposition 3.7 in Cont and Tankov (2003). □

Thus, a Lévy process with generating triplet (a_X, Σ_X, ν_X) can be expressed as the sum of a generalized Brownian motion $B = (B_t)_{t\geq 0}$ with covariance matrix $\Sigma_X = AA'$ and drift a_X, and two processes with discontinuous trajectories: a compound Poisson process $N = (N_t)_{t\geq 0}$, and a martingale $M = (M_t)_{t\geq 0}$. Unlike the compound Poisson process, a more general Lévy process can exhibit infinitely small jumps and their sum may not converge. For this reason, the jump measure $J_X\,(ds \times dx)$ in the integral in equation (3.11) has been compensated by $\nu_X(dx)$. Thus, $M = (M_t)_{t\geq 0}$ can be seen as an infinite superposition of independent compensated Poisson processes.

Here we give a basic classification related to the Lévy generating triplet (a_X, Σ_X, ν_X) of a d-dimensional Lévy process.

If all the elements of the matrix Σ_X are equal to zero, then $X = (X_t)_{t\geq 0}$ is called *purely non-Gaussian*. If also the elements of the vector a_X are equal to zero, then $X = (X_t)_{t\geq 0}$ is called a *pure jump* Lévy process.

Let $X = (X_t)_{t\geq 0}$ be a Lévy process on $\mathbb{R}^d$ with generating triplet (a_X, Σ_X, ν_X). If

$$\int_{|x|\leq 1} \nu_X(dx) < \infty,$$

then the process is of *finite activity.* If

$$\int_{\mathbb{R}^d} \nu_X(dx) = \infty,$$

then the process is of *infinite activity.* A finite activity Lévy process has finitely many jumps in any finite interval. Conversely, an infinite activity Lévy process is able to capture both rare large moves and frequent small moves. High activity is accounted for by a large (in most cases infinite) number of small jumps. See Schoutens (2003).

Let $X = (X_t)_{t \geq 0}$ be a d-dimensional Lévy process. Then X is a compound Poisson process with a drift if and only if its generating triplet (a_X, Σ_X, ν_X) satisfies

$$\Sigma_X = 0 \quad \text{and} \quad \nu_X(\mathbb{R}^d) < \infty.$$

We recall the definition of finite variation (see Revuz and Yor (2005)). Let Δ be a subdivision of the interval $[0, t]$ with $0 = t_0 < t_1 < \cdots < t_n = t$ and f a function $f : [a, b] \to \mathbb{R}^d$; the number $|\Delta| = \sup_i |t_{i+1} - t_i|$ is called the modulus or mesh of Δ. We consider the sum

$$S_t^\Delta = \sum_t |f(t_{i+1}) - f(t_i)|$$

The function f is of *finite variation* if for every t

$$S_t = \sup_\Delta S_t^\Delta < +\infty$$

The function $t \to S_t$ is called the total variation of f and S_t is the variation of f on $[0, t]$. The function S is obviously positive and increasing and if

$$\lim_{t \to \infty} S_t < +\infty,$$

the function f is said to be of *bounded variation.*

The same notions could be extended on any interval $[a, b]$. For example, trajectories of Brownian motion are almost surely of infinite variation. Therefore, a Lévy process with a Brownian component is of infinite variation. As proven in Revuz and Yor (2005), the

Brownian paths are almost surely of infinite variation on any interval. Consequently, a Lévy process is of finite variation if and only if its generating triplet (a_X, Σ_X, ν_X) satisfies

$$\Sigma_X = 0 \quad \text{and} \quad \int_{|x|\leq 1} |x|\nu_X(dx) < \infty. \tag{3.12}$$

Note that the finiteness of the integral in (3.12) also allows one to write the characteristic exponent of the process as

$$\psi(u) = id'_X u + \int_{\mathbb{R}^d} \exp\left(iu'x - 1\right)\nu_X(dx), \tag{3.13}$$

where the vector d_X relates to the vector a_X and the measure ν_X via

$$d_X = a_X - \int_{|x|<1} x\nu_X(dx).$$

In this case, the Lévy–Itô representation of the process reduces to

$$X_t = d_X t + \int_{[0,t]\times\mathbb{R}^d} xJ_X\left(ds \times dx\right) \tag{3.14}$$

with

$$\int_{[0,t]\times\mathbb{R}^d} xJ_X\left(ds \times dx\right) = \sum_{s\in[0,t]}^{\Delta X_s \neq 0} \Delta X_s. \tag{3.15}$$

Now we introduce a special class of finite variation Lévy processes called *subordinators*. These processes will be a key concept in Chapter 7 in the construction of multivariate Lévy process through the technique of the *stochastic time change*. A *subordinator* is a Lévy process with non-negative jumps, a non-negative drift, a null Gaussian component, and it has finite variation with almost surely non-decreasing trajectories.

A real-valued Lévy process $X = (X_t)_{t\geq 0}$ is called a *subordinator* if the following conditions are satisfied:

- $\nu_X(-\infty, 0) = 0$;
- $d_X \geq 0$;
- $\Sigma_X = 0$;
- $\int_0^\infty (1 \wedge x)\nu_X(dx) < \infty$.

When $X = (X_t)_{t\geq 0}$ is a subordinator, the Laplace transform of its distribution is more convenient than the characteristic function. The general form is

$$E[e^{-uX_t}] = \exp\left[-ud_X t + t\left(\int_0^\infty (e^{-ux} - 1)\nu_X(dx)\right)\right],$$

where $u \geq 0$.[1]

The tail behavior of the distribution of a Lévy process and consequently its moments are determined by the Lévy measure.

Proposition 3.5. *Let $X = (X_t)_{t\geq 0}$ be a Lévy process on $\mathbb{R}$ with generating triplet (a_X, σ_X, ν_X). The kth absolute raw moment of X_t, that is $E\left[|X_t|^k\right]$, is finite for some t or, equivalently, for all $t > 0$, if and only if*

$$\int_{|x|\geq 1} |x|^k \nu_X(dx) < \infty. \tag{3.16}$$

Note that it is not always simple to find the analytical behavior of the tails of an infinitely divisible distribution. Now, we recall another useful property associated with the Lévy measure ν_X (see Theorem 25.17 in Sato (1999)).

Proposition 3.6. *Let $X = (X_t)_{t\geq 0}$ be a Lévy process on $\mathbb{R}$ with generating triplet (a_X, σ_X, ν_X) and let $u \in \mathbb{R}$. The exponential moment $E[e^{uX_t}]$ is finite for some t or, equivalently, for all $t > 0$, if and only if*

$$\int_{|x|\geq 1} e^{ux} \nu_X(dx) < \infty. \tag{3.17}$$

In this case

$$E[e^{uX_t}] = e^{t\psi_X(-iu)},$$

where ψ_X is the characteristic exponent of the Lévy process.

---

[1]See Remark 21.6 in Sato (1999).

As discussed in Chapter 2, if the characteristic function is *entire* on $\mathbb{C}$, then the integral in equation (3.17) is finite for all u.

If they are finite, the first four cumulants can be written as follows:

$$
\begin{aligned}
c_1(X_t) &= t\left(a_X + \int_{|x|\geq 1} x\nu_X(dx)\right), \\
c_2(X_t) &= t\left(\sigma_X^2 + \int_{\mathbb{R}} x^2\nu_X(dx)\right), \\
c_3(X_t) &= t\int_{\mathbb{R}} x^3\nu_X(dx), \\
c_4(X_t) &= t\int_{\mathbb{R}} x^4\nu_X(dx).
\end{aligned}
$$

Since $c_4(X_t) > 0$, all infinitely divisible distributions have fatter tails than the normal. The Gaussian is the only infinitely divisible distribution with $c_4(X) = 0$. Furthermore, the cumulants are positive linear functions of time. The skewness and the excess kurtosis of the increments of the process X_{t-s} are, respectively, given by

$$
\text{skew}(X_{t-s}) = \frac{\text{skew}(X_1)}{\sqrt{t-s}}, \tag{3.18}
$$

and

$$
\text{ex.kurt}(X_{t-s}) = \frac{\text{ex.kurt}(X_1)}{t-s}. \tag{3.19}
$$

Thus, both skewness and kurtosis of the increments of the process decrease with the time scale over which increments are computed. For example, suppose we model the return dynamics of an asset through a Lévy process. As we move from daily returns to monthly returns, the skewness decays as $(t-s)^{-1/2}$ and the excess kurtosis as $(t-s)^{-1}$.

Proposition 3.7. *Let $(\Omega, \mathcal{F}, P)$ be a probability space and $X = (X_t)_{t\geq 0}$ a Lévy process on $\mathbb{R}$ with generating triplet (a_X, σ_X, ν_X).*

If for some $u \in \mathbb{R}$

$$\int_{|x|\geq 1} e^{ux}\nu_X(dx) < \infty, \tag{3.20}$$

then the process $Y = (Y_t)_{t\geq 0}$*, defined as*

$$Y_t = \frac{e^{uX_t}}{E[e^{uX_t}]} \qquad t \geq 0, \tag{3.21}$$

is a martingale.

Proof. By the definition of martingale and condition (3.20), we have Y_t is integrable (i.e., $Y_t \in L^1$) and thus we can calculate the conditional expectation. Since X is a Lévy process, we can write for $0 \leq s \leq t$

$$\begin{aligned} E_P\left[Y_t|\mathcal{F}_s\right] &= E_P\left[\frac{e^{uX_t}}{E[e^{uX_t}]}\middle|\mathcal{F}_s\right] \\ &= E_P\left[\frac{e^{u(X_t-X_s)}}{E[e^{u(X_t-X_s)}]}\frac{e^{uX_s}}{E[e^{uX_s}]}\middle|\mathcal{F}_s\right] \\ &= Y_s E_P\left[\frac{e^{u(X_t-X_s)}}{E[e^{u(X_t-X_s)}]}\middle|\mathcal{F}_s\right] = Y_s. \end{aligned}$$ □

3.2. Lévy Processes for Asset Log-Price

Lévy processes can be used to describe the evolution of the logarithm of the price of financial assets, such as stocks and stock indexes. Let $P = (P_t)_{t\geq 0}$ be the price process of an asset and $X = (X_t)_{t\geq 0}$ be a *one*-dimensional Lévy process describing the asset log-price dynamics, that is

$$X_t = \ln P_t.$$

Then, the increments

$$X_{t+\Delta t} - X_t = \ln\frac{P_{t+\Delta t}}{P_t} = r_{\Delta t}$$

are called asset log-returns and the price process

$$P_t = \exp(X_t)$$

is referred to as an *exponential Lévy model*. Lévy models for log-price dynamics can be divided into three categories:

- Brownian motions;
- jump–diffusion processes;
- infinite activity Lévy models.

3.2.1. Brownian motions

The log-price process $X = (X_t)_{t \geq 0}$ is given by a scaled Brownian motion with drift

$$X_t = \mu t + \sigma W_t,$$

where $W = (W_t)_{t \geq 0}$ is a standard Brownian motion, $\mu \in \mathbb{R}$ and $\sigma > 0$. Therefore, the characteristic function of X_t is

$$\phi_{X_t}(u) = \exp\left[\left(iu\mu - \frac{1}{2}u^2\sigma^2\right)t\right], \tag{3.22}$$

and log-returns are normally distributed

$$r_{\Delta t} \sim N(\mu\Delta t, \sigma^2\Delta t).$$

3.2.2. Jump–diffusion processes

The log-price dynamics is described through a Lévy process with a Gaussian diffusion part and a jump component represented by a compound Poisson process. This class of models has finitely many jumps in every time interval and jumps can be regarded as rare events. There are two well-known jump–diffusion models in finance. The Merton model considers jumps with normally distributed sizes (Merton (1976)). In the Kou model, there are jumps with double exponential distribution (Kou (2002)). In both cases, the log-price process $X = (X_t)_{t\geq 0}$ has the form

$$X_t = \mu t + \sigma W_t + \sum_{j=1}^{N_t} Y_j, \tag{3.23}$$

where $N = (N_t)_{t\geq 0}$ is a Poisson process with intensity parameter λ and Y_j are iid random variables. Since the diffusive and the jump components are independent, the characteristic function of a jump–diffusion process can be obtained by multiplying (3.22) and (3.4)

$$\phi_{X_t}(u) = \exp\left[\left(iu\mu + \lambda\left(\psi_Y(u) - 1\right) - \frac{1}{2}u^2\sigma^2\right)t\right]. \tag{3.24}$$

Assuming, for example, that $Y_j \sim N(a, b^2)$ for all j, (3.24) becomes

$$\phi_{X_t}(u) = \exp\left[\left(iu\mu + \lambda\left(\exp(iua - \frac{1}{2}u^2b^2) - 1\right) - \frac{1}{2}u^2\sigma^2\right)t\right].$$

Furthermore, since $r_{\Delta t}|(N_t = j) \sim N\left(\mu\Delta t + ja, \sigma^2\Delta t + jb^2\right)$, the log-returns density function can be written as

$$f_{r_{\Delta t}}(x) = \sum_{j=0}^{\infty} \frac{(\lambda\Delta t)^j \exp\left(-\lambda\Delta t\right)}{j!} \times \frac{1}{\sqrt{2\pi(\sigma^2\Delta t + jb^2)}} \exp\left[-\frac{1}{2}\frac{(x - (\mu\Delta t + ja))^2}{\sigma^2\Delta t + jb^2}\right].$$

3.2.3. Infinite activity Lévy models

Infinite activity Lévy models are characterized by an infinite number of jumps in every time interval. As has been shown by Carr *et al.* (2002), there is no need to introduce a Gaussian component in these models since the dynamics of jumps is able to give a realistic description of the price process at various time scales. Examples of such models are the variance gamma and the normal inverse Gaussian. According to Proposition 3.1, a Lévy process $X = (X_t)_{t\geq 0}$ can be defined by choosing an infinitely divisible distribution for X_1.

If $X_1 \sim \text{VG}(\mu, \theta, a, b, \sigma)$, then the process $X = (X_t)_{t \geq 0}$ is a Lévy process called *variance gamma*. The characteristic function of the log-price process is

$$\phi_{X_t}(u) = \exp(iu\mu t)\left[1 - \frac{1}{b}\left(iu\theta - \frac{1}{2}u^2\sigma^2\right)\right]^{-at} \tag{3.25}$$

and log-returns follows a variance gamma distribution

$$r_{\Delta t} \sim \text{VG}\left(\mu\Delta t, \theta, a\Delta t, b, \sigma\right).$$

If $X_1 \sim \text{NIG}(\mu, \theta, a, b, \sigma)$, then the process $X = (X_t)_{t \geq 0}$ is a Lévy process called *normal inverse Gaussian*. The characteristic function of the log-price process is

$$\phi_{X_t}(u) = \exp\left[iu\mu t + at\left(b - \sqrt{b^2 - 2\left(iu\theta - \frac{1}{2}u^2\sigma^2\right)}\right)\right], \tag{3.26}$$

and log-returns follows a normal inverse Gaussian distribution

$$r_{\Delta t} \sim \text{NIG}(\mu\Delta t, \theta, a\Delta t, b, \sigma).$$

In Chapter 4, the generalized hyperbolic distribution and some special cases among them, the variance gamma and the inverse Gaussian, will be presented. All those infinitely divisible distributions can be used as a model for the increments of a Lévy process with infinite activity.

3.2.4. Tempering the Lévy measure

The specification of an infinitely divisible law as the distribution of X_1 or of the increments $X_{t+\Delta t} - X_t$ is not the only convenient way to build a Lévy process to model financial returns. Indeed, a Lévy process can be defined by specifying directly its Lévy measure. Also, it is possible to obtain a new Lévy process by *tempering* or through an *exponential tilting* of the Lévy measure of a known process. To be precise, any transformation, $g(\nu_X)$, of a Lévy measure, ν_X, such that

$$\int_{|x|\leq 1} |x|^2 g(\nu_X(dx)) < \infty,$$

and

$$\int_{|x|\geq 1} g(\nu_X(dx)) < \infty$$

produces a new Lévy measure, $\nu_{gX} = g(\nu_X)$.

Let $X = (X_t)_{t\geq 0}$ be a Lévy process on $\mathbb{R}$ with characteristic triplet (a_X, Σ_X, ν_X). If there exists a real number θ such that

$$\int_{|x|\geq 1} \exp(\theta x)\nu_X(dx) < \infty,$$

then the transformation

$$\nu_{gX}(dx) = \exp(\theta x)\nu_X(dx),$$

called the Esscher transform, will lead to a new Lévy process with triplet $(a_{gX}, \Sigma_X, \nu_{gX})$.

Let $X = (X_t)_{t\geq 0}$ be a Lévy process on $\mathbb{R}$ with characteristic triplet (a_X, Σ_X, ν_X). Given two positive constants θ^+ and θ^-, the transformation

$$\nu_{gX}(dx) = \left(1_{x>0}\exp(-\theta^+ x) + 1_{x<0}\exp(-\theta^-|x|)\right)\nu_X(dx)$$

will lead to a new Lévy process with tails that decrease exponentially. We will make use of the exponential tilting technique in Chapter 11 to get the dynamics of the risk-neutral log-price process starting from the historical one through the Esscher change of measure. The Lévy measure tempering transformation will be used in Chapter 6 to dampen the tails of stable distribution to get the tempered stable distribution.

3.2.5. Subordination

Another convenient approach to build a Lévy process is based on *subordination*. In particular, it is possible to get a Lévy process by time-changing a Brownian motion with an independent increasing

Lévy process. In general, performing a stochastic time-change on given a Lévy process through a subordinator allows one to get a new Lévy process as stated in the following theorem.

Theorem 3.8 (Lévy process subordination). *Let $X = (X_t)_{t\geq 0}$ be a Lévy process on $\mathbb{R}^d$ with characteristic triplet (a_X, Σ_X, ν_X) and characteristic exponent $\psi_X(u)$. Let $S = (S_t)_{t\geq 0}$ be a one-dimensional subordinator with Lévy triplet $(a_S, 0, \nu_S)$ and Laplace exponent $l_S(u)$. Then the process $Y = (Y_t)_{t\geq 0}$, defined as $Y_t = X_{S_t}$, obtained through the stochastic time change of X_t with the subordinator S_t, is a Lévy process on $\mathbb{R}^d$, with characteristic triplet (a_Y, Σ_Y, ν_Y) where*

$$\begin{aligned} a_Y &= a_S a_X + \int_0^{+\infty} \nu_S(ds) \int_{|x|\leq 1} x f_{X_t}(dx), \\ \nu_Y(B) &= a_S \nu_X(B) + \int_0^{+\infty} f_{X_t}(B)\nu_S(ds), \quad \forall B \in \mathcal{B}(\mathbb{R}^d), \\ \Sigma_Y &= a_S \Sigma_X, \end{aligned}$$

and f_{X_t} is the probability distribution of X_t. Furthermore, the characteristic function of Y_t is given by

$$\phi_{Y_t}(u) = \exp(t l_S(\psi_X(u))).$$

Proof. See Theorem 4.2 in Cont and Tankov (2003). □

As an application of this theorem, we show how to get a variance gamma process through Brownian motion subordination. Let G be a gamma random variable with parameters a and b. Since the gamma distribution is infinitely divisible, it is possible to build a Lévy process with gamma distributed increments. The process $G = (G_t)_{t\geq 0}$ is called a *gamma process*. Let $B = (B_t)_{t\geq 0}$ be an independent generalized Brownian motion

$$B_t = \theta t + \sigma W_t.$$

Then, the process $X = (X_t)_{t\geq 0}$, defined as $X_t = B_{G_t}$, is a variance gamma process. In particular, since

$$\phi_{X_t}(u) = \left[1 - \frac{1}{b}\left(iu\theta - \frac{1}{2}u^2\sigma^2\right)\right]^{-at},$$

then $X_t \sim \mathrm{VG}(0, \theta, at, b, \sigma)$. To get the same variance gamma process described by (3.25), we should have defined the process $X = (X_t)_{t \geq 0}$ as $X_t = \mu t + B_{G_t}$.

In Chapter 7, we will construct four multivariate Lévy processes through Brownian motion subordination: the variance gamma, the normal inverse Gaussian, the generalized hyperbolic, and the normal tempered stable. Also, we will show that the resulting distributions for asset returns can be represented as multivariate normal variance–mean mixtures and, in Chapter 8, an expectation–maximization algorithm will be proposed to estimate these models.

3.2.6. Linear combination

Finally, it is possible to obtain new Lévy processes through the linear combination of other known Lévy processes, as it is established by the following theorem.

Theorem 3.9 (Linear combination of Lévy processes). *Let $X = (X_t)_{t \geq 0}$ be a Lévy process on $\mathbb{R}^d$ with characteristic triplet (a_X, Σ_X, ν_X) and C an $n \times d$ matrix. Then the process $Y = (Y_t)_{t \geq 0}$, defined as $Y_t = CX_t$, is a Lévy process on $\mathbb{R}^n$ with characteristic triplet (a_Y, Σ_Y, ν_Y) where*

$$
\begin{aligned}
a_Y &= Ca_X + \int_{\mathbb{R}^n} y(1_{|y| \leq 1}(y) - 1_{\{Cx:|x| \leq 1\}}(y))\nu_Y(dy), \\
\nu_Y(B) &= \nu_X(\{x : Cx \in B\}), \quad \forall B \in \mathcal{B}(\mathbb{R}^n), \\
\Sigma_Y &= C\Sigma_X C^T.
\end{aligned}
$$

Proof. See Theorem 4.1 in Cont and Tankov (2003). □

As an application of this theorem, we show how to get a variance gamma process as a linear combination of other Lévy processes. In particular, consider two independent gamma random variables: $G \sim \Gamma(a, b)$ and $S \sim \Gamma(c, d)$. Since $G \perp S$, the correspondent gamma processes $G = (G_t)_{t \geq 0}$ and $S = (S_t)_{t \geq 0}$ are independent. The process

$X = (X_t)_{t \geq 0}$, defined as

$$X_t = \mu t + G_t - S_t,$$

by Theorem 3.9 is a Lévy process since it is a linear combination of other Lévy processes, and its characteristic function is given by

$$\phi_{X_t}(u) = \exp(iu\mu t)\phi_{G_t}(u)\phi_{S_t}(-u),$$

where

$$\phi_{G_t}(u) = \left(1 - \frac{iu}{b}\right)^{-at},$$

and

$$\phi_{S_t}(u) = \left(1 + \frac{iu}{d}\right)^{-ct}.$$

Setting $a = c$, $b = 2(1-\theta)/\sigma^2$ and $d = 2/\sigma^2$, the characteristic function of X_t becomes

$$\phi_{X_t}(u) = \exp(iu\mu t)\left[1 - \frac{\sigma^2}{2(1-\theta)}\left(iu\theta - \frac{1}{2}u^2\sigma^2\right)\right]^{-at},$$

which is the characteristic function of a variance gamma process. In particular,

$$X_t \sim \mathrm{VG}(\mu t, \theta, at, 2(1-\theta)/\sigma^2, \sigma).$$

3.3. Change of Measure for Lévy Process

The following is the initial part of the seminal work of Delbaen and Schachermayer (1994), one of the most important papers in mathematical finance[2]:

> A basic result in mathematical finance, sometimes called the fundamental theorem of asset pricing, is that for a stochastic process $(\tilde{S}_t)_{t\geq 0}$, the existence of an equivalent martingale measure is essentially equivalent to the absence of arbitrage opportunities. In finance the process $(\tilde{S}_t)_{t\geq 0}$ describes the random evolution of the discounted price of one or several financial

[2]See also Delbaen and Schachermayer (2006) for further details.

> assets. The equivalence of no-arbitrage with the existence of an equivalent probability martingale measure is at the basis of the entire theory of *pricing by arbitrage*. Starting from the economically meaningful assumption that $\tilde{S}$ does not allow arbitrage profits, the theorem allows the probability P on the underlying probability space $(\Omega, \mathcal{F}, P)$ to be replaced by an equivalent measure Q such that the process $\tilde{S}$ becomes a martingale under the new measure. This makes it possible to use the rich machinery of martingale theory. In particular, the problem of fair pricing of contingent claims is reduced to taking expected values with respect to the measure Q. This method of pricing contingent claims is known to actuaries since the introduction of actuarial skills, centuries ago and known by the name of equivalence principle. (p. 463)

Let $X = (X_t)_{t\in[0,T]}$ be a Lévy process. Then, an exponential Lévy market model is arbitrage free if the log-price process of the asset satisfies one of the following (not mutually exclusive) conditions[3]:

- $\sigma_X > 0$;
- X has infinite variation;
- X has both positive and negative jumps;
- X has positive jumps and a negative drift or negative jumps and a positive drift.

While the existence of an equivalent martingale measure is related to the absence of arbitrage, its uniqueness is related to market completeness (i.e., perfect hedging). However, in arbitrage-free exponential Lévy models the risk due to jumps cannot be hedged and therefore there is no longer a unique risk-neutral measure. This is always the case when the price process is an arbitrage-free geometric Lévy process, except, of course, in the geometric Brownian motion case. In an arbitrage-free Lévy market model, there are many different equivalent measures under which the discounted asset price process is a martingale. Many candidates for a suitable martingale measure equivalent to the real-world probability measure have been suggested and, from a theoretical point of view, different criteria can be chosen based on hedging arguments or distance minimization: Hellinger distance, L^2 distance, entropy or Kullback–Leibler

[3]For further details see Proposition 9.9 in Cont and Tankov (2003).

distance. The following is only a list of possible equivalent martingale measures[4]:

(1) minimal martingale measure (see Föllmer and Schweizer (1991));
(2) variance optimal martingale measure (see Schweizer (1995));
(3) mean correcting martingale measure;
(4) utility based martingale measure;
(5) minimal entropy martingale measure (see Miyahara (1996), Fujiwara and Miyahara (2003), and Frittelli (2000));
(6) Esscher martingale measure (see Gerber and Shiu (1994)).

In this section, after having reviewed the change of measure for random variables, we will discuss the change of probability measure for Lévy processes starting with the following three special cases: Brownian motions, Poisson processes, and compound Poisson processes. Then, the general theorem on the measure change for any Lévy process will be provided. Finally, we will focus our attention on the special change of measure called the *Esscher transform*. The results provided by the theorems discussed in this section will be applied in Chapter 11. In particular, Girsanov's theorem for Brownian motions will be necessary in the review of the Black–Scholes option pricing model. The Esscher transform will be applied in the exponential generalized hyperbolic, exponential normal tempered stable, and exponential classical tempered stable option pricing models to get the risk-neutral log-price dynamics of the underlying asset. Furthermore, we will also show a direct application of the general theorem on the change of measure for Lévy processes in the classical tempered stable case.

3.3.1. Change of measure for random variables

Let P and Q be two probability measures on the same sample space Ω and σ-field $\mathcal{F}$. Q is *absolutely continuous* with respect to P, denoted

[4]The interested reader can find a good survey of martingale measures for geometric Lévy models in Miyahara (2005).

as $Q \ll P$, if $Q(A) = 0$ whenever $P(A) = 0$. P and Q are called *equivalent* if $Q \ll P$ and $P \ll Q$. Thus, two probability measures P and Q are equivalent if they have the same null sets, that is $P(A) = 0 \Leftrightarrow Q(A) = 0$.

Theorem 3.10. (Radon–Nikodym). *Let $Q \ll P$. Then there exists a non-negative random variable ξ such that*

$$E_P[\xi] = \int_\Omega \xi dP = 1,$$

$$Q(A) = E_P[1_A \xi] = \int_A \xi dP, \quad A \in \mathcal{F}. \tag{3.27}$$

ξ is P-almost surely unique. Conversely, if there exists a non-negative random variable ξ such that $E_P[\xi] = 1$ and Q is defined by (3.27), then $Q \ll P$.

The random variable ξ is referred to as *Radon–Nikodym derivative* or density of Q with respect to P and it is denoted as

$$\xi = \frac{dQ}{dP}.$$

If X is a random variable such that $E_Q[X]$ exists and $Q \ll P$, then

$$E_Q[X] = \int_\Omega X dQ = \int_\Omega X \xi dP = E_P[X\xi].$$

Let Q and P be equivalent probability measure and let X be a random variable with probability density functions $f_X^Q(x)$ on Q and $f_X^P(x)$ on P. Then the change of measure is defined by the Radon–Nikodym derivative

$$\xi = \frac{dQ}{dP} = \frac{f_X^Q(X)}{f_X^P(X)}.$$

3.3.2. Change of measure for continuous-time processes

In financial models, we often deal with continuous-time stochastic processes to describe the evolution of uncertain variables (like, for example, asset prices), over some time period $[0, T]$. The state space Ω is the set of possible paths of the relevant processes over the period $[0, T]$. Thus, a stochastic process can be thought of as a random variable on the sample space of paths. Specifically, a Brownian motion can be seen as a random variable on the space $\Omega = C[0, T]$ of continuous paths, while a Lévy process with jumps in the space $\Omega = D[0, T]$ of discontinuous cadlag paths. Two equivalent probability measures define the same set of scenarios (i.e., sample paths) but assign them different probabilities. As explained in Cont and Tankov (2003), even if the processes defined by two equivalent measures have in common the same trajectories, they can have very different distributional properties. In particular, the process defined by P can be a Lévy process while the one defined by Q does not necessarily have stationary and independent increments. A *structure preserving* change of measure does not modify the class of the process after the change.

At time T all the relevant uncertainties have been resolved and the values of all random variables, including the Radon–Nikodym derivative ξ, will be known. For this reason, the Radon–Nikodym derivative is usually written as

$$\xi_T = \frac{dQ}{dP}.$$

It is also possible to define a process $\xi = (\xi_t)_{\in[0,T]}$ by setting

$$\xi_t = \frac{dQ}{dP}|\mathcal{F}_t,$$

which is a P-martingale.

Theorem 3.11. *Let $\xi = (\xi_t)_{t\in[0,T]}$ be a positive P-martingale such that $E_P[\xi_T]$=1. Define the probability measure Q by the relation*

$$Q(A) = \int_A \xi_T dP.$$

Then $Q \ll P$ and for any random variable X

$$E_Q[X] = E_P[X\xi_T],$$

$$E_Q[X|\mathcal{F}_t] = E_P\left[X\frac{\xi_T}{\xi_t}|\mathcal{F}_t\right],$$

and if X is $\mathcal{F}_t$-measurable, then

$$E_Q[X|\mathcal{F}_s] = E_P\left[X\frac{\xi_t}{\xi_s}|\mathcal{F}_s\right], \quad \textit{for } s \leq t.$$

From this theorem, it follows that a process $X = (X_t)_{t\in[0,T]}$ is a Q-martingale if and only if the process $X\xi = (X_t\xi_t)_{t\in[0,T]}$ is a P-martingale.

Theorem 3.12 (Girsanov's theorem for Brownian motions). *Let $W=(W_t)_{t\in[0,T]}$ be ax P-standard Brownian. Consider the P-Brownian motion $B = (B_t)_{t\in[0,T]}$ defined as $B_t = \mu t + W_t$. There exists a probability measure Q equivalent to P, such that $B = (B_t)_{t\in[0,T]}$ is a Q-standard Brownian motion. The change of measure is provided by the Radon–Nikodym derivative*

$$\xi_T = \frac{dQ}{dP} = \exp\left(-\mu W_T - \frac{1}{2}\mu^2 T\right).$$

This theorem can be extended to arithmetic Brownian motions with drift.

Proposition 3.13. *Let $B = (B_t)_{t\in[0,T]}$ be a real-valued P-arithmetic Brownian motion with parameters μ and σ. Then, there exists a probability measure Q equivalent to P, such that $B = (B_t)_{t\in[0,T]}$ is a real-valued Q-arithmetic Brownian motion with parameters γ and σ, and the change of measure is given by*

$$\xi_T = \frac{dQ}{dP} = \exp\left[\frac{\gamma-\mu}{\sigma}W_T - \frac{1}{2}\left(\frac{\gamma-\mu}{\sigma}\right)^2 T\right],$$

where $W = (W_t)_{t\in[0,T]}$ is a P-standard Brownian motion with $W_t = (B_t - \mu t)/\sigma$. That is, the process $B_t = \gamma t + \sigma\tilde{W}_t$ is a Q-arithmetic Brownian motion, and $\tilde{W}_t$ is a Q-Wiener process given by $\tilde{W}_t = W_t - (\gamma-\mu)t/\sigma$.

Thus, P and Q are equivalent if a generalized Brownian motion has the same volatility σ under both measures. In other words, the change of measure modifies the drift but leaves unchanged the volatility parameter of the Brownian motion.

Now, consider the following application of Girsanov's theorem for Brownian motion. Let $B = (B_t)_{t\in[0,T]}$ be a P-generalized Brownian motions with drift μ and volatility σ. This process is not a P-martingale. However, it is possible to find a measure Q such that the process becomes a martingale. By defining the Radon–Nikodym derivative as

$$\xi_T = \frac{dQ}{dP} = \exp\left[-\frac{\mu}{\sigma}W_T - \frac{1}{2}\left(\frac{\mu}{\sigma}\right)^2 T\right],$$

we get the process $B_t = \sigma\tilde{W}_t$ which is a Q-martingale, where $\tilde{W}_t = W_t + \mu t/\sigma$ and W_t are standard Wiener processes under the measures Q and P, respectively.

Theorem 3.14 (Girsanov's theorem for Poisson processes). *Let $N = (N_t)_{t\in[0,T]}$ be a P-Poisson process with intensity 1. Given a positive constant λ, define the probability measure Q through the Radon–Nikodym derivative*

$$\xi_T = \frac{dQ}{dP} = \exp[(1-\lambda)T + N_T \ln\lambda].$$

Then, $N = (N_t)_{t\in[0,T]}$ is a Q-Poisson process with intensity λ.

The Girsanov's theorem for Poisson processes is extended by the following proposition.

Proposition 3.15. *Let $N = (N_t)_{t\in[0,T]}$ be a Poisson process with parameter λ under the probability measure P. Then it is a Poisson process with intensity η under the equivalent probability measure Q defined by the Radon–Nikodym derivative*

$$\xi_T = \frac{dQ}{dP} = \exp\left[(\lambda-\eta)T + N_T \ln\frac{\eta}{\lambda}\right].$$

Thus, an equivalent change of measure modifies the intensity parameter of a Poisson process but does not affect the size of its jumps.

Theorem 3.16. *Let (X_t, P) and (X_t, Q) be compound Poisson processes with Lévy measures ν_X^P and ν_X^Q. Then, P and Q are equivalent probability measures if and only if ν_X^P is equivalent to ν_X^Q. In this case, the change of measure is defined by the Radon–Nikodym derivative*

$$\xi_T = \frac{dQ}{dP} = \exp\left[(\lambda - \eta)T + \sum_{t \leq T} \varphi(\Delta X_t)\right],$$

where $\lambda = \nu_X^P(\mathbb{R})$, $\eta = \nu_X^Q(\mathbb{R})$, and $d\nu_X^Q / d\nu_X^P = \exp(\varphi(x))$.

Proof. See Proposition 9.6 in Cont and Tankov (2003). □

Now, we are ready to introduce the general theorem for equivalent probability measures change for general Lévy processes.

Theorem 3.17. *Let (X_t, P) and (X_t, Q) be Lévy processes on $\mathbb{R}$ with generating triplets (a_X, σ_X, ν_X) and $(a_X^Q, \sigma_X^Q, \nu_X^Q)$, respectively. Then, P and Q are equivalent for each t if and only if*

(1) $\sigma_X = \sigma_X^Q$;
(2) *the following integral is finite*

$$\int_{\mathbb{R}} (\exp(\varphi(x)/2) - 1)^2 \nu_X(dx) < \infty,$$

with the function $\varphi(x)$ defined by

$$\frac{d\nu_X^Q}{d\nu_X} = \exp(\varphi(x));$$

(3) *additionally if $\sigma_X = 0$, the following condition must hold:*

$$a_X^Q - a_X = \int_{|x| \leqslant 1} x(\nu_X^Q - \nu_X)(dx).$$

When the probability measures P and Q are equivalent, the Radon–Nikodym derivative is

$$\frac{dQ}{dP}|\mathcal{F}_t = \exp(\varepsilon_t),$$

where $\varepsilon = (\varepsilon_t)$ is a Lévy process with characteristic triplet $(a_\varepsilon, \sigma_\varepsilon, \nu_\varepsilon)$ given by

$$a_\varepsilon = -\frac{\sigma_\varepsilon^2}{2} - \int_{\mathbb{R}} \left(\exp(y) - 1 - y1_{|y|\leqslant 1}\right) \nu_\varepsilon(dx),$$

$$\sigma_\varepsilon = b\sigma_X,$$

$$\nu_\varepsilon = \nu_X * \varphi^{-1},$$

and the constant b is such that

$$a_X^Q - a_X - \int_{|x|\leqslant 1} x(\nu_X^Q - \nu_X)(dx) = b\sigma_X^2$$

if $\sigma_X > 0$ and zero if $\sigma_X = 0$.

Proof. See Theorems 33.1 and 33.2 in Sato (1999). □

Thus, for Lévy processes with jumps, the class of equivalent probability measure to a given measures is large. However, unless a Lévy process has a Gaussian component, we cannot change simply the drift to get an equivalent probability measure.

Here, we briefly review the Esscher transform.[5] In finance, Gerber and Shiu (1994) proposed the Esscher transform to construct equivalent martingale measures for processes with independent and stationary increments. The Esscher equivalent martingale measure sometimes is called the compound return Esscher transformed martingale measure to distinguish it from the simple return Esscher transformed martingale measure, which is known as the *minimal entropy martingale measure* (see for example Miyahara (2005)).

The *Esscher transform* is introduced by the following lemma.

[5]The Esscher transform has been introduced by Esscher in actuary theory for the calculation of collective risk (Esscher (1932)).

Lemma 3.18. *Let $X = (X_t)_{t\geq 0}$ be a real-valued Lévy process with Lévy triplet (a_X, σ_X, ν_X) under the measure P. If there exists a real number θ satisfying*

$$\int_{|x|\geq 1} \exp(\theta x)\nu_X(dx) < \infty,$$

then we can find the equivalent measure Q_θ whose Radon–Nikodym derivative is given by

$$\xi_t = \frac{dQ}{dP}|\mathcal{F}_t = \frac{e^{\theta X_t}}{E_P[e^{\theta X_t}]} = e^{\theta X_t - l_X(\theta)t},$$

where $l_X(\theta) = \log E_P[e^{\theta X_1}]$. That is,

$$Q(A)|_{\mathcal{F}_t} = \int_A \xi_t dP, \quad A \in \mathcal{F}_t$$

is equivalent to $P|_{\mathcal{F}_t}$ for all $t \geq 0$. Moreover, the process X is a Lévy process with Lévy triplet $(a_X^{Q_\theta}, \sigma_X, \nu_X^{Q_\theta})$ under the measure Q_θ, where

$$\nu_X^{Q_\theta}(dx) = e^{\theta x}\nu_X(dx)$$

and

$$a_X^{Q_\theta} = a_X + \int_{|x|\leq 1} x(e^{\theta x} - 1)\nu(dx).$$

Key Points

- The Brownian motion is a Lévy process with continuous paths.
- There is a link between infinitely divisible distributions and Lévy processes: given an infinitely divisible distribution one can build a Lévy process with infinitely divisible distributed increments. Conversely, given a Lévy process, the increments are infinitely divisible distributed.
- The Lévy–Khintchine formula describes what the characteristic function of an infinitely divisible distribution should be.
- The tail behavior of the distribution of a Lévy process and, as a consequence, its moments are determined by the Lévy measure.

- To define proper stock price dynamics within the Lévy framework, the existence of exponential moments is a necessary condition.
- Lévy models for log-price dynamics can be divided into three categories: Brownian motions, jump–diffusion processes, and infinite activity Lévy processes.
- Infinite activity Lévy models without a Gaussian component gives a realistic description of the price process at various time scales.
- Brownian motion subordination is a useful technique to build Lévy processes.
- The class of Lévy processes is invariant with respect to linear transformations, subordination, and exponential tilting of the Lévy measure.
- Even restricting attention to structure-preserving change of measures, the class of probabilities equivalent to a given one is surprisingly large.
- In order to find a change of measure between Lévy processes, one cannot freely change the drift if a diffusion component is not present.
- The Esscher transform is a convenient change of measure for processes with stationary and independent increments.

References

S.I. Boyarchenko and S.Z. Levendorskii. *Non-Gaussian Merton–Black–Scholes Theory.* World Scientific Publishing, River Edge, NJ, 2002.

P. Carr, H. Geman, D.B. Madan, and M. Yor. The fine structure of asset returns: An empirical investigation. *The Journal of Business*, 75(2): 05–332, 2002.

R. Cont and P. Tankov. *Financial Modelling with Jump Processes.* Chapman & Hall/CRC, Boca Raton, FL, 2003.

F. Delbaen and W. Schachermayer. A general version of the fundamental theorem of asset pricing. *Mathematische Annalen*, 300(1):463–520, 1994.

F. Delbaen and W. Schachermayer. *The Mathematics of Arbitrage.* Springer, Berlin, 2006.

F. Esscher. On the probability function in the collective theory of risk. *Scandinavian Actuarial Journal*, 15:175–195, 1932.

H. Föllmer and M. Schweizer. Hedging of contingent claims under incomplete information. In M.H.A. Davis and R.J. Elliot, editors,

Applied Stochastic Analysis. Gordon and Breach, New York, 1991, pp. 389–414.

M. Frittelli. The minimal entropy martingale measure and the valuation problem in incomplete markets. *Mathematical Finance*, 10(1):39–52, 2000.

T. Fujiwara and Y. Miyahara. The minimal entropy martingale measures for geometric Lévy processes. *Finance and Stochastics*, 7(4):509–531, 2003.

H.U. Gerber and E.S. Shiu. Option pricing by Esscher transform. *Transactions of the Society of Actuaries*, 46:99–144, 1994.

J. Jacod and P. Protter. *Probability Essentials.* Springer, Berlin, 2004.

F.C. Klebaner. *Introduction to Stochastic Calculus with Applications.* Imperial College Press, London, 3rd edition, 2012.

L. Koralov and Y.G. Sinai. *Theory of Probability and Random Processes.* Springer, Berlin, 2007.

S.G. Kou. A jump–diffusion model for option pricing. *Management Science*, 48(8):1086–1101, 2002.

R.C. Merton. Option pricing when underlying stock returns are discontinuous. *Journal of Financial Economics*, 3(1–2):125–144, 1976.

Y. Miyahara. Canonical martingale measures of incomplete assets markets. *Probability Theory and Mathematical Statistics — Proceedings of the Seventh Japan–Russia Symposium* (1995), 1996, pp. 343–352.

Y. Miyahara. Martingale measures for the geometric Lévy process models. Discussion papers in Economics, Nagoya City University, 431:1–14, 2005.

D. Revuz and M. Yor. *Continuous Martingales and Brownian Motion.* Springer, New York, 3rd edition, 2005.

G. Samorodnitsky and M.S. Taqqu. *Stable Non-Gaussian Random Processes.* Chapman & Hall, New York, 1994.

K.I. Sato. *Lévy Processes and Infinitely Divisible Distributions.* Cambridge University Press, Cambridge, 1999.

W. Schoutens. *Lévy Processes in Finance.* Wiley, Chichester, 2003.

M. Schweizer. On the minimal martingale measure and the Föllmer–Schweizer decomposition. *Stochastic Analysis and Applications*, 13(5):573–599, 1995.

Part II

Theory and Methodology

Chapter 4

The Generalized Hyperbolic Distribution

The main topics covered in this chapter are:

- the definition and the main properties of the class of generalized hyperbolic distributions including different parameterizations, tail behavior, and central and non-central moments common in applications in finance;
- special subclasses which represent popular models in finance including the hyperbolic, the normal inverse Gaussian, the hyperbolic skewed t, the variance gamma, and other distributions;
- a random number generator for the general case and the most important subclasses;
- the method of maximum likelihood for parameter estimation for the general case and the most important subclasses;
- an empirical comparison of the fits of the generalized hyperbolic class and the most important subclasses based on the daily returns of the S&P500 and Nikkei 225 indices;
- the intuition behind the multivariate generalized hyperbolic distribution.

4.1. Introduction

The class of generalized hyperbolic (GH) distributions is a large family of distributions suggested as a model for financial asset returns. Introduced by Barndorff-Nielsen (1977) in a study of the mass-size distribution of sand deposits laid down under steady wind conditions, applications of the general family in finance were suggested by Eberlein (2001), Eberlein and Özkan (2003). Various subclasses of the GH distribution had been used in the literature prior to these publications, see, for example, Rachev and Mittnik (2000) and the references therein.

The attractive properties of GH distributions for finance-related applications include the potential to model skewness, excess kurtosis, as well as heavy tails which are stylized facts about financial asset returns. The class includes representative distributions characterized by tails decaying according to a power law and representative distributions characterized by semi-heavy tails. Theoretically, these distributions arise as location-scale mixtures of normals allowing for a straightforward multivariate extension and implying that they can be used as a building block for Lévy processes. The general class includes many subclasses of other popular models used in finance such as the Gaussian distribution, Student's t-distribution, the hyperbolic skewed t-distribution, the normal inverse Gaussian distribution, the variance gamma distribution, the Laplace distribution, and the hyperbolic distribution.

Applications of GH distributions to the field of finance are not limited to modeling one-dimensional and multidimensional asset return distributions but also include applications to portfolio theory, derivative pricing, and risk management, see, for example, Eberlein and Keller (1995), Prause (1999), Eberlein (2001), Hellmich and Kassberger (2011), Badescu *et al.* (2011), and Kassberger and Kiesel (2006).

This chapter is organized as follows. Section 4.2 defines the class and discusses its basic properties. Section 4.3 describes the most popular special cases of the GH distribution and Section 4.4 discusses a random number generator for Monte Carlo simulations. Section 4.5 discusses parameter estimation and, the concluding section, Section 4.6, describes the multivariate case. In the appendix to this chapter, we provide Matlab functions for scenario generation and parameter estimation.

4.2. Definition and Basic Properties

The GH distributions arise as location-scale mixtures of normals. This section begins with a general discussion of the mixture model and then introduces the GH class. We also discuss general properties of the GH class including moments, parameterization issues, and tail behavior. A detailed technical description of the GH family and its various properties is available in Paolella (2007).

A random variable X is said to be a location-scale mixture of normals[1] if it has the following structure:

$$X = \mu + \beta W + \sqrt{W} Z \tag{4.1}$$

in which $Z \sim N(0,1)$ and W is a non-negative random variable independent of Z and is sometimes called the mixing variable. From this representation, it follows that the conditional distribution of X given W is Gaussian with a mean equal to $\mu + \beta W$ and variance equal to W, $X|W \sim N(\mu + \beta W, W)$.

4.2.1. The GH class in the standard parameterization

The class of GH distributions arises from (4.1) when the mixing random variable W follows the generalized inverse Gaussian (GIG) distribution. The random variable W is said to have a GIG distribution, $W \sim \text{GIG}(\lambda, \chi, \psi)$, if the density of W is given by

$$f_W(x) = \frac{1}{k_\lambda(\chi, \psi)} x^{\lambda - 1} \exp\left(-\frac{1}{2}(\chi x^{-1} + \psi x) \right), \quad x > 0 \tag{4.2}$$

[1]Location-scale mixtures of normals are also referred to as mean–variance mixtures of normals.

in which

$$k_\lambda(x,y) = 2\left(\frac{x}{y}\right)^{\lambda/2} K_\lambda(\sqrt{xy}), \quad x,y>0,$$
$$k_\lambda(0,y) = \left(\frac{y}{2}\right)^{-\lambda}\Gamma(\lambda), \quad y>0, \tag{4.3}$$
$$k_\lambda(x,0) = \left(\frac{x}{2}\right)^{\lambda}\Gamma(-\lambda), \quad x>0,$$

where $\Gamma(z) = \int_0^\infty t^{z-1}e^{-t}dt$ denotes the gamma function, and $K_\lambda(z)$ denotes the modified Bessel function of the second kind,

$$K_\lambda(z) = \frac{1}{2}\int_0^\infty t^{\lambda-1}e^{-\frac{1}{2}z(t+t^{-1})}dt, \quad z>0. \tag{4.4}$$

The parameter space is defined in the following way:

(a) if $\psi > 0$ and $\chi > 0$, then $\lambda \in \mathbb{R}$;
(b) if $\psi = 0$ and $\chi > 0$, then $\lambda < 0$;
(c) if $\psi > 0$ and $\chi = 0$, then $\lambda > 0$.

The special cases of the $k_\lambda(x,y)$ function given in (4.3) concern the boundary of the parameter space in (b) and (c) and represent the limits of $k_\lambda(x,y)$ as the corresponding argument converges to zero.

Because the GH distribution is a location-scale mixture of normals with a GIG mixing distribution, it can be defined through the parameters of the GIG distribution and the ones of the mixture in (4.1). However, the standard parameterization in the literature is given in terms of a transformation of these parameters.[2] A random variable X is said to have a GH distribution, $X \sim \mathrm{GH}(\lambda,\alpha,\beta,\delta,\mu)$, if the density of X is given by

$$f_X(x) = \frac{k_{\lambda-\frac{1}{2}}((x-\mu)^2+\delta^2,\alpha^2)}{\sqrt{2\pi}k_\lambda(\delta^2,\alpha^2-\beta^2)}e^{\beta(x-\mu)}, \quad x\in\mathbb{R} \tag{4.5}$$

in which $k_\lambda(x,y)$ is provided in (4.3). The connection between the new parameters and ones previously used to define the GIG family

[2]See, for example, Chapter 9 in Paolella (2007).

is: $\chi = \delta^2$ and $\psi = \alpha^2 - \beta^2$. The parameters $\lambda, \mu \in \mathbb{R}$, $\alpha, \delta \geq 0$, and $\beta \in [-\alpha, \alpha]$ satisfy the following constraints which follow directly from (a), (b), and (c) above:

(i) if $|\beta| < \alpha$ and $\delta > 0$, then $\lambda \in \mathbb{R}$;
(ii) if $|\beta| = \alpha$ and $\delta > 0$, then $\lambda < 0$;
(iii) if $|\beta| < \alpha$ and $\delta = 0$, then $\lambda > 0$.

If the vector of parameters is away from the boundary of the feasible parameter set, i.e., if (i) holds, then the density takes the following form which is commonly used to define the GH distribution directly,

$$f_X(x) = C(\sqrt{\delta^2 + (x-\mu)^2})^{\lambda - \frac{1}{2}} \times K_{\lambda - \frac{1}{2}}(\alpha\sqrt{\delta^2 + (x-\mu)^2})e^{\beta(x-\mu)}, \quad x \in \mathbb{R}, \tag{4.6}$$

where

$$C = \frac{(\alpha^2 - \beta^2)^{\frac{\lambda}{2}}}{K_\lambda(\delta\sqrt{\alpha^2 - \beta^2})\sqrt{2\pi}\delta^\lambda \alpha^{\lambda - \frac{1}{2}}}.$$

If this definition is adopted for the entire class, then on the boundary of the parameter set the density function needs to be calculated by computing the appropriate limits. In contrast, (4.5) provides the limit cases directly.

Matlab code — GH density, distribution function, and quantile function

In this section, we provide a Matlab implementation of the density function in (4.5) which is straightforward. The implementation does not rely on a separate implementation of the function $k_\lambda(x, y)$ in (4.3).

```
% function pdf = ghpdf(x, lambda, alpha, beta, delta, mu)
%
% Calculates the density function of the generalized ...
    hyperbolic distribution.
%
```

```
%     pdf: the density evaluated at x. The dimension of pdf is ...
      the same as the dimension of x
%

function pdf = ghpdf(x, lambda, alpha, beta, delta, mu)

if(abs(beta) < alpha) && (delta > 0)
    k_num = 2*(((x - mu).^2 + delta^2)/alpha^2).^(lambda/2 - ...
        1/4).*...
        besselk(lambda - 1/2, alpha*sqrt((x - mu).^2 + delta^2));
    k_denom = 2*(delta^2/(alpha^2 - beta^2))^(lambda/2)*...
        besselk(lambda, delta*sqrt(alpha^2 - beta^2));
elseif(abs(beta) == alpha) && (delta > 0) && (lambda < 0) && ...
    (alpha > 0)
    k_num = 2*(((x - mu).^2 + delta^2)/alpha^2).^(lambda/2 - ...
        1/4).*...
        besselk(lambda - 1/2, alpha*sqrt((x - mu).^2 + delta^2));
    k_denom = (delta^2/2)^lambda*gamma(-lambda);
elseif(abs(beta) < alpha) && (delta == 0) && (lambda > 0)
    k_num = 2*(((x - mu).^2 + delta^2)/alpha^2).^(lambda/2 - ...
        1/4).*...
        besselk(lambda - 1/2, alpha*sqrt((x - mu).^2 + delta^2));
    k_denom = ((alpha^2 - beta^2)/2)^(-lambda)*gamma(lambda);
elseif(beta == 0) && (alpha == 0) && (delta > 0) && (lambda < 0)
    k_num = (((x - mu).^2 + delta^2)/2).^(lambda - ...
        1/2)*gamma(-lambda + 1/2);
    k_denom = ((delta^2)/2)^(lambda)*gamma(-lambda);
else error('Wrong parameter values')
end

pdf = k_num.*exp(beta*(x - mu))/(sqrt(2*pi)*k_denom);
```

Unfortunately, no closed-form expressions exist for the cumulative distribution function (cdf) and the inverse distribution functions. We provide a Matlab implementation which integrates numerically the density function in (4.5) to compute the cdf.

```
function cdf = ghcdf(x, lambda, alpha, beta, delta, mu)

cdf = integral(@(x) ghpdf(x, lambda, alpha, beta, delta, mu), ...
    -Inf, x);
```

The implementation of the inverse cdf solves numerically a nonlinear equation and relies on the implementation of the cdf.

```
function q = ghinv(p, lambda, alpha, beta, delta, mu)

% Solve F(x) = p for x starting from x0 = 0
x0 = 0;
q = fzero(@(x) ghcdf(x, lambda, alpha, beta, delta, mu) - p, x0);
```

An implementation of the function in (4.3) is used subsequently in the chapter and is provided below.

```
function f = ghk(lambda, x, y)

  if(x > 0 && y > 0)
    f = 2*(x/y)^(lambda/2)*besselk(lambda, sqrt(x*y));
  elseif (x == 0 && y > 0)
    f = (y/2)^(-lambda)*gamma(lambda);
  elseif (x > 0 && y == 0)
    f = (x/2)^(lambda)*gamma(-lambda);
  else
      error('x and y should be non-negative and (x,y) != (0, 0)')
  end
```

Tail behavior and interpretation of the parameters

One of the attractive properties of the GH class is that it can account for skewness and fat tails. The tail decay of the density of the GH family is quite diverse. Some representatives have a semi-heavy tail and some have a fatter tail decaying according to a power law. The tail decay of the density in the general case when $0 \leq |\beta| < \alpha$ is given by[3]

$$\begin{aligned} f_X(x) &\approx |x|^{\lambda-1} \exp\left(-(\alpha+\beta)x\right), \quad x \to \infty, \\ f_X(x) &\approx |x|^{\lambda-1} \exp\left((\alpha-\beta)x\right), \quad x \to -\infty. \end{aligned} \tag{4.7}$$

This type of tail behavior is sometimes referred to as a semi-heavy tail decay. If $\alpha = 0$, then $\beta = 0$ and the tail behavior changes to a

[3]See Prause (1999) and the references therein.

power decay

$$f_X(x) \sim |x|^{-(|\lambda|+1)}, \quad x \to \pm\infty, \tag{4.8}$$

which represents a decay slower than the one described by (4.7). Finally, on the boundary of the parameter space when $0 < |\beta| = \alpha$, one of the tails has the behavior in (4.7) and the other one has the behavior in (4.8) depending on the sign of β.

The parameters of the GH distribution in the standard parameterization have the following interpretations:

- The parameter α is the tail parameter. It determines the type of tail behavior and also influences the tail thickness to a large degree. This is evident from (4.7) and (4.8).
- The parameter $\beta \in [-\alpha, \alpha]$ is a skewness parameter. If $\beta = 0$, then the distribution becomes symmetric around μ. However, it also plays a role in the tail behavior in that it removes probability mass from one of the tails and adds it to the other (see (4.7)) or its sign determines which tail has a power decay and which is semi-heavy when $|\beta| = \alpha$.
- The parameter δ is a shape parameter which controls how peaked the distribution is around the mode of the pdf. The larger δ is, the flatter the density becomes.
- The parameter μ is a location parameter.
- The parameter λ is known as a shape parameter with no straightforward interpretation.

A practical problem with the standard parameterization is that α, β, and δ are not scale invariant. It is possible to demonstrate that if $X \sim \text{GH}(\lambda, \alpha, \beta, \delta, \mu)$, then

$$aX + b \sim \text{GH}(\lambda, \alpha/|a|, \beta/a, \delta|a|, a\mu + b) \tag{4.9}$$

in which $a, b \in \mathbb{R}$ and $a \neq 0$. For statistical work, it is more convenient to change the parameterization so that one parameter can be interpreted as a scale parameter and the other parameters can become scale invariant. Even though δ seems to be a natural candidate for a scale parameter, it cannot play that role in the general case because on one of the boundaries of the parameter space δ can be equal to zero with X still being a well-defined random variable. However, under the additional assumption $\delta > 0$, the parameterization can be modified. Consider the simple transformations $\alpha_1 = \alpha\delta$ and $\beta_1 = \beta\delta$. Then, a scale transformation of $X \sim \text{GH}(\lambda, \alpha_1, \beta_1, \delta, \mu)$ does not modify the new parameters α_1 and β_1.

The parameter β is a skewness parameter but its range depends on α. Thus, for a fixed β the relative skewness depends on α — as α increases, the distribution becomes more symmetric. In order to have a skewness parameter independent of α, we can set $\beta_2 = \beta/\alpha$ and use it instead.

Some references define the density of the GH distribution excluding the boundary of the feasible parameter set as in (4.6) but in terms of the GIG parameters in (4.2). Using the inverse transformation $\delta^2 = \chi$ and $\alpha^2 = \psi + \beta^2$, the pdf of a $\text{GH}(\lambda, \psi, \beta, \chi, \mu)$ distributed random variables equals

$$f_X(x) = C\frac{K_{\lambda-\frac{1}{2}}(\sqrt{(\chi+(x-\mu)^2)(\psi+\beta^2)})}{(\sqrt{(\chi+(x-\mu)^2)(\psi+\beta^2)})^{\frac{1}{2}-\lambda}}e^{\beta(x-\mu)}, \quad x \in \mathbb{R}, \tag{4.10}$$

where

$$C = \frac{\psi^\lambda(\psi+\beta^2)^{\frac{1}{2}-\lambda}}{\sqrt{2\pi}(\sqrt{\chi\psi})^\lambda K_\lambda(\sqrt{\chi\psi})}.$$

Other parameterizations offering different advantages exist in the academic literature, see Chapter 9 in Paolella (2007) for additional details.

4.2.2. Moments of the GH distribution

From a practical perspective, the most important moments of a random variable modeling the return of a financial asset include the mean, the standard deviation, the skewness, and the kurtosis. They can be calculated explicitly using the stochastic representation in

(4.1) through the moments of the mixing variable W:

$$\begin{aligned}
E[X] &= \mu + \beta E[W], \\
\mathrm{var}[X] &= E[W] + \beta^2 \,\mathrm{var}[W], \\
\mathrm{skew}[W] &= \frac{3\beta \,\mathrm{var}[W] + \beta^3 \mu_3(W)}{(\mathrm{var}[W])^{3/2}}, \\
\mathrm{kurt}[W] &= \frac{3E[W]^2 + 6\beta^2(\mu_3(W) + E[W]\,\mathrm{var}[W]) + \beta^4 \mu_4(W)}{(\mathrm{var}[W])^2},
\end{aligned} \tag{4.11}$$

where $\mu_n(W) = E[W - E[W]]^n$ denotes the nth central moment, $\mathrm{var}[W] = \mu_2(W)$ denotes variance, $\mathrm{skew}[X] = E[X - E[X]]^3/(\mathrm{var}[X])^{3/2}$ and $\mathrm{kurt}[X] = E[X - E[X]]^4/(\mathrm{var}[X])^2$ denote skewness and kurtosis, respectively. If X has a GH distribution, then the corresponding moments can be computed by plugging in the corresponding moment of the GIG distribution.

Paolella (2007) provides a general formula for the moments of a GIG distribution with the density parameterized as in (4.2). If $W \sim \mathrm{GIG}(\lambda, \chi, \psi)$, then the rth moment of W equals

$$E[W^r] = \frac{k_{\lambda+r}(\chi, \psi)}{k_\lambda(\chi, \psi)} \tag{4.12}$$

in which $k_\lambda(x, y)$ is defined in (4.3). To compute the skewness and kurtosis, however, we need the first few central moments of W. They can be calculated using the following general formula:

$$\mu_n(W) = E[W - E[W]]^n = \sum_{k=0}^{n} \binom{n}{k} (-1)^{n-k} E[W^k](E[W])^{n-k},$$

where the moments on the right-hand side are provided in (4.12).

Finally, we use the substitution $\chi = \delta^2$ and $\psi = \alpha^2 - \beta^2$ to translate these expressions in the standard GH parameterization. We provide the expressions for the mean and variance only. If $X \sim \mathrm{GH}(\lambda, \alpha, \beta, \delta, \mu)$, then the mean and variance of X equal

$$E[X] = \mu + \beta \frac{k_{\lambda+1}(\delta^2, \alpha^2 - \beta^2)}{k_\lambda(\delta^2, \alpha^2 - \beta^2)},$$

$$\operatorname{var}[X] = \frac{k_{\lambda+1}(\delta^2, \alpha^2 - \beta^2)}{k_\lambda(\delta^2, \alpha^2 - \beta^2)} + \beta^2 \left(\frac{k_{\lambda+2}(\delta^2, \alpha^2 - \beta^2)}{k_\lambda(\delta^2, \alpha^2 - \beta^2)} - \left(\frac{k_{\lambda+1}(\delta^2, \alpha^2 - \beta^2)}{k_\lambda(\delta^2, \alpha^2 - \beta^2)} \right)^2 \right),$$

where $k_\lambda(x, y)$ is defined in (4.3).

A Matlab function computing the four moments in (4.11) is provided below.

```
% function [m, stdev, skew, kurt] = ghmom_4(lambda, alpha, ...
    beta, delta, mu)
%
% Calculates the mean, the standard deviation, the skewness, ...
    and the kurtosis of a generalized hyperbolic distribution ...
    given the distribution parameters.

function [m, stdev, skew, kurt] = ghmom_4(lambda, alpha, beta, ...
    delta, mu)

psi = alpha^2 - beta^2;
chi = delta^2;
EW = ghk(lambda + 1, chi, psi)/ghk(lambda, chi, psi);
EWsq = ghk(lambda + 2, chi, psi)/ghk(lambda, chi, psi);
varW = EWsq - EW^2;

mu3W = gig_mom(3, lambda, chi, psi, EW);
mu4W = gig_mom(4, lambda, chi, psi, EW);

m = mu + beta*EW;
var = EW + beta^2*varW;
stdev = sqrt(var);
skew = (3*beta*varW + beta^3*mu3W)/((var)^(3/2));
kurt = 3*EWsq + (beta^4*mu4W + 6*beta^2*(mu3W + EW*varW));
kurt = kurt/(var)^2;

% An auxiliary function computing the n-th central moment of a ...
    GIG distribution
function mom = gig_mom(n, lambda, chi, psi, EW)

mom = 0;
for k = 0:n
  mom = mom + nchoosek(n, k)*(-1)^(n - k)*...
    (ghk(lambda + k, chi, psi)/ghk(lambda, chi, psi))*(EW)^(n-k);
end
```

It is possible to provide a general formula for $E[X-\mu]^n$ where $X \sim \text{GH}(\lambda, \psi, \beta, \chi, \mu)$. Even though this is not the central moment $E[X - E[X]]^n$ because the mean of X is not equal to μ, a similar strategy can be adopted for the central moment of order n. The formula follows easily from the stochastic representation in (4.1):

$$\begin{aligned}
E[X-\mu]^n &= E[\beta W + \sqrt{W}Z]^n \\
&= \sum_{k=0}^{n} \binom{n}{k} \beta^{n-k} E[W^{n-\frac{k}{2}}]E[Z^k] \\
&= \sum_{m=0}^{\lfloor n/2 \rfloor} \binom{n}{2m} \beta^{n-2m} \frac{k_{\lambda+n-m}(\delta^2, \alpha^2-\beta^2)}{k_\lambda(\delta^2, \alpha^2-\beta^2)} \frac{(2m)!}{2^m m!} \\
&= \sum_{m=0}^{\lfloor n/2 \rfloor} \frac{n!\beta^{n-2m}}{(n-2m)!m!2^m} \frac{k_{\lambda+n-m}(\delta^2, \alpha^2-\beta^2)}{k_\lambda(\delta^2, \alpha^2-\beta^2)},
\end{aligned}$$

where $\lfloor x \rfloor$ means rounding to the nearest integer smaller than x. We used the formula in (4.12) and the fact that if $Z \sim N(0,1)$, then $E[Z^k] = k!/(2^{k/2}(k/2)!)$ if k is even and $E[Z^k] = 0$ if k is odd. Even though this formula is difficult to use analytically, computationally it is very easy to work with this formula. We provide a Matlab implementation.

```
% function mom = ghmom(n, lambda, alpha, beta, delta, mu)
%
% Calculates the moment E(X - mu)^n of a GH distribution.

function mom = ghmom(n, lambda, alpha, beta, delta, mu)

  chi = delta^2;
  psi = alpha^2 - beta^2;
  m = floor(n/2);
  mom = 0;
  for ii = 0:m
    mom = mom + factorial(n)*beta^(n - 2*ii)*ghk(lambda + n - ...
        ii, chi, psi)/...
      (factorial(n - 2*ii)*factorial(ii)*2^ii*ghk(lambda, chi, ...
        psi));
  end
```

If the central moment $E[X - E[X]]^n$ is of interest, then from the stochastic representation in (4.12) it follows that

$$E[X - E[X]]^n = E[\beta(W - E[W]) + \sqrt{W}Z]^n,$$

and the moment can be calculated using the approach above. General formulate for the absolute moments $E|X - \mu|^n$ of GH distributions are provided in Barndorff-Nielsen and Stelzer (2005).

4.3. Special Cases

The class of GH distributions contains many subclasses which have been suggested for modeling financial asset returns. They include the hyperbolic distribution, the normal inverse Gaussian distribution, a hyperbolic skewed Student's t- and symmetric Student's t-distributions, the Laplace and the skewed Laplace distributions, the variance-gamma distribution, and the Gaussian distribution. In this section, we describe each of these subclasses.

To illustrate the impact of the parameters on the shape of the density, we provide density plots using the Matlab function from Section 4.2. Because the parameter μ only shifts the distribution to the left or to the right, it has no impact on the shape and is set equal to zero on all plots. Finally, when illustrating the impact of the tail parameter α, we calculate the density by setting the skewness parameter $\beta = b\alpha$ in order to have the same relative skewness as measured in terms of $\beta_2 = \beta/\alpha = b$.

4.3.1. The hyperbolic distribution

The hyperbolic distribution arises from the GH family when $\lambda = 1$ and the other parameters are in the interior of the parameter space, $\alpha > 0$, $|\beta| < \alpha$, $\delta > 0$, and $\mu \in \mathbb{R}$. The random variable X has a hyperbolic distribution if $X \sim H(\alpha, \beta, \delta, \mu) = \text{GH}(1, \alpha, \beta, \delta, \mu)$.

The density function of X is given by

$$f_X(x) = \frac{\sqrt{\alpha^2-\beta^2}}{2\alpha\beta K_1(\delta\sqrt{\alpha^2-\beta^2})} \times \exp(-\alpha\sqrt{\delta^2+(x-\mu)^2}+\beta(x-\mu)), \tag{4.13}$$

where $K_\lambda(x)$ is given in (4.4). The functional form of the density is derived directly from (4.6) with $\lambda = 1$ and using the property that the Bessel function can be computed explicitly when the index parameter equals 1/2,

$$K_{1/2}(x) = \sqrt{\frac{\pi}{2x}}e^{-x}, \quad x > 0, \tag{4.14}$$

see, for example, Chapter 9 in Paolella (2007). From the general tail asymptotics in (4.7), it follows that the decay of the left and the right tail of the density function is exponential,

$$f_X(x) \sim \exp\left(-(\alpha+\beta)x\right), \quad x \to \infty,$$
$$f_X(x) \sim \exp\left((\alpha-\beta)x\right), \quad x \to -\infty.$$

Plots of the density function of the hyperbolic distribution for different parameter values are provided in Figure 4.1. The top-left plot shows the impact of α — smaller values of α imply fatter tails — and the top-right plot illustrates the impact of the skewness parameter. As β decreases, the density becomes more skewed to the left and shifts to the left because the mean depends on β as well as μ.

Finally, the bottom plot demonstrates that δ controls the shape of the distribution around the center. Smaller values of δ lead to a sharper mode. As δ converges to zero, the distribution does not become singular but instead converges to the asymmetric Laplace distribution, see Section 4.3.4.

4.3.2. The normal inverse Gaussian distribution

The normal inverse Gaussian (NIG) distribution arises from GH family when $\lambda = -1/2$ and the other parameters are in the interior of

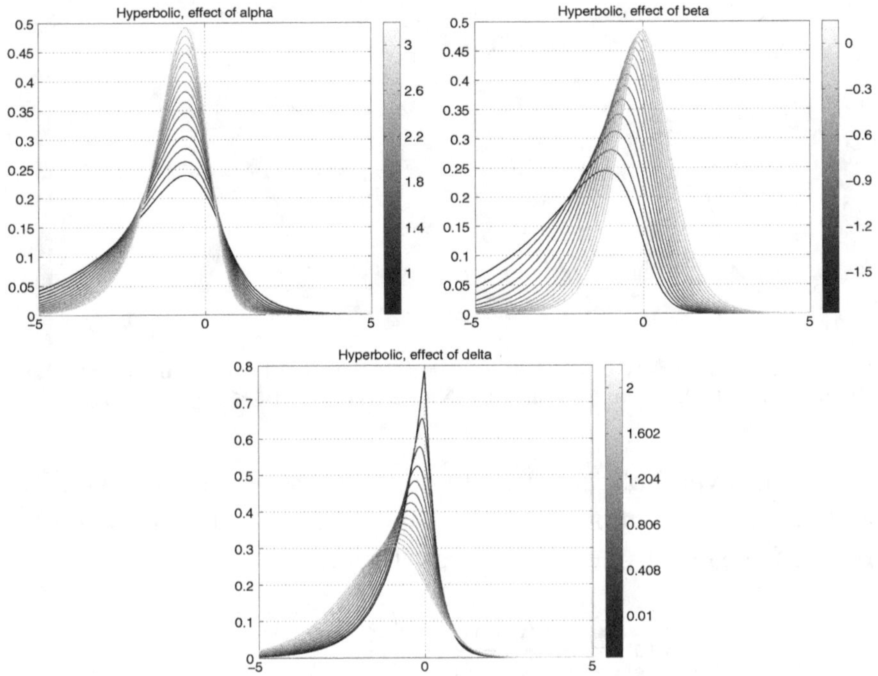

Figure 4.1. The density of the hyperbolic distribution with the following parameter values: $H(\alpha, -0.5\alpha, 1, 0)$, $1 \leq \alpha \leq 3$ (top left), $H(2, \beta, 1, 0)$, $-1.5 \leq \beta \leq 0$ (top right), and $H(2, -0.9, \delta, 0)$, $0.01 \leq \delta \leq 2$ (bottom).

the parameter space, $\alpha > 0$, $|\beta| < \alpha$, $\delta > 0$, and $\mu \in \mathbb{R}$. The random variable X has an NIG if $X \sim \text{NIG}(\alpha, \beta, \delta, \mu) = \text{GH}(-1/2, \alpha, \beta, \delta, \mu)$.

The density function of X is given by

$$f_X(x) = \frac{\alpha\delta K_1(\alpha\sqrt{\delta^2 + (x-\mu)^2})}{\pi\sqrt{\delta^2 + (x-\mu)^2}} \times \exp(\delta\sqrt{\alpha^2 - \beta^2} + \beta(x - \mu)). \tag{4.15}$$

The functional form of the density is derived directly from (4.6) with $\lambda = -1/2$ and the property in (4.14). The decay of the left and the right tail is as in (4.7) with $\lambda = -1/2$.

Plots of the density function of the NIG distribution are provided in Figure 4.2. The impact of the tail parameter α and the skewness parameter β on the shape of the distribution is the same as in the

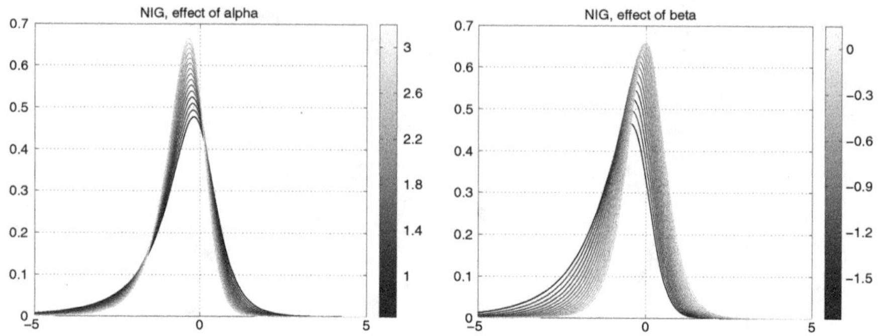

Figure 4.2. The density of the NIG distribution with the following parameter values: NIG($\alpha, -0.5\alpha, 1, 0$), $1 \le \alpha \le 3$ (top left), NIG($2, \beta, 1, 0$), $-1.5 \le \beta \le 0$ (top right).

case of the hyperbolic subclass. In this subclass, δ is a scale parameter and has no other impact on the shape. Indeed, setting $\alpha_1 = \alpha\delta$ and $\beta_1 = \beta\delta$ the density in (4.15) takes the form

$$f_X(x) = \frac{\alpha_1 K_1(\alpha_1\sqrt{1+((x-\mu)/\delta)^2})}{\delta\pi\sqrt{1+((x-\mu)/\delta)^2}} \times \exp\left(\sqrt{\alpha_1^2-\beta_1^2}+\beta_1\frac{x-\mu}{\delta}\right),$$

which demonstrates that δ is a scale parameter.

4.3.3. The hyperbolic skewed t-distribution

In contrast to the hyperbolic and NIG subfamilies which arise when the shape parameter λ takes a specific value, the hyperbolic skewed t-distribution (HSt) arises along one of the boundaries of the parameter space. A random variable X has a hyperbolic skewed t-distribution when $|\beta| = \alpha$; that is, $X \sim \text{HSt}(\lambda, \beta, \delta, \mu) = \text{GH}(\lambda, |\beta|, \beta, \delta, \mu)$. Because the parameters are on the boundary of the parameter set, the following conditions should also hold: $\lambda < 0$ and $\delta > 0$.

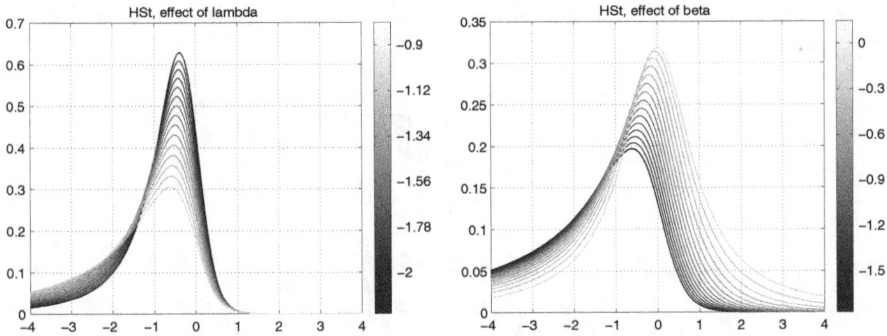

Figure 4.3. The density of the hyperbolic skewed t-distribution with the following parameter values: $\mathrm{HSt}(\lambda, -2, 1, 0)$, $-2 \le \lambda \le -0.9$ (top left), $\mathrm{HSt}(-1/2, \beta, 1, 0)$, $-1.5 \le \beta \le 0$ (top right).

The density of the hyperbolic skewed t-distribution equals

$$f_X(x) = \frac{2(\delta^2/2)^{-\lambda}}{\sqrt{2\pi}\Gamma(-\lambda)} \left(\frac{\sqrt{\delta^2 + (x-\mu)^2}}{|\beta|} \right)^{\lambda - \frac{1}{2}} \times K_{\lambda - \frac{1}{2}}(|\beta| \sqrt{\delta^2 + (x-\mu)^2}) e^{\beta(x-\mu)} \tag{4.16}$$

which is derived from (4.5). The decay of the left and the right tail of the density is not the same. Depending on the sign of beta, one of the tails has a power decay and the other one is semi-heavy. For example, if $\beta > 0$

$$f_X(x) \sim |x|^{\lambda-1} e^{-2\beta x}, \quad x \to \infty,$$
$$f_X(x) \sim |x|^{\lambda-1}, \quad x \to -\infty.$$

Plots of the density function of the hyperbolic skewed t-distribution are provided in Figure 4.3. The impact of the λ parameter is illustrated on the top-left plot and that of the skewness parameter is shown on the top-right plot.

Like the NIG distribution, the δ parameter can be interpreted as a scale parameter and has no other effect on the density shape.

If we set $\beta_1 = \beta\delta$, we obtain the following formula for the density function:

$$f_X(x) = \frac{2^{1-\lambda}}{\delta\sqrt{2\pi}\Gamma(-\lambda)} \left(\frac{\sqrt{1+\left(\frac{x-\mu}{\delta}\right)^2}}{|\beta_1|} \right)^{\lambda-\frac{1}{2}}$$

$$\times K_{\lambda-\frac{1}{2}} \left(|\beta_1| \sqrt{1+\left(\frac{x-\mu}{\delta}\right)^2} \right) e^{\beta_1 \frac{x-\mu}{\delta}}$$

which proves that δ is a scale parameter.

Student's *t*-distribution

Student's t-distribution can be viewed as a special case of the hyperbolic symmetric t. For this reason, the HSt class can be viewed as a generalization of the Student's t-distribution which is the reason behind the name of the subfamily.

The density in (4.16) with $\beta = 0$ takes a rather different functional form because the Bessel function disappears from the expression. If $X \sim \mathrm{HSt}(\lambda, 0, \delta, \mu)$, then

$$f_X(x) = \frac{\Gamma\left(\frac{-2\lambda+1}{2}\right)}{\Gamma\left(\frac{-2\lambda}{2}\right)} \frac{1}{\delta\sqrt{\pi}} \left(1+\left(\frac{x-\mu}{\delta}\right)^2\right)^{-\frac{-2\lambda+1}{2}}, \tag{4.17}$$

which is derived directly from the pdf of the GH distribution in (4.5) with $\beta = \alpha = 0$. The density in (4.17) turns into the density of the Student's t-distribution if $\mu = 0$ and $\delta^2 = -2\lambda = \nu$ where ν denotes the degrees of freedom parameter,

$$f_X(x) = \frac{\Gamma\left(\frac{\nu+1}{2}\right)}{\sqrt{\nu\pi}\Gamma\left(\frac{\nu}{2}\right)} \left(1+\frac{x^2}{\nu}\right)^{-\frac{\nu+1}{2}}.$$

Thus, adopting the classical notation for Student's t-distribution, the connection with the hyperbolic skewed t-distribution and the

GH class is

$$t(\nu) = \text{HSt}(-\nu/2, 0, \sqrt{\nu}, 0) = \text{GH}(-\nu/2, 0, 0, \sqrt{\nu}, 0).$$

The density of the hyperbolic symmetric t-distribution in (4.17) implies that the parameter δ can be interpreted as a scale parameter. As a result, if we set $\delta = \sigma\sqrt{\nu}$ where $\sigma > 0$ and $-2\lambda = \nu$, then the corresponding distribution becomes a t location-scale distribution. That is, if $X \sim t(\nu)$, then

$$\sigma X + \mu \sim \text{HSt}(-\nu/2, 0, \sigma\sqrt{\nu}, \mu).$$

For this reason, the hyperbolic symmetric t-distribution is indeed a generalization of the Student's t-distribution. Unfortunately, δ cannot be interpreted as a scale parameter in the case of hyperbolic skewed t.

4.3.4. The variance gamma distribution

Apart from HSt, the variance gamma (VG) distribution is the other class which arises on a boundary of the parameter space. The random variable X has a VG distribution if $\delta = 0$; that is, $X \sim \text{VG}(\lambda, \alpha, \beta, \mu) = \text{GH}(\lambda, \alpha, \beta, 0, \mu)$. The remaining parameters should satisfy the following conditions: $\lambda > 0$ and $|\beta| < \alpha$. The density function is derived from (4.5) and equals,

$$f_X(x) = \frac{2(\frac{\alpha^2-\beta^2}{2})^\lambda}{\sqrt{2\pi}\Gamma(\lambda)} \left(\frac{|x-\mu|}{\alpha}\right)^{\lambda-\frac{1}{2}} K_{\lambda-\frac{1}{2}}(\alpha|x-\mu|)e^{\beta(x-\mu)}. \quad (4.18)$$

Figure 4.4 provides plots of the impact of the distribution parameters on the density shape. Since the tail decay of the density this subclass coincides is given by the general asymptotics in (4.7), it is affected by both λ and α. The parameter λ also affects the shape of the density around the mode — the smaller its value is, the sharper the peak at the mode becomes.

The variance gamma distribution does not have a natural scale parameter. However, from the general property of linear transformations in (4.9) it follows that if $X \sim \text{VG}(\lambda, \alpha, \beta, 0)$, then

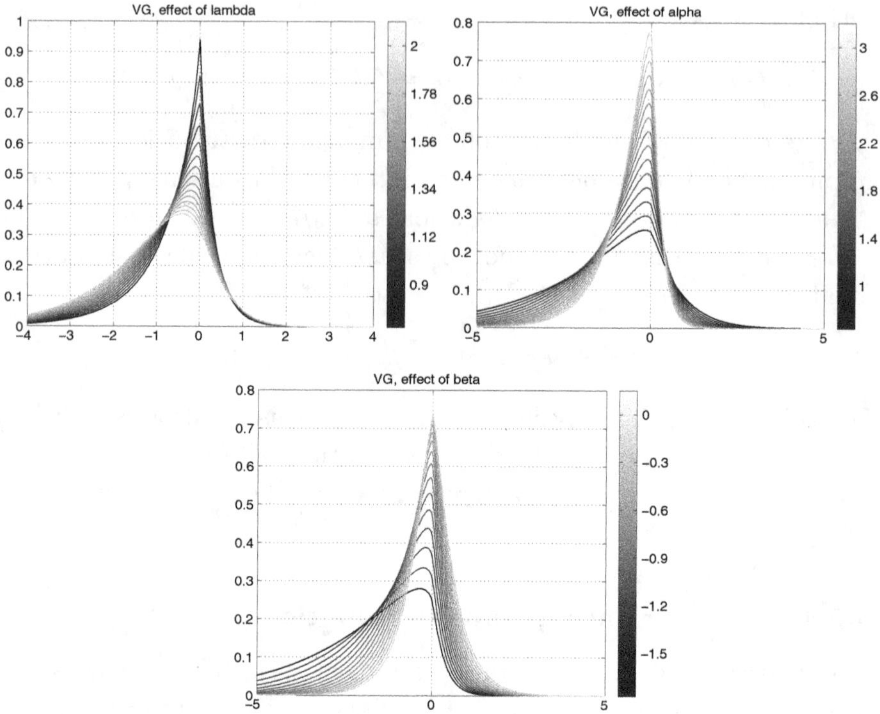

Figure 4.4. The density of the VG distribution with the following parameter values: $VG(\lambda, 2, -0.9, 0)$, $\lambda \in [0.9, 2]$ (top left), $VG(1.3, \alpha, -0.5\alpha, 0)$, $\alpha \in [1, 3]$ (top right), and $VG(1.3, 2, \beta, 0)$, $\beta \in [-0.5, 0]$ (bottom).

$\sigma X + \mu \sim VG(\lambda, \alpha/\sigma, \beta/\sigma, \mu)$. Therefore, a change in the scale of X is absorbed by both α and β.

The asymmetric Laplace distribution

The density of the VG distribution takes a much simpler form when $\lambda = 1$ due to the closed-form expression of the Bessel function in this case given in (4.14). This special case is known as the asymmetric Laplace distribution with a density function provided by the following expression:

$$f_X(x) = \frac{\alpha^2 - \beta^2}{2\alpha} e^{-\alpha|x-\mu|+\beta(x-\mu)}. \tag{4.19}$$

The asymmetric Laplace distribution arises also as a limit case in the hyperbolic subclass when δ approaches zero. Finally, if $\beta = 0$, then we obtain the Laplace distribution with a location parameter μ and a scale parameter equal to $1/\alpha$.

4.3.5. The Gaussian distribution

The GH class includes also the Gaussian distribution as a limit case. This is evident from the general mixture in (4.1) and also from the fact that Student's t-distribution is a special case in the GH family and that the Gaussian distribution arises at the limit when the degrees of freedom parameter approaches infinity. More generally, if $X \sim \text{GH}(\lambda, \alpha, \beta, \delta, \mu)$ then for all $\lambda \in \mathbb{R}$, when $\alpha \to \infty$, $\delta \to \infty$, and $\delta/\alpha \to \sigma^2$, the distribution of X converges to a Gaussian distribution with a mean equal to μ and a standard deviation equal to σ.

4.4. Random Number Generation

In this section, we discuss a Matlab implementation of a random number generator for the GH distribution. The general method is simple — given the GH distribution parameters, generate independently a random number from the corresponding GIG distribution and a random number from the standard Gaussian distribution and then use the formula in (4.1). An algorithm for GIG random number generation when the parameter values are in the interior of the parameter space ($|\beta| < \alpha$ and $\delta > 0$) is developed in Devroye (2014). Our Matlab implementation is provided in an appendix.

If the parameter values are on the border, then the distribution belongs to one of the special cases discussed in Section 4.3. In particular, the distribution is either a hyperbolic skewed t or a variance gamma. The general approach through equation (4.1) still works but one needs to identify the corresponding subclass in the GIG family because the GIG algorithm in Devroye (2014) does not apply. We provide these subclasses of the mixing variable below.

4.4.1. The hyperbolic skewed t-distribution

If $X \sim \text{HSt}(\lambda, \beta, \delta, \mu)$, then the mixing distribution in the representation in (4.1) is inverse gamma. The density function of W is derived from (4.2) with $\psi = \alpha^2 - \beta^2 = 0$,

$$f_W(x) = \frac{\left(\frac{\chi}{2}\right)^{-\lambda}}{\Gamma(-\lambda)} x^{\lambda-1} e^{-\frac{\chi}{2x}}, \quad x > 0. \tag{4.20}$$

Random numbers from an inverse gamma distribution can be generated in the following way. Suppose that V has a gamma distribution, $V \sim \Gamma(a, b)$, with a density function

$$f_V(x) = \frac{b^a}{\Gamma(a)} x^{a-1} e^{-bx}, \quad x > 0.$$

Then, $W = 1/V$ where $V \sim \Gamma(-\lambda, \chi/2)$ has an inverse gamma distribution with a density function provided in (4.20).

4.4.2. The variance gamma distribution

If $X \sim \text{VG}(\lambda, \alpha, \beta, \mu)$, then the random variable W in the representation in (4.1) has a gamma distribution. From the general form of the GIG density in (4.2), it follows that the density of W in this case equals,

$$f_W(x) = \frac{\left(\frac{\psi}{2}\right)^{\lambda}}{\Gamma(\lambda)} x^{\lambda-1} e^{-\frac{\psi}{2}x}, \quad x > 0, \tag{4.21}$$

which is the density of a gamma distribution with parameters $a = \lambda$ and $b = \psi/2$.

4.5. Parameter Estimation

GH distributions satisfy closed-form expressions of their density functions which make the method of maximum likelihood (ML) estimation a natural choice for parameter estimation. ML estimation of GH distribution parameters and various practical aspects are discussed at length in Prause (1999).

Let $(x_1, \ldots, x_n)$ be a sample of iid observations on a random variable $X \sim \mathrm{GH}(\lambda, \alpha, \beta, \delta, \mu)$. Denote the density of X by $f(x, \theta)$, where $\theta = (\lambda, \alpha, \beta, \delta, \mu)$. The ML estimator of the distribution parameters is the solution of the following optimization problem:

$$\widehat{\theta} = \arg\max_{\theta \in \Theta} \sum_{i=1}^{n} \log f(x_i, \theta), \tag{4.22}$$

in which Θ denotes the set of feasible parameter values. The logarithm of the density in the objective can be computed from the general formula for the density function in (4.5). The log-likelihood function equals

$$\begin{aligned}\sum_{i=1}^{n} \log f(x_i, \theta) = \beta \sum_{i=1}^{n} (x_i - \mu) \\ + \sum_{i=1}^{n} \log(k_{\lambda - \frac{1}{2}}((x_i - \mu)^2 - \delta^2, \alpha^2)) \\ - n \log(\sqrt{2\pi} k_\lambda(\delta^2, \alpha^2 - \beta^2)).\end{aligned}$$

The optimization problem given by (4.22) cannot be solved analytically. To solve the problem numerically, it is necessary to provide a starting point $\theta_0 = (\lambda_0, \alpha_0, \beta_0, \delta_0, \mu_0) \in \Theta$ which is best to be in the interior of the parameter space. Because financial asset returns can be skewed but the skewness is typically not very pronounced, it makes sense to provide a starting point that corresponds to a symmetric GH distribution; that is $\beta_0 = 0$. Under a symmetric hypothesis, the first four moments in (4.11) simplify a lot

$$\begin{aligned}E[X] &= \mu, \\ \mathrm{var}[X] &= E[W], \\ \mathrm{skew}[X] &= 0, \\ \mathrm{kurt}[X] &= \frac{3E[W]^2}{(\mathrm{var}[X])^2} = \frac{3E[W]^2}{(E[W])^2}.\end{aligned}$$

As a consequence, the parameter μ equals the mean of the distribution and, therefore, can be estimated by the sample average.

The higher-order moments, however, depend on the remaining three parameters and the only way to apply the method of moments is to solve a system of nonlinear equations. We adopt a simpler strategy. The parameters that control the tail thickness are set arbitrarily to $\lambda_0 = -1/2$ and $\alpha_0 = 1$. Finally, conditional on λ_0 and α_0, the parameter δ_0 can be estimated through the sample variance, see (4.11). Thus, the starting point is chosen to be

$$\theta_0 = (-1/2, 1, 0, \delta_0, \mu_0),$$

where

$$\mu_0 = \frac{1}{n}\sum_{i=1}^{n} x_i$$

and δ_0 is such that the following equality holds:

$$\frac{1}{n}\sum_{i=1}^{n}(x_i - \mu_0)^2 = \frac{k_{1/2}(\delta_0^2, 1)}{k_{-1/2}(\delta_0^2, 1)}.$$

The appendix to this chapter contains a Matlab implementation of the ML estimation of the parameters of the GH distribution and also of its various subclasses. The problem is specialized for a subclass by imposing a constraint on the set of feasible parameter values which makes Θ consistent with the corresponding subclass. For example, the ML estimator of the parameters of the hyperbolic subclass is the problem in (4.22) where

$$\Theta = \{\theta : \lambda = 1, |\beta| < \alpha, \delta > 0, \mu \in \mathbb{R}\}$$

and the starting point is modified so that $\theta_0 \in \Theta$. In some cases, it is possible to arrive at moment-based initial values of all parameters. We provide an illustration for the HSt case. The same method works for the VG distribution.

In the HSt case, $\psi = \alpha^2 - \beta^2 = 0$ which leads to further simplification of the first four moments in the symmetric case. Using the formulae (4.12) and (4.3), we find that

$$E[W] = \frac{\delta^2}{2}\frac{\Gamma(-\lambda-1)}{\Gamma(-\lambda)} \qquad \text{and} \qquad E[W^2] = \left(\frac{\delta^2}{2}\right)^2 \frac{\Gamma(-\lambda-2)}{\Gamma(-\lambda)},$$

which result in the following expressions for the variance and the kurtosis of the distribution,

$$\begin{aligned}
\mathrm{var}[X] &= \frac{\delta^2}{2}\frac{\Gamma(-\lambda-1)}{\Gamma(-\lambda)},\\
\mathrm{kurt}[X] &= \frac{3\Gamma(-\lambda-2)\Gamma(-\lambda)}{\Gamma^2(-\lambda-1)}.
\end{aligned}$$

As a consequence, the kurtosis of the distribution depends only on the parameter λ and it can be estimated from the sample kurtosis. Plugging in the estimated value of λ in the first equation leads to the estimator of δ which is proportional to the standard deviation of X.

As noted by Prause (1999), the GH distribution usually fits well to stock returns data. However, the shape of the log-likelihood function in (4.22) at the optimal solution is usually flat implying that statistically it is hard to distinguish between diverse sets of parameters in a neighborhood of the optimal solution. For this reason, some of the subclasses of the GH family may also fit very well to the data.

Figure 4.5 illustrates this phenomenon with real data — the daily returns of the S&P500 index in the period of the financial crisis, from January 1, 2008 to January 1, 2009. The figure includes four plots which compare a fitted GH distribution and fitted subclasses of the GH family to a fitted Gaussian and a kernel estimate of the density of the observed returns. The plot at the bottom shows the fitted GH distribution. The remaining subclasses are (clockwise): the hyperbolic, the HSt, the NIG, and the VG distributions. A visual inspection of the four plots shows that the Gaussian distribution is inadequate in describing the observed returns in that time period and that the GH distribution and the corresponding subclasses fit much better and perform similarly.

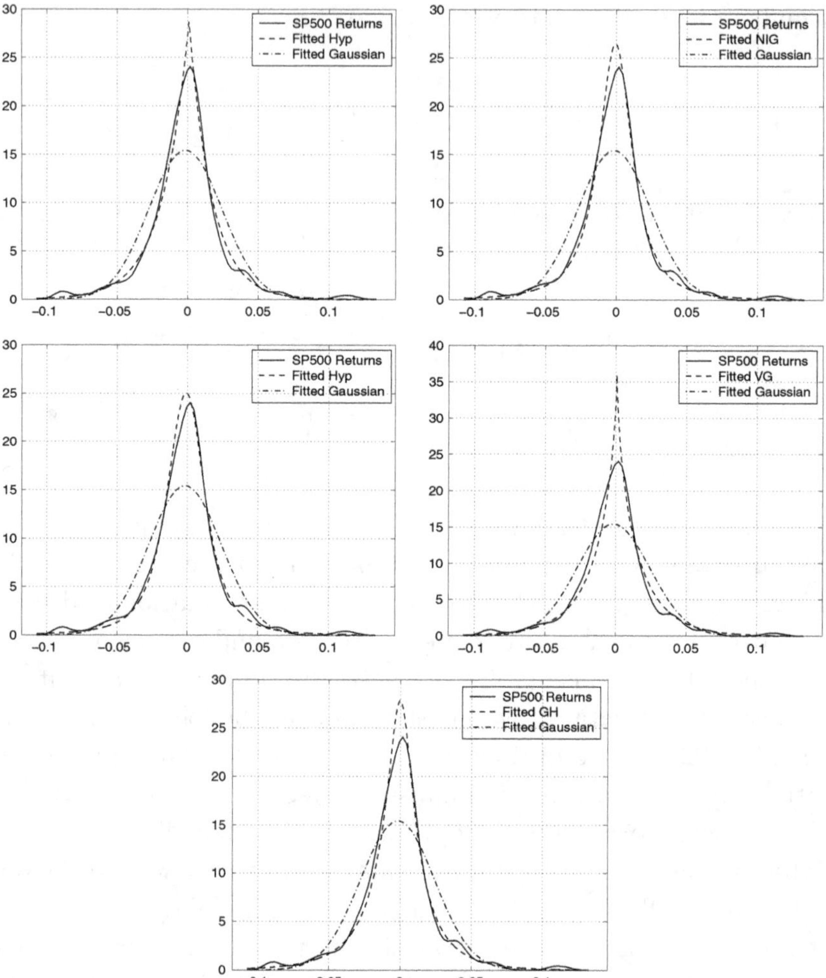

Figure 4.5. A comparison of distributions from the GH family fitted to the daily returns of the S&P500 index from January 1, 2008 to January 1, 2009 to a fitted Gaussian distribution and a kernel estimate of the density of the observed returns.

Table 4.1 reports the fitted parameter values, the value of the log-likelihood function at the solution point, and also the Anderson–Darling (AD) statistic for each of the fitted distributions. The values of the log-likelihood function at the solution point illustrate the relative impact of adding parameter constraints. For example, the value corresponding to the GH case is the highest because the subclasses are fitted by maximizing the same function under additional parameter constraints. Even though higher, the log-likelihood value is only marginally different from those corresponding to the subclasses which implies that in a statistical sense the quality of the fits is very similar.[4]

The AD statistic is used to assess the fit in the tails and is defined in the following way:

$$\mathrm{AD}(F, F_n) = \sup_{x\in\mathbb{R}} \frac{|F(x) - F_n(x)|}{\sqrt{F(x)(1 - F(x))}},$$

Table 4.1. The estimated parameters, the log-likelihood value (LL), and the Anderson–Darling (AD) statistic. The data used are the daily returns of the S&P500 index in the period from January 1, 2008 to January 1, 2009.

	AD	LL	$\widehat{\lambda}$	$\widehat{\alpha}$	$\widehat{\beta}$	$\widehat{\delta}$	$\widehat{\mu}$
GH	0.1338	594.5402	0.0656	33.1668	−2.4533	0.0092	0.0001
Hyp	0.3024	592.8092	1.0000	57.8216	−4.0720	0.0000	0.0009
NIG	0.2883	594.1007	−0.5000	21.3287	−1.6076	0.0150	−0.0005
HSt	0.1039	592.0798	−1.1508	0.8507	−0.8507	0.0217	−0.0009
VG	0.1619	592.5166	0.7731	48.0442	0.0000	0.0000	0.0006

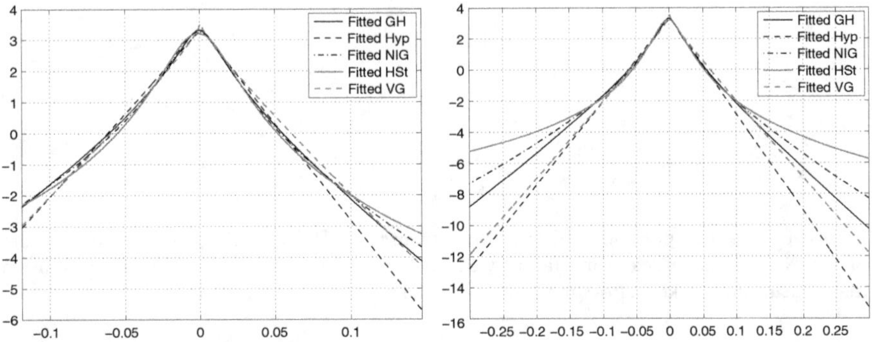

Figure 4.6. The logarithm of the fitted densities within the sample range (left) and beyond the sample range (right). The data used are the daily returns of the S&P500 index in the period from January 1, 2008 to January 1, 2009.

[4]The log-likelihood ratio test can be used to run a formal test, see Prause (1999).

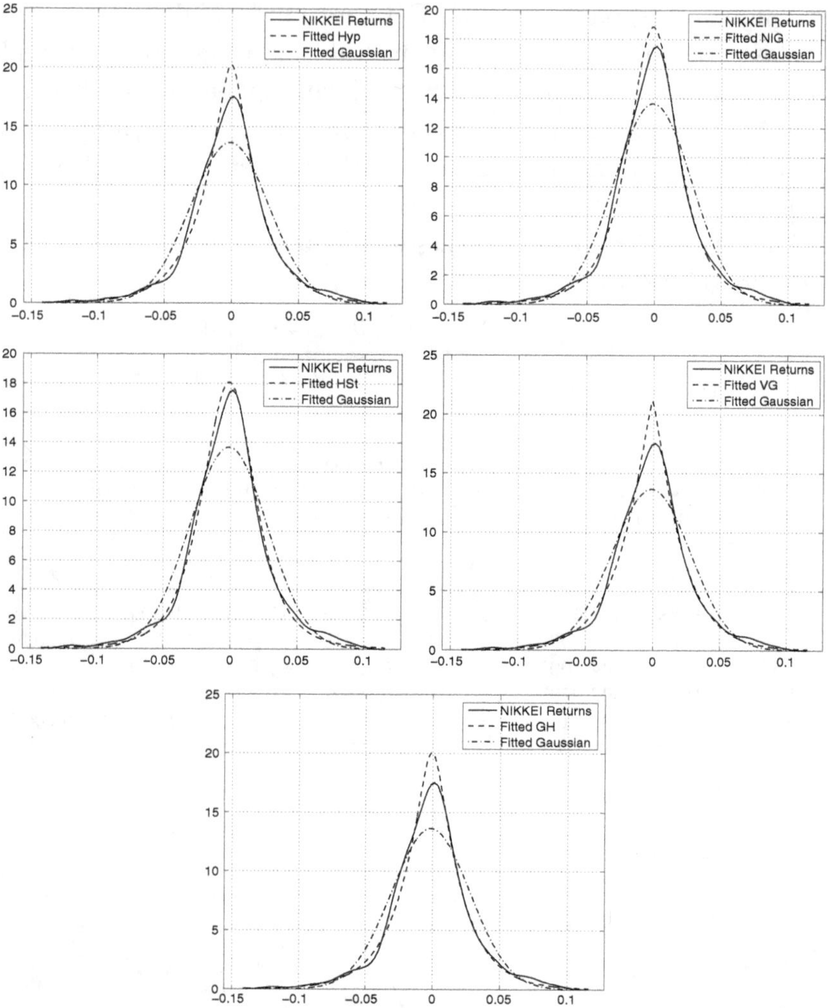

Figure 4.7. A comparison of distributions from the GH family fitted to the daily returns of the Nikkei 225 index from January 1, 2008 to January 1, 2009 to a fitted Gaussian distribution and a kernel estimate of the density of the observed returns.

where $F_n(x)$ denotes the sample cdf and $F(x)$ is the cdf of the model. The numbers in Table 4.1 indicate that the HG and the HSt distributions fit slightly better. It is not possible to test a formal hypothesis because the distribution of the AD statistic is generally unknown.

A visual comparison of the tails of the fitted distribution is provided on Figure 4.6. The plots compare the logarithms of the fitted densities because it reveals more clearly the differences in the tail decay — the logarithm of a density function decreases to $-\infty$ as $x \to \pm\infty$. The left plot shows the logarithm of the density of the fitted distributions for values of the argument within the minimal loss and the maximal gain of S&P500 in the corresponding period. The right plot shows the same functions but when the argument is allowed to go beyond the limits of the observed data points. Clearly within the range of the observed data, the fitted hyperbolic distribution tends to have lighter tails on a relative basis and the other distributions appear very similar. Extrapolating the tail beyond the data range indicates that the fitted HSt distribution has heavier tails than the rest.

As another example, we provide the same information for the daily returns of the Nikkei 225 index for the same period. Figure 4.7 compares the fitted distributions to the data and Table 4.2 provides the corresponding statistics. Similar to the S&P500 case, the differences in the log-likelihood at the optimal solution point are marginal and the four models seem indistinguishable even though there are substantial differences in the estimated parameter values. In contrast to the S&P500 case, the differences in the AD statistics are also marginal. The comparison in the left plot on Figure 4.8 is consistent with this observation — the logarithms of all fitted densities virtually coincide. Beyond the minimal loss and the maximal gain

Table 4.2. The estimated parameters, the log-likelihood value (LL), and the Anderson–Darling (AD) statistic. The data used are the daily returns of the Nikkei 225 index in the period from January 1, 2008 to January 1, 2009.

	AD	LL	$\widehat{\lambda}$	$\widehat{\alpha}$	$\widehat{\beta}$	$\widehat{\delta}$	$\widehat{\mu}$
GH	0.0834	544.8916	0.8691	47.8754	−0.9347	0.0080	−0.0005
Hyp	0.0868	544.8854	1.0000	49.9372	−1.0890	0.0059	−0.0004
NIG	0.0782	544.5225	−0.5000	28.7705	−0.0842	0.0253	−0.0013
HSt	0.0842	543.3365	−1.7613	0.1167	0.1167	0.0386	−0.0015
VG	0.0885	544.7887	1.2015	53.1756	−0.4666	0.0000	−0.0009

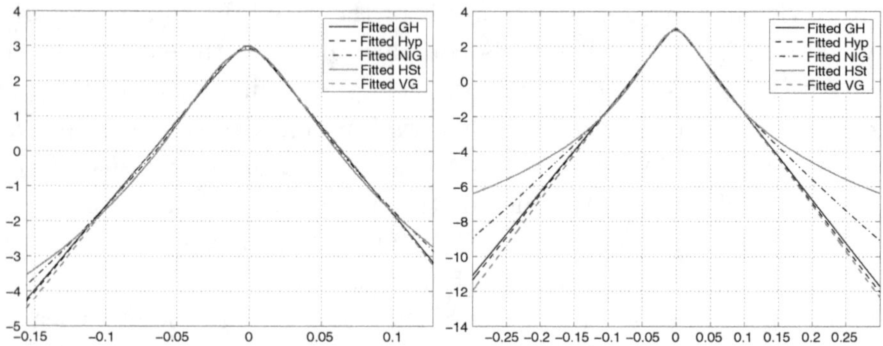

Figure 4.8. The logarithm of the fitted densities within the sample range (left) and beyond the sample range (right). The data used are the daily returns of the Nikkei 225 index in the period from January 1, 2008 to January 1, 2009.

however, the tails do diverge. The fitted HSt has the heaviest tail followed by the fitted NIG.

4.6. The Multivariate GH Distribution

An appealing feature of location-scale mixture models is that it is straightforward to extend the model to a multivariate setting. A detailed discussion of the properties of the multivariate case and how to estimate parameters is provided in Chapters 7 and 8. In this section, our objective is only to describe the intuition. For additional details, see also Chapter 3 in McNeil *et al.* (2005).

The main idea behind the multivariate extension is to use a multivariate normal Z in equation (4.1). A random vector X is said to have a multivariate GH distribution $X \sim \text{GH}(\lambda, \alpha, \boldsymbol{\beta}, \delta, \boldsymbol{\mu}, \Sigma)$,

$$X = \boldsymbol{\mu} + \boldsymbol{\beta} W + \sqrt{W} A Z \tag{4.23}$$

in which $\boldsymbol{\mu}, \boldsymbol{\beta} \in \mathbb{R}^n$, A is an $n \times n$ lower triangular matrix, $Z \sim N(0, I)$ where I is the identity matrix, and W and Z are independent. The product AZ has a multivariate normal distribution with a covariance matrix $\Sigma = AA'$, $AZ \sim N(0, \Sigma)$. To distinguish between the scalar β and μ and the vector $\boldsymbol{\beta}$ and $\boldsymbol{\mu}$, we use the same notation in bold.

When $W \sim \text{GIG}(\lambda, \chi, \psi)$, then X has a multivariate GH distribution. The components of X have one-dimensional GH distributions which are slightly different from the main definition in (4.5). The difference is that each component of AZ has a univariate Gaussian distribution with variance possibly different from one and, therefore, the corresponding univariate GH distributions have one additional parameter.

Because W is one and the same random variable for all vector elements, the components of X share the same (λ, χ, ψ), or $(\lambda, \alpha, \delta)$ respectively. The skewness parameter $\boldsymbol{\beta}$, however, is a vector and therefore each component of X can have a different skewness. As a consequence, any difference in the tail behavior of the components (see (4.7)) is only due to differences in the corresponding elements of $\boldsymbol{\beta}$. The multivariate distribution is, thus, not rich enough to accommodate different tail decays of the elements of X. For example, in a multivariate GH distribution it is not possible to have one Gaussian and one non-Gaussian component. If one component is Gaussian, then necessarily X has a multivariate Gaussian distribution because W must be a constant.

By a similar argument, because λ and δ are the same for all components of X, it is impossible to have X_i with a HSt distribution $(\lambda < 0)$ and X_j with a hyperbolic distribution $(\lambda = 1)$. That is, some of the special cases cannot co-exist in a multivariate distribution defined by (4.23).

Finally, a general property of the mixture model in (4.23) is that it is impossible for X to have independent components unless W is a constant in which case X is a multivariate Gaussian. Intuitively, the reason is the common scaling factor $\sqrt{W}$. For instance, if A is a diagonal matrix, then the components of Z are independent by construction but because of the common random scaling factor, the components of X become dependent.

The multivariate GH distribution has been applied to portfolio theory. If we build a portfolio of financial assets the return of which follows a multivariate GH distribution, it turns out that the portfolio return follows a univariate GH distribution. Denote by $v = (v_1, \ldots, v_n)$ a vector of portfolio weights and suppose that $X \sim \mathrm{GH}(\lambda, \alpha, \boldsymbol{\beta}, \delta, \boldsymbol{\mu}, \Sigma)$. Then, the portfolio return $v'X$ has the following mixture representation:

$$\begin{aligned} v'X &= v'\boldsymbol{\mu} + v'\boldsymbol{\beta} W + \sqrt{W} v' A Z \\ &= \mu_v + \beta_v W + \sqrt{W} \sigma_v Z. \end{aligned} \tag{4.24}$$

As a result, comparing (4.1) and (4.24) shows that the normalized portfolio return distribution is GH with the following parameters:

$$\frac{v'X - \mu_v}{\sigma_v} \sim \mathrm{GH}(\lambda, \alpha, \beta_v/\sigma_v, \delta, 0).$$

For a more general result on linear transformations of multivariate GH distributions, see Chapter 3 in McNeil *et al.* (2005).

Key Points

- The generalized hyperbolic distribution is a location-scale mixture of normals with a generalized inverse Gaussian mixing distribution. It represents a flexible model capable of capturing skewness and fat tails.
- Some of the main properties including closed-form expression of the density function, interpretation of parameters, moments, random number generation, and multivariate extensions follow directly from the mixture representation.
- The class of generalized hyperbolic distributions includes representatives with light tails, exponential tails, semi-heavy tails, and heavy tails with a power decay.
- Many classes of distributions popular for modeling financial asset returns arise as special cases of the generalized hyperbolic family such as the hyperbolic, the normal inverse Gaussian, the hyperbolic skewed *t*, Student's *t*, the variance gamma, the asymmetric Laplace, and the Gaussian distributions.
- Maximum likelihood estimation is straightforward with initial conditions which can be provided by the method of moments.
- The generalized hyperbolic distribution represents a very flexible family which tends to overfit the data. It is common to obtain

fits of similar quality when the general class is compared to some of the most popular subclasses such as the hyperbolic, the normal inverse Gaussian, the hyperbolic skewed t, and the variance gamma subclasses.
- The return of a portfolio of financial assets the returns of which follow a multivariate GH law has a univariate GH distribution.

References

A. Badescu, R. Elliott, R. Kulperger, J. Miettinen, and T.K. Siu. A comparison of different kernels for GARCH option pricing with generalized hyperbolic distributions. *International Journal of Theoretical and Applied Finance*, 14(5):669–708, 2011.

O.E. Barndorff-Nielsen. Exponentially decreasing distributions for the logarithm of particle size. *Proceedings of the Royal Society of London A: Mathematical, Physical and Engineering Sciences*, 353(1674): 401–419, 1977.

O.E. Barndorff-Nielsen and R. Stelzer. Absolute moments of generalized hyperbolic distributions and approximate scaling of normal inverse Gaussian–Lévy processes. *Scandinavian Journal of Statistics*, 32(4): 617–637, 2005.

L. Devroye. Random variate generation for the generalized inverse Gaussian distribution. *Statistics and Computing*, 24(2):239–246, 2014.

E. Eberlein. Application of generalized hyperbolic Lévy motions to finance. In O.E. Barndorff-Nielsen, T. Mikosch, and S. Resnick, editors, *Lévy Processes: Theory and Applications*, Birkhäuser, Boston, 2001, pp. 319–336.

E. Eberlein and U. Keller. Hyperbolic distributions in finance. *Bernoulli*, 1:281–299, 1995.

E. Eberlein and F. Özkan. Time consistency of Lévy models. *Quantitative Finance*, 3(1):40–50, 2003.

M. Hellmich and S. Kassberger. Efficient and robust portfolio optimization in the multivariate generalized hyperbolic framework. *Quantitative Finance*, 11(10):1503–1516, 2011.

S. Kassberger and R. Kiesel. A fully parametric approach to return modelling and risk management of hedge funds. *Financial Markets and Portfolio Management*, 20:472–491, 2006.

A. McNeil, R. Frey, and P. Embrechts. *Quantitative Risk Management: Concepts, Techniques, and Tools.* Princeton University Press, Princeton, NJ, 2005.

M. Paolella. *Intermediate Probability: A Computational Approach.* Wiley, Chichester, 2007.

K. Prause. *The generalized hyperbolic model: Estimation, financial derivatives, and risk measures.* PhD thesis, Universität Freiburg i. Br., Freiburg i. Br, 1999.

S.T. Rachev and S. Mittnik. *Stable Paretian models in finance.* Wiley, Chichester, 2000.

Appendix — Matlab Functions

This appendix includes a Matlab implementation of the random number generator discussed in Section 4.4 and the maximum likelihood estimator discussed in Section 4.5.

A.1. Generation of random numbers

We first provide a general implementation of the random number generator following the mixture formula. The general implementation uses a stand-alone vectorized implementation of the random number generator of GIG distributions given in Devroye (2014).

```
% function rnd = ghrnd(m, n, lambda, alpha, beta, delta, mu)
%
% Generates random numbers from the generalized hyperbolic ...
    distribution.
%
% Input parameters:
%     m and n: the dimension of the matrix to be filled with ...
    independent random numbers lambda, alpha, beta, delta, mu: ...
    the GH distribution parameters
%
% Output parameters:
%     out: an m x n matrix with iid random numbers

function rnd = ghrnd(m, n, lambda, alpha, beta, delta, mu)

chi = delta^2;
psi = alpha^2 - beta^2;

% generate the GIG subordinator
if(delta == 0 && abs(beta) < alpha)
  gig = gamrnd(lambda, 2/psi, m, n);
elseif(delta > 0 && abs(beta) == alpha)
  gig = 1./gamrnd(-lambda, 2/chi, m, n);
```

```
else
  gig = gigrnd(m, n, lambda, chi, psi);
end

% use the Gaussian mean-variance mixture
rnd = mu + gig*beta + sqrt(gig).*randn(m, n);
```

```
% function out = gigrnd(m, n, lambda, chi, psi)
%
% Generates random numbers from the generalized inverse ...
    Gaussian (GIG) distribution with a density function
%
%   f(x) = C*x^(lambda - 1)*exp(-(psi*x + chi/x)/2), x > 0
%
%   where C = (psi/chi)^(lambda/2)/(2*K_lambda(sqrt(psi*chi)))
%
% Input parameters:
%      m and n: the dimension of the matrix to be filled with ...
    independent random numbers lambda, psi, and chi: the GIG ...
    distribution parameters
%
% Output parameters:
%      out: an m x n matrix with iid random numbers
%
% The function has been vectorized to speed up calculations.
%
% The function implements the method published in:
% L. Devroye, Random variate generation for the generalized ...
    inverse Gaussian distribution Statistics and Computing, ...
    Vol. 24, pp. 239-246, 2014.

function out = gigrnd(m, n, lambda, chi, psi)

% pass to the two-parameter family
omega = sqrt(psi*chi);
alpha = sqrt(omega^2 + lambda^2) - lambda;

% define a few functions
f_psi = @(x, alpha, lambda) -alpha*(cosh(x) - 1) - ...
    lambda*(exp(x) - x - 1);
f_dpsi = @(x, alpha, lambda) -alpha*sinh(x) - lambda*(exp(x) - 1);
f_chi = @(x, s, t, sd, td, zeta, eta, theta, xi) (x >= -sd & x ...
    <= td) + ...
  (x > td).*exp(-eta - zeta*(x - t)) + (x < -sd).*exp(-theta + ...
      xi*(x + s));

% initialization step
x = -f_psi(1, alpha, lambda);
```

```
if((x >= 0.5) && (x <= 2))
  t = 1;
elseif(x > 2)
  t = sqrt(2 / (alpha + lambda));
elseif(x < 0.5)
  t = log(4/(alpha + 2*lambda));
end

x = -f_psi(-1, alpha, lambda);
if((x >= 0.5) && (x <= 2))
  s = 1;
elseif(x > 2)
  s = sqrt(4/(alpha*cosh(1) + lambda));
elseif(x < 0.5)
  s = min(1/lambda, log(1 + 1/alpha + sqrt(1/alpha^2+2/alpha)));
end

eta = -f_psi(t, alpha, lambda);
zeta = -f_dpsi(t, alpha, lambda);
theta = -f_psi(-s, alpha, lambda);
xi = f_dpsi(-s, alpha, lambda);
p = 1/xi;
r = 1/zeta;
td = t - r*eta;
sd = s - p*theta;
q = td + sd;

% prepare for the vectorized acceptance-rejection
notGood = true;
out = zeros(m, n);
chng = (out == 0);
ready = ~chng;
while(notGood)
  U = rand(m, n);
  V = rand(m, n);
  W = rand(m, n);

  out(chng) = -sd + p*log(V(chng));
  indU1 = (U < (q/(p + q + r)));
  indU2 = (U < ((q + r)/(p + q + r)));
  cond1 = chng & indU1;
  cond2 = chng & ~indU1 & indU2;

  out(cond1) = -sd + q*V(cond1);
  out(cond2) = td - r*log(V(cond2));

  ready(chng) = (W(chng).*f_chi(out(chng), s, t, sd, td, zeta, ...
      eta, theta, xi) <= ...
    exp(f_psi(out(chng), alpha, lambda)));
  chng = ~ready;
```

```
  if all(all(ready))
    notGood = false;
  end
end

% return to the three-parameter family
out = exp(out)*(lambda/omega + sqrt(1 + (lambda/omega)^2));
out = out./sqrt(psi/chi);
```

A.2. Maximum likelihood estimation

```
% function [mu, sigma, beta, alpha, lambda, info] = ghfit(sample)
%
% Estimates the parameters of the generalized hyperbolic ...
    distribution given a sample of observations. The function ...
    implements the maximum likelihood estimator.


function [lambda, alpha, beta, delta, mu, fval] = ...
    ghfit(sample, subclass)

if nargin == 1
    subclass = [];
end

% set the tolerance level for the optimization
str = optimset('TolX', 1.0e-9, 'TolFun', 1.0e-9, ...
    'MaxFunEvals', 10000, 'Display', 'Iter');
if isempty(subclass)
    % set initial values for the parameters, use the sample ...
        mean and volatility transform the parameter space to R^5
    d = fzero(@(x) ghk(1/2, x, 1)/ghk(-1/2, x, 1) - ...
        std(sample)^2, [0,1000]);
    d = sqrt(d);
    theta0 = [-0.5, log(1), 0, log(d), mean(sample)];

    [thetaMax, fval] = fminunc(@(x) -ghlogL(x, sample), ...
        theta0, str);

    % transform the solution back to the parameter space
    lambda = thetaMax(1);
    alpha = exp(thetaMax(2));
    beta = thetaMax(3);
    delta = exp(thetaMax(4));
    mu = thetaMax(5);
elseif strcmpi(subclass, 'Hyp')
```

```
    d = std(sample)*2;
    theta0 = [log(1), 0, log(d), mean(sample)];
    [thetaMax, fval] = fminunc(@(x) -ghlogL([1, x], sample),...
        theta0, str);

    % transform the solution back to the parameter space
    lambda = 1;
    alpha = exp(thetaMax(1));
    beta = thetaMax(2);
    delta = exp(thetaMax(3));
    mu = thetaMax(4);
elseif strcmpi(subclass, 'NIG')
    d = fzero(@(x) ghk(1/2, x, 1)/ghk(-1/2, x, 1) - ...
        std(sample)^2, [0,1000]);
    d = sqrt(d);
    theta0 = [log(1), 0, log(d), mean(sample)];
    [thetaMax, fval] = fminunc(@(x) -ghlogL([-1/2, x], ...
        sample), ...
        theta0, str);

    % transform the solution back to the parameter space
    lambda = -1/2;
    alpha = exp(thetaMax(1));
    beta = thetaMax(2);
    delta = exp(thetaMax(3));
    mu = thetaMax(4);
elseif strcmpi(subclass, 'HSt')
    lambda0 = -fzero(@(x) gamma(x)*gamma(x - 2)/(gamma(x - ...
        1)^2) - kurtosis(sample)/3, [2.00001, 100]);
    d = std(sample)*sqrt(2*gamma(-lambda0)/gamma(-lambda0 - 1));
    theta0 = [log(-lambda0), 0, log(d), mean(sample)];
    [thetaMax, fval] = fminunc(@(x) -HStlogL(x, sample), ...
        theta0, str);

    % transform the solution back to the parameter space
    lambda = -exp(thetaMax(1));
    alpha = abs(thetaMax(2));
    beta = thetaMax(2);
    delta = exp(thetaMax(3));
    mu = thetaMax(4);
elseif strcmpi(subclass, 'VG')
    lambda0 = fzero(@(x) gamma(x)*gamma(x + 2)/(gamma(x + ...
        1)^2) - kurtosis(sample)/3, [0.00001, 50]);
    a = (1/std(sample))*sqrt(2*gamma(lambda0 + 1)/gamma(lambda0));
    theta0 = [log(lambda0), log(a), 0, mean(sample)];
    [thetaMax, fval] = fminunc(@(x) -VGlogL(x, sample), ...
        theta0, str);

    % transform the solution back to the parameter space
    lambda = exp(thetaMax(1));
```

```
    alpha = exp(thetaMax(2));
    beta = thetaMax(3);
    delta = 0;
    mu = thetaMax(4);
else error('The subclass can be Hyp, NIG, HSt, or VG')
end

fval = -fval;

function l = VGlogL(theta, sample)
% An auxiliary function implementing the log-likelihood of the ...
    variance-gamma

% transform the input to the parameter space
lambda = exp(theta(1));
alpha = exp(theta(2));
beta = theta(3);
mu = theta(4);

% compute the log-likelihood and add a logarithmic barrier
l = beta*sum(sample - mu) + sum(log(ghk_v(lambda - 1/2, ...
    (sample - mu).^2, alpha^2))) -...
   length(sample)*log(sqrt(2*pi)*ghk_v(lambda, 0, alpha^2 - ...
       beta^2));

l = l + 0.01*log(alpha - abs(beta));

function l = HStlogL(theta, sample)
% An auxiliary function implementing the log-likelihood of the ...
    hyperbolic skewed t.

% transform the input to the parameter space
lambda = -exp(theta(1));
alpha = abs(theta(2));
beta = theta(2);
delta = exp(theta(3));
mu = theta(4);

% compute the log-likelihood, no need for a barrier
l = beta*sum(sample - mu) + sum(log(ghk_v(lambda - 1/2, ...
    (sample - mu).^2 + delta^2, alpha^2))) -...
    length(sample)*log(sqrt(2*pi)*ghk_v(lambda, delta^2, 0));

function l = ghlogL(theta, sample)
% An auxiliary function implementing the log-likelihood of the ...
    generalized hyperbolic distribution.

% transform the input to the parameter space
lambda = theta(1);
alpha = exp(theta(2));
```

```
beta = theta(3);
delta = exp(theta(4));
mu = theta(5);

% compute the log-likelihood and add a logarithmic barrier to ...
    make sure the inequality between beta and alpha is satisfied.
l = beta*sum(sample - mu) + sum(log(ghk_v(lambda - 1/2, ...
    (sample - mu).^2 + delta^2, alpha^2))) -...
    length(sample)*log(sqrt(2*pi)*ghk_v(lambda, delta^2, ...
        alpha^2 - beta^2));

l = l + 0.01*log(alpha - abs(beta));

function f = ghk_v(lambda, x, y)
% An auxiliary vectorized function computing ghk. The case (x ...
    == 0, y == 0) returns zero which translates into -Inf of ...
    the log-likelihood.

if(length(x) > 1) && (length(y) == 1)
    y = y*ones(size(x));
elseif (length(x) == 1) && (length(y) > 1)
    x = x*ones(size(y));
end

ind11 = (x > 0) & (y > 0);
ind01 = (x == 0) & (y > 0);
ind10 = (x > 0) & (y == 0);

f = zeros(size(x));
f(ind11) = 2*(x(ind11)./y(ind11)).^(lambda/2).*besselk(lambda, ...
    sqrt(x(ind11).*y(ind11)));
f(ind01) = (y(ind01)/2).^(-lambda)*gamma(lambda);
f(ind10) = (x(ind10)/2).^(lambda)*gamma(-lambda);
```

Chapter 5

The Class of Stable Distributions

The main topics covered in this chapter are:

- the history of stable distributions and why they represent an important building block for econometric models in finance;
- the main properties of the class of stable distributions;
- numerical techniques for approximating the densities and the distribution functions of stable laws;
- an algorithm for drawing efficiently random numbers from a stable distribution;
- several methods for estimation of the parameters of stable laws;
- a method for computing average value-at-risk when losses follow a stable distribution.

5.1. Introduction

As explained in Chapter 1, a central question in financial economics because of its significant theoretical and practical implications is: what is the distribution of financial returns? There is clearly no fundamental theory in finance that can suggest a distributional model and, therefore, the question remains largely an empirical one. Extensive empirical studies carried out since the 1950s have discovered several stylized facts: financial returns exhibit skewness, excess kurtosis, and fat tails. That is, realistic parametric models need to capture these empirical phenomena. On the other hand, the stylized facts are insufficient to identify even a broad class of models and that raises questions about *model risk*; that is, how sensitive the results are to the fact that we may have assumed a wrong parametric form. As a consequence, in many applications a parametric model is assumed directly without a specific motivation. Depending on the goal of the analysis, one then performs in-sample and out-of-sample tests to justify the parametric assumption.

It is also possible to argue differently; instead of hypothesizing a particular functional form, one can look at the problem of modeling the return distribution as an *approximation problem*. That is, given assumptions on the structure of the return process, one can look for a theory that provides an approximate model which is *close* in some sense to the unknown true distribution. This chapter is about such an approach which is based on the central limit theorem (CLT) and its generalizations.[1]

The CLT has an interesting history. In 1733, the English mathematician Abraham de Moivre published an article in which he calculated approximately the probability of the number of heads resulting from many independent tosses of a fair coin. In this calculation, he used the normal distribution as approximation. In the century that followed, the discovery of de Moivre was almost forgotten. It was rediscovered and extended in 1812 by the French mathematician Pierre-Simon Laplace. Nowadays, this is known as the theorem of de Moivre–Laplace of the normal approximation to the binomial distribution. It has many applications in different areas. In the field of finance, for example, it is the theoretical foundation behind the construction of binomial trees for option pricing.

---

[1]The common formal measure of *closeness* is the Kolmogorov metric $\rho(F, G) = \sup_x |F(x) - G(x)|$ but other choices are also possible, see Rachev *et al.* (2013).

The theorem of de Moivre–Laplace is a special case of the CLT but it was not until the beginning of the twentieth century that the importance of the CLT was fully recognized. In 1901, the Russian mathematician Aleksandr Lyapunov gave a more abstract formulation and showed that the limit result holds under certain very general conditions known as Lyapunov's conditions. Later, other conditions were established which generalized Lyapunov's conditions. A final solution to the problem was given by Bernstein, Lindeberg, and Feller who specified necessary and sufficient conditions for the CLT.

In the 1930s, the CLT was extended and the new result became known as the generalized CLT. Generally, the distribution of a sum of independent and identically distributed (iid) infinite variance random variables does not converge to the Gaussian distribution but to a *stable distribution.* The Gaussian distribution appears as a special case from the family of stable distributions. Stable non-Gaussian laws are also called stable Paretian because of the Pareto-like power decay of the tail of the distribution, or Lévy stable after the name of Paul Lévy who carried out the fundamental research of characterizing the family of non-Gaussian stable laws. As pointed out in Rachev and Mittnik (2000), stable distributions are attractive because they have an important desirable property — domains of attraction — which arises from the generalized CLT. Loosely speaking, according to this property, if a distribution is in the domain of attraction of a stable law, it has properties which are close to those of the corresponding stable law. The domain of attraction is completely determined by the tail behavior of the distribution and rationalizes the use of stable distributions as approximations.

In the pioneering work of Mandelbrot (1963) and Fama (1963), the class of stable laws was proposed as a model for financial asset returns. There were two reasons: (i) the domains of attraction property, and (ii) the empirical argument that they are heavy-tailed and can model skewness and excess kurtosis. Their work led to the consolidation of the stable Paretian hypothesis which underwent empirical scrutiny in the 1970s and 1980s. It was found out that even though stock returns exhibit tails thicker than those of the Gaussian

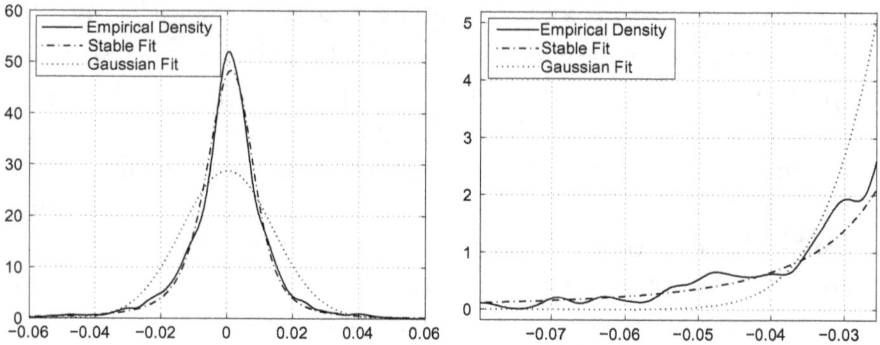

Figure 5.1. A comparison of a stable distribution and a Gaussian distribution fitted on the daily returns of the S&P 500 index from January 1, 2008 to January 1, 2016. The left plot shows the corresponding densities while the right plot shows the left tail zoomed in.

distribution, there are indications for a few inconsistencies with the stable Paretian hypothesis. The empirical inconsistencies are related mainly to a failure of the temporal aggregation property (i.e., that tail thickness does not change with return frequency) and the infinite variance property of stable laws.[2]

Nevertheless, stable distributions remain an important building block of econometric models. Grabchak and Samorodnitsky (2010) argue that stable distributions can be "close" to a sum of iid random variables with finite variance when the sum is truncated.[3] This property can explain why stable distributions seem to fit well-empirical data in spite of their infinite variance property. Efforts to obtain distributions that have a shape similar to stable laws and a finite variance led to the class of *tempered stable distributions* which are the main topic of Chapter 6.

The difference between the stable Paretian and the Gaussian distribution is illustrated in Figure 5.1. It compares a stable distribution and a Gaussian distribution estimated from the daily returns of the S&P 500 index in the period from January 1, 2008 to January 1, 2016.

---

[2]See, for example, Officer (1972), Akgiray and Booth (1988), and Akgiray and Lamoureux (1989). However, some of the statistical tests have been criticized, see Paolella (2016).

[3]In probability theory, this is known as a *pre-limit theorem*.

The left plot shows the fitted densities together with the empirical density and the right plot shows the left tail. Empirical examples with different asset classes in the context of different econometric models can be found in Rachev and Mittnik (2000). Mathematical models with application of stable laws in finance, economics and other areas can be found also in Adler *et al.* (1998), Zolotarev and Uchaikin (1999), and Embrechts *et al.* (1997).

This chapter is organized in the following way. Section 5.2 begins with a brief introduction to the classical and the generalized CLT because the domains of attraction property plays an important role in the motivation. Section 5.3 defines the class of stable distributions and provides some basic properties. Section 5.4 discusses numerical work with the densities and the distribution functions of the class. Section 5.5 provides an efficient random number generator. Section 5.6 discusses three parameter estimation methods which differ in computational complexity and estimation accuracy. Finally, Section 5.7 provides a formula for conditional value-at-risk when the losses have a stable distribution.

The appendix to this chapter includes efficient Matlab implementations of routines computing approximations of the density functions, the distribution functions, the quantile functions of stable distributions, and also provides routines for parameter estimation.

5.2. The Classical and the Generalized CLT

We begin with the result due to de Moivre and Laplace which shows that the normal distribution can be adopted as an approximate model for the binomial distribution which arises from n independent coin flips. We then proceed with the general case of a sum of iid summands and discuss the classical necessary conditions for the CLT. Finally, we discuss the generalized CLT.

Before we go into the details, note that the problem setup is that of a sum of random variables. How is this relevant for the practical problem of providing a model for financial returns? In finance, the distributional hypothesis concerns usually the logarithmic returns or the price changes which are also known as increments. Choose a time interval $[t_0, t_1]$ and take any $\tau \in (t_0, t_1)$. The value of τ separates $[t_0, t_1]$ into two subintervals. Denote by S_t the price of a financial security. The log-return $r_{(t_0,t_1)} = \log(S_{t_1}/S_{t_0})$ for the full period equals the sum of the log-returns for the subperiods,

$$r_{(t_0,t_1)} = \log(S_\tau/S_{t_0}) + \log(S_{t_1}/S_\tau) = r_{(t_0,\tau)} + r_{(\tau,t_1)}.$$

The same argument can be repeated for any arbitrary subdivision of the longer time period. Thus, for example, the monthly log-return equals the sum of the corresponding daily log-returns which are sums of the corresponding intra-day log-returns.

Likewise, the price increment $\Delta S_{(t_0,t_1)} = S_{t_1} - S_{t_0}$ can be represented as

$$\Delta S_{(t_0,t_1)} = S_{t_1} - S_\tau + S_\tau - S_{t_0} = \Delta S_{(t_0,\tau)} + \Delta S_{(\tau,t_1)}$$

and the same statement holds for the price increments. It is therefore natural to think of the lower frequency log-returns (increments) as sums of higher-frequency log-returns (increments) and to look for a conceptual framework that captures this feature. This observation alone, however, does not lead to any particularly useful insights. Only if we make assumptions about the structure of the summands is it possible to take advantage of celebrated results in the theory of probability and gain intuition about the possible behavior of the sum under those assumptions. We return to this idea at the end of the next section.

5.2.1. The normal approximation to the binomial distribution

Consider the following simple experiment. We flip an unfair coin independently n times. The probability that a head occurs in a single experiment is $0 < p < 1$. Denote by S_n the random variable indicating the number of heads resulting from the experiment. The probability that k heads occur in a sequence of n trials is given by the binomial distribution,

$$P(S_n = k) = \binom{n}{k} p^k q^{n-k}, \quad k = 0, 1, \dots, n, \tag{5.1}$$

where $q = 1 - p$. The mean and the variance of S_n are provided by $E[S_n] = np$ and $\mathrm{var}[S_n] = npq$, respectively.

The result by de Moivre–Laplace states that (5.1) can be approximated by the density of the normal distribution for large values of n,

$$P(S_n = k) = \binom{n}{k} p^k q^{n-k} \approx \frac{1}{\sqrt{2\pi npq}} \exp\left(-\frac{(k-np)^2}{2npq}\right). \tag{5.2}$$

Furthermore, probabilities of the type $P(a \leq S_n \leq b)$ can be approximated by summing the normal probabilities in (5.2),

$$P(a \leq S_n \leq b) \approx \int_a^b \frac{1}{\sqrt{2\pi npq}} \exp\left(-\frac{(x-np)^2}{2npq}\right) dx. \tag{5.3}$$

Both (5.2) and (5.3) indicate that the probabilistic properties of the binomial distribution for large n can be approximated by the normal distribution, i.e., the normal distribution with mean np and variance npq can be used as an approximate model.[4] By centering and normalizing S_n, we can also write the following version the last relationship,

$$\lim_{n\to\infty} P\left(\frac{S_n - np}{\sqrt{npq}} \leq b\right) = \frac{1}{\sqrt{2\pi}} \int_{-\infty}^b e^{-\frac{x^2}{2}} dx. \tag{5.4}$$

The above equation shows that the cumulative distribution function (cdf) of the standard normal distribution approximates the cdf of the

[4]In finance, this result is also used in the opposite direction. To approximate the price of a derivative contract written on an underlying asset the price of which is supposed to follow a geometric Brownian motion, we can construct a recombining binomial tree that allows two possible movements of the price of the underlying asset at every discrete time instant: up and down with values uS and dS, respectively, where u and d are positive multipliers that depend on the step size. The probabilities of the possible prices of the asset after n discrete steps are described by (5.1). The limit distribution as $n \to \infty$ is the log-normal distribution for a suitable choice of u, d, and the probability p.

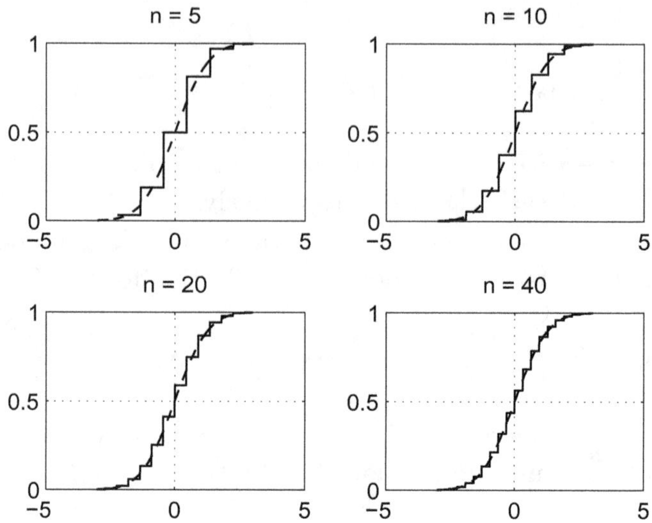

Figure 5.2. The binomial cdf converges to the Gaussian cdf when the number of trials n increases.

properly centered and normalized binomial distribution for large n. A numerical illustration is provided in Figure 5.2.

5.2.2. The classical CLT

The coin-flip experiment can be restated in the following way. We represent coin flips as a sequence of independent random variables $\delta_1, \delta_2, \ldots$ that can take two values: 0 and 1 with probabilities q and p, respectively. Thus, the number of successes in a sequence of n coin flips equals,

$$S_n = \sum_{i=1}^{n} \delta_i.$$

In this context, the limit relation in (5.4) is essentially a statement about the behavior of a sum of independent δ_i when the number of summands n tends to infinity.

It is quite remarkable that a result similar to (5.4) holds for sums of iid random variables in which the summands can have rather general distributions. That is, the Gaussian distribution can be accepted as an approximate model for a sum of iid random variables the common distribution of which we may not know. It only needs to satisfy some general regularity conditions and n needs to be large enough. We state the main result below.

Suppose that the random variables $X_1, X_2, \ldots, X_n, \ldots$ are independent copies of a random variable X with mean $E[X] = \mu$ and a finite variance $\mathrm{var}[X] = \sigma^2 < \infty$. Consider their sum

$$S_n = \sum_{i=1}^{n} X_i.$$

The classical CLT states that the centered and normalized sequence of S_n converges in distribution to the standard Gaussian distribution as n approaches infinity,

$$\lim_{n\to\infty} P\left(\frac{S_n - n\mu}{\sigma\sqrt{n}} \leq x\right) = \frac{1}{\sqrt{2\pi}} \int_{-\infty}^{x} e^{-\frac{u^2}{2}} du. \tag{5.5}$$

The centering is by the mean of the sum $E[S_n] = n\mu$ and the normalization is by the standard deviation of the sum, $\sqrt{\mathrm{var}[S_n]} = (n\sigma^2)^{1/2}$. In a practical application, if only the sum S_n can be observed, the CLT can be applied by first estimating the mean and the variance of S_n from the available sample and then by accepting a Gaussian distribution with the estimated mean and variance as a model that describes approximately the probabilistic behavior of the sum.

In every approximation problem it is reasonable to ask what error one is making by accepting an approximation. Suppose that the number of summands is large but fixed. Then, how close is the limit distribution to the distribution of the sum? Because we are asking a general question about the entire distribution, it makes sense to measure the error in terms of the maximal difference of the two distribution functions,

$$\rho(F_{\tilde{S}_n}, F_Z) = \sup_{x\in\mathbb{R}} |F_{\tilde{S}_n}(x) - F_Z(x)| \tag{5.6}$$

in which $\tilde{S}_n = (S_n - n\mu)/\sigma\sqrt{n}$ denotes the standardized distribution of the sum and $Z \in N(0,1)$ denotes the standard Gaussian distribution. A celebrated result known as the Berry–Esseen theorem states that if S_n is a sum of independent copies of a random variable X with a finite absolute third moment, $E|X|^3 < \infty$, then

$$\rho(F_{\tilde{S}_n}, F_Z) \leq C \frac{E|X|^3}{\sigma^3\sqrt{n}} \tag{5.7}$$

in which C is an absolute constant which does not depend on the distribution of X. The more irregular the distribution of X and the higher $E|X|^3$ is, the larger the error in the approximation for any n fixed.

In the classical CLT, the three critical assumptions are: (i) that of independence of the summands, (ii) that they have a common distribution, and (iii) that the variance of the common distribution is finite. Extensions of the classical CLT exist in which some of the three assumptions are relaxed. For example, (ii) can be relaxed: the summands may not have a common distribution but different distributions. Then, the CLT still holds as long as they are *asymptotically negligible*,

$$\max_{1\leq j\leq n} P\left(\frac{|X_j - \mu_j|}{\sigma_n} > \epsilon\right) \to 0, \; n \to \infty \quad \text{for each } \epsilon > 0,$$

where $\mu_j = E[X_j]$ and $\sigma_n^2 = \text{var}[S_n]$. That is, the CLT holds if the variability of each summand is negligible compared to the variability of the sum. In this more general setting $\mu_n = E[S_n]$ is used to center the sum S_n and σ_n is used to normalize it. The assumption of asymptotic negligibility holds if the summands are assumed iid because $\sigma_n = \sigma\sqrt{n}$ and therefore when n increases, then $|X_j - \mu_j|/\sigma\sqrt{n}$ becomes indeed negligible at the limit because the denominator explodes.

Another example is the assumption of independence (i). The CLT has been shown to hold assuming the summands are *weakly dependent* instead of independent. The assumption of weak dependence implies that X_j and X_k are nearly independent if the indices k and j are far apart. In a financial application, if S_n is used to describe a lower frequency log-return, the summands are higher frequency log-returns, and the assumption of weak dependence implies near independence only if the higher frequency log-returns X_j and X_k are observed far apart in time.[5]

Finally, the CLT may hold even if (iii) is relaxed. In particular, if the left tail $F_X(x)$ and the right tail $1 - F_X(x)$ of the random variable X decay asymptotically like $|x|^{-2}$, then $\text{var}[X] = \infty$ but the CLT holds albeit with a different normalizing sequence. Instead of choosing $n^{1/2}$ in (5.5), one should work with $(n\log(n))^{1/2}$. See Zolotarev and Uchaikin (1999) for additional details.

5.2.3. The generalized CLT

The conditions discussed in Section 5.2.2 represent only *sufficient* conditions under which the limit of properly centered and normalized sums of iid random variables converges in distribution to a Gaussian random variable. A necessary and sufficient condition is provided, for example, by Zolotarev and Uchaikin (1999). A sum $S_n = \sum_{i=1}^n X_i$ of independent copies of a random variable X converges in distribution to a Gaussian random variable Z under proper centering and normalization,

$$\frac{S_n - a_n}{b_n} \xrightarrow{d} Z \quad \text{as } n \to \infty,$$

where a_n and b_n are sequences of normalizing constants and $\xrightarrow{d}$ denotes convergence in distribution if and only if the cdf $F_X(x)$ of the random variable X satisfies the following condition:

$$\lim_{x\to\infty} \frac{x^2 \int_{|y|>x} dF_X(y)}{\int_{|y|<x} y^2 dF_X(y)} = 0. \tag{5.8}$$

[5]See Louhichi (2002) for rates of convergence under weak dependence.

If X has a finite variance, then the denominator in (5.8) converges to the second moment of X but the numerator decays and, therefore, the condition is satisfied. The second moment may be infinite and (5.8) may still hold. This happens when $\int_{|y|>x} dF_X(y)$ behaves approximately like x^{-2} for large x.[6]

Obviously, because (5.8) is a necessary and sufficient condition, if we relax it the CLT no longer holds. It is, however, still reasonable to ask whether the distribution of a sum of iid random variables converges to some limit distribution if (5.8) does not hold. Research into such questions lead to the discovery of the generalized CLT. Essentially, the idea is to keep only the assumptions of independence of the summands and that they have a common distribution.

The generalized CLT appears easier to state. Any properly centered and normalized sum of iid random variables converges at the limit to a distribution known as a *stable distribution*. This means that the stable distributions are the *only* distributions which can arise as limits of sums of iid random variables. This feature makes the stable distributions very attractive for modeling quantities which accumulate many independent random shocks. Understandably, the normal distribution is a special case of the stable distributions, just as the CLT is a special case of the generalized CLT. In contrast to the normal distribution, which is symmetric and cannot account for the heavy-tailed nature of the returns of financial variables, the class of non-Gaussian stable distributions has skewed and heavy-tailed representatives.

Questions about the rate of convergence, which are related to the error made by adopting the limit distribution as an approximate model, have been addressed. Results in terms of the Kolmogorov metric, i.e., along the lines of (5.7), and other metrics are available in Rachev *et al.* (2013).

[6] An example of such a distribution is Student's t with the degrees of freedom parameter equal to two. For more details on Student's t-distribution, see Chapter 4.

5.3. Definition and Basic Properties of Stable Distributions

The class of stable distributions can be introduced in three equivalent ways. We provide all of them because they also describe three important properties of the class.[7]

- *The stability property.* The stable distributions are the only class of distributions with the following property: if we take $n \geq 2$ independent copies of one member of the class and sum them up, we get the same distribution up to centering and normalization,

$$Y_1 + Y_2 + \cdots + Y_n \stackrel{d}{=} c_n Y + d_n,$$

  where $Y_1, Y_2, \ldots, Y_n$ are independent copies of Y, $c_n > 0$ and $d_n \in \mathbb{R}$ are constants that depend on the value of n, and $\stackrel{d}{=}$ denotes equality in distribution.
- *The domains of attraction property.* The class of stable distributions is the only class that arises as a limit of sums of iid random variables; this is essentially the statement of the generalized CLT. More formally, Y has a stable distribution if there is a sequence of iid random variables $X_1, X_2, \ldots, X_n, \ldots$ such that

$$\frac{X_1 + X_2 + \cdots + X_n - a_n}{b_n} \stackrel{d}{\to} Y,$$

  where a_n and b_n are sequences of normalizing constants.
- *Explicit form of the characteristic function.* Stable distributions can be introduced directly by means of their characteristic function $\phi_Y(u) = E[e^{iuY}]$. A random variable Y is said to have a stable distribution if there are parameters $0 < \alpha \leq 2$, $\sigma > 0$, $-1 \leq \beta \leq 1$, $\mu \in \mathbb{R}$ such that its characteristic function has the

[7]See Zolotarev (1986), Samorodnitsky and Taqqu (1994), and Zolotarev and Uchaikin (1999) for additional information.

following form:

$$\phi_Y(u) = \begin{cases} \exp\{-\sigma^\alpha |u|^\alpha \left(1 - i\beta \frac{u}{|u|} \tan\left(\frac{\pi\alpha}{2}\right)\right) + i\mu u\}, & \alpha \neq 1, \\ \exp\{-\sigma|u| \left(1 + i\beta \frac{2}{\pi} \frac{u}{|u|} \ln(|u|)\right) + i\mu u\}, & \alpha = 1, \end{cases} \tag{5.9}$$

where $\frac{u}{|u|} = 0$ if $u = 0$.

For numerical work, the definition in terms of the characteristic function is the most useful one. The four parameters that appear have important interpretations. The parameter α is also called *the index of stability* and determines the tail thickness of the distribution. Due to the important effect of this parameter on the properties of the stable distributions, they are often called *α-stable* or *alpha stable.*

The parameter β determines the skewness of the distribution while σ is a scale parameter and μ is a location parameter. If $\beta = 0$, then the random variable is symmetric with respect to the location parameter μ. The distribution is said to be skewed to the right if $\beta > 0$ and to the left if $\beta < 0$. It is said to be totally skewed to the right if $\beta = 1$ and totally skewed to the left if $\beta = -1$. Since stable distributions are uniquely determined by the four parameters, the common notation is $S_\alpha(\sigma, \beta, \mu)$. If Y is said to have stable distribution, we write $Y \sim S_\alpha(\sigma, \beta, \mu)$. The impact of the index of stability and the skewness parameter on the stable distribution density shape is illustrated in Figures 5.3 and 5.4.

The Gaussian distribution appears as a special case with $\alpha = 2$ and has variance equal to $2\sigma^2$. The value of β is irrelevant in this special case and the distribution is symmetric with respect to μ.

The parameterization of the characteristic function in (5.9) has the advantage that the parameters can be interpreted geometrically. However there is a serious disadvantage when it comes to numerical or statistical work: it is discontinuous at $\alpha = 1$ and $\beta \neq 0$; that is, if $\alpha \to 1, \beta \to \beta^* \neq 0, \sigma \to \sigma^*$ and $\mu \to \mu^*$ then the limit function is not the characteristic function of the stable random variable $S_1(\sigma^*, \beta^*, \mu^*)$. This drawback is not an inherent property of the class

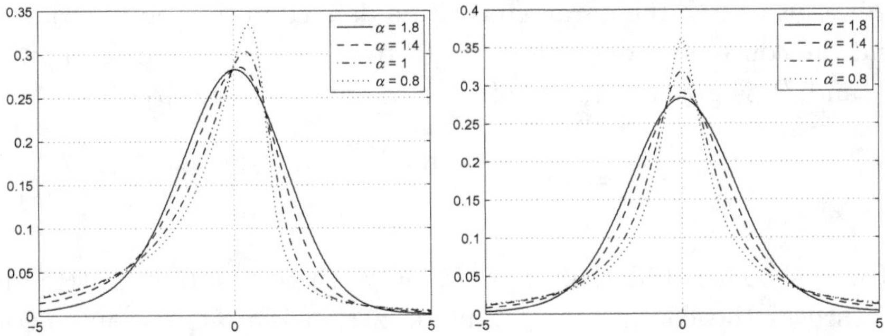

Figure 5.3. Stable distribution densities for different choices of α, $\beta = -0.6$ (left) and $\beta = 0$ (right). The scale and location are $\sigma = 1$ and $\mu = 0$ in both cases.

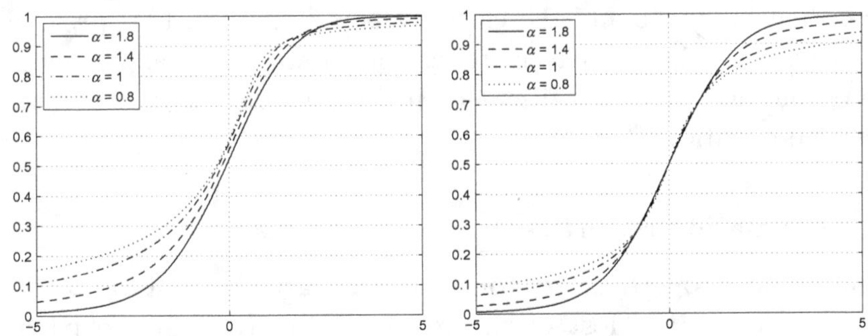

Figure 5.4. Stable distribution functions for different choices of α, $\beta = -0.6$ (left) and $\beta = 0$ (right). The scale and location are $\sigma = 1$ and $\mu = 0$ in both cases.

of α-stable laws and appears because of the special form of the classical characteristic function. As noted in Samorodnitsky and Taqqu (1994), it is possible to change the parameterization in order to have convergence in distribution when $\alpha \to \alpha^*, \beta \to \beta^*, \sigma \to \sigma^*$ and $\mu \to \mu^*$. An alternative parametric representation of the characteristic function equipped with this property as follows:

$$\phi_Y(u) = \begin{cases} \exp\{-|\sigma u|^\alpha \\ \quad + i\sigma u\beta(|\sigma u|^{\alpha-1} - 1)\tan\left(\frac{\pi\alpha}{2}\right) + i\mu_1 u\}, & \alpha \neq 1, \\ \exp\{-|\sigma u| + i\sigma u\beta\frac{2}{\pi}\ln|\sigma u| + i\mu_1 u\}, & \alpha = 1, \end{cases} \tag{5.10}$$

where $0 < \alpha \leq 2$, $-1 \leq \beta \leq 1$, $\sigma > 0$ and $\mu_1 \in R$.

Let us denote the parameterization defined in equation (5.9) as P_0 and the continuous one defined in (5.10) as P_1. The relation between P_0 and P_1 is given in terms of the parameters μ and μ_1

$$\mu_1 = \begin{cases} \mu + \beta\sigma \tan \dfrac{\pi\alpha}{2}, & \alpha \neq 1, \\ \mu, & \alpha = 1. \end{cases} \tag{5.11}$$

Obviously P_0 is different from P_1 only when $\beta \neq 0$. An attractive feature of the continuous parameterization is that it is not necessary to consider the case $\alpha = 1$ separately, i.e., it can be defined by means of the limit $\alpha \to 1$. Convergence in distribution follows because of the one-to-one relationship between the cumulative distribution functions and the characteristic functions. It is preferable to have a continuous parameterization in statistical and numerical work and all algorithms discussed in the chapter use the continuous parameterization P_1.

5.3.1. Basic properties

The basic properties we consider are easier to establish when working with P_0. Most of them follow directly from the particular form of equation (5.9), the corresponding proofs can be found in Samorodnitsky and Taqqu (1994).

From the stability property it is clear that when we sum up independent copies of a stable distribution, the sum has the same stable distribution. It turns out, if we sum up two independent stable random variables with the same α parameter but with possibly different skewness, scale, and location parameters, we get a stable random variable the parameters of which can be expressed through the parameters of the summands. If $Y_1 \sim S_\alpha(\sigma_1, \beta_1, \mu_1)$ and $Y_2 \sim S_\alpha(\sigma_2, \beta_2, \mu_2)$, then $Y_1 + Y_2 \sim S_\alpha(\sigma, \beta, \mu)$, with

$$\sigma = (\sigma_1^\alpha + \sigma_2^\alpha)^{\frac{1}{\alpha}}, \quad \beta = \frac{\beta_1\sigma_1^\alpha + \beta_2\sigma_2^\alpha}{\sigma_1^\alpha + \sigma_2^\alpha}, \quad \mu = \mu_1 + \mu_2.$$

The interpretation of σ and μ as scale and location parameters respectively follows from the fact that if we rescale and shift

$Y \sim S_\alpha(\sigma, \beta, \mu)$, only these two parameters change depending on the scale and shift factors. More formally, $cY + d \sim S_\alpha(|c|\sigma, \text{sign}(c)\beta, c\mu + d)$ if $\alpha \neq 1$ where $\text{sign}(c)$ denotes the sign of c. A linear transformation can normalize the distribution such that $\sigma = 1$ and $\mu = 0$.[8] If $c = -1$ and $d = 0$, then only β changes which is consistent with its interpretation as a skewness parameter.

We use the standard notation for the cumulative distribution function and probability density function: $P(X \leq x) = F_X(x; \alpha, \beta)$ specifies the cdf and $f_X(x; \alpha, \beta) = F'_X(x; \alpha, \beta)$ denotes the pdf of a random variable $X \sim S_\alpha(1, \beta, 0)$. The fact that we consider only standardized random variables is not limiting because if $Y \sim S_\alpha(\sigma, \beta, \mu)$, then $(Y - \mu)/\sigma \sim S_\alpha(1, \beta, 0)$. Furthermore, it is sufficient to examine only the cdf and the pdf of $X \sim S_\alpha(1, \beta, 0)$, with $\beta \geq 0$ and then take advantage of the expressions:

(a) if $\beta < 0$,

$$f_X(x; \alpha, \beta) = f_X(-x; \alpha, -\beta), \quad F_X(x; \alpha, \beta) = 1 - F_X(-x; \alpha, -\beta); \tag{5.12}$$

(b) if $\sigma \neq 1$ or $\mu \neq 0$,

$$f_Y(x) = \frac{1}{\sigma} f_X\left(\frac{x - \mu}{\sigma}; \alpha, \beta\right), \quad F_Y(x) = F_X\left(\frac{x - \mu}{\sigma}; \alpha, \beta\right), \tag{5.13}$$

where $f_Y(x)$ and $F_Y(x)$ are the pdf and the cdf of $Y \sim S_\alpha(\sigma, \beta, \mu)$, respectively.

Stable distributions can be used as a theoretical model when empirical data is heavy-tailed. As mentioned in the introduction, empirical studies confirm that financial time series possess this property. The application of stable laws in this aspect is motivated by the fact that the tail of the stable law approaches zero as a power function. This is usually referred to as "Pareto-like" behavior of the

[8] The case $\alpha = 1$ and $\beta \neq 0$ is different, $cY + d \sim S_1(|c|\sigma, \text{sign}(c)\beta, c\mu + d - 2\log(|c|)\beta/\pi)$.

tail because of the same power decay of the tail of the Pareto distribution. More formally, if $Y \sim S_\alpha(\sigma, \beta, \mu)$ and $0 < \alpha < 2$, then

$$\lim_{\lambda \to \infty} \lambda^\alpha P(Y > \lambda) = C_\alpha \frac{1+\beta}{2} \sigma^\alpha,$$

$$\lim_{\lambda \to \infty} \lambda^\alpha P(Y < -\lambda) = C_\alpha \frac{1-\beta}{2} \sigma^\alpha,$$

where the constant C_α can be calculated explicitly, see Samorodnitsky and Taqqu (1994) for additional details.

The tail behavior of the Gaussian distribution as $\lambda \to \infty$ is specified by

$$P(Y < -\lambda) = P(Y > \lambda) \sim \frac{1}{2\sqrt{\pi}\sigma\lambda} e^{-\frac{\lambda^2}{4\sigma^2}}$$

which is quite different from a power decay. The difference in the asymptotic behavior of the tails of α-stable distributions, with $\alpha < 2$ and the Gaussian distribution motivates the distinction between Gaussian and non-Gaussian stable distributions.

The power decay of the tail of Pareto stable distributions implies that they do not possess a finite second moment. More exactly, if $Y \sim S_\alpha(\sigma, \beta, \mu)$ and $0 < \alpha < 2$, then $E|Y|^p < \infty$, $0 < p < \alpha$ and $E|Y|^p = \infty$, $\alpha \leq p$. As a consequence, if $\alpha \leq 1$, then the corresponding α-stable distribution does not have a finite first absolute moment. In fact, if $\alpha > 1$ then the shift parameter μ equals the mean of Y, $\mu = E[Y]$. This interpretation of the shift parameter is not true under the parameterization P_1 but this does not present any problems in numerical work.

5.3.2. Is the iid setting relevant for financial asset returns?

Because of the domains of attraction property, stable distributions can be considered an approximate probabilistic description of quantities which arise as sums of iid random variables with an unknown distribution. We noted already that logarithmic returns and price increments of lower frequencies can be represented as sums of log-returns or price increments of higher frequencies,

$$\log\left(\frac{S_T}{S_0}\right) = \sum_{i=1}^{n} \log\left(\frac{S_{t_i}}{S_{t_{i-1}}}\right),$$

where $0 = t_0 \leq t_1 \leq \cdots \leq t_n = T$ is an arbitrary subdivision of a time interval $[0, T]$ and S_{t_i} denotes the price of a financial security at time t_i. This is, however, insufficient to claim that a stable distribution would be a reasonable approximation because we have not verified if the iid assumption holds for the summands. Generally, if the iid assumption is relaxed completely in the sense that the summands are allowed to be arbitrary, then no meaningful limit theory can exist.

In fact, financial asset returns at any frequency are neither independent, nor identically distributed. However, they are not arbitrary either. It is an empirical fact that lower frequency returns are less heavy-tailed than higher-frequency returns which indicates that the aggregation process leads to some regularity. In Figure 5.5, we illustrate this phenomenon by estimating the α parameter of stable laws using returns data of 1231 U.S. stocks in the period from January 1, 1990 to January 1, 2015. The return frequency varies from 1 day to

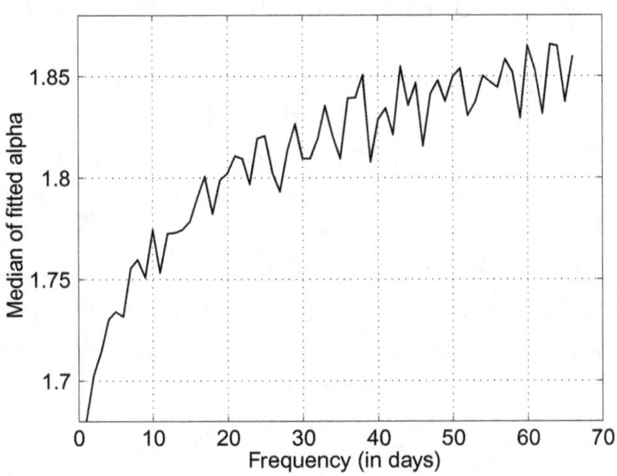

Figure 5.5. The median of the estimated α parameter from the returns of 1231 U.S. stocks from January 1, 1990 to January 1, 2015. The return frequency ranges from 1 day to 66 days.

66 days (approximately a quarter) with a step of 1 day. The plot shows the median of the estimated α for each return frequency.[9] Clearly, the higher frequency returns are characterized by a lower α than the lower frequency returns which implies heavier tails for higher frequencies.

From the standpoint of the CLT and its generalizations, the implication is that the higher frequency returns are very likely in the domain of attraction of the Gaussian distribution but the rate of convergence to the Gaussian distribution is very slow. If this is the case, one can still argue that a stable distribution can be good a approximation for the distribution of the partial sum for mid-size values of the number of summands but only in the center of the distribution and not in the tails. Such results in the iid setting are known as pre-limit theorems; further details can be found in Rachev and Mittnik (2000) and Grabchak and Samorodnitsky (2010). See also the discussion in Rachev *et al.* (2011, Chapter 7) for practical implications for risk estimation.

5.4. Probability Density and Distribution Functions

According to the generalized CLT, α-stable distributions are the only ones that have domains of attraction. Naturally one would expect that these distributions would play a central role in statistical theory and various applications. One representative, the Gaussian distribution, is really of primary importance which is due to the fact that all distributions with finite second moment belong to its domain of attractions. Undoubtedly, the main difficulty in the application of α-stable laws in practice is the lack of closed-form expressions for the pdf and the cdf, with the following exceptions:

[9] The estimation method is the regression-type procedure described in Section 5.6. We use the code provided in the appendix.

(a) The Gaussian distribution $S_2(\sigma, 0, \mu) = N(\mu, 2\sigma^2)$: the cdf $F_Y(x) = P(Y \leq x)$ and the pdf $f_Y(x)$ are well known:

$$F_Y(x) = \Phi\left(\frac{x-\mu}{\sigma\sqrt{2}}\right), \quad f_Y(x) = \frac{1}{2\sigma\sqrt{\pi}} e^{-\frac{(x-\mu)^2}{4\sigma^2}}, \quad x \in \mathbb{R},$$

where $\Phi(x)$ denotes the cdf of $N(0,1)$ distribution:

$$\Phi(x) = \frac{1}{\sqrt{2\pi}} \int_{-\infty}^{x} e^{-\frac{u^2}{2}} du. \tag{5.14}$$

(b) The Cauchy distribution $S_1(\sigma, 0, \mu)$: the cdf and the pdf are given by the following equations:

$$F_Y(x) = \frac{1}{2} + \frac{1}{\pi} \arctan\left(\frac{x-\mu}{\sigma}\right),$$

$$f_Y(x) = \frac{\sigma}{\pi((x-\mu)^2 + \sigma^2)}, \quad x \in \mathbb{R}.$$

(c) The Lévy distribution $S_{1/2}(\sigma, 1, \mu)$: the cdf and the pdf are given by

$$F_Y(x) = 2\left(1 - \Phi\left(\sqrt{\frac{\sigma}{x-\mu}}\right)\right),$$

$$f_Y(x) = \sqrt{\frac{\sigma}{2\pi}} \frac{1}{(x-\mu)^{3/2}} e^{-\frac{\sigma}{2(x-\mu)}}, \quad x \geq \mu.$$

The task of developing efficient numerical algorithms to overcome the lack of closed formulae is not trivial and much effort has been concentrated on it. We continue with a brief summary of some of the main approaches which include the series expansions of Bergström, the fast Fourier transform method, and Zolotarev's integral representations.[10]

5.4.1. Series expansions

Fama and Roll (1968) were the first to publish tabulated cdfs and percentiles of symmetric stable distributions. They took advantage

[10]See also Stoyanov and Racheva-Iotova (2004a) for a discussion of other methods.

of two infinite series expansions by Bergström (1952). The infinite series expansions are efficient only for either small or large values of $|x|$ where x denotes the argument of the density or the distribution function. For the middle range of $|x|$, one needs to adopt a different method but the infinite series expansions can provide an approximation of the tail of the distribution.

The Bergström expansion for the density function of stable laws when $\alpha \in (1,2)$ in the continuous parameterization P_1 is provided, for example, by Menn and Rachev (2006). Asymptotically, when $x \to \infty$, the following representation holds:

$$f_X(x;\alpha,\beta) = \frac{1}{\pi}\sum_{k=1}^{\infty} c a_k (cx)^{-k\alpha-1}, \tag{5.15}$$

where

$$\begin{aligned}
a_k &= (-1)^{k-1}\frac{\Gamma(1+k\alpha)}{k!}\sin\left(\frac{k\pi}{2}(\alpha+\beta^*(\alpha-2))\right),\\
c &= \left[\cos\left(\pi\beta^*\frac{\alpha-2}{2}\right)\right]^{\frac{1}{\alpha}},\\
\beta^* &= \frac{2}{\pi(\alpha-2)}\arctan\left(-\beta\tan\frac{\pi\alpha}{2}\right).
\end{aligned}$$

The fact that the formula applies asymptotically for positive x is not restrictive. Because of the symmetry property in (5.12), $f_X(-x;\alpha,\beta) = f_X(x;\alpha,-\beta)$ and, thus, formula (5.15) can be used with the sign of β flipped.

We provide a Matlab function which implements formula (5.15) in the following way. First, the value of the sum is calculated for $k = 1$ and we verify if this value is smaller than the floating-point relative accuracy equal to 2^{-52}. If this is not the case, new summands are added until either the contribution of a next summand is smaller than 2^{-52} or we reach 25 summands. The reason the number of summands should be limited is that the formula for a_k contains a ratio of two exploding terms which becomes hard to compute precisely for large k.

This is not restrictive because for large x the series in (5.15) converges quickly and only a few summands are sufficient.

```
function f = bergstroem(x, alpha, beta)

% function f = bergstroem(x, alpha, beta)
%
% Calculates the Bergstroem expansion of the density of a ...
    stable law.
% The parameterization of the characteristic function of ...
    normalized alpha-stable laws is:
%
% ln(phi(t)) = -|t|^(alpha)*(1 + ...
    i*beta*sign(t)*tan(pi*alpha/2)(|t|^(1 - alpha) - 1)), ...
    alpha ~= 1
% ln(phi(t)) = -|t|*(1 + i*beta*sign(t)ln(|t|)*2/pi), alpha = 1

b = atan(-beta*tan(pi*alpha/2))*2/(pi*(alpha - 2));
c = cos(pi*b*(alpha - 2)/2)^(1/alpha);

if x < 0
    f = bergstroem(-x, alpha, -beta);
    return
end

k = 1;
ak = (-1)^(k - 1)*gamma(1 + k*alpha)*sin((alpha + b*(alpha - ...
    2))*pi*k/2)/factorial(k);
Sk = c*ak*(c*x)^(-k*alpha-1);
f = Sk;
while (abs(Sk) > eps) && (k < 25)
    k = k + 1;
    ak = (-1)^(k - 1)*gamma(1 + k*alpha)*sin((alpha + b*(alpha ...
        - 2))*pi*k/2)/factorial(k);
    Sk = c*ak*(c*x)^(-k*alpha-1);
    f = f + Sk;
end
f = f/pi;
```

5.4.2. The Fast Fourier transform approach

As discussed in Chapter 2, distribution and density functions are derived from the characteristic function via the Fourier inversion integral. The most straightforward method to compute the pdf of a stable random variable X is direct numerical integration of the inversion formula:

$$f_X(x) = \frac{1}{2\pi}\int_{-\infty}^{\infty} e^{-itx}\phi_X(t)dt, \tag{5.16}$$

The general approach evaluates the density at the equally-spaced x values

$$x_k = \left(k - 1 - \frac{N}{2}\right)h, \quad k = 1, \ldots, N, \tag{5.17}$$

and computes the integral through a Riemann sum in which the integrand in (5.16) is also evaluated on a similar equally-spaced grid with space between nodes equal to $(hN)^{-1}$ and number of nodes equal to N. As a result, the integral in (5.16) is approximated by

$$f_X(x_k) \approx \frac{(-1)^{k-1+\frac{N}{2}}}{hN} \times \sum_{j=1}^{N}(-1)^{j-1}\phi_X\left(\frac{2\pi}{hN}\left(j - 1 - \frac{N}{2}\right)\right)e^{\frac{-i2\pi(j-1)(k-1)}{N}}, \tag{5.18}$$

where $k = 1, \ldots, N$. In the approximation, the upper and the lower bounds in (5.16) are not unbounded but equal $\frac{1}{2h}$ and $-\frac{1}{2h}$, respectively.

As will be explained in Chapter 6, the discrete Fast Fourier transform (FFT) is a numerical method developed for the calculation of $f(x_k)$ given the sequence

$$(-1)^{j-1}\phi_X\left(\frac{2\pi}{hN}\left(j - 1 - \frac{N}{2}\right)\right), \quad j = 1, \ldots, N. \tag{5.19}$$

In practice, we apply the discrete FFT to the sequence in (5.19).

Chapter 6 will analyze the impact of h and N on the quality of the approximation in the case of tempered stable distributions and summarize the three sources of approximation error. In the remainder of this section, we discuss an improvement of the FFT method unrelated to the parameters h and N arising from a different Riemann sum approximation. The improvement is generic and in the case of stable distributions Menn and Rachev (2006) demonstrate that the approximation error in the tails decreases substantially.

In general, integrals are numerically approximated by computing a Riemann sum

$$\int_a^b u(x)dx \approx \sum_{k=1}^{n-1} u(x_k^*)(x_{k+1} - x_k),$$

where $a = x_1 \leq x_2 \leq \cdots \leq x_n = b$ is a grid of points and $x_k^* \in [x_k, x_{k+1}]$, $k = 1, \ldots, n$. If $x_k* = x_k$ for all k, then the corresponding approximation is said to follow the left-point rule. If $x_k^* = (x_k + x_{k+1})/2$ for all k, then the approximation follows the midpoint rule. Finally, $x_k^* = x_{k+1}$ for all k defines the right-point rule. Generally, the error of the midpoint rule is smaller and the corresponding Riemann sum converges faster to the integral.

The classical FFT method outlined at the beginning of this section follows the left-point rule. Thus, we denote the approximation obtained in equation (5.18) by $f_X^{\mathrm{LPR}}(x)$. Following Menn and Rachev (2006), we describe a modification which corresponds to the midpoint rule consistent with the notation in this section.[11]

The grid of values for the argument of the density function is given by (5.17). The grid in the domain of the characteristic function is defined by

$$t_j = \frac{1}{hN}\left(j - 1 - \frac{N}{2}\right), \quad j = 1, \ldots, N,$$

and the midpoints are calculated from t_j,

$$t_j^* = \frac{t_j + t_{j+1}}{2} = \frac{1}{hN}\left(j - \frac{N}{2} + \frac{1}{2}\right), \quad j = 1, \ldots, N-1.$$

[11]Menn and Rachev (2006) construct the grids x_k and t_j differently and the meaning of the parameters h and N in their notation is not exactly the same as the meaning of h and N in (5.17).

The approximation is then constructed in the following way:

$$
\begin{aligned}
f_X(x_k) &= \int_{-\infty}^{\infty} \phi_X(2\pi t) e^{-i2\pi t x_k} dt \\
&\approx \int_{-\frac{1}{2h}}^{\frac{1}{2h}} \phi_X(2\pi t) e^{-i2\pi t x_k} dt \\
&\approx \frac{1}{hN} \sum_{j=1}^{N} \phi_X(2\pi t_j^*) e^{-i2\pi t_j^* x_k} \\
&= \frac{(-1)^{k-1} i e^{-i\pi \frac{k-1}{N}}}{hN} \sum_{j=1}^{N} (-1)^{j-1} \phi_X(2\pi t_j^*) e^{-i2\pi \frac{(k-1)(j-1)}{N}}.
\end{aligned}
$$

As a result,

$$
f_X^{\mathrm{MPR}}(x_k) = C_k \sum_{j=1}^{N} (-1)^{j-1} \phi_X(2\pi t_j^*) e^{-i2\pi \frac{(k-1)(j-1)}{N}}, \tag{5.20}
$$

where $C_k = \frac{(-1)^{k-1} i e^{-i\pi \frac{k-1}{N}}}{hN}$ and, thus, the discrete FFT is applied to the sequence

$$
(-1)^{j-1} \phi_X \left(\frac{2\pi}{hN} \left(j - \frac{N}{2} + \frac{1}{2} \right) \right), \quad j = 1, \ldots, N,
$$

which is similar to (5.19).

We compare the relative accuracy of the left-point rule and the midpoint rule in terms of the relative error (RE) of the approximation defined by

$$
\mathrm{RE}_X(x) = \frac{\tilde{f}_X(x) - f_X(x)}{f_X(x)}, \tag{5.21}
$$

where $\tilde{f}_X(x)$ denotes the density approximation obtained by the left-point rule $f_X^{\mathrm{LPR}}(x)$ in equation (5.18) or the midpoint rule $f_X^{\mathrm{MPR}}(x)$ in equation (5.20) and $f_X(x)$ denotes the true density function. Because the true density function is not known, we use the more precise integral approximation method described in Section 5.4.3 to construct a proxy.

Menn and Rachev (2006) note that the relative errors of the left-point and the right-point rule have different signs in the tails: the left-point rule overestimates the density in the tails while the midpoint rule underestimates it. It is, therefore, possible to construct a better approximation by a weighted average of the two approximations. Menn and Rachev (2006) recommend a weight of 1/3 for the left-point rule and a weight of 2/3 for the midpoint rule,

$$f_X^{\mathrm{WA}}(x) = \frac{1}{3} f_X^{\mathrm{LPR}}(x) + \frac{2}{3} f_X^{\mathrm{MPR}}(x). \tag{5.22}$$

The relative errors of the three approximations are plotted in Figure 5.6 for two different choices of the stable distribution parameters. Because the relative error does not depend on the scale and the location parameters, we set $\sigma = 1$ and $\mu = 0$. Both plots in Figure 5.6 illustrate the better precision gained with the weighted average approximation in the tails of the distribution.

Finally, as far as the choice of h and N is concerned, in the case of the left-point rule Rachev and Mittnik (2000) report that the choice $h = 0.01$ and $N = 2^{13}$ leads to an absolute approximation error of about 10^{-6}. The default choices in the Matlab implementation provided in the appendix to this chapter are $h = 0.01$ and $N = 2^{15}$ which lead to an even smaller error. The Matlab implementation includes the three approximation methods, the left-point rule, the right-point

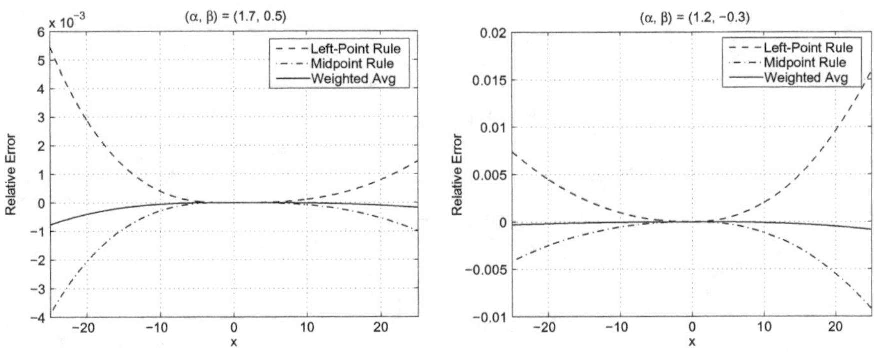

Figure 5.6. The relative error of the left-point rule approximation, the midpoint rule approximation, and the weighted average approximation in (5.22) for different choices of the stable distribution parameters. The distributions are normalized, $\sigma = 1$ and $\mu = 0$.

rule, and their weighted average with the weighted average approximation being the default choice.

5.4.3. The integral representations of Zolotarev

The density function and the distribution function of stable laws can be very accurately evaluated using integral representations derived by Zolotarev (1964). Nolan (1997) provides the representations in the continuous parameterization P_1. Following Nolan (1997), the density function and the distribution function of stable laws can be expressed as

$$f_X(x;\alpha,\beta) = c_2(x;\alpha,\beta)\int_{-\theta_0}^{\frac{\pi}{2}} g(\theta;x,\alpha,\beta)\exp(-g(\theta;x,\alpha,\beta))d\theta, \tag{5.23}$$

$$F_X(x;\alpha,\beta) = c_1(\alpha,\beta) + c_3(\alpha)\int_{-\theta_0}^{\frac{\pi}{2}} \exp(-g(\theta;x,\alpha,\beta))d\theta, \tag{5.24}$$

where

$$c_1(\alpha,\beta) = \begin{cases} \dfrac{1}{\pi}\left(\frac{\pi}{2}-\theta_0\right), & \alpha<1, \\ 0, & \alpha=1, \\ 1, & \alpha>1, \end{cases}$$

$$c_2(x;\alpha,\beta) = \begin{cases} \dfrac{\alpha}{\pi|\alpha-1|(x-\zeta)}, & \alpha\neq 1, \\ \dfrac{1}{2|\beta|}, & \alpha=1, \end{cases}$$

$$c_3(\alpha) = \begin{cases} \dfrac{\text{sign}(1-\alpha)}{\pi}, & \alpha\neq 1, \\ \dfrac{1}{\pi}, & \alpha=1, \end{cases}$$

$$g(\theta;x,\alpha,\beta) = \begin{cases} (x-\zeta)^{\frac{\alpha}{\alpha-1}}V(\theta;\alpha,\beta), & \alpha\neq 1, \\ e^{-\frac{\pi x}{2\beta}}V(\theta;\alpha,\beta), & \alpha=1, \end{cases}$$

and

$$\zeta = \zeta(\alpha, \beta) = \begin{cases} -\beta \tan \dfrac{\pi\alpha}{2}, & \alpha \neq 1, \\ 0, & \alpha = 1, \end{cases}$$

$$\theta_0 = \theta_0(\alpha, \beta) = \begin{cases} \dfrac{1}{\alpha} \arctan(\beta \tan \dfrac{\pi\alpha}{2}), & \alpha \neq 1, \\ \dfrac{\pi}{2}, & \alpha = 1, \end{cases}$$

$$V(\theta; \alpha, \beta) = \begin{cases} (\cos \alpha\theta_0)^{\frac{1}{\alpha-1}} \left(\frac{\cos\theta}{\sin \alpha(\theta_0+\theta)}\right)^{\frac{\alpha}{\alpha-1}} \frac{\cos(\alpha\theta_0+(\alpha-1)\theta)}{\cos\theta}, & \alpha \neq 1, \\ \frac{2}{\pi}\left(\frac{\frac{\pi}{2}+\beta\theta}{\cos\theta}\right) \exp\left(\frac{1}{\beta}\left(\frac{\pi}{2}+\beta\theta\right)\tan\theta\right), & \alpha = 1. \end{cases} \tag{5.25}$$

Equations (5.23) and (5.24) are valid for $x > \zeta$. The case $x < \zeta$ can be treated by taking advantage of the relations

$$f_X(x; \alpha, \beta) = f_X(-x; \alpha, -\beta),$$
$$F_X(x; \alpha, \beta) = 1 - F_X(-x; \alpha, -\beta),$$

which are the same as (5.12); the parameter ζ appears because of the change from a parameterization used by Zolotarev (1964) to P_1. Finally, the formula (5.24) for the distribution function $F(x; 1, \beta)$ holds when $\alpha = 1$ and $\beta > 0$. If $\beta < 0$, then one can take advantage of the symmetry in (5.12), $F_X(x; 1, \beta) = 1 - F_X(-x; 1, -\beta)$.

The advantage of the representations in (5.23) and (5.24) for numerical work is that the integration range is bounded and the integrands are also bounded functions. There is, however, one difficulty. The integrand in (5.23) is composed of two monotone segments, it is equal to zero at the lower and the upper bound of the integral and has a unique maximum for that value of θ for which $g(\theta; x, \alpha, \beta) = 1$. As remarked by Nolan (1997), for large values of $|x|$ the integrand in (5.23) has a sharp spike which can be an issue in the numerical integration; Figure 5.7 is an illustration of this point. As a result, for computational purposes it is more suitable to split the integration

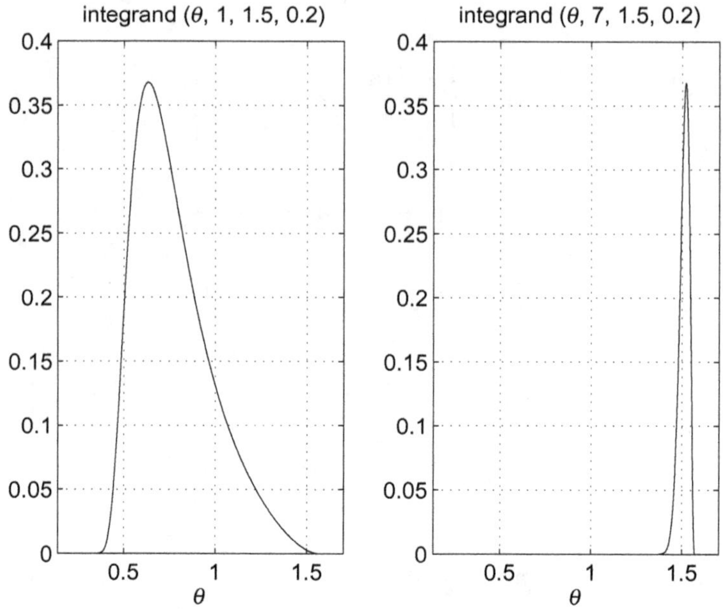

Figure 5.7. The integrand in (5.23) for $\alpha = 1.5$, $\beta = 0.2$ and two different values of the pdf argument: $x = 1$ (left) and $x = 7$ (right).

region into two parts:

$$f_X(x;\alpha,\beta) = c_2(x;\alpha,\beta)\left(\int_{-\theta_0}^{t} g(\theta;x,\alpha,\beta)\exp(-g(\theta;x,\alpha,\beta))d\theta \right.$$
$$\left. + \int_{t}^{\frac{\pi}{2}} g(\theta;x,\alpha,\beta)\exp(-g(\theta;x,\alpha,\beta))d\theta \right),$$

where

$$t = \{\theta : \ln\big(g(\theta;x,\alpha,\beta)\big) = 0\}. \tag{5.26}$$

The use of the logarithm in (5.26) is permitted because $g(\theta;x,\alpha,\beta)$ is positive (see Nolan (1997) and the references therein). Moreover, the logarithm transforms the product in (5.25) into a sum which is numerically more stable.

In contrast to (5.23), the expression for the distribution function in (5.24) is less problematic. The integrand is strictly monotonic (increasing or decreasing depending on the sign of β); at one of the integral bounds it equals 1 and at the other it equals zero.

The benefits of pdf and cdf estimation using Zolotarev's integral representations can be summarized as follows:

- the integral bounds are finite in contrast to the infinite ones in the Fourier inversion integral;
- the integrands are not periodic, they are bounded, continuous, and composed of monotonic segments;
- it is possible to approximate the pdf and the cdf for any x with no interpolation as in the FFT method.

It is important to remark that the accuracy and the computational burden significantly depend on the choice of quadrature to perform the numerical integration. In case we choose an adaptive quadrature, the speed of the pdf evaluation will depend on the particular value of x because for large x the integrand is not well behaved for numerical work and more iterations will be necessary before convergence.

5.4.4. Inverse distribution function

Many applications including finance rely on the inverse distribution function which is also known as the quantile function. In finance specifically, a risk measure known as value-at-risk is defined as a certain quantile in the left tail of the distribution and, therefore, its calculation relies on either knowing the inverse distribution function in closed form or on a numerical method for its calculation. In the case of stable distributions, unfortunately, no closed-form expressions are available.

Numerically, stable distribution quantiles can be obtained by solving a nonlinear equation. This is a general method which works as long as the distribution function $F_Y(x)$ can be reliably calculated. Denote by q_p the pth quantile of a random variable Y, then

$$F_Y(q_p) = p,$$

where $p \in (0,1)$. For numerical work, we use the approximation of the distribution function in (5.24) which leads to good precision if p is close to zero or one.

5.5. Random Number Generation

Generating random numbers from a stable distribution is computationally simple even though the random number generator is not trivial. The first algorithm for the general case was provided by Chambers *et al.* (1976) but in a parameterization convenient for theoretical work which is neither P_0 nor P_1. Additional details can be found in Samorodnitsky and Taqqu (1994), Janicki and Weron (1994), Weron (1996), and Zolotarev and Uchaikin (1999). In this section, we provide the algorithm in Weron (1996).

To generate a sample from random variable $Y \sim S_\alpha(\sigma,\beta,\mu)$ in the parameterization P_0, the following steps can be followed:

1. Generate independently a random variable V distributed uniformly on the interval $(-\frac{\pi}{2},\frac{\pi}{2})$ and a random variable W from an exponential distribution with mean equal to one; $W = -\log(U)$ where U is uniform in $[0,1]$ and is generated independently of V.
2. Depending on the value of α:

   (a) if $\alpha \neq 1$, calculate

   $$Z = a_{\alpha,\beta}\frac{\sin(\alpha(V+b_{\alpha,\beta}))}{\cos^{1/\alpha}(V)}\left(\frac{\cos(V-\alpha(V+b_{\alpha,\beta}))}{W}\right)^{(1-\alpha)/\alpha},$$

   where

   $$a_{\alpha,\beta} = \left(1+\beta^2\tan^2(\pi\alpha/2)\right)^{\frac{1}{2\alpha}},$$
   $$b_{\alpha,\beta} = \frac{1}{\alpha}\arctan(\beta\tan(\pi\alpha/2))$$

   are constants depending on the distribution parameters;

   (b) if $\alpha = 1$, calculate

   $$Z = \frac{2}{\pi}\left(\left(\frac{\pi}{2}+\beta V\right)\tan(V) - \beta\log\left(\frac{\frac{\pi}{2}W\cos(V)}{\frac{\pi}{2}+\beta V}\right)\right).$$

3. Depending on the value of α:

   (a) if $\alpha \neq 1$, calculate

$$Y = \sigma X + \mu;$$

   (b) if $\alpha = 1$, calculate

$$Y = \sigma X + \frac{2}{\pi}\beta\sigma\log(\sigma) + \mu.$$

We provide an implementation in Matlab below.[12] The function generates an $n \times m$ matrix of independent random numbers from $S_\alpha(\sigma, \beta, \mu)$. This convention is consistent with the built-in random number generators in Matlab. Our implementation does not depend on any of the Matlab toolboxes and all calculations are vectorized for better performance.

```
function r = srnd(n, m, alpha, beta, sigma, mu)

% function r = srnd(n, m, alpha, beta, sigma, mu)
%
% The function generates a matrix of independent random ...
    numbers from a stable distribution with a characteristic ...
    function:
%
% Eexp(itX) = exp(sigma^alpha(|t|^alpha + i*t*w(t, alpha, ...
    beta)) + i*mu*t)
%
% where if alpha ~= 1
%            w(t, alpha, beta) = beta(|t|^(alpha - ...
    1))tan(pi*alpha/2)
%       if alpha == 1
%            w(t, alpha, beta) = - beta*ln|t|*2/pi
%
% The input parameters n and m define the dimension of the ...
    output matrix.

v = (rand(n, m) - 1/2)*pi;
w = -log(rand(n, m));
a = (1 + beta^2*(tan(pi*alpha/2))^2)^(1/(2*alpha));
b = atan(beta*tan(pi*alpha/2))/alpha;
```

[12] Another Matlab implementation of the same random number generator is available on the web page of McCulloch: http://www.econ.ohio-state.edu/jhm/jhm.html.

```
if alpha ~= 1
  r = a*sin(alpha*(v + b))./(cos(v)).^(1/alpha);
  r = r.*(cos(v - alpha*(v + b))./w).^((1 - alpha)/alpha);
  r = r*sigma + mu;
else
  r = 2*((pi/2 + beta*v).*tan(v) - ...
       beta*log(pi*w.*cos(v)./(pi/2 + beta*v)))/pi;
  r = r*sigma + 2*beta*sigma*log(sigma)/pi + mu;
end
```

5.6. Parameter Estimation

The problem of parameter estimation of stable distributions was first tackled by Mandelbrot (1963) and Fama and Roll (1968, 1971). This is a non-trivial task because, with a few exceptions, there are no closed-form expressions for the probability density and distribution functions. In general, parameter estimation techniques for the class of stable laws belong to three categories: quantile methods, characteristic function-based methods, and maximum likelihood.[13] The quantile methods use predetermined empirical quantiles to estimate stable parameters. For example, the method of Fama and Roll (1968, 1971) for symmetric α-stable distribution and its modified version of McCulloch (1986) for the skewed case belong to this group.

Characteristic function-based methods include moment estimators suggested by Press (1972) and regression-type procedures proposed by Koutrouvelis (1981) and Kogon and Williams (1998). Simulation studies by Akgiray and Lamoureux (1989) suggest that the regression-type estimators are superior to quantile methods.[14]

The validity of maximum likelihood estimation (MLE) theory was demonstrated by DuMouchel (1973). Comparison studies

[13]Apart from these groups of methods, there is also the method of indirect inference where a different distribution which is easier to work with is used as a proxy. An application of this method for stable distributions is suggested in Garcia et al. (2011).

[14]See also Stoyanov and Racheva-Iotova (2004b) for a limited empirical study that suggests the method of Kogon and Williams can be comparable to MLE.

between MLE and the quantile method of McCulloch in Rachev and Mittnik (2000) recommend the maximum likelihood estimator. In terms of computational performance, MLE is the slowest method. Quantile methods are fastest and regression-type methods are in between.

In this section, we review McCulloch's quantile method, the regression-type estimator of Kogon and Williams, and the method of maximum likelihood.

5.6.1. The quantile method of McCulloch

The estimation procedure by McCulloch (1986) is a generalization of the quantile method by Fama and Roll (1968, 1971) for the symmetric stable case. The stable parameters estimator is provided in the parameterization P_0 and is consistent and asymptotically normal if $0.6 < \alpha \leq 2$. We adopt the standard notation for theoretical and empirical quantiles, namely x_p is the pth quantile if $F(x_p) = p$, where $F(x)$ is the distribution function of a random variable and given a sample of observations $x_1, x_2, \ldots, x_n$, then $\hat{x}_p$ is the sample quantile if $F_n(\hat{x}_p) = p$, where $F_n(x)$ is the sample distribution function.

Following McCulloch (1986), we define two functions of theoretical quantiles:

$$v_\alpha = \frac{x_{0.95} - x_{0.05}}{x_{0.75} - x_{0.25}},$$

$$v_\beta = \frac{x_{0.95} + x_{0.05} - 2x_{0.50}}{x_{0.95} - x_{0.05}}.$$

From (5.13) it follows that the functions v_α and v_β defined in this way do not depend on the scale and the location parameter and are functions only of α and β,

$$\begin{aligned} v_\alpha &= \varrho_1(\alpha, \beta), \\ v_\beta &= \varrho_2(\alpha, \beta). \end{aligned} \tag{5.27}$$

Employing equation (5.12), we have that $F(-x_p; \alpha, -\beta) = F(x_{1-p}; \alpha, \beta)$ and therefore the following relations hold:

$$\begin{aligned} \varrho_1(\alpha, \beta) &= \varrho_1(\alpha, -\beta), \\ \varrho_2(\alpha, \beta) &= -\varrho_2(\alpha, -\beta). \end{aligned} \tag{5.28}$$

The system of equations (5.27) can be inverted and the parameters α and β can be expressed as functions of the quantities v_α and v_β:

$$\begin{aligned} \alpha &= \tau_1(v_\alpha, v_\beta), \\ \beta &= \tau_2(v_\alpha, v_\beta). \end{aligned} \tag{5.29}$$

Replacing v_α and v_β in equations (5.29) with their sample counterparts $\hat{v}_\alpha$ and $\hat{v}_\beta$:

$$\begin{aligned} \hat{v}_\alpha &= \frac{\hat{x}_{0.95} - \hat{x}_{0.05}}{\hat{x}_{0.75} - \hat{x}_{0.25}}, \\ \hat{v}_\beta &= \frac{\hat{x}_{0.95} + \hat{x}_{0.05} - 2\hat{x}_{0.50}}{\hat{x}_{0.95} - \hat{x}_{0.05}} \end{aligned}$$

yields the following estimators $\hat{\alpha}$ and $\hat{\beta}$:

$$\begin{aligned} \hat{\alpha} &= \tau_1(\hat{v}_\alpha, \hat{v}_\beta), \\ \hat{\beta} &= \tau_2(\hat{v}_\alpha, \hat{v}_\beta). \end{aligned} \tag{5.30}$$

The functions $\tau_1(\cdot)$ and $\tau_2(\cdot)$ are tabulated in Tables 5.1 and 5.2. It should be noted that because of property (5.28), the following symmetry holds:

$$\begin{aligned} \tau_1(v_\alpha, v_\beta) &= \tau_1(v_\alpha, -v_\beta), \\ \tau_2(v_\alpha, v_\beta) &= -\tau_2(v_\alpha, -v_\beta). \end{aligned}$$

In other words, the sign of $\hat{v}_\beta$ determines the sign of β.

Since there are no closed-form expressions for the functions $\tau_1(\cdot)$ and $\tau_2(\cdot)$, we compute estimates of α and β from the statistics $\hat{v}_\alpha$ and $\hat{v}_\beta$ using the values in Tables 5.1 and 5.2 and linear interpolation for intermediate values. If it happens that $\hat{v}_\alpha$ is below 2.439, $\hat{\alpha}$ should be set equal to 2 and $\hat{\beta}$ equal to zero. Table 5.2 contains values

Table 5.1. $\alpha = \tau_1(v_\alpha, v_\beta) = \tau_1(v_\alpha, -v_\beta)$.

		v_β						
		0	0.1	0.2	0.3	0.5	0.7	1
v_α	2.439	2	2	2	2	2	2	2
	2.5	1.916	1.924	1.924	1.924	1.924	1.924	1.924
	2.6	1.808	1.813	1.829	1.829	1.829	1.829	1.829
	2.7	1.729	1.73	1.737	1.745	1.745	1.745	1.745
	2.8	1.664	1.663	1.663	1.668	1.676	1.676	1.676
	3	1.563	1.56	1.553	1.548	1.547	1.547	1.547
	3.2	1.484	1.48	1.471	1.46	1.448	1.438	1.438
	3.5	1.391	1.386	1.378	1.364	1.337	1.318	1.318
	4	1.279	1.273	1.266	1.25	1.21	1.184	1.15
	5	1.128	1.121	1.114	1.101	1.067	1.027	0.973
	6	1.029	1.021	1.014	1.004	0.974	0.935	0.874
	8	0.896	0.892	0.887	0.883	0.855	0.823	0.769
	10	0.818	0.812	0.806	0.801	0.78	0.756	0.691
	15	0.698	0.695	0.692	0.689	0.676	0.656	0.595
	25	0.593	0.59	0.588	0.586	0.579	0.563	0.513

Table 5.2. $\beta = \tau_2(v_\alpha, v_\beta) = -\tau_2(v_\alpha, -v_\beta)$.

		v_β						
		0	0.1	0.2	0.3	0.5	0.7	1
v_α	2.439	0	2.16	1	1	1	1	1
	2.5	0	1.592	3.39	1	1	1	1
	2.6	0	0.759	1.8	1	1	1	1
	2.7	0	0.482	1.048	1.694	1	1	1
	2.8	0	0.36	0.76	1.232	2.229	1	1
	3	0	0.253	0.518	0.823	1.575	1	1
	3.2	0	0.203	0.41	0.632	1.244	1.906	1
	3.5	0	0.165	0.332	0.499	0.943	1.56	1
	4	0	0.136	0.271	0.404	0.689	1.23	2.195
	5	0	0.109	0.216	0.323	0.539	0.827	1.917
	6	0	0.096	0.19	0.284	0.472	0.693	1.759
	8	0	0.082	0.163	0.243	0.412	0.601	1.596
	10	0	0.074	0.174	0.22	0.377	0.546	1.482
	15	0	0.064	0.128	0.191	0.33	0.478	1.362
	25	0	0.056	0.112	0.167	0.285	0.428	1.274

Table 5.3. $v_\sigma = \varrho_3(\alpha, \beta) = \varrho_3(\alpha, -\beta)$.

		β				
		0	0.25	0.5	0.75	1
α	0.5	2.588	3.073	4.534	6.636	9.144
	0.6	2.337	2.635	3.542	4.808	6.247
	0.7	2.189	2.392	3.004	3.844	4.775
	0.8	2.098	2.244	2.676	3.265	3.912
	0.9	2.04	2.149	2.461	2.886	3.356
	1	2	2.085	2.311	2.624	2.973
	1.1	1.98	2.04	2.205	2.435	2.696
	1.2	1.965	2.007	2.125	2.294	2.491
	1.3	1.955	1.984	2.067	2.188	2.333
	1.4	1.946	1.967	2.022	2.106	2.211
	1.5	1.939	1.952	1.988	2.045	2.116
	1.6	1.933	1.94	1.962	1.997	2.043
	1.7	1.927	1.93	1.943	1.961	1.987
	1.8	1.921	1.922	1.927	1.936	1.947
	1.9	1.914	1.915	1.916	1.918	1.921
	2	1.908	1.908	1.908	1.908	1.908

larger than 1 for more precise interpolation. If $\hat{\beta} > 1$, it should be reduced to 1.

McCulloch (1986) also provides an estimator for the scale parameter σ. Let us first define v_σ as

$$v_\sigma = \frac{x_{0.75} - x_{0.25}}{\sigma} = \varrho_3(\alpha, \beta).$$

The function $\tau_3(\alpha, \beta)$ is given in Table 5.3. Employing the same arguments that led us to equation (5.28) yields the relation $\varrho_3(\alpha, \beta) = \varrho_3(\alpha, -\beta)$. The estimator $\hat{\sigma}$ is obtained after replacing α and β with the corresponding estimators in (5.30):

$$\hat{\sigma} = \frac{\hat{x}_{0.75} - \hat{x}_{0.25}}{\varrho_3(\hat{\alpha}, \hat{\beta})}.$$

Estimation of the location parameter μ is a more involved affair because of the discontinuity of the parametric representation of the chf P_0 when $\alpha \to 1$ and $\beta \neq 0$. First we estimate the shifted location

Table 5.4. $\varrho_4(\alpha, \beta) = \varrho_4(\alpha, -\beta)$.

		β				
		0	0.25	0.5	0.75	1
	0.5	0	−0.061	−0.279	−0.659	−1.198
	0.6	0	−0.078	−0.272	−0.581	−0.997
	0.7	0	−0.089	−0.262	−0.52	−0.853
	0.8	0	−0.096	−0.25	−0.469	−0.742
	0.9	0	−0.099	−0.237	−0.424	−0.652
	1	0	−0.098	−0.223	−0.383	−0.576
	1.1	0	−0.095	−0.208	−0.346	−0.508
	1.2	0	−0.09	−0.192	−0.31	−0.447
α	1.3	0	−0.084	−0.173	−0.276	−0.39
	1.4	0	−0.075	−0.154	−0.241	−0.335
	1.5	0	−0.066	−0.134	−0.206	−0.283
	1.6	0	−0.056	−0.111	−0.17	−0.232
	1.7	0	−0.043	−0.088	−0.132	−0.179
	1.8	0	−0.03	−0.061	−0.092	−0.123
	1.9	0	−0.017	−0.032	−0.049	−0.064
	2	0	0	0	0	0

parameter ζ defined by:

$$\zeta = x_{0.50} + \sigma \ \mathrm{sign}(\beta)\varrho_4(\alpha, \beta),$$

where $\varrho_4(\alpha, \beta)$ is tabulated in Table 5.4 and has the property $\varrho_4(\alpha, \beta) = \varrho_4(\alpha, -\beta)$. The location parameter μ is related to ζ according to:

$$\mu = \begin{cases} \zeta - \beta\sigma \tan \dfrac{\pi\alpha}{2}, & \alpha \neq 1, \\ \zeta, & \alpha = 1. \end{cases} \tag{5.31}$$

Replacing the parameters in equations (5.6.1) and (5.31) with their sample counterparts yields the estimator $\hat{\mu}$:

$$\hat{\zeta} = \hat{x}_{0.50} + \hat{\sigma} sign(\hat{\beta})\varrho_4(\hat{\alpha}, \hat{\beta})$$

and

$$\hat{\mu} = \begin{cases} \hat{\zeta} - \hat{\beta}\hat{\sigma} \tan \dfrac{\pi\hat{\alpha}}{2}, & \hat{\alpha} \neq 1, \\ \hat{\zeta}, & \hat{\alpha} = 1. \end{cases}$$

It should be noted that a significant advantage of this method is the lack of heavy computations.[15]

5.6.2. Characteristic function-based methods

The characteristic function-based methods rely on the sample characteristic function for parameter estimation. The sample characteristic function is defined as

$$\hat{\phi}_X(u) = \frac{1}{n}\sum_{j=1}^{n} e^{iux_j}, \quad u \in \mathbb{R}, \tag{5.32}$$

where $x_1, x_2, \ldots, x_n$ is a sample of iid observations on a random variable X. Since $|\hat{\phi}_X(u)| \leq 1$, all moments of the random variable $\hat{\phi}_X(u)$ are finite and, according to equation (5.32), for any u it is the sample mean of the iid random variables e^{iux_j}. As a consequence, from the law of large numbers, it can be inferred that the sample characteristic function is a consistent estimator of the characteristic function $\phi_X(u) = E[e^{iuX}]$, $u \in \mathbb{R}$ of a random variable X.

Kogon and Williams (1998) suggest a method that reduces to running several regressions using the empirical characteristic function and the P_1. Their approach is similar to that of Koutrouvelis (1980) and Koutrouvelis (1981). The linear equations follow directly from the convenient form of the logarithm of the characteristic function,

$$\ln[-\Re(\ln\phi_X(u))] = \alpha\ln\sigma + \alpha\ln|u|, \tag{5.33}$$

$$\Im(\ln\phi_X(u)) = \mu_1 u + \beta\sigma u(|\sigma u|^{\alpha-1} - 1)\tan\frac{\pi\alpha}{2}. \tag{5.34}$$

As a result, estimators of the stable distribution parameters can be constructed using the method of least squares after replacing the characteristic function $\phi_X(u)$ with the sample characteristic function $\hat{\phi}_X(u)$.

---

[15] On the personal homepage of McCulloch http://www.econ.ohio-state.edu/jhm/jhm.html a GAUSS implementation of the algorithm is publicly available. In the appendix to this chapter, we provide a Matlab translation of the GAUSS code.

The main difficulty of this method is that $\hat{\phi}_X(u)$ should be evaluated at certain values of the argument u but the best choice might depend on the parameter values which are unknown. Koutrouvelis (1980, 1981) relate the values of u to the values of the index of stability α and the sample size. The main advantage of the procedure in Kogon and Williams (1998) is that the chosen values for u are invariant of any other parameters. Kogon and Williams (1998) report that an appropriate choice are 10 equally spaced values in the interval $[0.1, 1]$, $u_k = \{0.1 + 0.1k,\ k = 0, 1, \ldots, 9\}$. Certainly, the sample should be normalized before applying the method of least squares, otherwise the optimal choice of u would depend on the scale and the location parameters. For preliminary estimation of σ and μ, it is suggested to use the quantile method of McCulloch.

The algorithm is as follows:

(1) Given a sample of iid observations $x_1, x_2, \ldots, x_n$ first we find preliminary estimates σ_0 and μ_{01} using the quantile method of McCulloch and we normalize the observations by those values,

$$x'_j = \frac{x_j - \hat{\mu}_{01}}{\hat{\sigma}_0}, \quad j = 1, 2, \ldots, n.$$

(2) Next we consider the regression equation constructed from equation (5.33):

$$y_k = b + \alpha w_k + \epsilon_k, \quad k = 0, 1, \ldots, 9,$$

where $y_k = \ln[-\Re(\ln \hat{\phi}_X(u_k))]$, $w_k = \ln|u_k|$, $u_k = \{0.1 + 0.1k,\ k = 0, 1, \ldots, 9\}$ and ϵ_k denotes the error term. We find $\hat{\alpha}$ and $\hat{b}$ according to the method of least squares using the normalized sample $x'_1, x'_2, \ldots, x'_n$. The estimator $\hat{\sigma}_1$ of the scale parameter of the normalized sample equals

$$\hat{\sigma}_1 = \exp\left(\frac{\hat{b}}{\hat{\alpha}}\right).$$

(3) The estimators $\hat{\beta}$ and $\hat{\mu}_{11}$ of the skewness parameter and the modified location parameter, respectively, are derived from the

second regression equation (5.34):

$$z_k = \mu_{11} u_k + \beta v_k + \eta_k, \quad k = 0, 1, \ldots, 9,$$

where $z_k = \Im(\ln \hat{\phi}_X(u_k))$, $v_k = \hat{\sigma}_1 u_k (|\hat{\sigma}_1 u_k|^{\hat{\alpha}-1} - 1) \tan \frac{\pi \hat{\alpha}}{2}$, $u_k = \{0.1 + 0.1k,\ k = 0, 1, \ldots, 9\}$, and η_k is the error term.

(4) The final estimators $\hat{\sigma}$ and $\hat{\mu}_1$ follow from:

$$\hat{\sigma} = \hat{\sigma}_0 \hat{\sigma}_1, \quad \hat{\mu}_1 = \hat{\mu}_{01} + \hat{\sigma}_0 \hat{\mu}_{11}.$$

If we aim at estimating the location parameter μ, we need to take advantage of the connection between the two parametric forms P_0 and P_1:

$$\hat{\mu} = \hat{\mu}_1 - \hat{\beta} \hat{\sigma} \tan \frac{\pi \hat{\alpha}}{2}.$$

Kogon and Williams (1998) demonstrate in a Monte Carlo study that their method is computationally much more efficient than that of Koutrouvelis (1980, 1981). The approach by Koutrouvelis (1980, 1981) has better properties only in the estimation of β.

5.6.3. The method of maximum likelihood

The method of maximum likelihood is appealing because of the good large-sample properties of the estimator provided that certain general conditions are met. The likelihood function is defined by

$$L(x_1, x_2, \ldots, x_n | \theta) = \prod_{k=1}^{n} f_X(x_k | \theta),$$

where $x_1, x_2, \ldots, x_n$ is a sample of iid observations of a random variable X, $f_X(x|\theta)$ is the pdf of X, and θ is a vector of parameters. In the case of stable distributions, $\theta = (\alpha, \beta, \sigma, \mu)$. Maximum likelihood estimates are found by searching for that parameter values which maximize the likelihood function or, equivalently, the log-likelihood function:

$$\hat{\theta}_n = \arg\max_{\theta} \log(L(x_1, x_2, \ldots, x_n | \theta)).$$

DuMouchel (1973) verified that the likelihood function in the parameterization P_0 satisfies the general conditions if the following cases are excluded: $\alpha \leq \epsilon$, $\alpha = 2$ and $|\beta| = 1$, and $\alpha = 1$, $\beta_0 \neq 0$.

Clearly, if we intend to derive expressions for MLE analytically, we need to have closed-form expressions for the pdfs of stable laws. Because such expressions do not exist in the general case, the problem of MLE of stable parameters can be approached numerically by employing an approximation of the density function. Such an approximation can be based on the FFT method or on Zolotarev's integral representations.

5.6.4. A comparison of the estimation methods

In this section, we compare McCulloch's quantile method, the regression-type method, and the method of maximum likelihood in a Monte Carlo study using the Matlab code provided in the chapter. The comparison is carried out through the bootstrap method implemented according to the next steps.

(1) The stable distribution parameters α, β, σ, and μ are fixed to some predefined values.
(2) A random sample of size m is generated from a stable distribution with the parameters from the previous step.
(3) The three methods are used to estimate the parameters from the sample.
(4) Steps (2) and (3) are repeated 100 times. As a result, we have 100 sets of estimated parameters for each of the methods.

From the estimated parameters corresponding to each method, we calculate the bias and the variability which are characteristic of the estimation method. The method that results in a smaller bias and a smaller variability is preferred.

For a complete study, one needs to repeat the bootstrap procedure for each point in the parameter space. In this section, we provide an illustration for the standardized symmetric stable case, $\beta = 0$, $\sigma = 1$, and $\mu = 0$, while the tail index is varied from 1.4 to 2 with a step size of 0.1. The interval for the tail index $\alpha \in [1.4, 2]$ is typical for financial asset returns.[16] We set the sample size to $m = 1000$. Even though the four parameters are estimated jointly, we compare the methods only with respect to the α parameter.

Boxplots of the estimated values are provided in Figure 5.8. Each plot corresponds to one of the estimation methods discussed in the chapter and the boxplots reflect the relative spread of the estimated values. It is clear that the quantile method leads to relatively more dispersed estimates than the other two methods. The characteristic function-based method appears comparable to MLE.

This observation is supported by the data in Table 5.5 which contains two panels. The upper panel reports the average $\hat{\alpha}$ which shows the bias of the estimator relative to the true value. With a couple of exceptions, the biases are relatively small for all methods.

The lower panel reports the standard errors of the estimators. The numbers represent the standard deviation of the estimated parameter values. The standard errors of the quantile method are higher than those of the other methods, especially when α is close to 2. The characteristic function-based method is comparable to MLE for this sample size.

5.7. Risk Measures

We mentioned already that the quantile function has an important application in finance because it is directly related to the industry standard risk measure value-at-risk (VaR)which is defined through the quantile function of the distribution of the log-return of a given asset.

Another important function in finance is the conditional mean $E(Y|Y < q_p)$ where q_p is a given quantile. In finance, this function (changed by sign) is known as expected shortfall which coincides with the average value-at-risk (AVaR) or conditional value-at-risk

[16]See, for example, Rachev and Mittnik (2000).

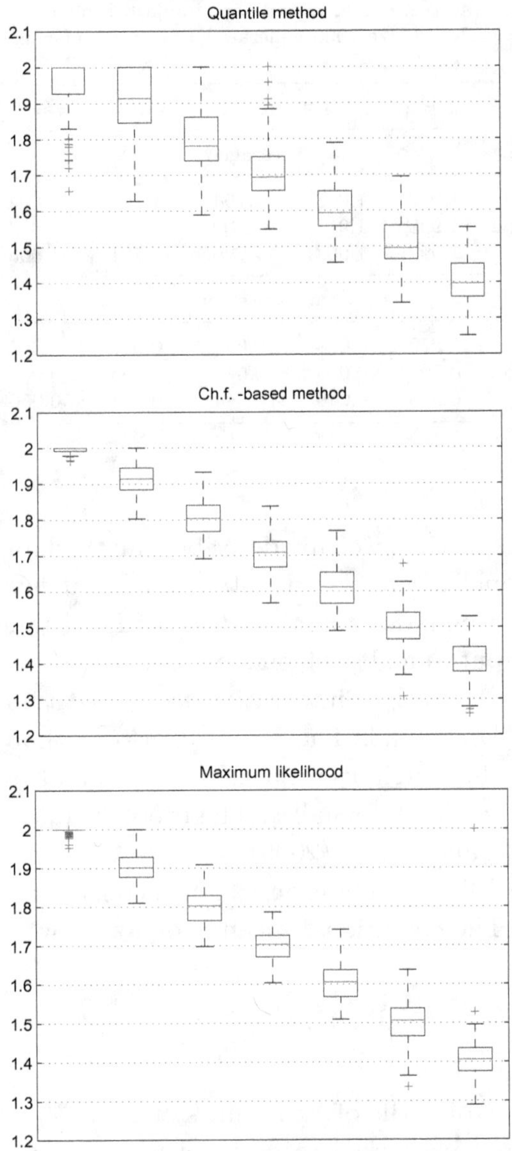

Figure 5.8. Boxplots of estimated α from 100 samples of size 1000 generated from a standardized symmetric stable distribution. The true value of α declines from 2 to 1.4 with a step of 0.1.

Table 5.5. The average estimated α and the standard error of the estimated α computed from 100 samples of size 1000 generated from a standardized symmetric stable distribution.

Estimation method	$\alpha = 2$	$\alpha = 1.9$	$\alpha = 1.8$	$\alpha = 1.7$	$\alpha = 1.6$	$\alpha = 1.5$	$\alpha = 4$
			Average $\hat{\alpha}$				
Quantile method	1.952	1.908	1.794	1.710	1.604	1.505	1.403
Ch.f.-based method	1.995	1.911	1.802	1.704	1.613	1.499	1.406
MLE	1.997	1.903	1.799	1.701	1.605	1.501	1.410
			Standard error of $\hat{\alpha}$				
Quantile method	0.079	0.083	0.090	0.082	0.068	0.070	0.061
Ch.f.-based method	0.008	0.043	0.049	0.055	0.061	0.059	0.054
MLE	0.008	0.040	0.044	0.043	0.050	0.052	0.075

(CVaR). In this section, we only describe a method for the calculation of the conditional mean. Its advantage is that it involves numerical integration of a bounded function in a bounded range which is not a difficult computational problem. In contrast, the definition of the conditional mean involves integration from $-\infty$ to q_p which can be problematic if p is small. The Monte Carlo method is not reliable either because the distributions are fat-tailed and the variance of the estimator can be high for small p. The proofs and further details can be found in Stoyanov *et al.* (2006).

We use the following notation which is consistent with the definition of AVaR. The conditional mean is denoted by

$$\text{AVaR}_\delta(Y) = -E(Y|Y < q_\delta),$$

where q_δ is the δ-quantile of Y. It turns out that Zolotarev's integral representations can be used to derive a formula for the conditional expectation. Suppose that $Y \in S_\alpha(\sigma, \beta, \mu)$ with $\alpha > 1$ has a standardized distribution in the continuous parameterization P_1. Then,

$$\text{AVaR}_\delta(Y) = \sigma A_{\delta,\alpha,\beta} - \mu.$$

The term $A_{\delta,\alpha,\beta}$ does not depend on the scale and the location parameters. If $q_\delta \neq 0$, then the term equals

$$A_{\delta,\alpha,\beta} = \frac{\alpha}{1-\alpha}\frac{|q_\delta|}{\pi\delta}\int_{-\overline{\theta}_0}^{\pi/2} g(\theta)\exp(-|q_\delta|^{\frac{\alpha}{\alpha-1}}v(\theta))d\theta, \qquad (5.35)$$

where

$$g(\theta) = \frac{\sin(\alpha(\overline{\theta}_0+\theta)-2\theta)}{\sin\alpha(\overline{\theta}_0+\theta)} - \frac{\alpha\cos^2\theta}{\sin^2\alpha(\overline{\theta}_0+\theta)},$$

$$v(\theta) = \left(\cos\alpha\overline{\theta}_0\right)^{\frac{1}{\alpha-1}}\left(\frac{\cos\theta}{\sin\alpha(\overline{\theta}_0+\theta)}\right)^{\frac{\alpha}{\alpha-1}}\frac{\cos(\alpha\overline{\theta}_0+(\alpha-1)\theta)}{\cos\theta},$$

$$\overline{\theta}_0 = \frac{1}{\alpha}\arctan\left(\overline{\beta}\tan\frac{\pi\alpha}{2}\right),$$

$$\overline{\beta} = -\mathrm{sign}(q_\delta)\beta$$

If $q_\delta = 0$, then we get

$$A_{\delta,\alpha,\beta} = \frac{2\Gamma\left(\frac{\alpha-1}{\alpha}\right)}{(\pi-2\theta_0)}\frac{\cos\theta_0}{(\cos\alpha\theta_0)^{1/\alpha}}, \qquad (5.36)$$

where $\Gamma(x)$ denotes the gamma function and $\theta_0 = \frac{1}{\alpha}\arctan(\beta\tan\frac{\pi\alpha}{2})$. Because of the power decay of the tails of α-stable laws, the case $\alpha \leq 1$ leads to an infinite conditional expectation.

We provide a Matlab routine that implements this formula. Because the formula is derived in the continuous parameterization but the shift parameter has an intuitive interpretation in the parameterization P_0, the implementation assumes the input parameters are in P_0.

```
function [AVaR, VaR] = stableavar(epsilon, alpha, beta, sigma, mu)

% Computes the average value-at-risk (AVaR) of stable laws. ...
   The parameterization used is
%
% ln Eexp(i*x*t) = i*delta*t + psi(c*t)
%
% where the standard (c = 1, delta = 0) log c.f. function is
%        if alpha ~= 1
```

```
%           psi(t) = -abs(t)^alpha*[1-i*b*sign(t)*tan(pi*alpha/2)]
%       elseif alpha == 1
%           psi(t) = -abs(t)*[1+i*beta*(2/pi)*sign(t)*ln(abs(t))
%
% The function returns both the conditional expectation and the
% corresponding quantile (VaR). The function uses ...
    stableq_int.m for quantile estimation

% modify a little the value of beta if abs(beta) = 1
if abs(beta) == 1
    beta = sign(beta)*0.999;
end

VaR = -stableq_int(epsilon, alpha, beta, 1, 0);

% this is necessary because the pdf is approximated in the ...
    continuous parameterization, now we are in the ...
    discontinuous case
VaR = -(-VaR + beta*tan(pi*alpha/2));

betaBar = -sign(VaR)*beta;
theta0Bar = atan(betaBar*tan(pi*alpha/2))/alpha;

AVaR = quadl(@integrand, -theta0Bar + eps, pi/2, 1.0e-15, [], ...
    alpha, betaBar, VaR);

AVaR = AVaR*alpha*abs(VaR);
AVaR = AVaR/(pi*epsilon*(1 - alpha));
AVaR = sigma*AVaR - mu;

% rescale the VaR if necessary
if nargout == 2
    VaR = sigma*VaR - mu;
end

%%%%%%%%%%%%%%%%%%%%%%%%%%%%%%%%%%%%%%%%%%%%%%%%%%%%%%%%%%%
%           Auxiliary function defined                    %
%%%%%%%%%%%%%%%%%%%%%%%%%%%%%%%%%%%%%%%%%%%%%%%%%%%%%%%%%%%

function f = integrand(theta, alpha, beta, VaR)
% computes the integrand

theta0 = atan(beta*tan(pi*alpha/2))/alpha;
g = sin(alpha*(theta0 + theta) - 2*theta)./sin(alpha*(theta0 + ...
    theta));
g = g - (alpha.*(cos(theta)).^2)./((sin(alpha*(theta0 + ...
    theta))).^2);

v = (cos(alpha*theta0))^(1/(alpha - 1));
```

```
v = v*((cos(theta)./sin(alpha*(theta0 + ...
    theta))).^(alpha/(alpha - 1)));
v = v.*(cos(alpha*theta0 + (alpha - 1)*theta)./cos(theta));

f = g.*exp(-(abs(VaR)^(alpha/(alpha - 1))).*v);
```

Key Points

- Stable distributions arise as the only distributions which can be limits of sums of iid random variables.
- The Gaussian distribution is a special case of the family of stable distributions. The stable non-Gaussian distributions, also known as Lévy stable, have Pareto-like tails and are heavy-tailed. They have an infinite variance.
- There are symmetric and asymmetric representatives of the Lévy stable distributions.
- The class of stable distributions is defined in terms of its characteristic function. Apart from a few special cases, no closed-form expressions for their densities and distribution functions are known to exist. Practical applications rely on numerical methods.
- The algorithm of Chambers–Mallows–Stuck provides a computationally easy random number generator.
- The stable distribution density and cdf can be approximated by the FFT method which is fast but imprecise in the tails. Zolotarev's integral representations are more precise but computationally more demanding.
- Parameter estimation can be carried out by different methods which vary in accuracy and computational complexity. The quantile method of McCulloch is fastest but has the highest variance, the regression-type method is in the middle in both accuracy and computational complexity, and the MLE has the lowest variance but is computationally demanding.
- The risk measure known as average value-at-risk can be calculated reliably by a formula that involves numerical integration of a bounded function in a bounded range.

References

R.J. Adler, R. Feldman, and M. Taqqu. *A Practical Guide to Heavy Tails: Statistical Techniques and Applications.* Birkhäuser, Boston, 1998.

A. Akgiray and G. Booth. The stable-law model of stock returns. *Journal of Business and Economic Statistics*, 6:51–57, 1988.

A. Akgiray and G. Lamoureux. Estimation of stable-law parameters: A comparative study. *Journal of Business and Economic Statistics*, 7:85–93, 1989.

H. Bergström. On some expansions of stable distributions. *Arkiv for Matematik*, II: 375–378, 1952.

M. Chambers, C.L. Mallows, and B.W. Stuck. A method for simulating stable random variables. *Journal of the American Statistical Association*, 71(354):340–344, 1976.

W.H. DuMouchel. On the asymptotic normality of the maximum-likelihood estimate when sampling from a stable distribution. *Annals of Statistics*, 1:948–957, 1973.

P. Embrechts, C. Klüppelberg, and T. Mikosch. *Modelling Extremal Events for Insurance and Finance.* Springer, Berlin, 1997.

E.F. Fama. Mandelbrot and the stable Paretian hypothesis. *Journal of Business*, 36:430–429, 1963.

E.F. Fama and R. Roll. Some properties of symmetric stable distributions. *Journal of the American Statistical Association*, 63:817–836, 1968.

E.F. Fama and R. Roll. Parameter estimates for symmetric stable distributions. *Journal of the American Statistical Association*, 66:331–338, 1971.

R. Garcia, E. Renault, and D. Veredas. Estimation of stable distributions by indirect inference. *Journal of Econometrics*, 161(2):325–337, 2011.

M. Grabchak and G. Samorodnitsky. Do financial returns have finite or infinite variance? A paradox and an explanation. *Quantitative Finance*, 10(8):883–893, 2010.

A. Janicki and A. Weron. *Simulation and Chaotic Behavior of Alpha-Stable Stochastic Processes.* Marcel Dekker, New York, 1994.

S. Kogon and D. Williams. Characteristic function based estimation of stable distribution parameters. In R. Adler, R. Feldman, and M. Taqqu, editors, *A Practical Guide to Heavy Tails: Statistical Techniques and Applications*, Birkhäuser, Boston, 1998, pp. 312–335.

I.A. Koutrouvelis. Regression-type estimation of the parameters of stable laws. *Journal of the American Statistical Association*, 75:918–928, 1980.

I.A. Koutrouvelis. An iterative procedure for the estimation of the parameters of stable laws. *Communications in Statistics — Simulation and Computation*, 10:17–28, 1981.

S. Louhichi. Rates of convergence in the CLT for some weakly dependent random variables. *Theory of Probability and Its Applications*, 46(2): 297–315, 2002.

B.B. Mandelbrot. The variation of certain speculative prices. *Journal of Business*, 36: 394–419, 1963.

J. H. McCulloch. Simple consistent estimators of stable distribution parameters. *Communications in Statistics — Simulation and Computation*, 15:1109–1136, 1986.

C. Menn and S. Rachev. Calibrated FFT-based density approximations for α-stable distributions. *Computational Statistics & Data Analysis*, 50: 1891–1904, 2006.

J. Nolan. Numerical computation of stable densities and distribution functions. *Stochastic Models*, 13:759–774, 1997.

R. Officer. The distribution of stock returns. *Journal of the American Statistical Association*, 76:807–812, 1972.

M. Paolella. Stable-GARCH models for financial returns: Fast estimation and tests for stability. *Econometrics*, 4(2), 2016.

S. Press. Estimation in univariate and multivariate stable distributions. *American Statistical Association*, 67:842–846, 1972.

S.T. Rachev and S. Mittnik. *Stable Paretian Models in Finance.* Wiley, Chichester, 2000.

S.T. Rachev, S.V. Stoyanov, and F.J. Fabozzi. *A Probability Metrics Approach to Financial Risk Measures.* Wiley-Blackwell, Chichester, 2011.

S.T. Rachev, L. Klebanov, S.V. Stoyanov, and F.J. Fabozzi. *The Methods of Distances in the Theory of Probability and Statistics.* Springer, New York, 2013.

G. Samorodnitsky and M.S. Taqqu. *Stable Non-Gaussian Random Processes.* Chapman & Hall, New York, 1994.

S.V. Stoyanov and B. Racheva-Iotova. Univariate stable laws in the field of finance – approximations of density and distribution functions. *Journal of Concrete and Applicable Mathematics*, 2(1):38–57, 2004a.

S.V. Stoyanov and B. Racheva-Iotova. Univariate stable laws in the field of finance – parameter estimation. *Journal of Concrete and Applicable Mathematics*, 2(4):24–49, 2004b.

S.V. Stoyanov, G. Samorodnitsky, S.T. Rachev, and S. Ortobelli. Computing the portfolio conditional value-at-risk in the α-stable case. *Probability and Mathematical Statistics*, 26:1–22, 2006.

R. Weron. On the Chambers–Mallows–Stuck method for simulating skewed stable random variables. *Statistics & Probability Letters*, 28(2):165–171, 1996.

V.M. Zolotarev. On the representations of stable laws by integrals. *Selected Translations in Mathematical Statistics and Probability*, 6:84–88, 1964.

V.M. Zolotarev. *One-Dimensional Stable Distributions*. Translation of Mathematical Monographs, Vol. 65. American Mathematical Society, 1986.

V.M. Zolotarev and V.V. Uchaikin. *Chance and Stability, Stable Distributions and Their Applications*. De Gruyter, 1999.

Appendix — Matlab Functions

Developing reliable routines for numerical work with stable distributions is not easy. The code is lengthy and often relies on other functions to carry out specific calculations. In this appendix, we provide Matlab code in two sections: one of them is dedicated to approximations to the density function, the distribution function, and the quantile function and the other — to parameter estimation. Because the code is lengthy, each function includes its own help with details on the particular input and output parameters and also on which parameterization is used.

In this appendix, we provide a short description of the function dependencies. Most of the functions do not require any Matlab toolboxes.

A.1. Approximations of the density function, the distribution function, and the quantile function

In this appendix, we provide Matlab code implementing approximations to the density function, the distribution function, and the quantile function.

Concerning density functions, we provide two methods: one based on the FFT method and one based on Zolotarev's integral representation. The FFT approximation is much faster but is less precise, especially in the tails. The function implements the left-point rule, the midpoint rule, and the weighted average approximation. The default method is the weighted average approximation but one can choose to work with any of the other two by setting the corresponding method in the function arguments.

Zolotarev's integral representation for the density in (5.23) requires solving a nonlinear equation and a procedure for numerical integration. To speed up calculations, we use specialized routines which are not a part of the standard Matlab toolboxes. The nonlinear equation is solved by an implementation of Newton's method (*fzerovec.m*) which has been vectorized to solve in parallel a set of nonlinear equations which differ by their parameter values. The application is very specific to the stable density approximation problem but the code is provided on a stand-alone basis.

Likewise, the numerical integration is performed by the adaptive Gauss–Kronrod quadrature (*adaptGKr.m*) which has been vectorized to solve simultaneously a number of integrals where the upper bound, the lower bound, and certain parameters of the integrand can be vectors. Also, in the adaptive part, the quadrature has been modified to split the integration range not in half but proportionately according to a pre-calculated ratio for each element of the vectorized parameters. The use of the quadrature is very specific to the integrand of the integral representations but the code is provided as a stand-alone function.

Concerning the distribution function, we provide an implementation of (5.24) which is numerically much better behaved than (5.23). Still, we use the vectorized quadrature to improve the computational performance.

Finally, the quantile function uses the vectorized Newton method to solve the nonlinear equation and also the FFT density approximation for a quick initial guess.

```
function [f, pdf, x]  = stablepdf_fft(arg, alpha, beta, sigma, ...
    mu, method)

% function f  = stablepdf_fft(arg, alpha, beta, sigma, mu, method)
%
% Approximates the probability density function of a stable law
% with parameters (alpha, beta, sigma, mu) using the FFT approach.
%
```

```
% The parameterization used is the following:
%
% Eexp(itX) = exp(-sigma^alpha*|t|^alpha*(1 + i*sign(t)*w(t, ...
    alpha, beta)) + i*mu*t)
%
% where if alpha ~= 1
%           w(t, alpha, beta, sigma) = beta*(|sigma*t|^(1 - ...
    alpha) - 1)*tan(pi*alpha/2)
%       elseif alpha == 1
%           w(t, alpha, beta, sigma) = beta*ln|sigma*t|*2/pi
%
% The input variable method is a string: 'left-point' or ...
    'mid-point'. If omitted, the function constructs the ...
    left-point and the mid-point approximation and applies ...
    weights 1/3 and 2/3 respectively.

% define the closed-form cases (Gaussian, Cauchy and Levy)
if alpha == 2 % (Gaussian case)
    f = normpdf(arg, mu + sigma^2*beta*tan(pi), sqrt(2)*sigma);
    return
elseif alpha == 1 && beta == 0 % (Cauchy case)
    f = pi*((arg - mu).^2 + sigma^2);
    f = sigma./f;
    return
elseif alpha == 1/2 && abs(beta) == 1 % (Levy case)
    index1 = find(arg <= mu - sigma^0.5);
    index2 = find(arg > mu - sigma^0.5);
    f(index2) = (sigma/(2*pi))^0.5*((arg(index2) - mu + ...
        sigma^0.5).^(-1.5))...
        .*exp(-(sigma./(2*(arg(index2) - mu + sigma^0.5))));
    f(index1) = 0;
    if beta < 0 % (Symmetric Levy case)
        f = stablepdf_fft(-arg, alpha, -beta, sigma, mu);
    end
    return
end

% define default values
if nargin == 5
    method = [];
end

if isempty(method)
    f = stablepdf_fft(arg, alpha, beta, sigma, mu, ...
        'left-point')/3 + ...
        2*stablepdf_fft(arg, alpha, beta, sigma, mu, ...
            'mid-point')/3;
    return
end
```

```
if alpha == 1
   alpha = 1 + 1.0e-5;
end

h = 0.01;
N = 2^15;

switch method
    case 'left-point'
        s = 1/(h*N);
        t1 = 1:1:N;
        t1 = t1(:);
        t2 = 2*pi*(t1 - 1 - N/2)*s;

        cfvalues = chf(t2, alpha, beta);
        x1 = (-1).^(t1 - 1).*cfvalues;

        % compute pdf values
        pdf = real(fft(x1, N)).*(s.*(-1).^(t1 - 1 - N/2));
        pdf = pdf/sigma;

        x = (t1 - 1 - N/2)*h*sigma + mu;
    case 'mid-point'
        t = ((0:N) - N/2)/(h*N);
        t_mid = (t(1:end-1) + t(2:end))/2;
        k = 0:(N-1);
        Ck = 1i*(-1).^k.*exp(-1i*pi*k/N)/(h*N);
        xk = h*(k - N/2);
        yj = chf(2*pi*t_mid, alpha, beta).*(-1).^k;

        % compute pdf values
        pdf = real(fft(yj, N).*Ck);
        pdf = pdf/sigma;

        x = xk*sigma + mu;
end

indexout = find(arg > max(x) | arg < min(x));
indexin = find(arg < max(x) & arg > min(x));

% approximate the indexin and indexout values
if isempty(indexout)
    f = interp1(x, pdf, arg, 'PCHIP');
else g(indexin) = interp1(x, pdf, arg(indexin), 'PCHIP');
    for j = 1:length(indexout)
        g(indexout(j)) = bergstroem((arg(indexout(j)) - ...
            mu)/sigma, alpha, beta)/sigma;
    end
    f = g;
end
```

```
f = f(:);

%%%%%%%%%%%%%%%%%%%%%%%%%%%%%%%%%%%%%%%%%%%%%%%%%%%%%%%%%%%%
% DEFINITION OF SOME AUXILIARY FUNCTIONS                   %
%%%%%%%%%%%%%%%%%%%%%%%%%%%%%%%%%%%%%%%%%%%%%%%%%%%%%%%%%%%%

function phi = chf(arg, alpha, beta)

% The stable distribution characteristic function
abs_t = abs(arg);
sign_t = sign(arg);
phi = -abs_t.^(alpha).*(1 + ...
    i*beta*tan(pi*alpha/2)*sign_t.*(abs_t.^(1-alpha) - 1));
phi(arg == 0) = 0;
phi = exp(phi);
```

```
function f = stablepdf_int(arg, alpha, beta, sigma, mu)

% function f = stablepdf_int(arg, alpha, beta, sigma, mu)
%
% Calculates stable pdf using Zolotarev's Integral representations
%
% Numerical integration is performed with Gauss-Kronrod 10-21 ...
    rule and has been vectorized for faster performance. A ...
    vectorized Newton method for finding the maximum of the ...
    integrand is also used. The function does not rely on any ...
    Matlab toolboxes.
%
% The parameterization of the characteristic function of ...
    normalized alpha-stable laws is:
%
% ln(phi(t)) = -|t|^(alpha)*(1 + ...
    i*beta*sign(t)*tan(pi*alpha/2)(|t|^(1 - alpha) - 1)), ...
    alpha ~= 1
% ln(phi(t)) = -|t|*(1 + i*beta*sign(t)ln(|t|)*2/pi), alpha = 1
%
% Based on "Numerical Calculation of Stable Densities and ...
    Distribution Functions", John P. Nolan


% Perform a consistency check of the parameter values
if alpha <= 0 || alpha > 2
    error('Alpha should be between 0 and 2')
elseif abs(beta) > 1
    error('Beta should be between -1 and 1')
elseif sigma <= 0
    error('Sigma should be non-negative')
end
```

```
% Define the closed form cases

if alpha == 2 % (Gaussian case)
    f = normpdf(arg, mu + sigma^2*beta*tan(pi), sqrt(2)*sigma);
    return
elseif alpha == 1 && beta == 0 % (Cauchy case)
    f = pi*((arg - mu).^2 + sigma^2);
    f = sigma./f;
    return
elseif alpha == 1/2 && abs(beta) == 1 % (Levy case)
    index1 = find(arg <= mu - sigma^0.5);
    index2 = find(arg > mu - sigma^0.5);
    f(index2) = (sigma/(2*pi))^0.5*((arg(index2) - mu + ...
        sigma^0.5).^(-1.5))...
        .*exp(-(sigma./(2*(arg(index2) - mu + sigma^0.5))));
    f(index1) = 0;
    if beta < 0 % (symmetric to Levy case)
        f = stablepdf_int(-arg, alpha, -beta, sigma, mu);
    end
    return
end

arg = arg(:);
arg = (arg - mu)/sigma;
f = zeros(size(arg));
if alpha ~= 1
    err = 1.0e-6;

    % Compute some constants for the two cases x < zeta or x > ...
        zeta
    power = alpha/(alpha - 1);
    zeta = -beta*tan(pi*alpha/2);
    theta = atan(-zeta)/alpha;
    below = find(arg < zeta - err);
    above = find(arg > zeta + err);
    middle = find((arg >= zeta - err) & (arg <= zeta + err));

    % COMPUTE THE DENSITY FOR THE "MIDDLE" VALUES
    f(middle) = gamma(1 + 1/alpha)*cos(theta);
    f(middle) = f(middle)/(pi*(1 + zeta*zeta)^(1/(2*alpha)));

    % COMPUTE THE DENSITY FOR THE "ABOVE" VALUES
    if ~isempty(above)
        xn = length(above);

        % Compute the logarithm for better numerical stability
        c1 = power*log((arg(above) - repmat(zeta, xn, 1)));
        c1 = c1 + repmat((power - 1)*log(cos(alpha*theta)), ...
            xn, 1);
```

```
        range_temp = atan((arg(above) - repmat(zeta, xn, ...
            1)).^alpha);

        % Apply a vectorized Newton procedure
        t1 = fzerovec(@range_det, -theta*(ones(xn, 1) - ...
            2*range_temp./pi) + ...
            range_temp, -theta, pi/2, 1.0e-8, [], alpha, ...
            theta, c1, power);
        t1
        % Get ready for a vectorized quadrature
        intError = [1  1.0e-10];
        ratio = (t1 + repmat(theta, xn, 1))./(repmat(pi/2, xn, ...
            1) + repmat(theta, xn, 1));
        ratio1 = max(ratio, 1 - ratio);
        ratio2 = min(ratio, 1 - ratio);
        I = adaptGKr(@integrand, -theta + eps, t1, intError, ...
            [], ratio1, ...
            alpha, theta, repmat(c1', 21, 1), power);
        I = I + adaptGKr(@integrand, t1, pi/2, intError, [], ...
            ratio2, ...
            alpha, theta, repmat(c1', 21, 1), power);
        f(above) = I.*alpha./(pi.*(arg(above) - ...
            zeta)*abs(alpha - 1));
    else f(above) = [];
    end

    % COMPUTE THE DENSITY FOR THE "BELOW" VALUES
    if ~isempty(below)
        f(below) = stablepdf_int(-arg(below), alpha, -beta, 1, 0);
    end

elseif alpha == 1
    xn = length(arg);
    range_temp = 1/2 - atan(arg)/pi;
    t1 = fzerovec(@range_det1, -pi*range_temp/2.5 + ...
        pi*(ones(xn, 1) - ...
        range_temp)/2.5, -pi/2, pi/2, 1.0e-8, [], beta, arg);

    intError = [1.0e-1  1.0e-10];
    ratio = (t1 + repmat(pi/2, xn, 1))/pi;

    I = adaptGKr(@integrand1, -pi/2, t1, intError, [], ...
        max(ratio, 1 - ratio), beta, repmat(arg', 21, 1));
    I = I + adaptGKr(@integrand1, t1, pi/2 - eps, intError, ...
        [], min(ratio, 1 - ratio), beta, repmat(arg', 21, 1));

    f = I/(2*abs(beta));
end
f = f/sigma;
```

```
%%%%%%%%%%%%%%%%%%%%%%%%%%%%%%%%%%%%%%%%%%%%%%%%%%%%%%%%%%%%%%%%
%     DEFINITION OF SOME AUXILIARY FUNCTIONS                   %
%%%%%%%%%%%%%%%%%%%%%%%%%%%%%%%%%%%%%%%%%%%%%%%%%%%%%%%%%%%%%%%%

function [g, dg] = range_det(t, alpha, theta, c1, power)

% This function is used when splitting the range over which
% the integration is performed when alpha ~= 1. The function ...
    is used to solve log(f(x)) = 0 instead of f(x) = 1 for ...
    better numerical stability.

temp1 = alpha.*theta + (alpha - 1).*t;
temp2 = alpha.*(theta + t);
g = c1 + (power - 1).*log(cos(t)) + log(cos(temp1)) - ...
    power.*log(sin(temp2));

dg = (1 - power).*tan(t) + (1 - alpha).*tan(temp1) -...
    alpha.*power.*cot(temp2);

function [g, dg] = range_det1(t, beta, arg)

% This function is used when splitting the range over which
% the integration is performed when alpha = 1. The function is ...
    used to solve log(f(x)) = 0 instead of f(x) = 1 for better ...
    numerical stability.

g = -pi.*arg./(2*beta) + log(2/pi) + log((pi/2 + ...
    beta.*t)./cos(t))...
    + (pi/2 + beta.*t).*tan(t)./beta;
dg = beta./(pi/2 + beta*t) + 2*tan(t) + (pi/2 + ...
    beta*t)./(beta*cos(t).*cos(t));

function g = integrand(t, alpha, theta, c1, power)
% This function is the integrand when alpha ~= 1

g = c1 + (power - 1).*log(cos(t));
g = g + log(cos(alpha.*theta + (alpha - 1).*t));
g = g - power.*log(sin(alpha.*(t + theta)));
g = exp(g);
g = g.*exp(-g);
g(isnan(g)) = realmin;

function g = integrand1(t, beta, arg)
% This function is the integrand when alpha = 1

g = (pi + 2*beta.*t).*exp((pi./(2*beta) + t).*tan(t) - ...
    pi*arg./(2*beta))./(pi*cos(t));
g = g.*exp(-g);
g(isnan(g)) = realmin;
```

```
function F = stablecdf_int(arg, alpha, beta, sigma, mu)

% function F = stablecdf_int(arg, alpha, beta, sigma, mu)
%
% Calculates the cumulative distribution function of a stable
% law, taking advantage of Zolotarev's integral representation.
%
% Numerical integration is performed with Gauss-Kronrod 10-21 rule
% The parameterization of the characteristic function of ...
    normalized alpha-stable laws is:
%
% ln(phi(t)) = -|t|^(alpha)*(1 + ...
    i*beta*sign(t)*tan(pi*alpha/2)(|t|^(1 - alpha) - 1)), ...
    alpha ~= 1
% ln(phi(t)) = -|t|*(1 + i*beta*sign(t)ln(|t|)*2/pi), alpha = 1
%
% Based on "Numerical Calculation of Stable Densities and ...
    Distribution Functions", John P. Nolan

% Define the closed-form cases
if alpha == 2
    F = normcdf(arg, mu + sigma^2*beta*tan(pi), sqrt(2)*sigma);
    F = F(:);
    return
elseif alpha == 1 && beta == 0
    F = 1/2 + atan((arg - mu)/sigma)/pi;
    F = F(:);
    return
end

arg = arg(:);
arg = (arg - mu)/sigma;
F = zeros(size(arg));

if alpha ~= 1
    err = 1.0e-6;

    % Compute some constants for the two cases x < zeta or x > ...
        zeta
    power = alpha/(alpha - 1);
    zeta = - beta*tan(pi*alpha/2);
    theta = atan(-zeta)/alpha;
    below = find(arg < zeta - err);
    above = find(arg > zeta + err);
    middle = find((arg >= zeta - err) & (arg <= zeta + err));

    if alpha < 1
        c1 = (pi/2 - theta)/pi;
    elseif alpha > 1
```

```
        c1 = 1;
    end

    % COMPUTE THE CDF FOR THE "MIDDLE" VALUES
    F(middle) = (pi/2 - theta)/pi;

    % COMPUTE THE CDF FOR THE "ABOVE" VALUES
    if ~isempty(above)
        xn = length(above);

        % Compute the logarithm for better numerical stability
        c = power*log((arg(above) - repmat(zeta, xn, 1)));
        c = c + repmat((power - 1)*log(cos(alpha*theta)), xn, 1);

        % Get ready for a vectorized quadrature
        intError = [1  1.0e-10];

        % compute a midpoint to split the integration. ...
            Increases precision for high |beta|
        range_temp = atan((arg(above) - repmat(zeta, xn, ...
            1)).^alpha);
        t1 = -theta*(ones(xn, 1) - 2*range_temp/pi) + range_temp;

        ratio = repmat(0.5, xn, 1);
        I = adaptGKr(@integrand, -theta + eps, t1, intError, ...
            [], ratio, ...
            alpha, theta, repmat(c', 21, 1), power);
        I = I + adaptGKr(@integrand, t1, repmat(pi/2, xn, 1), ...
            intError, [], ratio, ...
            alpha, theta, repmat(c', 21, 1), power);
        F(above) = c1 + (sign(1 - alpha)/pi)*I;
    else F(above) = [];
    end

    if ~isempty(below)
        F(below) = 1 - stablecdf_int(-arg(below), alpha, ...
            -beta, 1, 0);
    end

elseif alpha == 1
    xn = length(arg);

    intError = [1.0e-1  1.0e-10];
    ratio = repmat(0.5, xn, 1);

    % calculate a midpoint to split the integration. Increases ...
        precision for high |beta|
    range_temp = 1/2 - atan(arg)/pi;
    t1 = -pi*range_temp/2.5 + pi*(ones(xn, 1) - range_temp)/2.5;
```

```
    I = adaptGKr(@integrand1, -pi/2, t1, intError, [], ratio, ...
        beta, repmat(arg', 21, 1));
    I = I + adaptGKr(@integrand1, t1, repmat(pi/2, xn, 1), ...
        intError, [], ratio, beta, repmat(arg', 21, 1));
    F = I/pi;

    if beta < 0
        F = 1 - F;
    end
end

%%%%%%%%%%%%%%%%%%%%%%%%%%%%%%%%%%%%%%%%%%%%%%%%%%%%%%%%%%%
% DEFINITION OF SOME AUXILIARY FUNCTIONS                  %
%%%%%%%%%%%%%%%%%%%%%%%%%%%%%%%%%%%%%%%%%%%%%%%%%%%%%%%%%%%

function g = integrand(t, alpha, theta, c1, power)
% This function is the integrand when alpha ~= 1

g = c1 + (power - 1).*log(cos(t));
g = g + log(cos(alpha.*theta + (alpha - 1).*t));
g = g - power.*log(sin(alpha.*(t + theta)));
g = exp(-exp(g));
g(isnan(g)) = realmin;

function g = integrand1(t, beta, arg)
% This function is the integrand when alpha = 1

g = (pi + 2*beta.*t).*exp((pi./(2*beta) + t).*tan(t) - ...
    pi*arg./(2*beta))./(pi*cos(t));
g = exp(-g);
g(isnan(g)) = realmin;
```

```
function q = stableq_int(pvalue, alpha, beta, sigma, mu)

% function q = stableq_int(pvalue, alpha, beta, sigma, mu)
%
% Calculates quantiles using Zolotarev's integral representation
% of the cumulative distribution function of a stable law. The
% function needs stablecdf_int.m and stablepdf_fft.m
% pvalue should contain numbers in [0, 1]
% alpha, beta, sigma and mu are the stable parameters.
%
% The parameterization of the characteristic function of ...
    normalized alpha-stable laws is:
%
```

```
% ln(phi(t)) = -|t|^(alpha)*(1 + ...
    i*beta*sign(t)*tan(pi*alpha/2)(|t|^(1 - alpha) - 1)), ...
    alpha ~= 1
% ln(phi(t)) = -|t|*(1 + i*beta*sign(t)ln(|t|)*2/pi), alpha = 1

% Define the closed-form cases
if alpha == 2
    q = norminv(pvalue, mu + sigma^2*beta*tan(pi), sqrt(2)*sigma);
    return
elseif alpha == 1 && beta == 0
    q = sigma*tan((pvalue - 1/2)*pi) + mu;
    return
end

pvalue = pvalue(:);
% Calculate initial values from the fft-based approximation
x0 = stableq_fft_fast(pvalue, alpha, beta, sigma, mu);

% Find the zero of the function F(x) - pvalue using a ...
    vectorized Newton method
q = fzerovec(@stableq_int_df, x0, -Inf, Inf, 1.0e-8, [], ...
    pvalue, alpha, beta);
q = q(:);
q = q*sigma + mu;

%%%%%%%%%%%%%%%%%%%%%%%%%%%%%%%%%%%%%%%%%%%%%%%%%%%%%%%%%%%%%
%       DEFINITIONS OF SOME AUXILIARY FUNCTIONS             %
%%%%%%%%%%%%%%%%%%%%%%%%%%%%%%%%%%%%%%%%%%%%%%%%%%%%%%%%%%%%%

function [f, df] = stableq_int_df(arg, pvalue, alpha, beta)

% Defines the equation F(arg) = p. The function is called in a ...
    vectorized format in which arg and p have the same dimension.

f = stablecdf_int(arg, alpha, beta, 1, 0) - pvalue;
df = stablepdf_int(arg, alpha, beta, 1, 0);

function q = stableq_fft_fast(pvalue, alpha, beta, sigma, mu)

% Calculates quantiles using the fft approach to stable density
% calculation. The function needs stable_pdfb.m for stable pdf
% approximation.

arg = (-40.95*sigma + mu):sigma/500:(40.95*sigma + mu);
pdf = stablepdf_fft(arg, alpha, beta, sigma, mu, 'linear');
cdf = cumsum(pdf)*sigma/500;
index_zero = max(find(cdf < 1.0e-5));
index_one = min(find(cdf > 0.99999));

cdf(index_one + 1:end) = [];
```

```
arg(index_one + 1:end) = [];

cdf(1:index_zero - 1) = [];
arg(1:index_zero - 1) = [];

q = interp1(cdf, arg, pvalue);
```

```
function x = fzerovec(fun, x0, lower, upper, xtol, ftol, varargin)

% function x = fzerovec(fun, x0, xtol, ftol, varargin)
%
% Uses Newton's method to find the solution to F(x) = 0, x0 is
% the initial guess. If x0 is a vector column, some of the ...
    varargin elements can also be vectors with the dimension ...
    of x0.
% The function is used in stablepdf_int and stableq_int.
%

if nargin < 4 || isempty(xtol), xtol = 5*eps;  end
if nargin < 5 || isempty(ftol), ftol = 5*eps;  end
xeps = max(xtol,5*eps);  feps = max(ftol,5*eps);  % Smallest ...
    tols are 5*eps

x0 = x0(:);
x = x0;  k = 0;  maxit = 50;    % Initial guess, current and ...
    max iterations

ready = false(size(x0));
while sum(~ready) ~= 0

    % Check if the maximum number of iterations has not been ...
        reached
    if k >= maxit
        % Exit with a warning
        x = x(:);
        warning(sprintf('root not found within tolerance after ...
            %d iterations\n',k));
        return
    end

    % increment the counter
    k = k + 1;

    % find and use only non-NaN values and their corresponding
    % varargin elements
    n = find(~ready);
    x11 = x(n);
    for m = 1:length(varargin)
```

```
        if length(varargin{m}) > 1
            varargin_current{m} = varargin{m}(n);
        else varargin_current{m} = varargin{m};
        end
    end

    % Evaluate the function @fun
    [f, dfdx] = feval(fun, x11, varargin_current{:});    % ...
        Returns f( x(k-1) ) and f'( x(k-1) )
    dx = f./dfdx;
    x12 = x11 - dx;

    % correct the result if not within (theta, pi/2)
    x12(x12 < lower) = (lower + x11(x12 < lower))/2;
    x12(x12 > upper) = (upper + x11(x12 > upper))/2;

    % Return the values
    x(n) = x12;

    % Check for which elements the increment is small enough
    idx = (abs(f) < feps) | (abs(dx) < xeps);

    % indicate that those values as ready
    ready(n(idx)) = true;
end

x = x(:);
```

```
function I = adaptGKr(fun, a, b, tol, maxLevel, ratio, varargin)

% adaptGKr  Adaptive numerical integration using Gauss-Kronrod ...
    21 rule and automatic interval subdivision.  Interval over ...
    which the G-K rule is applied is bisected until the larger ...
    of an absolute or relative tolerance is met, or until ...
    recursion limit is reached.
%
% The function is used in stablepdf_int and stablecdf_int and ...
    has been
% vectorized to take the integral bounds a, b, and the ratio ...
    parameters as vectors of equal dimension.
%

if nargin<4,  tol = [5e-3  5e-6];        end  % Default tol and ...
    maxLevel
if nargin<5,  maxLevel=16;               end
if nargin<6, ratio = 0.5;                end
if isempty(maxLevel),  maxLevel = 30;   end
if isempty(tol)             % case of tol = [] on input
```

```
  tol = [5e-3  5e-6];
else
  if length(tol)==1        % True if only relative tolerance ...
      is supplied
    tol = [tol 0];
  end
end

% --- Call worker function with "level" and "I1" arguments that
%     are invisible to user of adaptGK
I = doAdaptGKr(fun,a,b,tol,maxLevel,ratio,0,0, [], 0, ...
    varargin{:});

% ====================

function I = doAdaptGKr(fun,a,b,tol,maxLevel,ratio,level,I1, ...
    index, n, varargin)
% doAdaptGK  Routine that does work of adaptive Gauss-Kronrod ...
    quadrature adaptGK is a simple interface to doAdaptGK that ...
    hides the level and I1 parameters from the user
%
% Synopsis:  I = doAdaptGK(fun,a,b,tol,maxLevel,level,I1, ...
    index, vararing)
%
% Input:  Same parameters as adaptGK except
%         level = local refinement level;  level = 0 on startup
%         I1    = current value of integral on this level;  I1 ...
    = 0 on startup

if length(a) ~= length(b) && length(b) ~= 1
    a = repmat(a, size(b));
elseif length(a) ~= length(b) && length(a) ~= 1
    b = repmat(b, size(a));
end

level = level + 1;         % Current refinement level
if level == 1              % on first call, I1 has not been
                                % evaluated
   I1 = gaussKronrod21(fun,a,b,varargin{:});
end

c = a + ratio.*(b-a);

I21 = gaussKronrod21(fun,a,c,varargin{:});
I22 = gaussKronrod21(fun,c,b,varargin{:});

I2 = I21 + I22;

if isempty(index)
    index = 1:1:length(c);
    n = length(c);
```

```
end

I = zeros(n, 1);
k1 = find(abs(I2 - I1) <= min(tol(1)*abs(I1), tol(2)));
I(index(k1)) = I2(k1);

if level > maxLevel+2
    I(index) = I2;
    return
end

index(k1) = [];
k2 = find(abs(I2 - I1) > min(tol(1)*abs(I1), tol(2)));

for i = 1:length(varargin)
    if size(varargin{i}, 2) > 1
        varargin{i} = varargin{i}(:, k2);
    end
end

if isempty(k2)
    return
else
    I = I + doAdaptGKr(fun,a(k2),c(k2),[tol(1) tol(2)/2],...
        maxLevel,ratio(k2),level,I21(k2), index, n, ...
            varargin{:})...        % I_left
        + doAdaptGKr(fun,c(k2),b(k2),[tol(1) tol(2)/2],...
        maxLevel,ratio(k2),level,I22(k2), index, n, ...
            varargin{:});          % I_rightend
end

function I = gaussKronrod21(fun,a,b,varargin)
% Synopsis:    I = gaussKronrod21(f,a,b)
%
% Input:       f = (string) name of m-file that evaluates the ...
    integrand, f(x)
%              a,b = lower and upper limits of the integral
%              arg1,arg2,... = optional arguments passed through ...
    to fun
%
% Output:      I = approximation to integral of f(x)*dx from a ...
    to b
%                  Obtained with 21 point Kronrod rule
xgk = [ 0.995657163025808080735527280689003;  % nodes of 21 ...
    point Kronrod rule
        0.973906528517171720077964012084452;  % xgk(2:2:21) ...
            are nodes of 10 point
        0.930157491355708226001207180059508;  % Gauss rule
        0.865063366688984510732096688423493;
        0.780817726586416897063717578345042;
```

```
        0.679409568299024406234327365114874;
        0.562757134668604683339000099272694;
        0.433395394129247190799265943165784;
        0.294392862701460198131126603103866;
        0.148874338981631210884826001129720;
        0                                    ];

wgk  =[ 0.011694638867371874278064396062192;   % weights of 21 ...
    point Kronrod rule
        0.032558162307964727478818972459390;
        0.054755896574351996031381300244580;
        0.075039674810919952767043140916190;
        0.093125454583697605535065465083366;
        0.109387158802297641899210590325805;
        0.123491976262065851077958109831074;
        0.134709217311473325928054001771707;
        0.142775938577060080797094273138717;
        0.147739104901338491374841515972068;
        0.149445554002916905664936468389821; ] ;

half = 0.5*(b-a);                                  % half ...
    length of the interval
n = length(half);
points = [-xgk(1:11); xgk(10:-1:1)];
points = repmat(points, 1, n);
half = repmat(half', 21, 1);
first = repmat(0.5*(a+b)', 21, 1);
x = first + half.*points;                          % Translate 21 ...
    nodes to interval (a,b)
f = feval(fun,x,varargin{:});                      % Evaluate ...
    f at all 21 nodes

I = half(1, :).*(wgk(1:11)'*f(1:11, :) + ...
    wgk(10:-1:1)'*f(12:21, :));  % 21 pt Kronrod
```

A.2. Parameter estimation — McCulloch's quantile method, the regression-type method, and MLE

In this appendix, we provide Matlab implementations of McCulloch's quantile method, the regression-type method described in the chapter, and the MLE method.

The implementation of McCulloch's method is based on the GAUSS code available on his website. The implementation of the regression-type method is straightforward but relies on the quantile method for initial estimates. Finally, the MLE relies on the

FFT density approximation and uses the regression-type method to get an initial guess. The MLE is the only function which relies on the optimization toolbox of Matlab. It uses unconstrained optimization to find the parameter values which maximize the log-likelihood function.

```
function [alpha, beta, c, delta, zeta] = mcculloch(sample)

% function [alpha, beta, c, delta, zeta] = mcculloch(sample)
%
% A matlab implementation of McCulloch's quantile method of
% stable parameters estimation. The original fortran code can
% be found at McCulloch's homepage:
% http://www.econ.ohio-state.edu/jhm/jhm.html
%
% The input variable of the function is 'sample'. The output
% variables are:
%       alpha is the index of stability
%       beta is skewness parameter
%       c is scale parameter
%       delta is location parameter
%       zeta is shifted location parameter
%           zeta = delta + beta*c*tan(pi*alpha/2), if alpha ~= 1
%           zeta = delta                          , if alpha == 1
%
% Parametetrization
% ln Eexp(i*x*t) = i*delta*t + psi(c*t)
%
% where the standard (c = 1, delta = 0) log c.f. function is
%       if alpha ~= 1
%           psi(t) = -abs(t)^alpha*[1-i*b*sign(t)*tan(pi*alpha/2)]
%       elseif alpha == 1
%           psi(t) = -abs(t)*[1+i*beta*(2/pi)*sign(t)*ln(abs(t))]
ena =      [2.4388     2.4388     2.4388     2.4388     2.4388;
            2.5120     2.5117     2.5125     2.5129     2.5148;
            2.6080     2.6093     2.6101     2.6131     2.6174;
            2.7369     2.7376     2.7387     2.7420     2.7464;
            2.9115     2.9090     2.9037     2.8998     2.9016;
            3.1480     3.1363     3.1119     3.0919     3.0888;
            3.4635     3.4361     3.3778     3.3306     3.3161;
            3.8824     3.8337     3.7199     3.6257     3.5997;
            4.4468     4.3651     4.1713     4.0052     3.9635;
            5.2172     5.0840     4.7778     4.5122     4.4506;
            6.3140     6.0978     5.6241     5.2195     5.1256;
            7.9098     7.5900     6.8606     6.2598     6.1239;
           10.4480     9.9336     8.7790     7.9005     7.6874;
           14.8378    13.9540    12.0419    10.7219    10.3704;
```

```
         23.4831    21.7682    18.3320    16.2163    15.5841;
         44.2813    40.1367    33.0018    29.1399    27.7822];

enb =    [0.0000    0.0000     0.0000     0.0000     0.0000;
         0.0000     0.0179     0.0357     0.0533     0.0710;
         0.0000     0.0389     0.0765     0.1133     0.1480;
         0.0000     0.0626     0.1226     0.1784     0.2281;
         0.0000     0.0895     0.1736     0.2478     0.3090;
         0.0000     0.1183     0.2282     0.3199     0.3895;
         0.0000     0.1478     0.2849     0.3942     0.4686;
         0.0000     0.1769     0.3422     0.4703     0.5458;
         0.0000     0.2062     0.3993     0.5473     0.6210;
         0.0000     0.2362     0.4561     0.6240     0.6934;
         0.0000     0.2681     0.5134     0.6993     0.7616;
         0.0000     0.3026     0.5726     0.7700     0.8248;
         0.0000     0.3415     0.6343     0.8339     0.8805;
         0.0000     0.3865     0.6994     0.8900     0.9269;
         0.0000     0.4408     0.7678     0.9362     0.9620;
         0.0000     0.5095     0.8381     0.9700     0.9847];

enc =    [1.9078    1.9078     1.9078     1.9078     1.9078;
         1.9140     1.9150     1.9160     1.9185     1.9210;
         1.9210     1.9220     1.9275     1.9360     1.9470;
         1.9270     1.9305     1.9425     1.9610     1.9870;
         1.9330     1.9405     1.9620     1.9970     2.0430;
         1.9390     1.9520     1.9885     2.0450     2.1160;
         1.9460     1.9665     2.0220     2.1065     2.2110;
         1.9550     1.9845     2.0670     2.1880     2.3330;
         1.9650     2.0075     2.1255     2.2945     2.4910;
         1.9800     2.0405     2.2050     2.4345     2.6965;
         2.0000     2.0850     2.3115     2.6240     2.9735;
         2.0400     2.1490     2.4610     2.8865     3.3565;
         2.0980     2.2445     2.6765     3.2650     3.9125;
         2.1890     2.3920     3.0040     3.8440     4.7755;
         2.3370     2.6355     3.5425     4.8085     6.2465;
         2.5880     3.0735     4.5340     6.6365     9.1440];
za =     [0.0000    0.0000     0.0000     0.0000     0.0000;
         0.0000    -0.0166    -0.0322    -0.0488    -0.0644;
         0.0000    -0.0302    -0.0615    -0.0917    -0.1229;
         0.0000    -0.0434    -0.0878    -0.1321    -0.1785;
         0.0000    -0.0556    -0.1113    -0.1699    -0.2315;
         0.0000    -0.0660    -0.1340    -0.2060    -0.2830;
         0.0000    -0.0751    -0.1542    -0.2413    -0.3354;
         0.0000    -0.0837    -0.1733    -0.2760    -0.3896;
         0.0000    -0.0904    -0.1919    -0.3103    -0.4467;
         0.0000    -0.0955    -0.2080    -0.3465    -0.5080;
         0.0000    -0.0980    -0.2230    -0.3830    -0.5760;
         0.0000    -0.0986    -0.2372    -0.4239    -0.6525;
         0.0000    -0.0956    -0.2502    -0.4688    -0.7424;
         0.0000    -0.0894    -0.2617    -0.5201    -0.8534;
```

```
            0.0000     -0.0779     -0.2718     -0.5807     -0.9966;
            0.0000     -0.0610     -0.2790     -0.6590     -1.1980];

% Transpose the matrices
ena = ena';
enb = enb';
enc = enc';
za = za';

q05=prctile(sample,5);
q25=prctile(sample,25);
q50=prctile(sample,50);
q75=prctile(sample,75);
q95=prctile(sample,95);

d = q95 - q05;
cn = q75 - q25;
an = d/cn;

if cn == 0
    disp ('Error: Interquantile range is 0')
elseif an < 2.4388
    alpha = 2;
    beta = 0;
    c = cn/1.9078;
    delta = q50;
    zeta = delta;
    return
end

bn = (q05 + q95 - 2 * q50)/d;
sign = 1.0;

if bn < 0
    sign = -1;
elseif bn == 0
    sign = 0;
end
bn = abs(bn);

ah = zeros(5,1);
for j = 1:5
    for i = 2:15
        if an <= ena(j,i)
            break
        end
    end
    t = (an - ena(j,i-1))/(ena(j,i)-ena(j,i-1));
    ah(j) = 2 - (i-2+t)*0.1;
end
```

```
bh = zeros(16,1);
for i = 2:16
    for j = 2:4
        if bn <= enb(j,i);
            break
        end
    end
    t = (bn - enb(j-1,i))/(enb(j,i) - enb(j-1,i));
    bh(i) = (j - 2 + t) * 0.25;
end
bh(1) = 2 * bh(2) - bh(3);

for j = 2:5
    jj = j;
    i = floor((2 - ah(j)) * 10 + 2);
    i = max(i, 2);
    i = min(i, 16);
    aa = 2 - 0.1*(i-2);
    s2 = -(ah(j) - aa) * 10;
    b = (1 - s2) * bh(i-1) + s2 * bh(i);
    if b < (j-1)/4
        break
    end
end

j = jj;
bb = 0.25 * (j-2);
t1 = (bh(i-1) - bb) / .25;
t2 = (bh(i) - bb) / .25;
s1 = - (ah(j-1) - aa) * 10;
dt = t2 - t1;
ds = s2 - s1;
s = (s1 + t1 * ds) / (1 - ds * dt);
t = t1 + s * dt;

alpha = aa - s * 0.1;
alpha = max(0.5, alpha);
alpha = min(2, alpha);

beta = bb + t * .25;
beta = min(beta, 1);
beta = beta * sign;
if alpha == 2
    beta = 0;
end

c = enc(j-1,i-1) * (1-s) * (1-t) + enc(j-1,i) * s * (1-t)...
    + enc(j,i-1) * t * (1-s) + enc(j,i) * t * s;
c = cn/c;
```

```
zeta = za(j-1,i-1) * (1-s) * (1-t) + za(j-1,i) * s * (1-t)...
      + za(j,i-1) * t * (1-s) + za(j,i) * t * s;
zeta = q50 + c * sign * zeta;

if alpha == 1;
  delta = zeta;
else
  delta = zeta - beta * c * tan(pi*alpha / 2);
end
```

```
function [alpha, beta, sigma, delta, mu] = cfestimate(sample)

% function [alpha, beta, sigma, delta, mu] = cfestimate(sample)
%
% A characteristic function based estimator of stable parameters.
% The function output is
%       alpha - the index of stability
%       beta - the skewness parameter
%       sigma - the scale parameter
%       delta - location parameter
%       mu - modified location parameter
%
% The parameterization used is the following:
%
% Eexp(itX) = exp(sigma^alpha(|t|^alpha + i*t*w(t, alpha, ...
    beta)) + i*mu*t)
%
% where if alpha ~= 1
%           w(t, alpha, beta) = beta(|t|^(alpha - 1) - ...
    1)tan(pi*alpha/2)
%       elseif alpha == 1
%           w(t, alpha, beta) = - beta*ln|t|*2/pi
%

% Set default range and number of points
if nargin == 1
    wk = 0.1:0.1:1;
else wk = range_l:(range_u - range_l)/(NofPoints - 1):range_u;
end

wk = wk(:);

% Calculate initial values for delta and zeta
[alpha0, beta0, c0, delta0, zeta0] = mcculloch(sample);

% Normalize and center the sample
sample = (sample - zeta0)/c0;
```

```
% Calculate the sample log c.f.
samplelcf = log(scf(wk, sample));

% Estimate the coeffs of the first regression
x1 = [ones(size(wk)), log(abs(wk))];
b1 = coeff(log(-real(samplelcf)), x1);

alpha = b1(2);

% Verify if the estimates are in the admissible bounds
if alpha > 2
    alpha = 2;
end

c1 = exp(b1(1)/alpha);

% Estimate the coeffs of the second regression
x2 = [wk, auxfun(wk, alpha, c1)];
b2 = coeff(imag(samplelcf), x2);
zeta1 = b2(1);
beta = b2(2);

% Verify if the estimates are in the admissible bounds
if abs(beta) > 1
    beta = sign(beta);
end

% Calculate final values
sigma = c0*c1;
mu = zeta0 + c0*zeta1;
if alpha ~= 1
    delta = mu - beta*sigma*tan(pi*alpha/2);
else delta = mu;
end

%%%%%%%%%%%%%%%%%%%%%%%%%%%%%%%%%%%%%%%%%%%%%%%%%%%%%%%%%%%%%%%%%%%%%%%%%%%%%%
%            DEFINITION OF SOME AUXILIARY FUNCTIONS                         %
%%%%%%%%%%%%%%%%%%%%%%%%%%%%%%%%%%%%%%%%%%%%%%%%%%%%%%%%%%%%%%%%%%%%%%%%%%%%%%

function f = scf(t, sample)

% Calculates the sample characteristic function at t
for k = 1:length(t)
    f(k) = mean(exp(i*t(k)*sample));
end
f = f(:);

function g = auxfun(t, alpha, c)
```

```
% An auxiliary function used in the estimation of beta
if alpha ~= 1
    g = c*t*tan(pi*alpha/2).*(abs(c*t).^(alpha - 1) - 1);
elseif alpha == 1
    g = -2*c*t.*log(abs(c*t))/pi;
else error('Error: Estimated alpha is outside [0, 2]')
end
g = g(:);

function b = coeff(y, x)
% Calculates the regression coefficients
b = inv(x'*x)*x'*y;
```

```
function [alpha, beta, sigma, mu] = stablefitmax(sample)

% function [alpha, beta, sigma, mu] = stablefitmax(sample)
%
% Performs a numerical maximum likelihood fit of the
% parameters of a stable distribution given a sample (vector)
% uses stablepdf_fft.m for density approximation
%

[alphai, betai, sigmai, mui] = cfestimate(sample);

if alphai > 1
    mui = mean(sample);
end

if alphai == 2
    alphai = 1.99;
    warning('Initial Alpha altered additionally');
elseif abs(betai) == 1
    betai = sign(betai)*0.95;
    warning('Initial Beta altered additionally');
end

if alphai < 0.5
    alphai = 0.7;
end

% transform [0.5, 2] -> [-1, 1] where atanh is defined
% alpha is constrained in [0.5, 2]
initial = [atanh((alphai - 1.25)/0.75), atanh(betai), ...
    log(sigmai), mui]';

[result, fval] = fminunc(@(x)likelifun(x, sample), initial, ...
    optimset('Display', 'Off'));
```

```
% apply the inverse transform
alpha = tanh(result(1))*0.75 + 1.25;
beta = tanh(result(2));
sigma = exp(result(3));
mu = result(4) - beta*sigma*tan(pi*alpha/2);

%%%%%%%%%%%%%%%%%%%%%%%%%%%%%%%%%%%%%%%%%%%%%%%%%%%%%%%%%%%%
%            Auxiliary function defined                    %
%%%%%%%%%%%%%%%%%%%%%%%%%%%%%%%%%%%%%%%%%%%%%%%%%%%%%%%%%%%%

function f = likelifun(x, s)

% Computes the negative of the logarithm of the likelihood ...
    function

alpha = tanh(x(1))*0.75 + 1.25;
beta = tanh(x(2));
sigma = exp(x(3));
mu = x(4);

f = -sum(log(stablepdf_fft(s, alpha, beta, sigma, mu)));
```

Chapter 6

Tempered Stable Distributions

The main topics covered in this chapter are:

- the history of the tempered stable distribution and why this distribution has been applied in finance;
- the main properties and formulas of tempered stable laws;
- the evaluation of the characteristic function in a non-trivial case (i.e., for the rapidly decreasing tempered stable law);
- probability density and cumulative distribution function evaluation when only the characteristic function is available in closed form while the density function is not;
- how to generate sample draws from a tempered stable distribution;
- how to estimate the parameters of a tempered stable distribution from a sample;
- how to evaluate well-known risk measures for a tempered stable stock market model.

6.1. Introduction

As noted in earlier chapters, Mandelbrot (1963) presented empirical evidence about the distribution of commodity returns and interest rates. He reported that observed distributions do not follow the normal law but that of a non-normal distribution within the family of the stable law. Mandlebrot's findings were devastating for the financial community because much of finance theory at the time, as well as today, begins with the following: "assume returns are normally distributed". Shortly after the publication of the Mandelbrot paper, Paul Cootner, a prominent financial economist at MIT, expressed his concern regarding the implications by warning that: "Almost without exception, past econometric work is meaningless. Surely, before consigning centuries of work to the ash pile, we should like to have some assurance that all our work is truly useless. If we have permitted ourselves to be fooled for as long as this into believing that the Gaussian assumption is a workable one, is it not possible that the Paretian revolution is similarly illusory?" (Cootner, 1964, p. 337).

Fast forward 50 years and the empirical findings of Mandlebrot continue to hold for a wide-range of financial assets as well as in non-U.S. financial markets (see Rachev *et al.*, 2005, for a review). While still rejecting the normal distribution to describe observed returns for financial assets, more recent findings suggest that the stable distribution is not an ideal fit despite being better than the normal distribution. More specifically, a study by Grabchak and Samorodnitsky (2010) found that the tails of the distribution for asset returns are heavier than the normal distribution but thinner than the stable distribution.

Moreover, chiefly because the stable law does not have finite variance and finite exponential moments, it is difficult to apply the stable law to two important applications in finance: (1) mean–variance portfolio construction, and (2) modeling of option prices. It is clear that if the variance is infinite, a mean–variance portfolio cannot be defined. Then, it is crucial for exponential Lévy option pricing models to have finite exponential moments, otherwise the driving process is not a martingale and an option pricing model cannot be defined.

This is the reason why the study of the tempered stable (TS) law has become of increased interest in mathematical finance.

The TS family was introduced by Boyarchenko and Levendorskii (2000, 2002) from both a univariate and multivariate perspective and further studied on a theoretical level by Rosinski (2007). We recall that non-normal infinitely divisible distributions are usually defined by their Lévy measure, from which it is possible to derive the tail behavior of the distribution as well as its moments. The formal and elegant definition of TS distributions and processes has been proposed in the seminal work of Rosinski (2007) where a completely monotone function is chosen to transform the Lévy measure of a stable distribution (various parametric classes have been discussed by Terdik and Woyczyński, 2006).

The main idea is to modify the Lévy measure of a stable law by tempering (or tilting) both the left and right tails through an exponential function. Subsequently, Bianchi *et al.* (2010b) introduced the class of tempered infinitely divisible (TID) distributions by multiplying the Lévy measure of a stable distribution with a positive definite radial function (see Schoenberg, 1938) instead of with a completely monotone function as in Rosinski (2007). Here the main idea is to modify the Lévy measure of a stable law by tempering (or tilting) both the left and right tails through an exponential quadratic function, so that both the left and right tails approach zero faster than a TS tail. For this reason, Bianchi *et al.* (2010b) defined a parametric example in the TID class and referred to it as a rapidly decreasing TS distribution.

In Figure 6.1 we show the behavior of the probability density function of two similar slightly asymmetric classical tempered stable (CTS) and rapidly decreasing tempered stable (RDTS) laws. While the CTS belongs to the TS class, the RDTS belongs to the TID class. It is clear that the RDTS goes to zero faster. Although the TID class of distributions has the same desirable properties as the TS class, the advantage of the TID is that it may admit exponential moments of any order. More specifically, in some cases, the characteristic function of a TID random variable is extendible to an entire function on $\mathbb{C}$ (i.e., it admits any exponential moment). This last property allows one

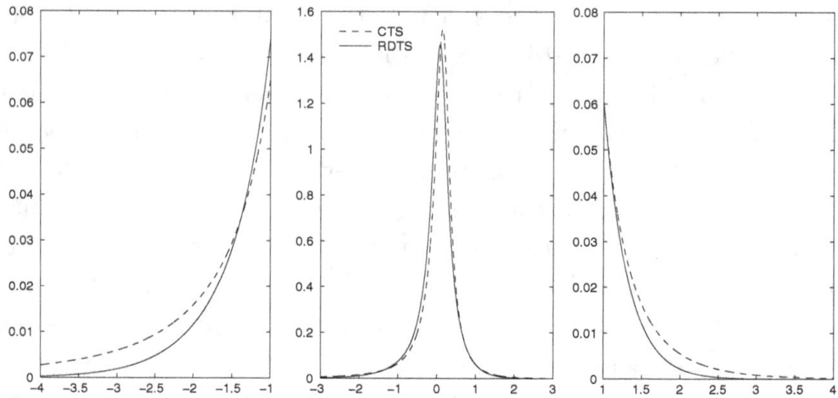

Figure 6.1. Shape of the probability density function of a CTS and an RDTS slightly asymmetric law with comparable parameters with a focus on the left and right tails.

to define a discrete-time option pricing model of the GARCH type with heavy-tailed innovations, as proposed by Kim *et al.* (2010a). Bianchi *et al.* (2017) studied the Ornstein–Uhlenbeck-based model with marginal law given by TS and TID distributions. In particular, they analyzed three different parametric examples belonging to these classes and evaluated the characteristic function of integrated tempered Ornstein-Uhlenbeck (OU) processes with a view toward practical applications.

Recently, two classes of distributions broader than the TS class were proposed: Rosiński and Sinclair (2010) introduced the generalized tempered stable class and Grabchak (2012) proposed the p-tempered stable class. Again the main idea is to modify the Lévy measure of a stable law with a function in order to temper (or tilt) both the left and right tails. Note that not all real functions can be tempering functions because for a measure to be a well-defined Lévy measure it has to satisfy some properties, as briefly shown in Section 6.2. It is important to observe that while if one looks at the infinitely divisible distribution the Lévy measure gives the weights in the tails; if one looks at the corresponding Lévy process, this measure can be viewed as the measure of jumps of the process.

The chapter is organized as follows. In Section 6.2 we review the definition of the characteristic function with a look toward infinitely divisible distributions and some formulas needed to compute moments. Then we define (1) the CTS distribution, (2) the normal tempered stable (NTS) distribution defined as a normal mean–variance mixture, where the mixing distribution is a one-sided tempered stable distribution, and (3) the RDTS distribution. For the RDTS, we describe and provide the Matlab code for computing the characteristic function of this law which involves the evaluation of a Kummer's or confluent hypergeometric function, a particular mathematical function whose evaluation is non-trivial. In Section 6.3 we provide some formulas that can be used in practical applications to compare these three non-normal laws with the normal one. In practice, given the sample mean and standard deviation, we show how to cast these estimates in a non-normal framework. In Section 6.4 we briefly review and analyze the fast Fourier transform approach for evaluating the probability density and the cumulative distribution function given a closed-form expression for the characteristic function. We provide the Matlab code and a numerical assessment of the algorithm. To use the TS model in practical applications, it is necessary to find efficient simulation algorithms to draw random samples from this type of probability law. The simulation of a TS random draw is not a simple task. Based on this observation, we discuss in Section 6.5 a method based on the inversion of the cumulative distribution function, that can be efficiently implemented in Matlab to simulate thousands of random draws. In Section 6.6 we briefly review how to estimate the parameters of a CTS law using the maximum likelihood estimation (MLE) method. The same approach can be used in the NTS and RDTS cases. We discuss a stock market model assuming that the driving law is tempered stable and empirically illustrate how to compute the value-at-risk (VaR) and the average value-at-risk (AVaR).

6.2. Characteristic Functions

In this section, we analyze three different parametric examples belonging to the TS family: the *classical tempered stable* (CTS) distribution, the *normal tempered stable* (NTS) distribution, and the *rapidly decreasing tempered stable* (RDTS) distribution (see Bianchi *et al.* (2010b)). All three examples are defined through their characteristic function, because the density function does not exist for all values of the parameters, but only for some particular cases. For this reason, before analyzing the features and properties of each distribution, we begin this section with some useful definitions.

The characteristic function of a random variable X with density function f_X is defined by its Fourier transform, that is

$$\phi_X(u) = E[\exp(iuX)] = \int_{-\infty}^{\infty} e^{iux} f_X(x)dx. \tag{6.1}$$

Conversely, given the characteristic function, the density function can be obtained using the inverse Fourier transform, that is

$$f_X(x) = \frac{1}{2\pi}\int_{-\infty}^{\infty} e^{-iux}\phi_X(u)du. \tag{6.2}$$

The following relationship holds for any characteristic function

$$\phi_X(-u) = \overline{\phi_X(u)}, \tag{6.3}$$

where $\overline{z}$ means the complex conjugate of z. As we will see in Section 6.4, equality (6.3) is useful in practical applications because it allows one to evaluate the function ϕ_X only on the negative part of the real line and obtain the evaluation on the positive part of the real line simply by conjugating.

The characteristic function of laws we analyze in this chapter has a specific form, because the three TS laws belong to the class of one-dimensional infinitely divisible distribution. The characteristic function of an infinitely divisible law X is given by the Lévy–Khintchine formula, that is

$$\phi_X(u) = \exp\left(i\gamma_c u - \frac{1}{2}\sigma^2u^2 + \int_{-\infty}^{\infty}(e^{iux} - 1 - iuxc(x))\nu(dx)\right), \tag{6.4}$$

where $c(x)$ is usually defined as follows:

$$c(x) = 1_{|x|\leq 1}.$$

In the formula, the measure ν is referred to as the *Lévy measure*. This measure is a Borel measure on the real line $\mathbb{R}$ satisfying the conditions that

$$\nu(0) = 0$$

and

$$\int_{\mathbb{R}} (1 \wedge |x^2|)\nu(dx) < \infty. \tag{6.5}$$

The parameters γ_c and σ are real numbers. The parameter γ_c depends on the function $c(x)$. Under some condition on ν it is possible to select $c(x) = 1$ (finite mean) or $c(x) = 0$ (if $\sigma = 0$, finite variation). The parameter γ_c is referred to as the *center*, if $c(x) = 1$, or *drift*, if $c(x) = 0$ and determines the location. This triplet $(\gamma_c, \sigma^2, \nu)$, referred to as the Lévy triplet, is uniquely defined for each infinitely divisible distribution. While σ and ν are invariant from the choice of the function $c(x)$, γ_c is affected by the choice. In both the CTS and the RDTS cases, we select $c(x) = 1$, and in the NTS case we select $c(x) = 1_{|x|\leq 1}$. If $\sigma = 0$, then the distribution X is referred to as a *purely non-Gaussian distribution*, and in the case when its mean is finite the characteristic function is computed by

$$\phi_X(u) = \exp\left(i\gamma_c u + \int_{-\infty}^{\infty} (e^{iux} - 1 - iux1_{|x|\leq 1})\nu(dx)\right).$$

Given an infinitely divisible law X_1, one can define a process $X = (X_t)_{t\geq 0}$ which we refer to as a Lévy process. While if one looks at the infinitely divisible distribution, the Lévy measure can be viewed as a measure of tails, and if one looks at the corresponding Lévy process, this measure can be viewed as the measure of jumps of the process. A well-known distribution is obtained by assuming $\nu(z) = 0$ for each $z \in \mathbb{R}$, that is

$$\phi_X(u) = \exp\left(i\gamma_c u - \frac{1}{2}\sigma^2 u^2\right), \tag{6.6}$$

and we have only the Gaussian part, no heavy tails in the distribution and no jumps in the process; that is, we have the normal distribution and the corresponding Brownian motion. It can be shown that

Brownian motion is the only Lévy process with continuous but non-differentiable (at any point) trajectories. The proof of this property is not trivial and can be found in Revuz and Yor (2005).

By equation (6.5) it follows that not all real functions q can be tempering functions. To be more precise, given the Lévy measure $\nu_{S_\alpha}(z)$ of an alpha stable law,[1] the function

$$\nu(z) = q(z)\nu_{S_\alpha}(z)$$

is a Lévy measure of a TS distribution only for some functions $q(z)$ for which the inequality in equation (6.5) is satisfied. Additionally, the function $q(z)$ has some further properties in order to be defined as a tempering function.

In the following, we will consider only purely non-Gaussian infinitely divisible distributions. Given a purely non-Gaussian distribution D one can define a pure jump process $X = (X_t)_{t\geq 0}$ with $X_0 = 0$ such that the distribution of the increments $X_{t+1} - X_t$ is D. We say that a stochastic process is of finite variation if the sample paths are of finite variation with probability 1, that is if

$$\int_{-1}^{1} |x|\nu(dx) < \infty.$$

If this is not the case, we say that the process is of infinite variation.

6.2.1. Classical tempered stable distributions

The CTS model is well-known in finance and statistical physics. In the physics literature it appeared with the name of truncated Lévy flight model in the work of Koponen (1995) and later in Novikov (1994), by considering ideas contained in Mantegna and Stanley (1994). It was introduced to model turbulence. In contrast to the stable distribution, a CTS distribution may have finite variance. It has the local behavior of the alpha stable distribution, but tails are tempered. Then, the importance of TS processes comes from the

[1]The stable law has been discussed in Chapter 5.

fact that they combine both the alpha stable and Gaussian trends. In mathematical finance, it was introduced with the name of KoBoL in Boyarchenko and Levendorskii (2000), further studied under the name CGMY in Carr *et al.* (2002), see also Rachev *et al.* (2011) for other applications to finance. We can find the mathematical formalization of TS distributions and processes, also called tempering stable processes, in Rosinski (2007), where the entire class is defined and a view toward simulation is given. In the following, we will refer principally to the simplest parametric example, even if we note that the class introduced by Rosinski (2007) is much wider: Terdik and Woyczyński (2006) and Bianchi *et al.* (2010a) proposed further parametric examples.

As already observed in Section 6.1, the class of TS distributions and processes was proposed in the work of Rosinski (2007); subsequently, Bianchi *et al.* (2010b) introduced the class of tempered infinitely divisible (TID) distributions. Note that while the CTS law described in this section is a parametric example in the class proposed by Rosinski (2007), the RDTS law, described in Section 6.2.3, is a parametric example in the class proposed by Bianchi *et al.* (2010b).

Let α, C, λ_+, and λ_- be positive constants, $m \in \mathbb{R}$, and $\alpha \in (0,2)\backslash\{1\}$. For both CTS and RDTS cases, we do not discuss the case $\alpha = 1$. The law X is said to have a CTS distribution with parameters (α, C, λ_+, λ_-, m) if the characteristic function of X is given by

$$\begin{aligned}\phi_X(u) = \exp(iu(m - C\Gamma(1-\alpha)(\lambda_+^{\alpha-1} - \lambda_-^{\alpha-1}))\\ + C\Gamma(-\alpha)((\lambda_+ - iu)^\alpha - \lambda_+^\alpha + (\lambda_- + iu)^\alpha - \lambda_-^\alpha))\end{aligned} \tag{6.7}$$

and its Lévy measure can be written as

$$\begin{aligned}\nu(z) &= q(z)\nu_{S_\alpha}(z)\\ &= \frac{Ce^{-\lambda_+ z}}{z^{1+\alpha}}1_{z>0} + \frac{Ce^{-\lambda_-|z|}}{|z|^{1+\alpha}}1_{z<0},\end{aligned} \tag{6.8}$$

where $q(z)$ is the tempering function of the CTS law and $\nu_{S_\alpha}(z)$ is the Lévy measure of the stable law, that is,

$$\nu_{S_\alpha}(z) = \frac{C}{z^{1+\alpha}} 1_{z>0} + \frac{C}{|z|^{1+\alpha}} 1_{z<0}. \tag{6.9}$$

Given a CTS law X_1, one can define a Lévy process $X = (X_t)_{t\geq 0}$ which we refer to as a CTS process. By equation (6.7), it follows that the increments $\Delta X = X_{t+\Delta t} - X_t$ are distributed as a CTS law with parameters $(\alpha,\ C\Delta t,\ \lambda_+,\ \lambda_-,\ m\Delta t)$. It can be proven that if $\alpha < 1$, the process has finite variation. By simple calculations, one can write the Laplace exponent of X that is given by

$$\begin{aligned} l_X(u) = u(m - C\Gamma(1-\alpha)(\lambda_+^{\alpha-1} - \lambda_-^{\alpha-1})) \\ + C\Gamma(-\alpha)((\lambda_+ - u)^\alpha - \lambda_+^\alpha + (\lambda_- + u)^\alpha - \lambda_-^\alpha)), \\ -\lambda_- < u < \lambda_+. \end{aligned} \tag{6.10}$$

In Figure 6.2, we show the probability density function of a CTS law for different values of the parameters: (a) $\alpha \in (0.25, 0.75, 1, 25, 1.75)$, $C = 1$, and $\lambda_+ = \lambda_- = 1$; (b) $\alpha = 1.5$,

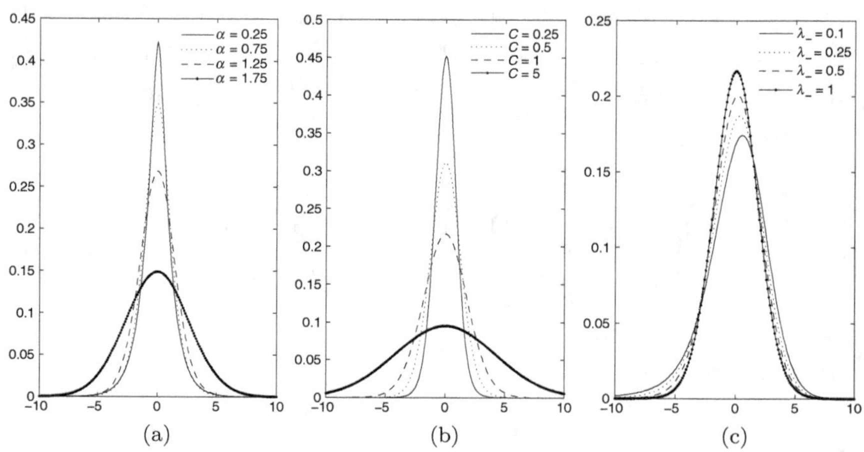

Figure 6.2. Shape of the probability density function of a CTS law for different values of the parameters: (a) $\alpha \in (0.25, 0.75, 1, 25, 1.75)$, $C = 1$, and $\lambda_+ = \lambda_- = 1$; (b) $\alpha = 1.5$, $C \in (0.25, 0.5, 1, 5)$, and $\lambda_+ = \lambda_- = 1$; (c) $\alpha = 1.5$, $C = 1$, $\lambda_+ = 1$, and $\lambda_- \in (0.1, 0.25, 0.5, 1)$.

$C \in (0.25, 0.5, 1, 5)$, and $\lambda_+ = \lambda_- = 1$; (c) $\alpha = 1.5$, $C = 1$, $\lambda_+ = 1$, and $\lambda_- \in (0.1, 0.25, 0.5, 1)$. Note that α is the main tail parameter, and λ_+ and λ_- control the rate of decay on the positive and negative tails. Given the Lévy measure in equation (6.8), we obtain asymmetric CTS laws only if λ_+ and λ_- are different. If $\lambda_+ > \lambda_-$ ($\lambda_+ < \lambda_-$), then the distribution is skewed to the left (right), and if $\lambda_+ = \lambda_-$, then it is symmetric. The parameter C is mainly a scale parameter, and m is a location parameter.

Using the derivatives of the Laplace exponent evaluated at 0, we get the first four cumulants and from equation (2.33) the mean, variance, skewness, and kurtosis are given as follows:

$$
\begin{aligned}
E[X] &= c_1(X) = m, \\
\text{var}[X] &= c_2(X) = C\Gamma(2-\alpha)(\lambda_+^{\alpha-2} + \lambda_-^{\alpha-2}), \\
\text{skew}[X] &= \frac{c_3(X)}{c_2(X)^{3/2}} = \frac{\Gamma(3-\alpha)C\left(\lambda_+^{\alpha-3} - \lambda_-^{\alpha-3}\right)}{\left(C\Gamma(2-\alpha)(\lambda_+^{\alpha-2} + \lambda_-^{\alpha-2})\right)^{3/2}}, \\
\text{kurt}[X] &= 3 + \frac{c_4(X)}{c_2(X)^2} = 3 + \frac{C\Gamma(4-\alpha)\left(\lambda_+^{\alpha-4} + \lambda_-^{\alpha-4}\right)}{\left(C\Gamma(2-\alpha)(\lambda_+^{\alpha-2} + \lambda_-^{\alpha-2})\right)^2}.
\end{aligned}
$$

Matlab code — CTS characteristic function evaluation

In this section, we show the Matlab code needed to evaluate the characteristic function of a CTS random variable. In this case the implementation is simple. We point out that u can be a vector. Since we are considering the distribution of the increments of a stochastic CTS process $X = (X_t)_{t\geq 0}$ we consider also a possible time scale dt, in order to work with the increments $\Delta X = X_{t+\Delta t} - X_t$. Recall that in the CTS case a time scale does not modify the characteristics of the driving distribution. This means in practice that if one selects $dt = 1/250$ and casts the code into an estimation algorithm having as input daily log-returns (e.g., the maximum likelihood estimation method we describe in Section 6.6), then the parameters of the distribution are annualized. The same idea is also implemented in the code in Sections 6.2.2 and 6.2.3.

```
% implementation of the characteristic function of
% a CTS random variable

function phi = chf_CTS(u, param, dt)

C = param(1);
lm = param(2);
lp = param(3);
alpha = param(4);
m = param(5);

phi = exp( dt*( ...
          i*u*(m - C*gamma(1-alpha)*(lp^(alpha - 1) - ...
              lm^(alpha - 1))) ...
      + C*gamma(-alpha)*( (lp - i*u).^alpha - lp^alpha ...
              + (lm + i*u).^alpha   - lm^alpha ) ) );
return;
```

6.2.2. Normal tempered stable distributions

The NTS law can be defined as a normal mean–variance mixture where the mixing distribution is a one-sided CTS distribution (see Rachev *et al.* (2011) and Kim *et al.* (2012)) or alternatively it can be seen as the law of the increments of a Brownian motion with tempered stable stochastic time (see Bianchi *et al.* (2016)).

Let α, λ, and C be positive constants, and $0 < \alpha < 2$. The law S is said to be a one-sided CTS law with parameters $(\alpha,\ \lambda,\ C)$ if $0 < \alpha < 1$ and the characteristic function of S is given by

$$\phi_S(u) = E[\exp(iuS)] = \exp(C\Gamma(-\alpha)((\lambda - iu)^\alpha - \lambda^\alpha)), \qquad (6.11)$$

where $u \in \mathbb{R}$ and $\Gamma(\cdot)$ is the gamma function. From equation (6.11) it is possible to compute the Laplace exponent of the CTS subordinator

$$l_S(u) = C\Gamma(-\alpha)((\lambda - u)^\alpha - \lambda^\alpha), \quad u < \lambda \qquad (6.12)$$

and its moments

$$\begin{aligned} E[S] &= -\alpha C\Gamma(-\alpha)\lambda^{\alpha-1}, \\ \mathrm{var}[S] &= \alpha(\alpha-1)C\Gamma(-\alpha)\lambda^{\alpha-2}, \\ \mathrm{skew}[S] &= (2-\alpha)\left[\alpha(\alpha-1)C\Gamma(-\alpha)\lambda^\alpha\right]^{-\frac{1}{2}}, \\ \mathrm{kurt}[S] &= 3 + (\alpha-2)(\alpha-3)\left[\alpha(\alpha-1)C\Gamma(-\alpha)\lambda^\alpha\right]^{-1}. \end{aligned} \qquad (6.13)$$

Then we define the normal mean–variance mixture X, where the mixing distribution is a one-sided tempered stable distribution, as

$$X = \mu + \theta S + \sqrt{S}\sigma Z, \tag{6.14}$$

where

- Z is a normal distribution with zero mean and unit variance;
- S is the mixing distribution with Laplace exponent $l_S(u)$ defined in equation (6.11) with $\alpha = a/2$ and S is independent of Z;
- μ and θ are constants in $\mathbb{R}$.

Then X is referred to as the normal tempered stable (NTS) law with parameters (a, λ, C, θ, μ, σ). It is possible to prove that the NTS law X has characteristic function given by

$$\phi_X(u) = \exp\left\{iu\mu + C\Gamma\left(-\frac{a}{2}\right)\left(\left(\lambda - iu\theta + \frac{1}{2}u^2\sigma^2\right)^{\frac{a}{2}} - \lambda^{\frac{a}{2}}\right)\right\} \tag{6.15}$$

and compute its Laplace exponent

$$l_X(u) = u\mu + C\Gamma\left(-\frac{a}{2}\right)\left(\left(\lambda - u\theta - \frac{1}{2}u^2\sigma^2\right)^{\frac{a}{2}} - \lambda^{\frac{a}{2}}\right). \tag{6.16}$$

We point out that if one selects $\alpha = a/2$ and $C = \lambda^{1-\alpha}/\Gamma(1-\alpha)$, equation (6.11) can be written as

$$\phi_S(u) = \exp\left(-\frac{2\lambda^{1-\frac{a}{2}}}{a}\left((\lambda - iu)^{\frac{a}{2}} - \lambda^{\frac{a}{2}}\right)\right) \tag{6.17}$$

and the distribution Y defined as

$$Y = \mu' + \theta(S-1) + \sqrt{S}\sigma Z,$$

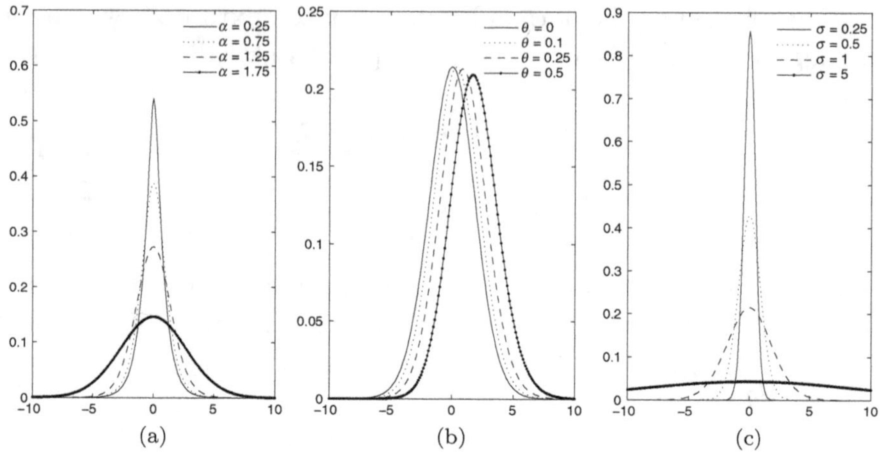

Figure 6.3. Shape of the probability density function of an NTS law for different values of the parameters: (a) $a \in (0.25, 0.75, 1, 25, 1.75)$, $C = 1$, $\lambda = 1$, $\mu = 0$, $\theta = 0$, and $\sigma = 1$; (b) $a = 1.5$, $C = 1$, $\lambda = 1$, $\mu = 0$, $\theta \in (0, 0.1, 0.25, 0.5)$, and $\sigma = 1$; (c) $a = 1.5$, $C = 1$, $\lambda = 1$, $\mu = 0$, $\theta = 0$, and $\sigma \in (0.25, 0.5, 1, 5)$.

is the NTS distribution with characteristic function

$$\phi_Y(u) = \exp\left\{iu(\mu - \theta) - \frac{2\lambda^{1-\frac{a}{2}}}{a}\left(\left(\lambda - iu\theta + \frac{1}{2}u^2\sigma^2\right)^{\frac{a}{2}} - \lambda^{\frac{a}{2}}\right)\right\} \tag{6.18}$$

analyzed by Kim *et al.* (2012).

Given an NTS law X_1, one can define a Lévy process $X = (X_t)_{t\geq 0}$ which we refer to as an NTS process. By equation (6.15), it follows that the increments $\Delta X = X_{t+\Delta t} - X_t$ are distributed as an NTS law with parameters $(a, \lambda, C\Delta t, \theta, \mu\Delta t, \sigma)$. In Figure 6.3, we show the probability density function of an NTS law for different values of the parameters: (a) $a \in (0.25, 0.75, 1, 25, 1.75)$, $C = 1$, $\lambda = 1$, $\mu = 0$, $\theta = 0$, and $\sigma = 1$; (b) $a = 1.5$, $C = 1$, $\lambda = 1$, $\mu = 0$, $\theta \in (0, 0.1, 0.25, 0.5)$, and $\sigma = 1$; (c) $a = 1.5$, $C = 1$, $\lambda = 1$, $\mu = 0$, $\theta = 0$, and $\sigma \in (0.25, 0.5, 1, 5)$. The parameter a controls the tail behavior of the distribution and the parameter θ is related to the distribution's skewness. If $\theta < 0$ ($\theta > 0$), the distribution is skewed to the left (right). Moreover, if $\theta = 0$, then it is symmetric. Lastly, C and σ are scale parameters, and μ is a location parameter.

From (6.16) it is possible to compute moments of X and express them as functions of the moments of the mixing variable S

$$\begin{aligned}
E[X] &= \mu + E\left[S\right]\theta, \\
\mathrm{var}[X] &= \mathrm{var}[S]\left(\theta^2 + \frac{\sigma^2\lambda}{1-\alpha}\right), \\
\mathrm{skew}[X] &= \mathrm{skew}[S]\left(\theta^3 + \frac{3\theta\sigma^2\lambda}{2-\alpha}\right)\left(\theta^2 + \frac{\sigma^2\lambda}{1-\alpha}\right)^{-\frac{3}{2}}, \\
\mathrm{kurt}[X] &= 3 + (\mathrm{kurt}[S]-3)\left[\theta^4 + \frac{3\sigma^2\lambda}{3-\alpha}\left(2\theta^2 + \frac{\sigma^2\lambda}{2-\alpha}\right)\right] \\
&\quad\times\left(\theta^2 + \frac{\sigma^2\lambda}{1-\alpha}\right)^{-2}.
\end{aligned} \tag{6.19}$$

Matlab code — NTS characteristic function evaluation

In this section, we show the Matlab code needed to evaluate the characteristic function of a NTS random variable. Also in this case the implementation is simple. We point out that u can be a vector.

```
% implementation of the characteristic function of
% a NTS random variable

function phi = chf_NTS(u, param, dt)

a = param(1);
lambda = param(2);
C  = param(3);
sigma = param(4);
theta  = param(5);
mu = param(6);

g = 1i*theta*u - sigma^2*u.^2/2;
phi = exp( dt*( 1i*mu.*u + C*gamma(-a/2) ...
            *((lambda - g).^(a/2) - lambda^(a/2) ) ) );

return;
```

6.2.3. Rapidly decreasing tempered stable distributions

The TID class of distributions has the same desirable properties as the TS class, but with the advantage that it may admit exponential moments of any order. More precisely in some cases, the characteristic function of a TID random variable is extendible to an entire function on $\mathbb{C}$; that is, it admits any exponential moment. As observed in Bianchi *et al.* (2010b), some practical problems in the field of mathematical finance have motivated the study of TID. Specifically, Bianchi *et al.* (2010b) sought a discrete-time option pricing model that accounts for well-known stylized facts observed about real-world stock market returns: asymmetry, time-varying variance, and heavy tails. As shown in Kim *et al.* (2010a), since an RDTS random variable admits any exponential moment, it is possible to define an RDTS-GARCH option pricing model without artificially restricting the variance process as in the CTS case. Furthermore, Kim *et al.* (2010a) and Bianchi (2015) empirically studied the shape of the RDTS distribution, demonstrating that the RDTS distribution is much closer to the CTS distribution than to the normal distribution, even if it has exponential moments of any order.

Let α, C, λ_+, λ_- positive constants, $m \in \mathbb{R}$ and $\alpha \in (0,2)\backslash\{1\}$. The law X is said to have an RDTS distribution if the characteristic function of X is given by

$$\phi_X(u) = \exp(ium + CG(iu; \alpha, \lambda_+) + CG(-iu; \alpha, \lambda_-)) \quad (6.20)$$

with

$$\begin{aligned} G(x;\alpha,\lambda) = 2^{-\alpha/2-1}\lambda^{\alpha} &\left(\Gamma\left(-\frac{\alpha}{2}\right) M\left(-\frac{\alpha}{2}, \frac{1}{2}; \left(\frac{\sqrt{2}x}{2\lambda}\right)^2\right)\right. \\ &+ \frac{\sqrt{2}x}{\lambda}\Gamma\left(\frac{1-\alpha}{2}\right) M\left(\frac{1}{2}-\frac{\alpha}{2}, \frac{3}{2}; \left(\frac{\sqrt{2}x}{2\lambda}\right)^2\right) \\ &\left. - \frac{\sqrt{2}x}{\lambda}\Gamma\left(\frac{1}{2}-\frac{\alpha}{2}\right) - \Gamma\left(-\frac{\alpha}{2}\right)\right), \end{aligned} \quad (6.21)$$

where $M(a, b; z)$ is the Kummer's or confluent hypergeometric function of the first kind as defined in equation (13.1.2) in Abramowitz and Stegun (1974) (see also Tricomi (1954)) that is

$$M(a, b; z) = 1 + \frac{az}{b} + \frac{(a)_2 z^2}{(b)_2 2!} + \cdots + \frac{(a)_n z^n}{(b)_n n!}, \tag{6.22}$$

where

$$(m)_n = m(m+1)(m+2)\dots(m+n-1),$$

with $m \in \mathbb{R}$.

For a more simple derivatives computation, by the definition of the function M, the function G can be written as

$$G(x; \alpha, \lambda) = \frac{1}{2}\sum_{n=2}^{\infty} \frac{x^n}{n!} \left(\frac{\lambda}{\sqrt{2}}\right)^{\alpha-n} \Gamma\left(\frac{n-\alpha}{2}\right). \tag{6.23}$$

The Lévy measure of an RDTS law can be written as

$$\begin{aligned} \nu(z) &= q(z)\nu_{S_\alpha}(z) \\ &= \frac{Ce^{-\lambda_+^2 z^2/2}}{z^{1+\alpha}} 1_{z>0} + \frac{Ce^{-\lambda_-^2 z^2/2}}{|z|^{1+\alpha}} 1_{z<0}, \end{aligned} \tag{6.24}$$

where $q(z)$ is the tempering function of the RDTS law and $\nu_{S_\alpha}(z)$ is the Lévy measure of the stable law described in equation (6.9). Given an RDTS law X_1, one can define a Lévy process $X = (X_t)_{t\geq 0}$ which we refer to as an RDTS process. By equation (6.20), it follows that the increments $\Delta X = X_{t+\Delta t} - X_t$ are distributed as an RDTS law with parameters (α, $C\Delta t$, λ_+, λ_-, $m\Delta t$). If $\alpha < 1$, it can be demonstrated that the process has finite variation. By simple calculations, one can write the Laplace exponent of X as

$$l_X(u) = um + CG(u; \alpha, \lambda_+) + CG(-u; \alpha, \lambda_-). \tag{6.25}$$

Note that the first derivative of the function G is given by

$$G'(x; \alpha, \lambda) = \frac{1}{2}\sum_{n=2}^{\infty} \frac{x^{n-1}}{n!n} \left(\frac{\lambda}{\sqrt{2}}\right)^{\alpha-n} \Gamma\left(\frac{n-\alpha}{2}\right). \tag{6.26}$$

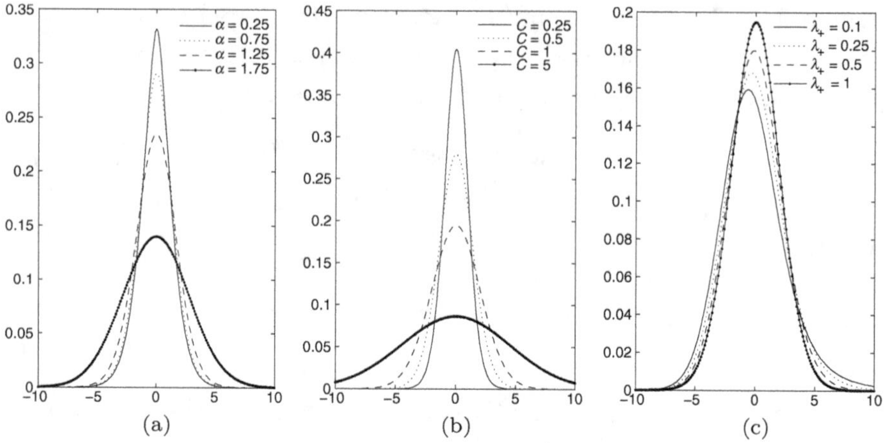

Figure 6.4. Shape of the probability density function of an RDTS law for different values of the parameters: (a) $\alpha \in (0.25, 0.75, 1, 25, 1.75)$, $C = 1$, and $\lambda_+ = \lambda_- = 1$; (b) $\alpha = 1.5$, $C \in (0.25, 0.5, 1, 5)$, and $\lambda_+ = \lambda_- = 1$; (c) $\alpha = 1.5$, $C = 1$, $\lambda_- = 1$, and $\lambda_+ \in (0.1, 0.25, 0.5, 1)$.

In Figure 6.4, we show the probability density function of an RDTS law for different values of the parameters: (a) $\alpha \in (0.25, 0.75, 1, 25, 1.75)$, $C = 1$, and $\lambda_+ = \lambda_- = 1$; (b) $\alpha = 1.5$, $C \in (0.25, 0.5, 1, 5)$, and $\lambda_+ = \lambda_- = 1$; (c) $\alpha = 1.5$, $C = 1$, $\lambda_- = 1$, and $\lambda_+ \in (1, 2, 5, 10)$. As in the CTS case, α is the main tail parameter, and λ_+ and λ_- control the rate of decay on the positive and negative tails. Given the Lévy measure in equation (6.8), we obtain asymmetric RDTS laws only if λ_+ and λ_- are different. If $\lambda_+ > \lambda_-$ ($\lambda_+ < \lambda_-$), then the distribution is skewed to the left (right), and if $\lambda_+ = \lambda_-$, then it is symmetric. The parameter C is mainly a scale parameter.

Using the derivatives of the Laplace exponent evaluated at 0 we get the first four cumulants and from equation (2.33) the mean, variance, skewness, and kurtosis are given as follows:

$$E[X] = c_1(X) = m,$$

$$\text{var}[X] = c_2(X) = 2^{-\frac{\alpha}{2}} C\Gamma\left(1 - \frac{\alpha}{2}\right)\left(\lambda_+^{\alpha-2} + \lambda_-^{\alpha-2}\right),$$

$$\text{skew}[X] = \frac{c_3(X)}{c_2(X)^{3/2}} = 2^{\frac{1}{2}+\frac{\alpha}{4}} \frac{\Gamma(\frac{3-\alpha}{2})C\left(\lambda_+^{\alpha-3} - \lambda_-^{\alpha-3}\right)}{\left(C\Gamma(1-\frac{\alpha}{2})(\lambda_+^{\alpha-2} + \lambda_-^{\alpha-2})\right)^{3/2}},$$

$$\text{kurt}[X] = 3 + \frac{c_4(X)}{c_2(X)^2} = 3 + 2^{\frac{\alpha}{2}+1} \frac{C\Gamma(\frac{4-\alpha}{2})\left(\lambda_+^{\alpha-4} + \lambda_-^{\alpha-4}\right)}{\left(C\Gamma(1-\frac{\alpha}{2})(\lambda_+^{\alpha-2} + \lambda_-^{\alpha-2})\right)^2}.$$

While the calculus of the characteristic function in both the CTS and the NTS cases involves only elementary functions, more interesting is the evaluation of the characteristic function in the RDTS case that is connected with the Kummer's or confluent hypergeometric function. Equation (6.20) implies the evaluation of the $G(x;\alpha,\lambda)$ function only on the straight line represented by the subset $I = \{iy \mid y \in \mathbb{R}\}$ of the complex plane $\mathbb{C}$. We do not need a general algorithm to evaluate the function on the entire complex plane $\mathbb{C}$, but just on a subset of it. An efficient algorithm to compute the characteristic function in equation (6.20) can be constructed (see Gil *et al.* (2007), and Bianchi *et al.* (2010a)) by means of the analytic continuation, without having recourse either to numerical integration or to numerical solution of a differential equation. The Kummer's or confluent hypergeometric functions belongs to the special functions class and often occurs in many practical computational problems. It is defined by the power series (6.23), and the following equalities hold:

$$\begin{aligned} M(a,b;z) &= e^z M(b-a,a;-z), \\ M(a,b;z) &= \frac{\Gamma(b)}{\Gamma(b-a)}(-z)^{-a}\sum_{n=0}^{\infty}(-1)^n\frac{(a)_n(a-b+1)_n}{zn!} \\ &\quad + \cos\pi(b-a)\frac{\Gamma(b)}{\Gamma(a)}\exp(z)z^{a-b}\sum_{n=0}^{\infty}\frac{(b-a)_n(1-a)_n}{zn!}. \end{aligned} \tag{6.27}$$

By the equality (6.3), one can determine the values of $G(x;\alpha,\lambda)$ only for the subset $I_- = \{iy \mid y \in \mathbb{R}_-\}$ and then simply consider the conjugate for the set $I_+ = \{iy \mid y \in \mathbb{R}_+\}$. Second, in order to obtain a fast convergence of the series (6.23), we split the negative imaginary line into two subsets without intersection,

$$\begin{aligned} I_-^1 &= \{iy \mid -30 < y \leq 0\}, \\ I_-^2 &= \{iy \mid y < -30\}, \end{aligned}$$

and we use the first equality in (6.27) and (6.22) to evaluate $M(a, b; z)$ in I_-^1. Then, the second equality of (6.27) is enough to evaluate $M(a, b; z)$ in I_-^2. This subdivision allows one to truncate the series (6.22) to the integer $N = 1000$ and obtain the same results as other algorithms to evaluate special functions. We point out that the value of y ranges in the interval $[-M, M]$. This method, together with the Matlab vector calculus, increase considerably the speed with respect to algorithms based on the numerical solution of the differential equation. Our method is grounded only on basic summations and multiplication. As a result the computational effort in the RDTS density evaluation is satisfactory and it can be applied in practice.

Matlab code — G function evaluation

In this section, we show the Matlab code needed to evaluate the function G in equation (6.21). In this case the implementation is not trivial. The function M is computed by analytic continuation and only for the subset I_-. More in detail the code is divided in two parts: in the first part M is evaluated on I_-^1, and in the second part in I_-^2.

```
% implementation of the function G

function y = G(x,alpha,lambda)

        z = x*sqrt(2)/lambda;
        w = (z/2).^2;

        t = gamma(-alpha/2)*Mneg(-alpha/2,1/2,w) ...
            + z.*gamma((1-alpha)/2).*Mneg((1-alpha)/2,3/2,w);
        y = ...
            2^(-1-alpha/2)*lambda^alpha*(t-z*gamma(1/2-alpha/2) ...
            - gamma(-alpha/2));
```

```
return

function y = Mneg(a,b,z)

% this function computes the value of M for a real vector z, ...
    where z has to be a negative row vector of real numbers

n = 1000;
power = 1:n;

z1 = z(z>=-30);
z2 = z(z<-30);

y1=[];

    if numel(z1)>0
    % for small z we use the series approximation
    c = b - a;
    an_over_bn = (c:1:(c+n-1))./(b:1:(b+n-1));
    num1 = cumprod(an_over_bn);

    matrix = (-z1)'*(1./power);
    zn = cumprod(matrix,2);

    zn = zn*(num1)';

    y = 1 + sum(zn,2);

    y1 = exp(z1).*y';
    end

y2=[];

    if numel(z2)>0
    % for large z we use the asymptotic formula
    a = b-a;
    z2 = -z2;
    k = 10;
    power=1:k;
    an_times_factorial = ((b-a):1:(b-a+k-1))./power;

    num1 = cumprod(an_times_factorial);

    matrix = (1./z2)'*((1-a):1:(k-a));
    zn = cumprod(matrix,2);

    zn = zn*(num1)';

    sum1 = 1 + sum(zn,2);
```

```
    an_times_factorial = (a:1:(a+k-1))./power;
    num1 = cumprod(an_times_factorial);

    matrix = (-1./z2)'*((1+a-b):1:(k+a-b));
    zn = cumprod(matrix,2);

    zn = zn*(num1)';

    sum2 = 1 + sum(zn,2);

    y2 = gamma(b)/gamma(a)*z2.^(a-b).*sum1' ...
    + cos(pi*a)*gamma(b)/gamma(b-a)*exp(-z2).*z2.^(-a).*sum2';
    end

% return a row vector
y=[y2 y1];

return;
```

Matlab code — RDTS chf evaluation

In this section, we show the Matlab code needed to evaluate the characteristic function of a RDTS random variable. We point out that u can be a vector and that the function G is evaluated only on I_- and then equation (6.3) is considered.

```
% implementation of the characteristic function of
% a RDTS random variable

function y = chf_RDTS(u, param, dt)

%C = param(1);
%lp = param(2);
%lm = param(3);
%alpha = param(4);
%m = param(5);

y = exp(dt*chExp_RDTS(u,param));

return;

function y = chExp_RDTS(u,param)

% characteristic exponent of a RDTS distribution
% The code cannot be for any complex number u
```

```
C = param(1);
lp = param(2);
lm = param(3);
alpha = param(4);
m = param(5);

    if (0.0>=alpha || alpha>=2.0)
        printf( 'alpha must be in the open interval (0,2)\n' )
    end;


    if ( alpha == 1.0 )
       printf( 'alpha cannot be 1\n')
    end;

% we want consider only negative values of u
% this part is needed to evaluate the pdf with the fft algorithm

um = u(u<=0);

ym = C*G(1i*um, alpha, lp) + conj(C*G(1i*um, alpha, lm)) + ...
    1i*um*m ;

l = length(ym);
yp = conj(ym);
yp = yp(l:-1:1);
ym = ym(2:(l-1));
y = [ym yp];

return;
```

6.3. Characteristic Functions Given Sample Mean and Standard Deviation

In this section, we provide the characteristic function of a random variable X defined as

$$X = \mu + \sigma Z,$$

where μ and σ are parameters, and Z is a random variable with zero mean and unit variance. We consider different distributions for the standardized variable Z: Normal, CTS, NTS, and RDTS. Thus, the random variable X has the same expectation μ and the same variance

σ^2 independently by the distribution of Z. Only moments of order greater than two may differ among the four distributions considered. In practice, given an estimate of the mean and of the standard deviation obtained from a time series of log-returns, we show how to cast these values in a non-normal framework. Therefore, in this section, the parameter μ represents the mean and σ represents the standard deviation and not the parameters of the NTS distribution defined in Section 6.2.2.

Given the characteristic function in equation (6.6), a law X is said to have a normal distribution with parameters μ and σ, if the characteristic function of X is given by

$$\phi_X(u) = E[\exp(iuX)] = \exp\left(i\mu - \frac{1}{2}\sigma^2 u^2\right), \tag{6.28}$$

where i is the imaginary unit.

Let α, λ_+, and λ_- be positive constants, and $\alpha \in (0,2)\backslash\{1\}$ (in all TS cases, we do not discuss the case $\alpha = 1$). Law X is said to have a CTS distribution with parameters (α, λ_+, λ_-, μ, σ) if the characteristic function of X is given by

$$\begin{aligned}\phi_X(u) = \exp\left(iu(\mu - C\Gamma(1-\alpha)({\lambda'_+}^{\alpha-1} - {\lambda'_-}^{\alpha-1}))\right.\\ \left. + C\Gamma(-\alpha)((\lambda'_+ - iu)^\alpha - {\lambda'_+}^{\alpha} + (\lambda'_- + iu)^\alpha - {\lambda'_-}^{\alpha})\right)\end{aligned} \tag{6.29}$$

with

$$C = \sigma^2\left(\Gamma(2-\alpha)({\lambda'_+}^{\alpha-2} + {\lambda'_+}^{\alpha-2})\right)^{-1} \quad \text{and} \quad \lambda'_\pm = \lambda_\pm/\sigma.$$

The parameters α, λ_+, and λ_- are related to tail weights: α is the main tail parameter, and λ_+ and λ_- control the rate of decay on the positive and negative tails. Additionally, λ_+ and λ_- are also skewness parameters. If $\lambda_+ > \lambda_-$ ($\lambda_+ < \lambda_-$), then the distribution is skewed to the left (right), and if $\lambda_+ = \lambda_-$, then it is symmetric. Then, μ and σ are location and scale parameter, respectively.

Let a and λ be positive constants, $\theta \in \mathbb{R}$ and $a \in (0,2)\backslash\{1\}$. Law X is said to have an NTS distribution with parameters $(a, \lambda, \theta, \mu, \sigma)$

if the characteristic function of X is given by

$$\phi_X(u) = \exp\left(iu(\mu - \theta') - \frac{2\lambda^{1-a/2}}{a} \times \left(\left(\lambda - iu\theta' + \frac{\gamma^2 u^2}{2}\right)^{a/2} - \lambda^{a/2}\right)\right) \quad (6.30)$$

with

$$\gamma = \left(\sigma^2 - \theta'^2 \frac{2-a}{2\lambda}\right)^{1/2} \quad \text{and} \quad \theta' = \theta\sigma.$$

The parameter a controls the tail behavior of the distribution and the parameter θ is related to the distribution's skewness. If $\theta < 0$ ($\theta > 0$), then the distribution is skewed to the left (right). Moreover, if $\theta = 0$, then it is symmetric. Then, λ and σ are scale parameters, and μ is a location parameter.

Let α, λ_+, and λ_- be positive constants, and $\alpha \in (0, 2)\backslash\{1\}$. Law X is said to have an RDTS distribution with parameters (α, λ_+, λ_-, μ, σ) if the characteristic function of X is given by

$$\phi_X(u) = \exp(iu\mu + CG(iu; \alpha, \lambda'_+) + CG(-iu; \alpha, \lambda'_-) \quad (6.31)$$

with the $G(x; \alpha, \lambda)$ function define in equation (6.21) and

$$C = 2^{\alpha/2}\sigma^2(\Gamma(1 - \alpha/2)(\lambda'_+{}^{\alpha-2} + \lambda'_+{}^{\alpha-2}))^{-1}, \qquad \lambda'_\pm = \lambda_\pm/\sigma.$$

The role of the parameters is the same as for the CTS distribution. The tails of an RDTS distribution are lighter than those of a CTS distribution.

6.4. The Fast Fourier Transform Approach

In theory, as discussed in Chapter 5, given the characteristic function $\phi_X(u)$ of a law X, density and cumulative distribution functions can be derived via the Fourier inversion formula, that is

$$f_X(x) = \frac{1}{2\pi}\int_{-\infty}^{\infty} e^{-iux}\phi_X(u)du \tag{6.32}$$

or

$$F_X\left(x+\frac{h}{2}\right) - F_X\left(x-\frac{h}{2}\right) = \frac{1}{2\pi}\int_{-\infty}^{\infty} e^{-iux}\phi_X(u)\frac{2}{u}\sin\frac{hu}{2}du, \tag{6.33}$$

where $h > 0$. Unfortunately, it is not always possible to find a closed-form solution for equations (6.32) and (6.33), and one has to then employ numerical integration of the inversion formulas. The algorithm described here draws from Rachev and Mittnik (2000) and Stoyanov and Racheva-Iotova (2004) (see Scherer *et al.* (2012) for further details on this topic).

To compute the density function for a large number of x values, the fast Fourier transform (FFT) algorithm can be efficiently employed, particularly when the likelihood function has to be computed and inserted into an optimization routine to perform an MLE. It is well known that the FFT approach is computationally efficient with only the disadvantage that the density function has to be evaluated on an equally spaced grid. As a consequence, one has to interpolate for intermediate values and when the arguments are outside the grid (this occurs rarely in practical applications).

The main idea is to calculate the integral in (6.32) for the following grid of equally spaced x values

$$x_k = \left(k - 1 - \frac{N}{2}\right)h,$$

where $k = 1, \ldots, N$. Then equation (6.32) can be rewritten as

$$f_X(x_k) = \int_{-\infty}^{\infty} e^{-i2\pi\omega(k-1-\frac{N}{2})h}\phi_X(2\pi\omega)d\omega. \tag{6.34}$$

Since the integral in (6.34) is convergent, we can choose a large enough upper and a small enough lower bound to approximately compute (6.34) through the Riemann sum; that is,

$$f_X(x_k) \approx s \sum_{n=1}^{N} \phi_X \left(2\pi s \left(n - 1 - \frac{N}{2}\right)\right) e^{-i2\pi(k-1-\frac{N}{2})(n-1-\frac{N}{2})sh}. \tag{6.35}$$

In this particular example, the upper and the lower bounds equal $\frac{sN}{2}$ and $-\frac{sN}{2}$, respectively. The integrand is evaluated for the equally spaced grid $n - 1 - \frac{N}{2}$, $n = 1, \ldots, N$ with distance s between them. The choice of the integral bounds is not arbitrary. If $s = (hN)^{-1}$, we have the following expression for the density:

$$f_X(x_k) \approx \frac{1}{hN} \sum_{n=1}^{N} \phi_X \left(2\pi \frac{1}{hN} \left(n - 1 - \frac{N}{2}\right)\right) e^{-i2\pi(k-1-\frac{N}{2})(n-1-\frac{N}{2})\frac{1}{N}}.$$

Having rearranged the terms in the exponent, the following approximation holds:

$$f_X(x_k) \approx \frac{(-1)^{k-1+\frac{N}{2}}}{hN} \sum_{n=1}^{N} (-1)^{n-1} \phi_X \left(\frac{2\pi}{hN} \left(n - 1 - \frac{N}{2}\right)\right) e^{\frac{-i2\pi(n-1)(k-1)}{N}}. \tag{6.36}$$

The discrete FFT is a numerical method developed for calculation of sequences such as $f_X(x_k)$ in (6.36) given the sequence

$$(-1)^{n-1} \phi_X \left(\frac{2\pi}{hN} \left(n - 1 - \frac{N}{2}\right)\right). \tag{6.37}$$

So by applying the discrete Fourier transform for the sequence (6.37), we approximately compute the density values $f_X(x_k)$. The benefit of this approach is that the FFT algorithm needs $N \log_2 N$ arithmetical operations. In comparison, the direct computation of the integral needs N^2. Obviously, the FFT approach reduces the computational burden enormously when N is a large number. It should be noted that the approximation error has three different sources: (1) the interchange of the infinite integral bounds in (6.34) with finite ones; (2) the approximation of (6.34) with the Riemann sum (6.35);

and (3) the interpolation for intermediate values if the function argument is not a grid node.

The parameters of the FFT method N (the number of summands in the Riemann sum) and h (the grid spacing) should be carefully chosen since there is a trade-off between accuracy and computational burden. From the construction of (6.35), it follows that, to reduce the approximation error, N should be as large as possible. A peculiarity of the numerical method is that efficiency is gained if N is expressed as 2^q. In addition the length of the integration region in terms of the original variable u is $2\pi/h$. Hence h should be as small as possible to increase accuracy.

By following the approach of Scherer *et al.* (2012), in both the CTS and RDTS cases, one can always approximate the density function in the standardized case (with zero mean and unit variance) with parameter $(\alpha, \lambda_+, \lambda_-, 0, 1)$ as defined in Section 6.3 and then estimate the more general case with parameter by properly scaling and shifting. In this way it is possible to improve the efficiency and the quality of the approximation.

Because of the symmetry of the grid nodes, further efficiency can be achieved in the computations of (6.37) if we use the following relationship valid for any characteristic function:

$$\phi_X(-u) = \overline{\phi_X(u)}, \tag{6.38}$$

where $\overline{z}$ means the complex conjugate of z. Due to (6.38), it is possible to compute the $\phi_X(u)$ only for positive (negative) grid nodes, then change the sign of their imaginary parts and achieve the characteristic function values for negative (positive) grid nodes. This approach is particularly useful in the approximation of the RDTS since in these cases a special function has to be evaluated.

Concerning the cumulative distribution function, there are two potential approaches, that is

(1) it is possible to use (6.33) and derive an expression similar to (6.36);

(2) one can work directly with the evaluated density according to (6.36) and

$$F_X(x) = \int_{-\infty}^{x} f_X(u)du = h \sum_{\{k:u_k \le x\}} f_X(u_k).$$

In this chapter we consider the second approach.

In Table 6.1, we report the error of evaluating the density of CTS, NTS and RDTS according to the FFT method with different choices of q and h. We also evaluate the computational burden that depends mainly on q and on the characteristic function of the model. The procedure was run on an 8-core AMD FX processor with 16 GB of RAM with a Linux-based 64-bit operating system. The computational time necessary to interpolate intermediate values is negligible with respect to the computational time for the discrete FFT. In the CTS and RDTS cases we select $\alpha = 0.75$, $C = \lambda_+ = \lambda_- = 1$, and $m = 0$, and in the NTS case we select $a = 0.75$, $C = \lambda = \sigma = 1$, and

Table 6.1. CTS, NTS, and RDTS percentage errors. Maximum absolute difference $e^{\mathrm{FFT}}(h, q; X(\Theta))$ between the density evaluated via the FFT method and the exact density (with $\log_{10} h^* = -5$ and $q^* = 25$). We consider the interval $[x_{\min}, x_{\max}] = [-10, 10]$ with $\Delta x = 10^{-3}$ for a total of 20 thousand values. In all cases $dt = 1$. In the CTS and RDTS cases we select $\alpha = 0.75$, $C = \lambda_+ = \lambda_- = 1$, and $m = 0$, and in the NTS case we select $a = 0.75$, $C = \lambda = \sigma = 1$, and $\theta = \mu = 0$.

		$q = 14$	$q = 15$	$q = 16$	$q = 17$	$q = 18$
				CTS		
$\log_{10} h$	−1	3.90e–4	3.90e–4	3.90e–4	3.90e–4	3.90e–4
	−2	3.88e–6	3.88e–6	3.88e–6	3.88e–6	3.88e–6
	−3	3.54e+4	6.86e–7	8.81e–12	4.13e–12	1.01e–11
	−4	3.81e+10	7.41e+10	8.46e+9	1.84e+7	8.80e–4
				NTS		
$\log_{10} h$	−1	4.81e–4	4.81e–4	4.81e–4	4.81e–4	4.81e–4
	−2	4.79e–6	4.79e–6	4.79e–6	4.79e–6	4.79e–6
	−3	9.81e+4	7.10e–9	7.97e–11	5.50e–11	5.37e–11
	−4	6.91e+11	8.88e+11	7.31e+10	7.92e+7	1.03e–4
				RDTS		
$\log_{10} h$	−1	1.70e–2	1.76e–2	1.77e–2	1.78e–2	1.78e–2
	−2	5.61e–2	8.62e–4	1.31e–2	1.66e–2	1.75e–2
	−3	1.00e+0	1.00e+0	4.15e–1	9.69e–2	1.13e–2
	−4	1.10e+0	1.00e+0	1.00e+0	1.00e+0	1.00e+0

$\theta = \mu = 0$. The percentage error is defined by

$$e^{\text{FFT}}(h, q; X(\Theta)) = \max_{x \in [x_{\min}, x_{\max}]} \frac{|f_X^{\text{FFT}}(x, h, q; X(\Theta)) - f_X^{\text{exact}}(x)|}{f_X^{\text{exact}}(x)}, \tag{6.39}$$

where $N = 2^q$ is the number of summands in the Riemann sum, h is the grid spacing, X is a one-dimensional law with parameters Θ (that may include dt), and f_X^{exact} represents the exact valuation of the density function of the law X. Since in all cases of interest, the density f_X cannot be computed exactly, we approximate it as follows:

$$f_X^{\text{exact}}(x) \approx f_X^{\text{FFT}}(x, h^*, q^*; X(\Theta)),$$

where $\log_{10} h^* = -5$ and $q^* = 25$ (in the algorithm considered here, the greatest possible number of integration steps depends on the maximum available RAM). In this last case (i.e., when $q^* = 25$) the computing time range from 25 s in the NTS case to 73 s in the RDTS case (35 s in the CTS case). The computing time is much smaller for smaller values of q ranging from 0.015 s ($q = 14$) to 0.3 s ($q = 18$) in the CTS and NTS from 0.03 s ($q = 14$) to 0.7 s ($q = 18$) in the RDTS case. Note that density is evaluated on 20,000 values. As expected, the optimal balance between accuracy and efficiency (i.e., the choice of the parameters h and q) depends on the model and the parameter choice. Specifically, one has to be careful about the choice of h and q for small values of α. In Figure 6.5, we show the shape of the probability density function of a CTS law with parameters $\alpha = 0.75$, $C = \lambda_+ = \lambda_- = 1$, and $m = 0$ for different choices of q and h in the FFT algorithm. It is clear that for some combinations of h and q one obtains a totally wrong density evaluation with large percentage errors (e.g., when $\log_{10} h = -3$ and $q = 14$ or $\log_{10} h = -4$ and $q = 14$). Note that in the implementation of the RDTS characteristic function we use the relation in equation (6.3).

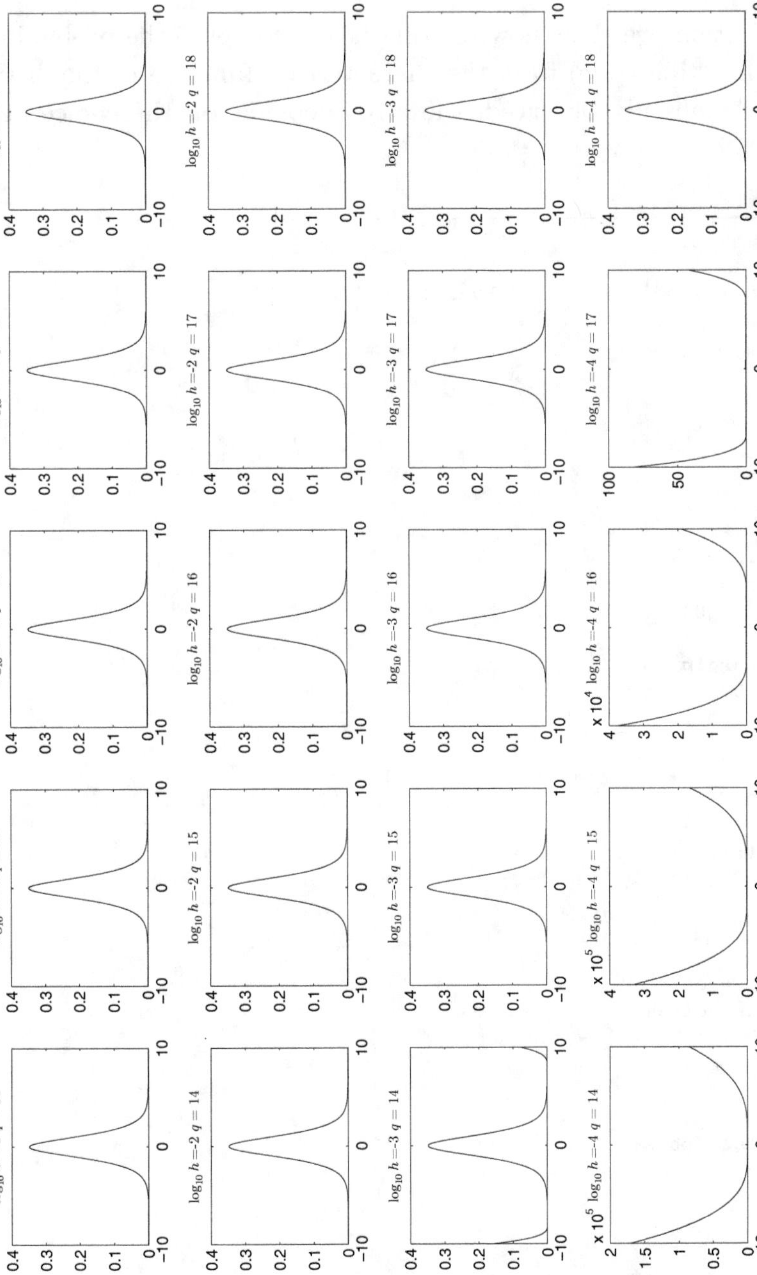

Figure 6.5. Shape of the probability density function of a CTS law with parameters $\alpha = 0.75$, $C = \lambda_+ = \lambda_- = 1$, and $m = 0$ for different choices of q and h in the FFT algorithm.

6.4.1. Matlab code — pdf evaluation

In this section, we show how to numerically compute the probability density function (pdf) from the characteristic function of the distribution. As already observed, the error depends on the choice of h and $N = 2^q$.

```
% implementation of the probability density function

function [f, pdf, x]  = pdfEvaluation(arg, chf, method, h, N)

% this function evaluate the pdf at points arg given the
% characteristic function 'chf' with e.g.
% chf = @(z) chf_CTS(z, param_CTS, dt)
% where param_CTS and dt are given
% WARNING: in the RDTS case
% chf = @(z) chf_RDTS(z', param, dt)
% has to be used

if nargin == 2
     method = 'pchip';
     h = 0.001;
     N = 2^17;
elseif nargin == 3
     h = 0.001;
     N = 2^17;
end

s = 1/(h*N);
t1 = 1:1:N;
t1 = t1(:);
t2 = 2*pi*(t1 - 1 - N/2)*s;

% compute the value of the characteristic function

cfvalues = chf(t2); % e.g. chf = @(z) chf_CTS(z, param_CTS, dt)
cfvalues = cfvalues(:);
x1 = (-1).^(t1 - 1).*cfvalues;

% compute pdf values

pdf = real(fft(x1, N)).*(s.*(-1).^(t1 - 1 - N/2));

% make sure the density has only positive values

negative = pdf<0;
pdf(negative) = 0;
```

```
% compute the properly transformed values of x, i.e. f(x) = ...
    pdf values

x = (t1 - 1 - N/2)*h;

% approximate the value of the pdf function at the specified ...
    argument value arg and according to the specified method

f = interp1(x, pdf, arg, method);

return;
```

6.4.2. Matlab code — cdf evaluation

In this section, we show how to numerically compute the cumulative distribution function (cdf) from a probability density function in closed form or from the characteristic function of the distribution. The interpolation is conducted by considering the interval $[-M, M]$. The selection of M depends on the data to analyze.

```
% implementation of the cumulative density function

function [cdf, cdfi, x] = cdfEvaluation(arg, ...
    randomVariableFunction, ...
                                          method, M)

% this function evaluate the cdf at points arg;
% (1) if method = 'pdf' then 'randomVariableFunction' should ...
    be the pdf in this case M = 1 and x = -1:dx:1;
% (2) if method = 'AVaR' then 'randomVariableFunction' should ...
    be the pdf in this case M = 20 (default) if it is not ...
    specified and x = -M:dx:M;

%this option should be used to evaluate the average ...
    value-at-risk (AVaR)
% (3) if method = 'chf' then 'randomVariableFunction' should ...
    be the chf e.g. chf = @(z) chf_CTS(z, param_CTS, dt) where ...
    param_CTS and dt are given and x is given as output of the ...
    function 'pdfEvaluation'

if strcmp(method,'pdf')

    dx = 1e-3;
    x = -1:dx:1;
```

```
    x = x(:);

    pdf = randomVariableFunction(x);

elseif strcmp(method,'AVaR')

    if nargin == 3
       M = 20;
    end

    dx = 1e-3;
    x = -M:dx:M;
    x = x(:);

    pdf = randomVariableFunction(x);

elseif strcmp(method,'chf')

    [~, pdf, x] = pdfEvaluation(0, randomVariableFunction);

end

cdfi = cumsum(pdf)*(x(2) - x(1));

if min(arg) < min(x) || max(arg) > max(x)
    in = find(arg > max(x));
    arg(in) = max(x);
    in = find(arg < min(x));
    arg(in) = min(x);
    warning('arg value outside x grid. Corrected.')
end

cdf = interp1(x, cdfi, arg);

return
```

6.5. Simulation

The simulation of a TS random draw is not a simple task. A series representation algorithm has been proposed in Rosinski (2007) and empirically studied in Bianchi *et al.* (2010a) and Imai and Kawai (2011). Since there exists an efficient algorithm to draw random samples from stable distributions (see Chambers *et al.* (1976)), the problem of generating random numbers from a TS law X can be solved by using a stable law Y possessing a probability density g similar to the probability density f of X. One can generate a value for Y and accept (reject) this value if a given condition is satisfied (not satisfied). More details on this method can be found in Gentle (2003) or Rachev *et al.* (2011). We refer to the probability of the acceptance event as the *acceptance rate.* This acceptance–rejection simulation method has been widely studied in the literature (see Kawai and Masuda (2011a,b, 2012), Jelonek (2012), and references therein). By applying this algorithm, one can sample TS random numbers in an exact (or approximate) way if the tail index α is less (or greater) than 1. The computational cost strictly depends on the parameters. In the case when $\alpha < 1$, a double rejection sampling algorithm that does not depend upon the model parameters has been proposed by Devroye (2009).

Furthermore, one can use the method proposed by Rosiński (2001) where the ratio between the Lévy measures of a TS and a stable law is used to construct an acceptance–rejection algorithm. Also in this case the probability of the acceptance event depends on the parameters of the TS distribution to simulate. However, this method does not seem to work efficiently and further investigation is needed to use it in practice.

Alternatively, one can consider the inverse transform algorithm. That is, given the cumulative distribution function F_X, one can use the following three-step procedure:

1. generate a sequence $u_1, \ldots, u_n$ of independent draws from a uniform random variable on $[0, 1]$;
2. find a root x_i of the equation $F_X(x_i) - u_i = 0$ for each $i = 1, \ldots, n$;
3. return the sequence $x_1, \ldots, x_n$.

In the TS case, it is easy to see that the function F_X is not available in closed form. To find values of F_X, first we have to invert the characteristic function in order to get both the density (f_X) and the cumulative distribution function (F_X), and then find the value

x_i satisfying the equality $F_X(x_i) - u_i = 0$. Here the term inversion means Fourier inversion. The details of this method were explained in Section 6.4. Even if this method may seem computationally demanding, an efficient procedure can be written in order to increase the speed and make the time necessary for the simulation of large matrices of minor concern. Further details on this method are provided in Section 6.5.1. This approach can be applied for all values of the parameters and it is more efficient than using the acceptance–rejection method when a huge matrix of TS random numbers has to be drawn. See also the work of Ballotta and Kyriakou (2014) and Bianchi *et al.* (2018).

However, as shown in Section 6.4, it sometimes happens that for certain values of the parameters the Fourier inversion does not work properly due to numerical errors. In these cases, to obtain the right simulation algorithm, one has to increase the integration limits or decrease the discretization step into the FFT algorithm in order to reduce the error in evaluating the density function and the cumulative distribution function. However, the error in evaluating these two functions decreases at the expense of computational time (see again Section 6.4). Similarly, for other values of the parameters it may happen that, if using the acceptance–rejection algorithm, the acceptance rate is so low that it cannot be used in real applications (as observed above, it is only for $\alpha < 1$ that a double rejection sampling algorithm that does not depend upon the model parameters has been developed). This in practice means that the optimal choice of the algorithm may depend on the parameters.

Bianchi *et al.* (2017) test both the acceptance–rejection and the inverse transform algorithm by drawing stdCTS random numbers. As observed in Section 6.3, it is possible to define a CTS law with zero mean and unit variance. Bianchi *et al.* (2017) analyze the CTS law with parameters $(\alpha, \lambda_+, \lambda_-, \mu, \sigma)$ with $\mu = 0$ and $\sigma = 1$ and considering three parameter sets for $(\alpha, \lambda_+, \lambda_-)$: (0.75, 0.5, 0.5), (1.5, 1, 0.5), and (1.75, 0.1, 0.05). The computational time needed for generating a matrix with dimension 20,000 $\times$ 1000 is 15 seconds (56 seconds, and 11 seconds) with the acceptance–rejection algorithm, slightly more than 5 seconds with the inverse algorithm in all the cases considered.

The procedure was run on an 8-core AMD FX processor with 16 GB of Ram with a Linux-based 64-bit operating system. According to the notation introduced in Section 6.4, $\log_{10} h = -3$ and $q = 17$ are considered, representing the grid spacing and the grid steps in the FFT inversion. We point out that, if using the acceptance–rejection algorithm, the computational time can be much greater if one selects other parameters. Conversely, the computational time of the inverse algorithm depends only on h and q.

6.5.1. Matlab code — CTS random numbers simulation

In this section, we show how to numerically simulate CTS random number starting from the CTS characteristic function and evaluating the cumulative distribution function implemented in Section 6.4.2. The selection of the parameter $xLim$ depends on the data one wants to simulate. In Figure 6.6, we show the empirical cumulative distribution function of simulated CTS random numbers and the theoretical one. The empirical cumulative distribution function is based on simulated data with different number of simulations, that is $NSim \in (50, 100, 250, 500)$, and the parameters of the CTS law are $\alpha = 0.75$, $m = 0$, and $C = \lambda_+ = \lambda_- = 1$. Furthermore, we select $dt = 1$, $MSim = 1$, and $xLim = 15$.

```
% Simulate a CTS random numbers (NSim x MSim)

function xSim = SimCTS_invcdf(param, dt, NSim, MSim, xLim, dx)

% C = param(1);
% lm = param(2);
% lp = param(3);
% alpha = param(4);
% m = param(5);

chf = @(z) chf_CTS(z, param, dt);
[cdf, cdfi, x] = cdfEvaluation(0, chf, 'chf');

if nargin == 3
    dx = 10;
    xLim = 1;
    MSim = 1;
```

```
elseif nargin == 4
    dx = 10;
    xLim = 1;
end

x        = x(1:dx:length(x));
cdfi     = cdfi(1:dx:length(cdfi));

index = cdfi >= 1 | cdfi <= 0 | abs(x) > xLim;

cdfi(index) = [];
x(index)    = [];

cdfi = [0; cdfi; 1];
x = [-xLim; x; xLim];

figure
plot(x, cdfi)

U = rand(NSim*MSim,1);
xSim = interp1(cdfi,x,U,'pchip');
xSim = reshape(xSim,NSim,MSim);

return
```

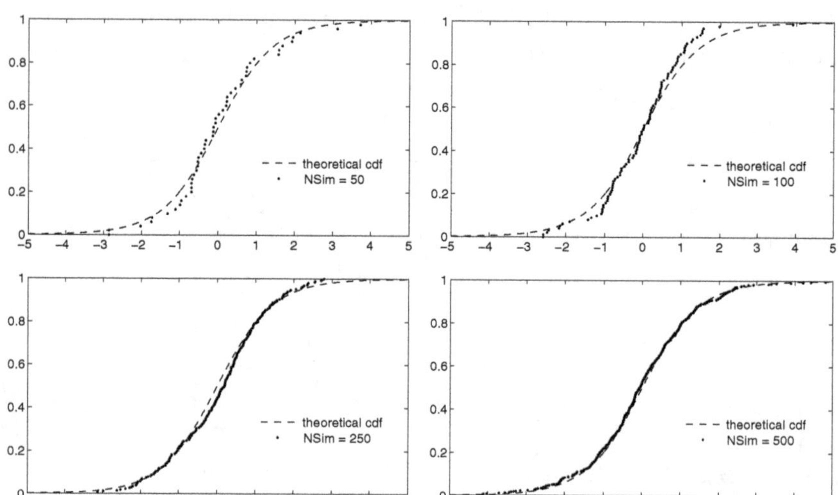

Figure 6.6. The empirical cumulative distribution function based on simulated data with different numbers of simulation (50, 100, 250, 500) is compared with the theoretical one.

6.6. Estimation

In this section, we show how to estimate parameters using the maximum likelihood estimation (MLE) method. This estimation method requires a procedure to determine a numerical approximation of the density function of the CTS (NTS and RDTS, respectively) law by using the FFT as shown in Section 6.4. Let X be a random variable with a parameter vector Θ. For example, if X is a CTS random variable with parameters α, C, λ_+, λ_-, and m, then

$$\Theta = (\alpha, C, \lambda_+, \lambda_-, m).$$

Let $\phi_X(u;\Theta)$ and $f_X(x;\Theta)$ be, respectively, the characteristic function and the density function of X. As already observed in equation (6.32), the inverse Fourier transform describes the relation between ϕ and f, that is

$$f_X(x;\Theta) = \frac{1}{2\pi}\int_{-\infty}^{\infty} e^{-iux}\phi_X(u;\Theta)\,du.$$

Given this relation and the characteristic function of a CTS (NTS and RDTS, respectively) law, we are able to calculate the value $f_X(x)$ of the density function at every given point x. This task can be performed in a very efficient way by using the FFT approach described in Section 6.4. Given a sample from a law X with observations $\{x_i\}_{0\le i\le N}$, the MLE is a solution $\hat{\Theta}$ of the optimization problem

$$\hat{\Theta} = \max_{\Theta} \log L(\Theta; \{x_i\}_{0\le i\le N}),$$

where the likelihood function is given by

$$L(\Theta) = \prod_{i=1}^{N} f_X(x_i;\Theta).$$

It is obvious that when the density function f_X can be computed in closed-form (e.g., for the normal distribution), the problem is less computationally demanding compared to when the density function has to be computed via the discrete FFT.

6.6.1. Matlab code — likelihood evaluation

In this section, we show how to implement the likelihood of a CTS-based model. The parameters dt, $hFFT$, and $nFFT$ must be provided.

```
% implementation of the likelihood function of
% a CTS random variable

function y = likelihoodCTS(returns, param, dt, hFFT, nFFT)

    C = param(1);
    lm = param(2);
    lp = param(3);
    alpha = param(4);
    if (alpha == 1) % alpha = 1 is not allowed
        param(4) = 0.99;
        alpha = param(4);
    end

    m = param(5);

    chf = @(z) chf_CTS(z, [C lm lp alpha m], dt);
    returnsPdf = pdfEvaluation(returns, chf, 'pchip', hFFT, nFFT);

    lhVec = returnsPdf;
    y = -sum(log(lhVec));

return;
```

6.6.2. Matlab code — MLE estimation

In this section, we show how to implement the simple optimization algorithm to find an MLE in the CTS case. In this script we consider the Matlab optimization function *fmincon*, and we select upper and lower bounds (*ub* and *lb*) for the parameters. We show some simple options that one can select in the optimization algorithm. Note that the choice of *ub* and *lb* may depend on the sample to fit.

```
% MLE algorithm for a CTS random variable

function [param, fval, exitflag] = mleCTS(returns, dt, param0,
                                          hFFT, nFFT, lb, ub)

% C = param(1);
```

```
% lm = param(2);
% lp = param(3);
% alpha = param(4);
% m = param(5);

if nargin == 2

    hFFT = 0.001;
    nFFT = 2^17;
    param0 = [1   10  10  1.5 0];
    lb = [0.001, 0.1,  0.1, 0.5 -1];
    ub = [10,   10, 10, 1.99 1];

    options = optimset('Display', 'off', ...
                       'Algorithm','interior-point',...
                       'UseParallel','always');

elseif nargin == 5

    lb = [0.001, 0.1,  0.1, 1.5 -1];
    ub = [10,   10, 10, 1.99 1];

    options = optimset('MaxFunEvals', 1500, ...
                       'TolFun', 1e-15, ...
                       'TolCon', 1e-15, ...
                       'TolX', 1e-15, ...
                       'Display', 'iter', ...
                       'Algorithm','interior-point');

end

fopt = @(x) likelihoodCTS(returns, x, dt, hFFT, nFFT);

[param, fval, exitflag] = fmincon(@(x) fopt(x), ...
                          param0, [],[],[],[],...
                          lb,ub,[],options);


return
```

6.6.3. Matlab code — A simple test

Thus, we show the Matlab script to simulate CTS random numbers and perform an MLE. The code is based on the simulation method proposed in Section 6.5.1 and the MLE algorithm shown in Section 6.6.2. The execution of the entire test may take some time, since each iteration run for some seconds. The computing time depends on different reasons (e.g., the processor, the Matlab version, the available RAM, the number of cores). If one does not want to wait for a long time, one has to select *numIteration* equal to 1.

```
% test of the simulation and the mle algorithm
% select a proper number for 'numIteration'

clc
clear all
close all

display(['The test takes some time ...'])

param = [0.1 1 1 1.5 0];
dt = 1;
NSim = 2500;
MSim = 1;
xLim = 25;

numIteration = 10;

for i = 1:numIteration

tic

xSim = SimCTS_invcdf(param, dt, NSim, MSim, xLim);
[param_mle(i,:), fval(i), exitflag(i)] = mleCTS(xSim, dt);

display(['iteration n. ', num2str(i),' - elapsed time ',...
         num2str(toc),' seconds'])
display(['estimated parameters: ',num2str(param_mle(i,:))])
end
```

6.7. Risk Measures

In this section, we apply the framework defined in the previous sections to stock log-returns to compute two well-known risk measures. Here the observations of our random variable X are given by observed log-returns. The risk measures we study in this chapter are value-at-risk (VaR) and average value-at-risk (AVaR), the average of VaRs greater than the VaR for a given tail probability. AVaR, also known as conditional value-at-risk (CVaR) and expected shortfall, is a superior risk measure to VaR because it satisfies all axioms of a coherent risk measure and it is consistent with preference relations of risk-averse investors (see Rachev *et al.* (2008)). Both measures have been adopted as standard risk measures in the financial industry and by bank regulators (Basel Committee on Banking Supervision (January 2016)).

In order to have a visual assessment of the differences between the three tempered stable and the normal distribution, in Figure 6.7 we report both the probability density and the cumulative distribution functions of the models estimated on the log-returns of a randomly selected Italian open-ended fund (this example is drawn from Bianchi (2015)). For each model estimation we assume that the fund log-returns are normal (CTS, RDTS, NTS, respectively) distributed. An MLE as described in Section 6.6 is conducted to find the model

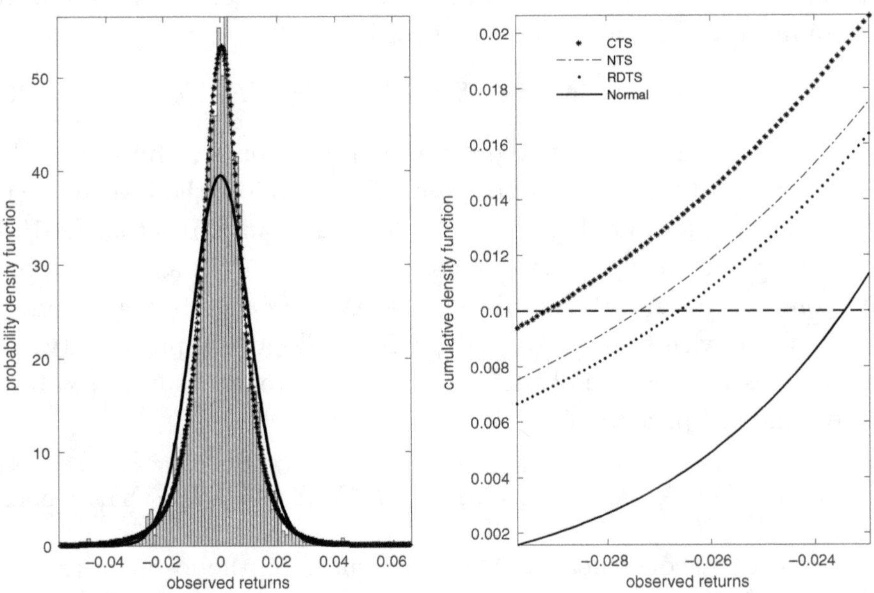

Figure 6.7. Normal, CTS, NTS, and RDTS estimation results of a randomly selected fund. On the left, the probability density functions are reported. On the right, a detail of the cumulative distribution functions around the left tail (1% level) is shown. The abscissa (changed in sign) of the point of intersection between the dashed line and the cumulative distribution function is the 1% VaR.

parameters. Examining the probability density function on the left side of the figure, we can assert that the normal model cannot capture the statistical properties of the observed returns, while all TS models do. Then, on a first look the probability density functions of the three TS models seem to differ only slightly. On the right side of the figure, the behavior of the cumulative distribution functions around the left tail is shown. For each model, the abscissa (changed in sign) of the point of intersection between the dashed line and the cumulative distribution function is the 1% VaR. The model assessment change under this different perspective which provides a focus on the left tail. Clearly there are differences in the tails and as a consequence in the measurement of VaR among the four models, at least for the fund on which the figure is based.

As shown in Section 6.4, by taking into consideration the works of Stoyanov and Racheva-Iotova (2004) and Scherer *et al.* (2012) it is possible to implement an MLE method based on the FFT and to compute the VaR at tail level δ; that is,

$$\mathrm{VaR}_\delta(X) = -\inf\{x|P(X \leq x) > \delta\} = -F_X^{-1}(\delta) \tag{6.40}$$

by inverting the cumulative distribution function F_X. In Section 6.4 we provided a framework to evaluate F_X, therefore the inversion can be easily implemented (see Section 6.7.1). From Kim *et al.* (2010b, 2011) it is possible to obtain a closed formula (up to an integration) to compute the AVaR. Recall that the AVaR of a continuous random variable X with finite mean (i.e., $E[X] < \infty$) at tail probability δ is defined as the average of the VaRs that are greater than the VaR at the same tail probability, that is

$$\mathrm{AVaR}_\delta(X) = \frac{1}{\delta}\int_0^\delta \mathrm{VaR}_p(X)dp = -E\big[X\big|X < -\mathrm{VaR}_\delta(X)\big]. \tag{6.41}$$

Therefore, by construction, AVaR is focused on the losses in the tail that are greater than the corresponding VaR level. As shown in Kim *et al.* (2010b), equation (6.41) is equivalent to

$$\mathrm{AVaR}_\delta(X) = \mathrm{VaR}_\delta(X) - \frac{e^{-VaR_\delta(X)\rho}}{\pi\delta}\int_0^\infty e^{-ix\mathrm{VaR}_\delta(X)}\frac{\phi_X(i\rho - x)}{(i\rho - x)^2}dx.$$

We refer to $\mathrm{VaR}_{0.01}(X)$ and $\mathrm{AVaR}_{0.01}(X)$ as 1% VaR and 1% AVaR, respectively.

6.7.1. Matlab code — VaR evaluation

In this section, we show how to numerically compute the VaR given the cumulative distribution function. Note that δ can be a vector. The inversion of the cumulative distribution function is obtained by finding the x such that the equation $\delta - F_X(x) = 0$. The cumulative distribution function is numerically evaluated by using the code in Section 6.4.2.

```
% implementation of the VaR with tail level delta

function q = VaR_given_cdf(delta, cdf)

    options = optimset('Display','off');
    q = zeros(size(delta));

    for j=1:length(delta)
        q(j) = fsolve( @(x)(delta(j) - cdf(x)), 0, options);
    end

    q = -q;

end
```

6.7.2. Matlab code — AVaR evaluation

In this section, we show how to numerically compute the AVaR given the cumulative distribution function. Note that δ is a constant (e.g., 0.01). The AVaR is evaluated by means of a numerical integration. Again, the cumulative distribution is numerically evaluated by using the code in Section 6.4.2.

```
function v = AVaR_quad( delta, VaR, chf, N, rho )

    if nargin <= 3
        N = 1000;
        rho = 1e-4;
```

```
    end

    F = @(x)(exp(-1i.*x.*VaR).*chf(-x+rho*1i)./(rho*1i-x).^2);
    v = VaR - exp(-VaR*rho)*real(quadl( F, 0, N ))/(pi*delta);

return;
```

6.7.3. Matlab code — A simple test

Here we show the Matlab script to compute the VaR and the AVaR in the CTS case given a set of parameters. The code is based on the method proposed in Sections 6.7.1 and 6.7.2. The execution of the entire test takes few seconds.

```
clc
clear all

param = [0.1 1 1 1.5 0];
dt = 1/250;
delta = 0.01;

chf = @(z) chf_CTS(z, param, dt);
cdf = @(x) cdfEvaluation(x, chf, 'chf');

tic
VaR  = VaR_given_cdf(delta, cdf);
AVaR = AVaR_quad(delta, VaR, chf);
time = toc;

display(['VaR = ',num2str(VaR),' AVaR = ',num2str(AVaR),...
        ' computed in ', num2str(time), ' seconds'])
```

Key Points

- A one-dimensional TS (TID) distribution is obtained by tempering the Lévy measure of a one-dimensional stable law through a completely monotone (positive definite radial) function.
- The CTS (RDTS) law is a parametric example in the TS (TID) class.

- The NTS law can be viewed as the law of the increments of a time-changed Brownian motion with one-side CTS stochastic time, or as a normal mean–variance mixture with one-side CTS mixing distribution.
- The characteristic function f_X of a random variable X with density function f_X is defined by its Fourier transform. Conversely, given the characteristic function ϕ_X, the density function f_X can be obtained using the inverse Fourier transform.
- Given a characteristic function it is possible to define the cumulant generating function (or Laplace exponent) and the kth cumulant of a random variable can be computed by taking the derivatives of the cumulant generating function evaluated at 0.
- An efficient method based on the discrete FFT can be used to compute the probability density function f_X and the cumulative distribution function F_X starting from the characteristic function ϕ_X.
- The FFT method to evaluate the density function given its characteristic function is a numerical procedure. The parameters N (the number of summands in the Riemann sum) and h (the grid spacing) should be carefully chosen since there is a trade-off between accuracy and computational burden.
- The use of TS and TID distributions is obstructed by the difficulty of calibrating and simulating them, mainly because the density function is not known in closed form and numerical methods have to be used.
- For CTS, NTS and RDTS processes a change of time scale leads to a change of spatial scale of the distribution of the increments.
- The evaluation of the characteristic function of an RDTS law is not trivial, since a Kummer's or confluent hypergeometric function has to be evaluated.
- To draw a random sample from a CTS (NTS and RDTS) distribution an efficient algorithm based on the FFT and the classical inverse transform algorithm can be implemented.
- The parameter of one-dimensional TS and TID distributions can be found by considering a standard MLE method, where the likelihood function is evaluated through an FFT method.

- For any infinitely divisible law, the VaR and the AVaR can be computed by assuming that a numerical procedure to compute the cumulative distribution function is available.

References

M. Abramowitz and I.A. Stegun. *Handbook of Mathematical Functions: With Formulas, Graphs, and Mathematical Tables.* Dover Publications, New York, 1974.

L. Ballotta and I. Kyriakou. Monte Carlo option pricing coupled with Fourier transform for the CGMY process. *Journal of Futures Markets*, 34(12):1095–1121, 2014.

Basel Committee on Banking Supervision. Minimum capital requirements for market risk, January 2016. https://www.bis.org/bcbs/publ/d352.pdf.

M.L. Bianchi. Are the log-returns of Italian open-end funds normally distributed? A risk assessment perspective. *Journal of Asset Management*, 16:437–449, 2015.

M.L. Bianchi, S.T. Rachev, and F.J. Fabozzi. Calibrating the Italian smile with time-varying volatility and heavy-tailed models. *Computational Economics*, 51:339–378, 2018.

M.L. Bianchi, S.T. Rachev, and F.J. Fabozzi. Tempered stable Ornstein–Uhlenbeck processes: A practical view. *Communications in Statistics - Simulation and Computation*, 46(1):423–445, 2017.

M.L. Bianchi, G.L. Tassinari, and F.J. Fabozzi. Riding with the four horsemen and the multivariate normal tempered stable model. *International Journal of Theoretical and Applied Finance*, 19(4):1–28, 2016.

M.L. Bianchi, S.T. Rachev, Y.S. Kim, and F.J. Fabozzi. Tempered stable distributions and processes in finance: numerical analysis. In M. Corazza and C. Pizzi, editors, *Mathematical and Statistical Methods for Actuarial Sciences and Finance.* Springer, Berlin, 2010a, pp. 33–42.

M.L. Bianchi, S.T. Rachev, Y.S. Kim, and F.J. Fabozzi. Tempered infinitely divisible distributions and processes. *SIAM Theory of Probability and Its Applications*, 55(1):59–86, 2010b.

S.I. Boyarchenko and S.Z. Levendorskii. The variation of certain speculative prices. *International Journal of Theoretical and Applied Finance*, 3(3):549–552, 2000.

S.I. Boyarchenko and S.Z. Levendorskii. *Non-Gaussian Merton–Black–Scholes theory.* World Scientific Publishing, River Edge, NJ, 2002.

P. Carr, H. Geman, D.B. Madan, and M. Yor. The fine structure of asset returns: An empirical investigation. *The Journal of Business*, 75(2): 05–332, 2002.

J.M. Chambers, C.L. Mallows, and B.W. Stuck. A method for simulating stable random variables. *Journal of the American Statistical Association*, 71:340–344, 1976.

P.H. Cootner. *The Random Character of Stock Market Prices.* MIT Press, Cambridge, 1964.

L. Devroye. Random variate generation for exponentially and polynomially tilted stable distributions. *ACM Transactions on Modeling and Computer Simulation*, 19(18), 2009.

J.E. Gentle. *Random number generation and Monte Carlo Methods.* Springer, New York, 2003.

A. Gil, J. Segura, and N.M. Temme. *Numerical Methods for Special Functions.* Society for Industrial and Applied Mathematics (SIAM), Philadelphia, PA, 2007.

M. Grabchak. On a new class of tempered stable distributions: moments and regular variation. *Journal of Applied Probability*, 49:1015–1035, 2012.

M. Grabchak and G. Samorodnitsky. Do financial returns have finite or infinite variance? A paradox and an explanation. *Quantitative Finance*, 10(8):883–893, 2010.

J. Imai and R. Kawai. On finite truncation of infinite shot noise series representation of tempered stable laws. *Physica A: Statistical Mechanics and its Applications*, 390:4411–4425, 2011.

P. Jelonek. Generating tempered stable random variates from mixture representation. Working paper 14, University of Leicester, Department of Economics, 2012.

R. Kawai and H. Masuda. Exact discrete sampling of finite variation tempered stable Ornstein-Uhlenbeck processes. *Monte Carlo Methods and Applications*, 17:279–300, 2011a.

R. Kawai and H. Masuda. On simulation of tempered stable random variates. *Journal of Computational and Applied Mathematics*, 235: 2873–2887, 2011b.

R. Kawai and H. Masuda. Infinite variation tempered stable Ornstein–Uhlenbeck processes with discrete observations. *Communications in Statistics - Simulation and Computation*, 41:125–139, 2012.

Y.S. Kim, S.T. Rachev, M.L. Bianchi, and F.J. Fabozzi. Tempered stable and tempered infinitely divisible GARCH models. *Journal of Banking & Finance*, 34(9):2096–2109, 2010a.

Y.S. Kim, S.T. Rachev, M.L. Bianchi, and F.J. Fabozzi. Computing VaR and AVaR in infinitely divisible distributions. *Probability and Mathematical Statistics*, 30(2):223–245, 2010b.

Y.S. Kim, S.T. Rachev, M.L. Bianchi, I. Mitov, and F.J. Fabozzi. Time series analysis for financial market meltdowns. *Journal of Banking & Finance*, 35(8):1879–1891, 2011.

Y.S. Kim, R. Giacometti, S.T. Rachev, F.J. Fabozzi, and D. Mignacca. Measuring financial risk and portfolio optimization with a non-Gaussian multivariate model. *Annals of Operations Research*, 201(1):325–343, 2012.

I. Koponen. Analytic approach to the problem of convergence of truncated Lévy flights towards the Gaussian stochastic process. *Physical Review E*, 52(1):1197–1199, 1995.

B.B. Mandelbrot. The variation of certain speculative prices. *Journal of Business*, 36:394–419, 1963.

R.N. Mantegna and H.E. Stanley. Stochastic process with ultraslow convergence to a Gaussian: The truncated Lévy flight. *Physical Review Letters*, 73(22):2946–2949, 1994.

E.A. Novikov. Infinitely divisible distributions in turbulence. *Physical Review E*, 50(5):3303–3305, 1994.

S.T. Rachev and S. Mittnik. *Stable Paretian Models in Finance.* Wiley, Chichester, 2000.

S.T. Rachev, C. Menn, and F.J. Fabozzi. *Fat-tailed and Skewed Asset Return Distributions: Implications for Risk Management, Portfolio Selection, and Option Pricing.* Wiley, Hoboken, NJ, 2005.

S.T. Rachev, S.V. Stoyanov, and F.J. Fabozzi. *Advanced Stochastic Models, Risk Assessment, and Portfolio Optimization: The Ideal Risk, Uncertainty, and Performance Measures.* Wiley, Hoboken, NJ, 2008.

S.T. Rachev, Y.S. Kim, M.L. Bianchi, and F.J. Fabozzi. *Financial Models with Lévy Processes and Volatility Clustering.* Wiley, Hoboken, NJ, 2011.

D. Revuz and M. Yor. *Continuous Martingales and Brownian Motion.* Springer, New York, 3rd edition, 2005.

J. Rosiński. Series representations of Lévy Processes from the Perspective of Point Processes. In O.E. Barndorff-Nielsen, T. Mikosch, and S.I. Resnick, editors, *Lévy Processes: Theory and Applications*, Birkhäuser, Boston, 2001, pp. 401–415.

J. Rosinski. Tempering stable processes. *Stochastic Processes and Their Applications*, 117(6):677–707, 2007.

J. Rosiński and J.L. Sinclair. Generalized tempered stable processes. In J.K. Misiewicz, editor, *Stability in Probability*, Banach Center Publication 90, Warsaw, 2010, pp. 153–170.

M. Scherer, S.T. Rachev, Y.S. Kim, and F.J. Fabozzi. Approximation of skewed and leptokurtic return distributions. *Applied Financial Economics*, 22:1305–1316, 2012.

I.J. Schoenberg. Metric spaces and completely monotone functions. *Annals of Mathematics*, 39:811–841, 1938.

S.V. Stoyanov and B. Racheva-Iotova. Numerical methods for stable modeling in financial risk management. In S.T. Rachev, editor, *Handbook of Computational and Numerical Methods in Finance*, Birkhäuser, Boston, MA, 2004, pp. 299–329.

G. Terdik and W.A. Woyczyński. Rosiński measures for tempered stable and related Ornstein-Uhlenbeck processes. *Probability and Mathematical Statistics*, 26:213–243, 2006.

F.G. Tricomi. *Funzioni ipergeometriche confluenti.* Edizioni Cremonese, Roma, 1954.

Chapter 7

Multivariate Time-Changed Brownian Motion

The main topics covered in this chapter are:

- the history of time-changed Brownian motion and why this process has been applied to finance;
- the main properties and formulas for different parametric examples of multivariate time-changed Brownian motion;
- the relation between multivariate time-changed Brownian motion and multivariate normal mean–variance mixture distributions;
- the derivation of the closed formula for the density function of some multivariate normal mean–variance mixture laws;
- the evaluation of the probability density function of a multivariate normal mean–variance mixture law when only the characteristic function of the mixing distribution is available in closed form while the density function is not;
- how to generate sample draws from a multivariate normal mean–variance mixture law.

7.1. Introduction

Many interesting problems in finance are multidimensional. These problems are usually solved by considering a multivariate normal distribution, mainly because the complexity of a model increases as soon as one moves from a normal to a non-normal distribution framework. The dependence model implied by the multivariate normal distribution is symmetric and it is not capable of capturing asymmetry of dependence during extreme market movements and the dependence of tail events. A more complex dependence structure of multivariate distributions is needed rather than a simple linear correlation. Additionally, the margins of a multivariate normal distribution are normally distributed and they do not assign enough weight to tail events and are not capable of explaining negative skew. A visual assessment of these empirically observed facts is provided in Figure 7.1 where we show the simulated returns drawn from a multivariate normal distribution fitted on market data and compare them to real market data. The number of simulations is equal to the number of observed historical data. More specifically, we consider two margins, the Deutsche Aktienindex 30 (ticker DAX) and Financial

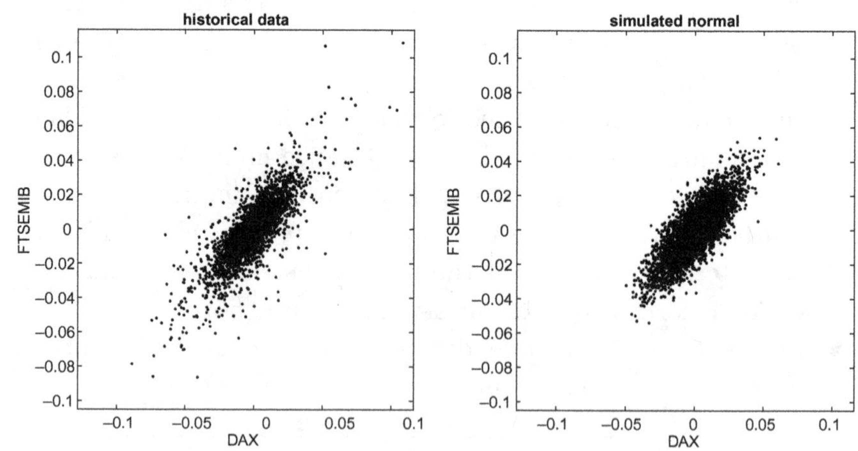

Figure 7.1. Bivariate scatterplot of DAX versus FTSEMIB by considering market data and simulated sample from the multivariate normal. Depicted market data are daily log-returns from January 2, 1998 to August 29, 2014.

Times Stock Exchange Milano Indice di Borsa (ticker FTSEMIB) of a five-dimensional distributions representing the log-returns of five major European indexes.

A large part of modern finance has been concerned with modeling the evolution of return processes over time. The time-changed Brownian motion construction we will describe in this chapter is well known from the theory of stochastic processes. In the theoretical work of Monroe (1978), it is shown under what conditions a stochastic process is equivalent to a time-changed Brownian motion. The application to finance is even older and it is based on the following evidence: "The different evolution of prices series on different days is due to the fact that information is available to traders at a varying rate. On days when no new information is available trading is slow, and the price process evolves slowly. On days when new information violates old expectation, trading is brisk, and the price process evolves much faster" (see Clark (1973)). Financial return series are asymmetric and heavy-tailed, where the tails are important because bad news is tail events. In periods of high volatility, time runs faster than in periods of low volatility. The physical time in the well-known Brownian motion is substituted by an intrinsic (or a business) time providing tail effects.

In practice, a non-decreasing process, which we refer to as the *subordinator* (or *stochastic clock*), can be used as the stochastic time of a Brownian motion. By subordination, it is possible to capture empirically observed anomalies that contradict the classical log-normality assumption for asset prices. The subordination approach has been widely studied in the literature, principally from a univariate perspective (see Hurst *et al.* (1997, 1999), Geman *et al.* (2001, 2002), Schoutens (2003), Cont and Tankov (2003), and Rachev *et al.* (2011)). The subordinator models operational time and provides the so-called fat-tail effects often observed in financial markets.

The time-changed Brownian motion construction can be extended to a multivariate framework. Indeed, it is possible to subordinate a multivariate Brownian motion by a univariate or a multivariate stochastic time. While in Luciano and Schoutens (2006), Guillaume

(2012, 2013), and Tassinari and Corradi (2013) a model with independent Brownian motions is proposed, in Leoni and Schoutens (2008), Tassinari (2009), Luciano and Semeraro (2010), Eberlein and Madan (2010), Tassinari and Corradi (2014), Tassinari and Bianchi (2014), and Bianchi *et al.* (2016) correlated Brownian motions are considered.

The distribution of the increments of the time-changed Brownian motions analyzed in this chapter can be also viewed as a generalization of the multivariate normal distribution known as a multivariate normal mean–variance mixture distribution (see, e.g., Tassinari and Bianchi (2014)). These models share much of the structure of the multivariate normal distribution, but they allow both asymmetry and heavy tails. Furthermore, according to Frahm (2004), these distributions belong to the class of elliptical variance–mean mixture. Elliptical and generalized elliptical heavy tail distributions have been widely studied and applied to finance (see e.g., Kring *et al.* (2009) and Dominicy *et al.* (2013)).

Most of the multivariate non-Gaussian financial models proposed in the academic literature have dealt with the portfolio allocation problem (Kawai (2009), Marfé (2012), and Kim *et al.* (2012)). Only few works have investigated option pricing under a multivariate non-Gaussian framework, mainly because the complexity of option pricing models increases as soon as one moves from a univariate to a multivariate setting (see Tassinari (2009), Wu *et al.* (2009), Kawai and Kohatsu-Higa (2010), Guillaume (2013), Tassinari and Corradi (2013), Tassinari and Bianchi (2014), and references therein).

In this chapter, we look closely at three models driven by subordinated Brownian motions with univariate (1) gamma, (2) normal inverse Gaussian, (3) and generalized inverse Gaussian. We extend the variance gamma (VG), the normal inverse Gaussian (NIG), and the generalized hyperbolic (GH) to a multivariate setting. In these three cases (VG, NIG and GH), we show how the density function can be computed in closed-form. All three models belong to the class of the GH model. More precisely, both the VG and NIG distributions are particular cases of the GH distribution (see McNeil *et al.*

(2005)). Then we focus on a fourth model that can be seen as a multivariate time-changed Brownian with one-sided classical tempered stable time or as a multivariate mean–variance models with a tempered stable mixture distribution, as proposed by Kim *et al.* (2012). In practice we extend the univariate normal tempered stable (NTS) model described in Chapter 6 to a multivariate setting. We recall that a formal elegant definition of tempered stable (TS) distributions and processes was proposed in the work of Rosinski (2007) and applied to finance in numerous empirical studies (see Rachev *et al.* (2011)), principally under a univariate framework. We show that both the VG and NIG distributions are also particular cases of the NTS distribution. In all four cases (VG, NIG, GH, and NTS), we further assume that the components of the underlying Brownian motion are correlated.

As shown in McNeil *et al.* (2005) for the GH class and in Kim *et al.* (2012) for the NTS case, the portfolio constructed as a linear combination of the multivariate GH (multivariate NTS) marginal distributions has a univariate GH (NTS) distribution. Thus, the models proposed in this chapter can be used to compute in a relatively simple way widely known portfolio risk measures. Furthermore, from a practical perspective, the simulation of random draws from these multivariate distributions is simple, mainly because they can be viewed as normal mean–variance mixtures and one only needs an efficient algorithm to simulate the univariate mixing distribution and a standard procedure to generate a multivariate normal distribution.

As discussed in Bianchi *et al.* (2016), from a theoretical standpoint, while for multivariate NTS (MNTS) processes a change of time scale leads to a change of spatial scale of the distribution of the increments, for multivariate GH (MGH) processes a change of time scale modifies the distributional properties of the increments (see Cont and Tankov (2003)). This means that if the distribution of the increments of a process on a daily basis is MGH, the distribution of the increments of the same process at other time scales is no longer MGH, thereby making it inconvenient for working with data on different time scales. This inconvenience comes from generalized inverse Gaussian processes and it may occur in practical applications such as in option pricing models that combine the information provided by daily log-returns observations and implied volatilities, usually computed on a yearly basis (see Tassinari and Bianchi (2014)). Furthermore, the corresponding Lévy processes are difficult to simulate, because the law depends on the time scale.

The chapter is organized as follows. After having defined the multivariate stock market model, in Sections 7.2.1–7.2.4 we define the multivariate time-changed Brownian motion with gamma, inverse Gaussian, generalized inverse Gaussian, and one-sided classical tempered stable stochastic time, respectively. In Section 7.3 we show that the increments of time-changed Brownian motions considered in this study can be seen as a normal mean–variance mixture, and three models we study are particular cases of the GH model. Thus, we show how to compute the probability density function for the four models. While in three cases it can be computed in closed form, the density of the multivariate NTS distribution does not have an analytic solution and therefore a numerical method is needed to evaluate it. In Section 7.4 we provide a simple algorithm to simulate random numbers from these multivariate distributions.

7.2. The Multivariate Time-Changed Brownian Motion

In our model we consider a market with n stocks, and we assume that the dynamics of asset log-returns are described through a Lévy process obtained by a multivariate Brownian motion, time-changed by a single one-dimensional non-negative non-decreasing Lévy process. We refer to this last Lévy process as a *subordinator* or a *stochastic clock*. We further assume that the components of the multivariate Brownian motion are correlated, and the subordinator is independent of the Brownian motion. Building a process in this way allows one to get two sources of dependence: (1) a jump in the subordinator produces a jump in the price processes and all jumps occur at the same time; (2) since we assume a correlated Brownian motion, the jump sizes are correlated.

The price at time t of stock j is given by the following equation:

$$P_t^j = P_0^j \exp(Y_t^j), \tag{7.1}$$

where P_t^j and P_0^j are the price of stock j at times t and 0, respectively, and Y_t^j is the log-return of the jth underlying asset over the interval $[0, t]$, for every $j = 1, \ldots, n$. Thus, the log-return process of the jth underlying asset is defined as

$$Y_t^j = \mu_j t + X_t^j = \mu_j t + \theta_j S_t + \sigma_j W_{S_t}^j, \tag{7.2}$$

where $X^j = (X_t^j)_{t\geq 0}$ is the pure jump part of the process, $S = (S_t)_{t\geq 0}$ is the subordinator, $W^j = (W_t^j)_{t\geq 0}$ and $W^k = (W_t^k)_{t\geq 0}$ are correlated one-dimensional Brownian motions with correlation coefficient ρ_{jk}, $W_S^j = (W_{S_t}^j)_{t\geq 0}$ is the jth Brownian motion evaluated at the common stochastic clock S_t. Then, μ, θ, and σ are vectors in $\mathbb{R}^n$, and the elements of the vector σ are strictly positive ($\sigma_j > 0$).

The characteristic function of the multivariate pure jump part process $X = (X_t)_{t\geq 0}$ can be computed by considering equation (4.6) in Cont and Tankov (2003), that is

$$\phi_{X_t}(u) = \exp\left(t l_{S_1}(\psi(u))\right), \tag{7.3}$$

where $l_{S_1}(\cdot)$ is the Laplace exponent of the common subordinator and $\psi(u)$ is the characteristic exponent of the multivariate Brownian motion, that is

$$\begin{aligned}\psi(u) &= iu'\theta - \frac{1}{2}u'\Sigma u \\ &= \sum_{j=1}^{n} iu_j\theta_j - \frac{1}{2}\sum_{j=1}^{n}\sum_{k=1}^{n} u_j u_k \sigma_j \sigma_k \rho_{jk},\end{aligned} \tag{7.4}$$

with $u \in \mathbb{R}^n$, and where the matrix Σ has elements $\sigma_{jk} = \sigma_j\sigma_k\rho_{jk}$. Since Σ is a variance-covariance matrix, we can rewrite equation (7.4) in the following form:

$$\psi(u) = iu'\theta - \frac{1}{2}u'D_\sigma \Omega D_\sigma u, \tag{7.5}$$

where D_σ is a diagonal matrix with diagonal $\sigma \in \mathbb{R}^n_+$, and Ω is the correlation matrix of the Brownian motions with elements ρ_{jk}. The characteristic function of the log-return process is given by

$$\phi_{Y_t}(u) = \exp\left(itu'\mu\right)\phi_{X_t}(u). \tag{7.6}$$

After having defined the multivariate model we want to apply, we consider four possible parametric examples, that is the *multivariate variance gamma* (MVG), the *multivariate normal inverse Gaussian* (MNIG), the *multivariate generalized hyperbolic* (MGH), and the *multivariate normal tempered stable* (MNTS) model.

7.2.1. The multivariate VG process

In this section we review the construction of the MVG process and provide some useful formulas. This process and its extension have been studied in different works. Luciano and Schoutens (2006) use the MVG model to describe the behavior of a series of stocks or indexes and to study a multifirm value-based default model. Leoni and Schoutens (2008) present an application of the VG distribution to price multivariate derivatives. Their work focuses on the practical implementation of the model in a multivariate setting and several calibration procedures are discussed and applied to the pricing of exotic structures. Semeraro (2008) subordinates a multivariate Brownian motion with independent components by a multivariate gamma subordinator. The dependence is introduced via a multivariate-dependent random vector. In this model, the parameters, which control the dependence, can be fitted using the sample covariance. Wang (2009) proposes a multivariate model based on the sum of independent VG. Guillaume (2012) extends the VG multivariate framework by considering processes with independent but time inhomogeneous increments (Sato processes). Guillaume (2013) propose an extension that allows a calibration not requiring the existence of actively traded multivariate derivatives. Hitaj and Mercuri (2013) cast the MVG model in a portfolio allocation problem. Tassinari and Corradi (2013) proposed a framework based on the Esscher transform to price equity and debt tranches of collateralized funds of hedge fund obligations. Tassinari and Bianchi (2014) introduce a joint calibration algorithm in which they simultaneously estimate the model parameters on the time series of log-returns by minimizing the Kolmogorov–Smirnov distance and calibrate the implied volatility surface by minimizing the average relative percentage error which is a measure of the distance between model and observed implied volatilities. Linders and Stassen (2016) propose a methodology for pricing basket options in the MVG framework and provide an approximate basket option pricing formula in order to have an efficient and fast calibration of the model. In this chapter, we will not analyze all alternative models proposed in the literature, but we provide only the simplest extension of a one-dimensional VG process to a multivariate framework: the MVG process.

Let $S = (S_t)_{t\geq 0}$ be a gamma process, that is, a process which starts at zero and has stationary and independent increments, in which the law of S_1 is gamma with parameters $\alpha > 0$ and $\beta > 0$. We refer to the law S_1 as $\Gamma(\alpha, \beta)$. The density function of the random variable S_1 is

$$f_{S_1}(x; \alpha, \beta) = \frac{\beta^\alpha}{\Gamma(\alpha)} x^{\alpha-1} \exp(-x\beta), \quad x > 0, \tag{7.7}$$

where

$$\Gamma(\alpha) = \int_0^\infty x^{\alpha-1} \exp(-x) dx, \tag{7.8}$$

and its characteristic function is

$$\phi_{S_1}(s) = \left(\frac{\beta}{\beta - is}\right)^\alpha. \tag{7.9}$$

From (7.9) it is possible to compute the Laplace exponent of the gamma subordinator

$$l_{S_1}(s) = \ln \phi_{S_1}\left(\frac{s}{i}\right) = \alpha \ln \frac{\beta}{\beta - s}, \qquad s < \beta, \tag{7.10}$$

and, from it, the cumulants

$$c_i\left[S_1\right] = \left[l_{S_1}^{(i)}(s)\right]_{s=0},$$

and the main moments

$$\begin{aligned}
E[S_1] &= c_1[S_1] = \frac{\alpha}{\beta}, \\
\mathrm{var}[S_1] &= c_2[S_1] = \frac{\alpha}{\beta^2}, \\
\mathrm{skew}[S_1] &= \frac{c_3[S_1]}{c_2[S_1]^{3/2}} = \frac{2}{\sqrt{\alpha}}, \\
\mathrm{kurt}[S_1] &= 3 + \frac{c_4[S_1]}{c_2[S_1]^2} = 3\left(1 + \frac{2}{\alpha}\right).
\end{aligned} \tag{7.11}$$

Finally, using (7.6) we get the characteristic function of the multivariate variance gamma (MVG) process with linear drift

$$\phi_{Y_t}(u) = \exp\left(itu'\mu\right)\left[1 - \frac{1}{\beta}\left(iu'\theta - \frac{1}{2}u'\Sigma u\right)\right]^{-\alpha t}. \tag{7.12}$$

Setting $u_i = 0$ into (7.12), $\forall i \neq j$, we get the characteristic function of the log-return process of the jth underlying asset

$$\phi_{Y_t^j}(u_j) = \exp\left(itu_j\mu_j\right)\left[1 - \frac{1}{\beta}(iu_j\theta_j - \frac{1}{2}u_j^2\sigma_j^2)\right]^{-\alpha t}. \tag{7.13}$$

From (7.12) it is possible to compute marginal and joint moments of log-return increments over the period $[0, t]$ and express them as functions of the moments of the subordinator. For the marginal moments the following equalities hold:

$$\begin{aligned}
E[Y_t^j] &= \left(\mu_j + \frac{\alpha}{\beta}\theta_j\right)t = \mu_j t + E[S_t]\theta_j, \\
\mathrm{var}[Y_t^j] &= \left(\beta\sigma_j^2 + \theta_j^2\right)\frac{\alpha}{\beta^2}t = \left(\beta\sigma_j^2 + \theta_j^2\right)\mathrm{var}[S_t],
\end{aligned}$$

$$\text{skew}[Y_t^j] = \frac{\theta_j\left(3\beta\sigma_j^2 + 2\theta_j^2\right)}{\sqrt{\alpha t}\left(\beta\sigma_j^2 + \theta_j^2\right)^{3/2}} = \frac{\theta_j\left(3\beta\sigma_j^2 + 2\theta_j^2\right)}{2\left(\beta\sigma_j^2 + \theta_j^2\right)^{3/2}}\text{skew}[S_t],$$

$$\begin{aligned}\text{kurt}[Y_t^j] &= 3\left[1 + \frac{2}{\alpha t} - \frac{\beta^2\sigma_j^4}{\alpha t(\beta\sigma_j^2 + \theta_j^2)^2}\right] \\ &= 3\left[1 + \frac{\text{kurt}[S_t] - 3}{3}\left(1 - \frac{\beta^2\sigma_j^4}{2(\beta\sigma_j^2 + \theta_j^2)^2}\right)\right], \qquad (7.14)\end{aligned}$$

and the correlation and the covariance are given by

$$\begin{aligned}\text{corr}[Y_t^j; Y_t^k] &= \frac{\theta_j\theta_k + \beta\sigma_j\sigma_k\rho_{jk}}{\sqrt{\left(\beta\sigma_j^2 + \theta_j^2\right)\left(\beta\sigma_k^2 + \theta_k^2\right)}}, \qquad (7.15) \\ \text{cov}[Y_t^j; Y_t^k] &= (\theta_j\theta_k + \beta\sigma_j\sigma_k\rho_{jk})\,\text{var}[S_t].\end{aligned}$$

From equation (7.15) it is clear that the correlation depends on both the correlation between the normal margins of the driving Brownian motion and the parameters of the stochastic time. From equation (7.15) it follows that we have null correlation if at least one margin has a symmetric distribution (i.e., θ_j or θ_k is zero) and the underlying Brownian motions are uncorrelated (i.e., ρ_{jk} is zero). However, due to the common stochastic clock null correlation does not imply independence.

Finally, assuming that $Y_0 = 0$ by equation (7.9) it follows that if Y_1 is MVG distributed with parameters (α, β, μ, θ, Σ) then the law of the increments $Y_{t+\Delta t} - Y_t$ is MVG with parameters ($\alpha\Delta t$, β, $\mu\Delta t$, θ, Σ). This means that a change of time scale leads to a change of spatial scale of the distribution of the increments.

7.2.2. The multivariate NIG process

In this section we review the construction of the MNIG process and provide some useful formulas. Also this process and its extensions have been studied in different works. Aas *et al.* (2005) use the MNIG as the conditional distribution for a multivariate GARCH model and they perform a value-at-risk (VaR) exercise on a portfolio of European, American, and Japanese equities. Then they show that for a one-day holding period the model proposed outperforms a Gaussian GARCH model and a Student's t-GARCH model, and it is slightly better than a skew Student's t-GARCH model. Oigard *et al.* (2005) prove that the MNIG model is useful for the modeling of impulsive signals, noise, and interference in multivariate sonar, radar, and communication systems. Wu *et al.* (2009) use an MNIG model to develop closed-form pricing formulas for both geometric and arithmetic basket options. For geometric basket options, an exact analytical solution is possible; for arithmetic basket options, the formula is an approximation. Ballotta and Bonfiglioli (2016) introduce a class of multivariate Lévy processes, investigate its general properties, and apply it to build multivariate VG, NIG, and jump diffusion processes. The models are tested by calibrating them to option prices. In this chapter we will not analyze all alternative models proposed in the literature, but provide only the simplest extension of a one-dimensional NIG process to a multivariate framework: the MNIG process.

Let $S_t = \{S_t,\ t \geq 0\}$ be an inverse Gaussian process, that is, a process which starts at zero and has stationary and independent increments, in which the law of S_1 is inverse Gaussian (IG) with parameters $\gamma > 0$ and $\eta > 0$. We refer to the law S_1 as IG(γ, η). The density function of the random variable S_1 is

$$f_{S_1}(x; \gamma, \eta) = \frac{\gamma}{\sqrt{2\pi}} \exp\left[-\frac{1}{2}\frac{(\eta x - \gamma)^2}{x}\right] x^{-3/2}, \quad x > 0, \tag{7.16}$$

and its characteristic function

$$\phi_{S_1}(s) = \exp[\gamma(\eta - \sqrt{\eta^2 - 2is})]. \tag{7.17}$$

From (7.17) it is possible to compute the Laplace exponent of the IG subordinator

$$l_{S_1}(s) = \ln \phi_{S_1}\left(\frac{s}{i}\right) = \gamma(\eta - \sqrt{\eta^2 - 2s}), \quad s \leq \eta^2/2, \tag{7.18}$$

and, using it, the following moments can be derived

$$
\begin{aligned}
E[S_1] &= \frac{\gamma}{\eta}, \\
\mathrm{var}[S_1] &= \frac{\gamma}{\eta^3}, \\
\mathrm{skew}[S_1] &= \frac{3}{\sqrt{\gamma\eta}}, \\
\mathrm{kurt}[S_1] &= 3\left(1 + \frac{5}{\gamma\eta}\right).
\end{aligned}
\tag{7.19}
$$

Finally, using (7.6) we get the characteristic function of the multivariate normal inverse Gaussian (MNIG) process with linear drift

$$
\phi_{Y_t}(u) = \exp\left\{t\left[iu'\mu + \gamma\left(\eta - \sqrt{\eta^2 - 2\left(iu'\theta - \frac{1}{2}u'\Sigma u\right)}\right)\right]\right\}. \tag{7.20}
$$

Setting $u_i = 0$, $\forall i \neq j$ into (7.20), we get the characteristic function of the log-return process of the jth underlying asset

$$
\phi_{Y_t^j}(u_j) = \exp\left\{t\left[iu_j\mu_j + \gamma\left(\eta - \sqrt{\eta^2 - 2\left(iu_j\theta_j - \frac{1}{2}u_j^2\sigma_j^2\right)}\right)\right]\right\}. \tag{7.21}
$$

From (7.20) it is possible to compute marginal and joint moments of log-return increments over the period $[0, t]$ and express them as functions of the moments of the subordinator. For the marginal moments, the following equalities hold:

$$
\begin{aligned}
E[Y_t^j] &= \left(\mu_j + \frac{\gamma}{\eta}\theta_j\right)t = \mu_j t + E[S_t]\theta_j, \\
\mathrm{var}[Y_t^j] &= \left(\sigma_j^2\eta^2 + \theta_j^2\right)\frac{\gamma}{\eta^3}t = \left(\sigma_j^2\eta^2 + \theta_j^2\right)\mathrm{var}[S_t],
\end{aligned}
$$

$$\text{skew}[Y_t^j] = \frac{3\theta_j}{\sqrt{\gamma\eta t\left(\sigma_j^2\eta^2+\theta_j^2\right)}} = \frac{\theta_j}{\sqrt{\sigma_j^2\eta^2+\theta_j^2}}\text{skew}[S_t], \quad (7.22)$$

$$\begin{aligned}\text{kurt}[Y_t^j] &= 3\left[1+\frac{5\theta_j^2+\sigma_j^2\eta^2}{\gamma\eta t\left(\sigma_j^2\eta^2+\theta_j^2\right)}\right]\\ &= 3\left[1+\frac{\text{kurt}[S_t]-3}{15}\left(\frac{5\theta_j^2+\sigma_j^2\eta^2}{\sigma_j^2\eta^2+\theta_j^2}\right)\right],\end{aligned}$$

and the correlation and the covariance are given by

$$\begin{aligned}\text{corr}[Y_t^j;Y_t^k] &= \frac{\theta_j\theta_k+\eta^2\sigma_j\sigma_k\rho_{jk}}{\sqrt{\left(\sigma_j^2\eta^2+\theta_j^2\right)\left(\sigma_k^2\eta^2+\theta_k^2\right)}},\\ \text{cov}[Y_t^j;Y_t^k] &= (\theta_j\theta_k+\eta^2\sigma_j\sigma_k\rho_{jk})\,\text{var}[S_t].\end{aligned} \quad (7.23)$$

From equation (7.23) it is clear that the correlation depends on both the correlation between the normal margins of the driving Brownian motion and the parameters of the stochastic time. From equation (7.23) it follows that we have null correlation if at least one margin has a symmetric distribution (i.e., θ_j or θ_k is zero) and the underlying Brownian motions are uncorrelated (i.e., ρ_{jk} is zero). However, due to the common stochastic clock, null correlation does not imply independence.

Finally, assuming that $Y_0 = 0$ by equation (7.17) it follows that if Y_1 is MNIG distributed with parameters $(\gamma, \eta, \mu, \theta, \Sigma)$, then the law of the increments $Y_{t+\Delta t} - Y_t$ is MNIG with parameters $(\gamma\Delta t, \eta, \mu\Delta t, \theta, \Sigma)$. This means that a change of time scale leads to a change of spatial scale of the distribution of the increments.

7.2.3. The multivariate GH model

The generalize hyperbolic (GH) distribution has received a lot of attention in the financial-modeling literature, particularly in the univariate case (see Chapter 4 and McNeil *et al.* (2005)). The GH parametric family includes many familiar distributions such as, for example, Student's t, the skew-t, the VG, and the NIG. In this section we study its multivariate extension (MGH) and provide some useful formulas. Barndorff-Nielsen *et al.* (1992) first introduced the class of MGH distributions. Prause (1999) studies MGH distributions with a look-forward estimation and option pricing, proposing an approach to estimate these distributions also for high-dimensional data efficiently, and discussing the generation of random variates and the pricing of derivatives following the Esscher approach. Protassov (2004) shows how it is possible to perform a fast historical estimation on the time series of log-returns through the so-called expectation–maximization (EM) maximum likelihood estimation method when a parameter is fixed. Hu (2005), McNeil *et al.* (2005), and Hu and Kercheval (2007) describe a way to stably calibrate MGH distributions for a wider range of parameters and provide a framework in which it is possible to evaluate portfolio risk measures and solve a portfolio optimization problem. Hellmich and Kassberger (2011) apply the MGH distribution to portfolio modeling, using average value-at-risk (AVaR) as a risk measure. In this chapter we review the model proposed in McNeil *et al.* (2005) from a multivariate perspective.

Let $S = (S_t)_{t\geq 0}$ be a generalized inverse Gaussian (GIG) process, i.e., a process which starts at zero and has stationary and independent increments. $S = (S_t)_{t\geq 0}$ is a Lévy process in which the defining distribution of the increments $S_{\Delta t} = S_{t+\Delta t} - S_t$ is GIG with parameters λ, ψ, χ. The parameters ψ and χ are both non-negative and not simultaneously 0. The density function of the random variable $S_{\Delta t}$ is

$$\begin{aligned} f_{S_1}(x;\lambda,\psi,\chi) &= \frac{1}{2K_\lambda\left(\sqrt{\chi\psi}\right)}\left(\frac{\psi}{\chi}\right)^{\frac{\lambda}{2}} x^{\lambda-1} \\ &\quad \times \exp\left[-\frac{1}{2}\left(\frac{\chi}{x}+\psi x\right)\right],\ x>0, \end{aligned} \tag{7.24}$$

and its characteristic function is

$$\phi_{S_{\Delta t}}(s) = \left(1-\frac{2is}{\psi}\right)^{-\frac{\lambda}{2}} \frac{K_\lambda(\sqrt{\chi(\psi-2is)})}{K_\lambda\left(\sqrt{\chi\psi}\right)}, \tag{7.25}$$

where K_λ is the modified Bessel function of the second kind of order λ. Even if the function K_λ is a special function, efficient algorithms for the evaluation of this function are available. For example, in Matlab it is implemented as *besselk*.

From (7.25) it is possible to compute the Laplace exponent of the GIG subordinator

$$\begin{aligned} l_{S_{\Delta t}}(s) &= \ln \phi_{S_{\Delta t}}\left(\frac{s}{i}\right) \\ &= -\frac{\lambda}{2}\ln\left(1-\frac{2s}{\psi}\right) + \ln\frac{K_\lambda\left(\sqrt{\chi(\psi-2s)}\right)}{K_\lambda\left(\sqrt{\chi\psi}\right)}, \quad s<\psi/2, \end{aligned} \tag{7.26}$$

evaluate its cumulants

$$\begin{aligned} c_1[S_{\Delta t}] &= \left(\frac{\chi}{\psi}\right)^{\frac{1}{2}} \frac{K_{\lambda+1}\left(\sqrt{\chi\psi}\right)}{K_\lambda\left(\sqrt{\chi\psi}\right)}, \\ c_2[S_{\Delta t}] &= \left(\frac{\chi}{\psi}\right)\left[\frac{K_{\lambda+2}\left(\sqrt{\chi\psi}\right)}{K_\lambda\left(\sqrt{\chi\psi}\right)} - \left(\frac{K_{\lambda+1}\left(\sqrt{\chi\psi}\right)}{K_\lambda\left(\sqrt{\chi\psi}\right)}\right)^2\right], \\ c_3[S_{\Delta t}] &= \left(\frac{\chi}{\psi}\right)^{\frac{3}{2}}\left[\frac{K_{\lambda+3}\left(\sqrt{\chi\psi}\right)}{K_\lambda\left(\sqrt{\chi\psi}\right)} - \frac{3K_{\lambda+2}\left(\sqrt{\chi\psi}\right)K_{\lambda+1}\left(\sqrt{\chi\psi}\right)}{K_\lambda^2\left(\sqrt{\chi\psi}\right)}\right. \\ &\qquad \left. +2\left(\frac{K_{\lambda+1}\left(\sqrt{\chi\psi}\right)}{K_\lambda\left(\sqrt{\chi\psi}\right)}\right)^3\right], \\ c_4[S_{\Delta t}] &= \left(\frac{\chi}{\psi}\right)^2\left[\frac{K_{\lambda+4}\left(\sqrt{\chi\psi}\right)}{K_\lambda\left(\sqrt{\chi\psi}\right)} - \frac{4K_{\lambda+3}\left(\sqrt{\chi\psi}\right)K_{\lambda+1}\left(\sqrt{\chi\psi}\right)}{K_\lambda^2\left(\sqrt{\chi\psi}\right)}\right. \\ &\qquad \left. -3\left(\frac{K_{\lambda+2}\left(\sqrt{\chi\psi}\right)}{K_\lambda\left(\sqrt{\chi\psi}\right)}\right)^2\right] \end{aligned} \tag{7.27}$$

$$+6\left(\frac{\chi}{\psi}\right)^2\left[\frac{2K_{\lambda+2}\left(\sqrt{\chi\psi}\right)K_{\lambda+1}^2\left(\sqrt{\chi\psi}\right)}{K_\lambda^3\left(\sqrt{\chi\psi}\right)}\right.$$
$$\left.-\left(\frac{K_{\lambda+1}\left(\sqrt{\chi\psi}\right)}{K_\lambda\left(\sqrt{\chi\psi}\right)}\right)^4\right],$$

and the first four main moments

$$E[S_{\Delta t}] = c_1[S_{\Delta t}],$$
$$\mathrm{var}[S_{\Delta t}] = c_2[S_{\Delta t}],$$
$$\mathrm{skew}[S_{\Delta t}] = \frac{c_3[S_{\Delta t}]}{c_2[S_{\Delta t}]^{3/2}},$$
$$\mathrm{kurt}[S_{\Delta t}] = 3 + \frac{c_4[S_{\Delta t}]}{c_2[S_{\Delta t}]^2}.$$

Finally, using (7.6) we get the characteristic function of the increments $Y_{\Delta t} = Y_{t+\Delta t} - Y_t$ of the multivariate generalized hyperbolic (MGH) process with linear drift

$$\phi_{Y_{\Delta t}}(u) = \exp\left(iu'\mu\right)\left[1-\frac{2}{\psi}\left(iu'\theta-\frac{1}{2}u'\Sigma u\right)\right]^{-\frac{\lambda}{2}}$$
$$\times\frac{K_\lambda\left(\sqrt{\chi\left(\psi-2\left(iu'\theta-\frac{1}{2}u'\Sigma u\right)\right)}\right)}{K_\lambda\left(\sqrt{\chi\psi}\right)}. \tag{7.28}$$

Setting $u_i = 0$, $\forall i \neq j$, into (7.28) we get the characteristic function of the log-return process of the jth underlying asset

$$\phi_{Y^j_{\Delta t}}(u_j) = \exp\left(iu_j\mu_j\right)\left[1-\frac{2}{\psi}\left(iu_j\theta_j-\frac{1}{2}u_j^2\sigma_j^2\right)\right]^{-\frac{\lambda}{2}}$$
$$\times\frac{K_\lambda\left(\sqrt{\chi\left(\psi-2\left(iu_j\theta_j-\frac{1}{2}u_j^2\sigma_j^2\right)\right)}\right)}{K_\lambda\left(\sqrt{\chi\psi}\right)}. \tag{7.29}$$

As already observed, both the MGV described in Section 7.2.1 and the MNIG models described in Section 7.2.2 are special cases of the MGH one. If we set $\chi = 0$ and $\lambda > 0$, $S_{\Delta t}$ follows a gamma distribution with parameters, $\alpha = \lambda$ and $\beta = \psi/2$. If we set $\lambda = -1/2$, $S_{\Delta t}$ follows an IG distribution with parameters $\gamma = \sqrt{\chi}$ and $\eta = \sqrt{\psi}$.

From (7.28) it is possible to compute marginal and joint moments of log-return increments over the period $[t, t+\Delta t]$ and express them as functions of the moments and cumulants of the subordinator. For the marginal moments the following equalities hold

$$\begin{aligned} E[Y^j_{\Delta t}] &= \mu_j + E[S_{\Delta t}]\theta_j, \\ \mathrm{var}[Y^j_{\Delta t}] &= E[S_{\Delta t}]\sigma_j^2 + \mathrm{var}[S_{\Delta t}]\theta_j^2, \\ \mathrm{skew}[Y^j_{\Delta t}] &= \frac{c_3[Y^j_{\Delta t}]}{\mathrm{var}[Y^j_{\Delta t}]^{3/2}}, \\ \mathrm{kurt}[Y^j_{\Delta t}] &= 3 + \frac{c_4[Y^j_{\Delta t}]}{\mathrm{var}[Y^j_{\Delta t}]^2}, \end{aligned} \tag{7.30}$$

where

$$\begin{aligned} c_3\left[Y^j_{\Delta t}\right] &= 3\,\mathrm{var}[S_{\Delta t}]\theta_j\sigma_j^2 + c_3\left[S_{\Delta t}\right]\theta_j^3, \\ c_4\left[Y^j_{\Delta t}\right] &= 3\,\mathrm{var}[S_{\Delta t}]\sigma_j^4 + 6c_3\left[S_{\Delta t}\right]\theta_j\sigma_j^2 + c_4\left[S_{\Delta t}\right]\theta_j^4, \end{aligned}$$

and the correlation and the covariance are given by

$$\begin{aligned} \mathrm{corr}[Y^j_{\Delta t}; Y^k_{\Delta t}] &= \frac{\theta_j\theta_k\Delta\sqrt{\chi} + \sqrt{\psi}\sigma_j\sigma_k\rho_{jk}}{\sqrt{\left[\theta_j^2\Delta\sqrt{\chi} + \sigma_j^2\sqrt{\psi}\right]\left[\theta_k^2\Delta\sqrt{\chi} + \sigma_k^2\sqrt{\psi}\right]}}, \\ \mathrm{cov}[Y^j_{\Delta t}; Y^k_{\Delta t}] &= E\left[S_{\Delta t}\right]\sigma_{jk} + \mathrm{var}[S_{\Delta t}]\theta_j\theta_k, \end{aligned} \tag{7.31}$$

where

$$\Delta = \left(\frac{K_{\lambda+2}\left(\sqrt{\chi\psi}\right)}{K_{\lambda+1}\left(\sqrt{\chi\psi}\right)} - \frac{K_{\lambda+1}\left(\sqrt{\chi\psi}\right)}{K_{\lambda}\left(\sqrt{\chi\psi}\right)}\right). \tag{7.32}$$

The correlation depends on both the correlation between the normal margins of the driving Brownian motion and the parameters of the stochastic time. From equation (7.31) it follows that if both χ and ψ are strictly positive, we have null correlation if at least one margin has a symmetric distribution (i.e., θ_j or θ_k is zero) and the underlying Brownian motions are uncorrelated (i.e., ρ_{jk} is zero). However, due to the common stochastic clock, null correlation does not imply independence.

In contrast to both the MVG and MNIG cases, assuming that $Y_0 = 0$ by equation (7.25) it follows that if Y_1 is MGH distributed with parameters (λ, χ, ψ, μ, θ, Σ), then the law of the increments $Y_{t+\Delta t} - Y_t$ is no longer MGH distributed. This means that a change of time scale modifies the distributional properties of the increments. More specifically, if for the time scale Δt the parameters of the MGH law are (λ, χ, ψ, μ, θ, Σ), then for a different time scale $\Delta t'$ the characteristic function of the increments is

$$\phi_{Y_{\Delta t'}}(u) = \exp\left(iu'\mu\Upsilon\right)\left(\left[1 - \frac{2}{\psi}\left(iu'\theta - \frac{1}{2}u'\Sigma u\right)\right]^{-\frac{\lambda}{2}}\right.$$

$$\left.\times\frac{K_\lambda\left(\sqrt{\chi\left(\psi - 2\left(iu'\theta - \frac{1}{2}u'\Sigma u\right)\right)}\right)}{K_\lambda\left(\sqrt{\chi\psi}\right)}\right)^{\Upsilon},$$

where

$$\Upsilon = \frac{\Delta t'}{\Delta t}.$$

This in practice means that for a different time scale we do not know the distribution of the increments but only their characteristic function.

By considering Proposition 3.13 in McNeil *et al.* (2005), it follows that if $w \in \mathbb{R}^n$, then the random variable $w'Y_{\Delta t}$ has a GH distribution with parameters (λ, χ, ψ, $\tilde{\theta}$, $\tilde{\mu}$, $\tilde{\sigma}$), where

$$\tilde{\theta} = w'\theta, \qquad \tilde{\mu} = w'\mu, \qquad \tilde{\sigma} = \sqrt{w'\Sigma w}.$$

A similar result also holds for the MVG and the MNIG case, since both laws are particular cases of the MGH law.

7.2.4. The multivariate NTS model

In this section we review the construction of the MNTS process and provide some useful formulas. Kim *et al.* (2012) describe an MNTS distributed market model and derive the closed-form solution for two important measures used by portfolio managers — the marginal VaR and the marginal AVaR. Guerard *et al.* (2013) consider the MNTS distribution to capture the dependence between different securities and to several portfolio construction and optimization techniques to US and global stock universes. Kim *et al.* (2015) present an approach to pricing European quanto options assuming that the underlying instruments follow an MNTS process. Bianchi *et al.* (2016) extend the EM-based maximum likelihood estimation algorithm to normal mean–variance mixture distributions in which only the characteristic function of the mixing distribution is known in closed form and, conversely, its density function is not. The authors show that the computational effort needed to evaluate the likelihood function of an MNTS distribution is equal to the computation of a univariate density with a fast Fourier transform (FFT) algorithm and a numerical integration of a one-dimensional integral. The method is applied to fit a five- and a 30-dimensional series of stock returns and to evaluate widely known portfolio risk measures. In their study, the MNTS model is compared with the MGH model. In this chapter, we follow the construction proposed in Bianchi *et al.* (2016).

Let $S = (S_t)_{t\geq 0}$ be a classical tempered stable subordinator (CTS), that is, an increasing process which starts at zero and has stationary and independent increments, in which the law of S_1 is classical tempered stable (CTS) with parameters $\lambda > 0$, $C > 0$, and $0 < \alpha < 1$. We refer to the law S_1 as CTS (α, λ, C). Sometimes it is convenient working with the alternative parameterization (a, λ, C), where $\alpha = \frac{a}{2}$.

The characteristic function is

$$\phi_{S_1}(s) = \exp\left(C\Gamma(-\alpha)((\lambda - is)^\alpha - \lambda^\alpha)\right), \tag{7.33}$$

while the density function of the random variable S_1 is not known in closed form.

From (7.33), it is possible to compute the Laplace exponent of the classical tempered stable subordinator

$$l_{S_1}(s) = \ln \phi_{S_1}\left(\frac{s}{i}\right) = C\Gamma(-\alpha)((\lambda - s)^\alpha - \lambda^\alpha), \quad s < \lambda, \tag{7.34}$$

and its moments

$$\begin{aligned} E[S_1] &= -\alpha C\Gamma(-\alpha)\lambda^{\alpha-1}, \\ \text{var}[S_1] &= \alpha(\alpha-1)C\Gamma(-\alpha)\lambda^{\alpha-2}, \\ \text{skew}[S_1] &= (2-\alpha)\left[\alpha(\alpha-1)C\Gamma(-\alpha)\lambda^\alpha\right]^{-\frac{1}{2}}, \\ \text{kurt}[S_1] &= 3 + (\alpha-2)(\alpha-3)\left[\alpha(\alpha-1)C\Gamma(-\alpha)\lambda^\alpha\right]^{-1}. \end{aligned} \tag{7.35}$$

Finally, using (7.6) we get the characteristic function of the multivariate normal classical tempered stable (MNTS) process with linear drift

$$\begin{aligned} \phi_{Y_t}(u) = \exp\Bigg\{t\Bigg[iu'\mu + C\Gamma(-\alpha) \\ \times\left(\left(\lambda - iu'\theta + \frac{1}{2}u'\Sigma u\right)^\alpha - \lambda^\alpha\right)\Bigg]\Bigg\}. \end{aligned} \tag{7.36}$$

Setting $u_i = 0$, $\forall i \neq j$, into (7.36) we get the characteristic function of the log-return process of the jth underlying asset

$$\begin{aligned} \phi_{Y_t^j}(u_j) = \exp\Bigg\{t\Bigg[iu_j\mu_j + C\Gamma(-\alpha) \\ \times\left(\left(\lambda - iu_j\theta_j + \frac{1}{2}u_j^2\sigma_j^2\right)^\alpha - \lambda^\alpha\right)\Bigg]\Bigg\}. \end{aligned} \tag{7.37}$$

Both the MGV described in Section 7.2.1 and the MNIG models described in Section 7.2.2 are special cases of the MNTS one. If $\alpha \to 0$ (here we refer to α of the CTS subordinator), S_1 follows a gamma distribution with parameters $\alpha = C$ and $\beta = \lambda$ (here we refer to α and β of the gamma distribution). If we set $\alpha = 1/2$ (here we refer to α of the CTS subordinator), S_1 follows an IG distribution with parameters $\gamma = C\sqrt{2\pi}$ and $\eta = \sqrt{2\lambda}$.

From (7.36) it is possible to compute marginal and joint moments of log-return increments over the period $[0, t]$ and express them as functions of the moments of the subordinator. For the marginal moments, the following equalities hold:

$$E[Y_t^j] = \mu_j t + E[S_t]\theta_j,$$

$$\mathrm{var}[Y_t^j] = \left(\theta_j^2 + \frac{\sigma_j^2\lambda}{1-\alpha}\right)\mathrm{var}[S_t],$$

$$\mathrm{skew}[Y_t^j] = \theta_j\left(\theta_j^2 + \frac{3\sigma_j^2\lambda}{2-\alpha}\right)\left(\theta_j^2 + \frac{\sigma_j^2\lambda}{1-\alpha}\right)^{-\frac{3}{2}}\mathrm{skew}[S_t],$$

$$\begin{aligned}\mathrm{kurt}[Y_t^j] &= 3 + (\mathrm{kurt}[S_t] - 3)\\ &\quad\times\left[\theta_j^4 + \frac{3\sigma_j^2\lambda}{3-\alpha}\left(2\theta_j^2 + \frac{\sigma_j^2\lambda}{2-\alpha}\right)\right]\left(\theta_j^2 + \frac{\sigma_j^2\lambda}{1-\alpha}\right)^{-2},\end{aligned} \tag{7.38}$$

and the correlation and the covariance are given by

$$\begin{aligned}\mathrm{corr}[Y_t^j; Y_t^k] &= \frac{\theta_j\theta_k(1-\alpha) + \lambda\sigma_j\sigma_k\rho_{jk}}{\sqrt{\left(\theta_j^2(1-\alpha) + \lambda\sigma_j^2\right)\left(\theta_k^2(1-\alpha) + \lambda\sigma_k^2\right)}},\\ \mathrm{cov}[Y_t^j; Y_t^k] &= \mathrm{var}[S_t]\left(\theta_j\theta_k + \frac{\sigma_{jk}\lambda}{1-\alpha}\right).\end{aligned} \tag{7.39}$$

The correlation depends on both the correlation between the normal margins of the driving Brownian motion and the parameters of the stochastic time. From equation (7.39) it follows that we have null correlation if at least one margin has a symmetric distribution (i.e., θ_j or θ_k is zero) and the underlying Brownian motions are uncorrelated (i.e., ρ_{jk} is zero). However, due to the common stochastic clock null correlation does not imply independence.

Assuming that $Y_0 = 0$ by equation (7.33) it follows that if Y_1 is MNTS distributed with parameters (α, C, λ μ, θ, Σ), then the law of the increments $Y_{t+\Delta t} - Y_t$ is MNTS with parameters (α, $C\Delta t$, λ, $\mu\Delta t$, θ, Σ). This means that a change of time scale leads to a change of spatial scale of the distribution of the increments.

By considering the result of Proposition 1 in Kim *et al.* (2012), it follows that if $w \in \mathbb{R}^n$, then the random variable $w'Y_1$ has characteristic function (7.37) with parameters $(a, \lambda, C, \tilde{\theta}, \tilde{\mu}, \tilde{\sigma})$, where

$$\tilde{\theta} = w'\theta, \qquad \tilde{\mu} = w'\mu, \qquad \tilde{\sigma} = \sqrt{w'\Sigma w}.$$

This result is useful for computing portfolio risk measures since one has the characteristic function of the one-dimensional portfolio returns. Given the characteristic function, numerical procedure based on the evaluation of the cumulative distribution function can be used to compute both the VaR and the AVaR.

7.3. A Normal Mean–Variance Perspective

The relation between subordinated multivariate Brownian motions and multivariate distributions defined as normal mean–variance mixture (see McNeil *et al.* (2005)) is the subject of this section. Consider the random process $Y = (Y_t)_{t\geq 0}$ in $\mathbb{R}^n$, where the jth component is defined in equation (7.2) as

$$Y_t^j = \mu_j t + \theta_j S_t + \sigma_j W_{S_t}^j, \tag{7.40}$$

where $S = (S_t)_{t\geq 0}$ is a subordinator, $W = (W_t)_{t\geq 0}$ is a multivariate Brownian motion in $\mathbb{R}^n$, $W^j = (W_t^j)_{t\geq 0}$ and $W^k = (W_t^k)_{t\geq 0}$ are correlated with correlation coefficient ρ_{jk}, $W_S^j = (W_{S_t}^j)_{t\geq 0}$ is the jth component of the Brownian motion evaluated at the common stochastic clock S_t. Then, μ, θ, and σ are vectors in $\mathbb{R}^n$, and the characteristic exponent of the multivariate Brownian motion is given by equation (7.4). The random vector Y_t defined in equation (7.40) can be rewritten,[1] at time $t = 1$, as

$$Y_1 = \mu + \theta S_1 + \sqrt{S_1} D_\sigma A Z, \tag{7.41}$$

where

[1]This formulation has been briefly described in Chapter 4.

- Z is a multivariate normal distribution with zero mean and unit variance–covariance matrix I of dimension n;
- assuming that $S_0 = 0$, S_1 is the law of the increments of the subordinator at time $t = 1$ with Laplace exponent $l_{S_1}(s)$ and S_1 is independent of Z;
- A is the lower Cholesky decomposition of the correlation matrix Ω, that is, $\Omega^{1/2} = A$; furthermore, $\Sigma^{1/2} = D_\sigma A$, where D_σ is the diagonal matrix with diagonal $\sigma \in \mathbb{R}^n_+$ defined in Section 7.2.

The equality between (7.40) and (7.41) follows by the fact that the characteristic function of the random vector in equation (7.41) is given by

$$\phi_{Y_1}(u) = \exp(iu'\mu)\exp\left(l_{S_1}(\psi(u))\right), \tag{7.42}$$

where $\psi(u)$ is defined in equation (7.4). This means that equation (7.42) coincides with equation (7.6) in $t = 1$. It follows that for each discrete-time step Δt the distribution of the increments of Y_t is given by

$$Y_{\Delta t} = \mu\Delta t + \theta S_{\Delta t} + \sqrt{S_{\Delta t}} D_\sigma A Z,$$

where we refer to $S_{\Delta t}$ as the law of the increments of the subordinator $S = (S_t)_{t\geq 0}$.

Additionally, if Σ is non-singular, it is possible to find the density function related to the characteristic function in equation (7.42). Thus, the density function $f_Y(y)$ can be written as

$$f_Y(y) = \int_0^\infty \frac{e^{(y-\mu)'\Sigma^{-1}\theta}}{(2\pi)^{n/2}|\Sigma|^{1/2}w^{n/2}} \times \exp\left\{-\frac{(y-\mu)'\Sigma^{-1}(y-\mu)}{2w} - \frac{\theta'\Sigma^{-1}\theta}{2/w}\right\} f_{S_1}(w)dw, \tag{7.43}$$

where f_{S_1} is the density of S_1 (see McNeil *et al.* (2005) and references therein). As we will see in the following examples in Sections 7.3.1–7.3.3, this finding is particularly useful in applications.

7.3.1. The multivariate VG density

Given equation (7.43), we compute the density function of an MVG random variable. In the MVG case, the density of the subordinator $S = (S_t)_{t\geq 0}$ at time $t = 1$ is given by equation (7.7). Thus, equation (7.43) becomes

$$\begin{aligned} f_{Y_{\mathrm{MVG}}}(y) &= \int_0^{\infty} \frac{e^{(y-\mu)'\Sigma^{-1}\theta}}{(2\pi)^{n/2}|\Sigma|^{1/2}w^{n/2}} \\ &\quad \times \exp\left\{-\frac{(y-\mu)'\Sigma^{-1}(y-\mu)}{2w} - \frac{\theta'\Sigma^{-1}\theta}{2/w}\right\} \\ &\quad \times \frac{\beta^{\alpha}}{\Gamma(\alpha)} w^{\alpha-1} \exp\left(-w\beta\right) dw \\ &= c\int_0^{\infty} \frac{1}{w^{n/2-\alpha+1}} \exp\left(-\frac{a}{w} - bw\right) dw, \end{aligned} \tag{7.44}$$

where

$$\begin{aligned} c &= \frac{\beta^{\alpha} e^{(y-\mu)'\Sigma^{-1}\theta}}{\Gamma(\alpha)(2\pi)^{n/2}|\Sigma|^{1/2}}, \\ a &= \frac{(y-\mu)'\Sigma^{-1}(y-\mu)}{2}, \\ b &= \frac{\theta'\Sigma^{-1}\theta}{2} + \beta. \end{aligned}$$

From the equality (3.471.9) in Gradshteyn and Ryzhik (2007), that is

$$\int_0^{\infty} x^{\nu-1} e^{-\frac{a}{x} - bx} dx = 2\left(\frac{a}{b}\right)^{\frac{\nu}{2}} K_{\nu}(2\sqrt{ab}), \tag{7.45}$$

the last term in (7.44) becomes

$$c\int_0^{\infty} w^{\alpha-n/2-1} \exp\left(-\frac{a}{w} - bw\right) dw = 2c\left(\frac{a}{b}\right)^{\frac{\nu}{2}} K_{\nu}(2\sqrt{ab}), \tag{7.46}$$

where $\nu = \alpha - n/2$ and, as in Section 7.3.2, K_ν is the modified Bessel function of the second kind of order ν.

Matlab code — MVG density

In this section we show how to numerically compute the probability density function from equation (7.44). We provide an implementation extensively based on the Matlab matrix calculus. Since the MVG is a special case of the MGH, we consider the parameters λ and ψ, and the equalities $\alpha = \lambda$ and $\beta = \psi/2$. In the code, the variable d refers to the number of stocks (i.e., $d = n$). The input x is the matrix of observations with dimension $N \times d$.

```
% implementation of the density function of the MVG

function y = densityMVG_matrix(x, lambda, psi, mu, theta, ...
    Sigma, dt)

lambda   = lambda*dt;
mu       = mu*dt;

% row vectors
mu       = mu(:)';
theta    = theta(:)';

[N, d] = size(x);

L = chol(Sigma,'lower');

alpha = lambda;
beta  = psi/2;

A = ((x - ones(N,1)*mu)/L');
A = sum(A.^2,2)/2;

B = theta*(Sigma\theta')/2 + beta;

C = beta^(alpha) * exp((x - ones(N,1)*mu)*(Sigma\theta')) ...
    / ((2*pi)^(d/2)*det(Sigma)^(1/2)*gamma(alpha));

Z = 2*sqrt(A*B);

y = 2*C.*(A/B).^(alpha/2 - d/4).*besselk(alpha - d/2, Z);

return
```

7.3.2. The multivariate NIG density

Given equation (7.43), we compute the density function of an MNIG random variable. In the MNIG case, the density of the subordinator $S = (S_t)_{t\geq 0}$ at time $t = 1$ is given by equation (7.16). Thus, equation (7.43) becomes

$$\begin{aligned}
f_{Y_{\text{MNIG}}}(y) &= \int_0^\infty \frac{e^{(y-\mu)'\Sigma^{-1}\theta}}{(2\pi)^{n/2}|\Sigma|^{1/2}w^{n/2}} \\
&\quad \times \exp\left\{-\frac{(y-\mu)'\Sigma^{-1}(y-\mu)}{2w} - \frac{\theta'\Sigma^{-1}\theta}{2/w}\right\} \\
&\quad \times \frac{\gamma}{\sqrt{2\pi}} \exp\left\{-\frac{1}{2}\frac{\eta^2 w^2 - 2\eta\gamma w + \gamma^2}{w}\right\} w^{-3/2} dw \\
&= \int_0^\infty \frac{\gamma e^{(y-\mu)'\Sigma^{-1}\theta}}{(2\pi)^{n/2+1/2}|\Sigma|^{1/2}w^{n/2+3/2}} \\
&\quad \times \exp\left\{-\frac{(y-\mu)'\Sigma^{-1}(y-\mu)}{2w}\right. \\
&\quad \left. - \frac{\theta'\Sigma^{-1}\theta}{2/w} - \frac{1}{2}\eta^2 w + \eta\gamma - \frac{\gamma^2}{2w}\right\} dw \\
&= c\int_0^\infty w^{-(n+1)/2-1} \exp\left(-\frac{a}{w} - bw\right) dw, \qquad (7.47)
\end{aligned}$$

where

$$\begin{aligned}
c &= \frac{\gamma e^{(y-\mu)'\Sigma^{-1}\theta+\eta\gamma}}{(2\pi)^{n/2+1/2}|\Sigma|^{1/2}}, \\
a &= \frac{(y-\mu)'\Sigma^{-1}(y-\mu)}{2} + \frac{\gamma^2}{2}, \\
b &= \frac{\theta'\Sigma^{-1}\theta}{2} + \frac{\eta^2}{2}.
\end{aligned}$$

From the equality (3.471.9) in Gradshteyn and Ryzhik (2007) (see equation (7.45)), the last term in (7.47) becomes

$$c\int_0^\infty w^{-(n+1)/2-1} \exp\left(-\frac{a}{w} - bw\right) dw = 2c\left(\frac{a}{b}\right)^{\frac{\nu}{2}} K_\nu(2\sqrt{ab}), \qquad (7.48)$$

where $\nu = -(n+1)/2$ and K_ν is the modified Bessel function of the second kind of order ν. As in Section 7.3.1, we consider the function implemented in the Matlab *besselk* function.

Matlab code — MNIG density

In this section we show how to numerically compute the probability density function from equation (7.47). Also in this case, we provide an implementation extensively based on the Matlab matrix calculus. Since the MNIG is a special case of the MGH, we consider the parameters χ and ψ, and then the equalities $\gamma = \sqrt{\chi}$ and $\eta = \sqrt{\psi}$. In the code, the variable d refers to the number of stocks (i.e., $d = n$). The input x is the matrix of observations with dimension $N \times d$.

```
% implementation of the density function of the MNIG

function y = densityMNIG_matrix(x, chi, psi, mu, theta, Sigma, dt)

chi      = chi*dt^2;
mu       = mu*dt;

% row vectors
mu       = mu(:)';
theta    = theta(:)';

[N, d] = size(x);

L = chol(Sigma,'lower');

gamma = sqrt(chi);
eta   = sqrt(psi);

A = ((x - ones(N,1)*mu)/L');
A = sum(A.^2,2)/2 + gamma^2/2;

B = theta*(Sigma\theta')/2 + eta^2/2;

C = gamma * exp((x - ones(N,1)*mu)*(Sigma\theta') + gamma*eta) ...
    / ((2*pi)^(d/2+1/2)*det(Sigma)^(1/2));

Z = 2*sqrt(A*B);

nu = -(d + 1)/2;
```

```
y = 2*C.*(A/B).^(nu/2).*besselk(nu, Z);

return
```

7.3.3. The multivariate GH density

Given equation (7.43), we compute the density function of an MGH random variable. In the MGH case, the density of the subordinator $S = (S_t)_{t\geq 0}$ at time $t = 1$ is given by equation (7.24). Thus, equation (7.43) becomes

$$\begin{aligned}
f_{Y_{\mathrm{MGH}}}(y) &= \int_0^\infty \frac{e^{(y-\mu)'\Sigma^{-1}\theta}}{(2\pi)^{n/2}|\Sigma|^{1/2}w^{n/2}} \\
&\quad \times \exp\left\{-\frac{(y-\mu)'\Sigma^{-1}(y-\mu)}{2w} - \frac{\theta'\Sigma^{-1}\theta}{2/w}\right\} \\
&\quad \times \frac{1}{2K_\lambda\left(\sqrt{\chi\psi}\right)}\left(\frac{\psi}{\chi}\right)^{\frac{\lambda}{2}} w^{\lambda-1}\exp\left[-\frac{1}{2}\left(\frac{\chi}{w}+\psi w\right)\right] dw \\
&= \int_0^\infty \left(\frac{\psi}{\chi}\right)^{\frac{\lambda}{2}} \frac{e^{(y-\mu)'\Sigma^{-1}\theta}}{2K_\lambda\left(\sqrt{\chi\psi}\right)(2\pi)^{n/2}|\Sigma|^{1/2}} w^{\lambda-n/2-1} \\
&\quad \times \exp\left\{-\frac{(y-\mu)'\Sigma^{-1}(y-\mu)+\chi}{2w} - \frac{\theta'\Sigma^{-1}\theta+\psi}{2/w}\right\} dw \\
&= c\int_0^\infty w^{\lambda-n/2-1}\exp\left(-\frac{a}{w} - bw\right) dw, \qquad (7.49)
\end{aligned}$$

where

$$\begin{aligned}
c &= \frac{(\sqrt{\chi\psi})^{-\lambda}\psi^{\lambda}e^{(y-\mu)'\Sigma^{-1}\theta}}{2K_\lambda\left(\sqrt{\chi\psi}\right)(2\pi)^{n/2}|\Sigma|^{1/2}}, \\
a &= \frac{(y-\mu)'\Sigma^{-1}(y-\mu)}{2} + \frac{\chi}{2}, \\
b &= \frac{\theta'\Sigma^{-1}\theta}{2} + \frac{\psi}{2}.
\end{aligned}$$

From the equality (3.471.9) in Gradshteyn and Ryzhik (2007) (see equation (7.45)), the last term in (7.49) becomes

$$c \int_0^\infty w^{\lambda - n/2 - 1} \exp\left(-\frac{a}{w} - bw\right) dw = 2c \left(\frac{a}{b}\right)^{\frac{\nu}{2}} K_\nu(2\sqrt{ab}), \quad (7.50)$$

where $\nu = \lambda - n/2$ and K_ν is the modified Bessel function of the second kind of order ν. As in Section 7.3.1, we consider the function implemented in the Matlab *besselk* function. After some rearrangements, equation (7.50) may be written as equation (3.30) in McNeil *et al.* (2005).

Matlab code — MGH density

In this section we show how to numerically compute the probability density function from equation (7.49). Also in this case, we provide an implementation extensively based on the Matlab matrix calculus. In the code, the variable d refers to the number of stocks (i.e., $d = n$). Note that the factors in the multiplications have to be properly arranged to avoid numerical errors. For MGH processes a change of time scale modifies the distributional properties of the increments, the distribution of the increments of the same process at other time scales is no longer MGH, and for this reason we do not include dt as an input parameter. The input x is the matrix of observations with dimension $N \times d$.

```
% implementation of the density function of the MGH

function y = densityMGH_matrix(x, lambda, chi, psi, mu, theta, ...
    Sigma)

% We do not consider dt in this function

% row vectors
mu      = mu(:)';
theta   = theta(:)';

[N, d] = size(x);

L = chol(Sigma,'lower');
```

```
epsilon = sqrt(chi*psi);

A = theta*(Sigma\theta');
B = (x - ones(N,1)*mu)*(Sigma\theta');
C = ((x - ones(N,1)*mu)/L');
C = sum(C.^2,2);

k = epsilon^(-lambda)*psi^lambda*(psi + A).^(d/2 - lambda) / ...
    ((2*pi)^(d/2)*det(Sigma)^(1/2)*besselk(lambda, epsilon));

Z = sqrt( (chi + C) * (psi + A));

y = k.*besselk(lambda - d/2, Z) .* exp(B) ./ Z.^(d/2 - lambda);

return
```

7.3.4. The multivariate NTS density

The integral in equation (7.51) representing the density of the MNTS distribution does not have an analytic solution and therefore a numerical method is needed to evaluate it. As observed above, in the MNTS case the density function h is computed by means of a FFT because only the characteristic function of the mixing distribution is known in closed form and, conversely, its density function is not. This means in practice that the computational effort is given by two different numerical procedures: (1) an FFT and (2) a numerical integration.

Note that the density function of an MNTS distribution can be written as

$$f_Y(y;\Theta) = \int_0^\infty f_{Y|S}(y|s;\mu,\theta,\Sigma)h(s;a,\lambda,C)ds, \tag{7.51}$$

where $Y|S \sim N(\mu+\theta S, S\Sigma)$ (see Hu (2005)), h is the density function of the subordinator defined in equation (7.33), and Θ is the set of model parameters (a, λ, C, θ, μ, Σ). Recall that equation (7.51) holds for all multivariate mean–variance mixture distributions where h is the selected mixing distribution (e.g., the GIG distribution in the MGH case). Because it is not possible to find a closed-form solution for f_Y and h, numerical integrations to evaluate these two functions must be employed. Since if one considers a CTS subordinator, only the characteristic function (7.33) is known, a discrete evaluation of the density function h together with an interpolation algorithm is used. That is, by means of an FFT procedure, the characteristic function (7.33) is inverted to calculate the density function h, as described in detailed in Chapter 5 which covers the univariate stable distribution and in Chapter 6 which covers the univariate tempered stable distribution. Then, since the function h cannot be written in closed-form, the density f_Y in equation (7.51) has to be found by numerical integration.

Matlab code — MNTS density

In this section we show how to numerically compute the probability density function from the integral in equation (7.51). As already observed, there are two different numerical procedures: (1) an FFT to evaluate the density function of the mixing distribution, and (2) a numerical integration. While in the first case the error depends on the choice of $hFFT$ and $NFFT = 2^q$, in the second case the error depends on the integration limits $s0$ and M and on the integration step ds. All these parameters may affect the value of the density. The proposed function allows one to play with the parameters. We implement the code by assuming that input *paramSubordinatorCTS* follows the alternative parameterization (a, λ, C), where $\alpha = \frac{a}{2}$. In the code, the variable d refers to the number of stocks (i.e., $d = n$). The input x is the matrix of observations with dimension $N \times d$. Note that the function *pdfEvaluation* is provided in Chapter 6.

```
% implementation of the density function of the MNTS

function y = densityMNTS_matrix(data, paramSubordinatorCTS, ...
    mu, theta, Sigma, dt, hFFT, NFFT, s0, ds, M)

%a          = paramSubordinatorCTS(1);
%lambda     = paramSubordinatorCTS(2);
%C          = paramSubordinatorCTS(3);

mu          = mu*dt;

s   = s0:ds:M;
```

```
f1  = conditionalNormalDensity(s, data, mu, theta, Sigma);
chf = @(u) chf_subordinatorCTS(u, paramSubordinatorCTS, dt);
f2  = pdfEvaluation(s, chf, 'pchip', hFFT, NFFT);

f2 = f2(:);

y = (f1*ds)*f2;

return

function y = conditionalNormalDensity(s, data, mu, theta, Sigma)

% row vectors
mu       = mu(:)';
theta    = theta(:)';
s        = s(:)';

[lr, lc] = size(data);
d = lc;
N = lr;

L = chol(Sigma,'lower');

data = data - ones(lr, 1)*mu;

rho = data/(L');
rho = sum(rho.^2,2);

v1 = data*(Sigma\theta'); % column vector
v1 = repmat(v1,1,length(s));

k = theta*(Sigma\theta'); % theta*inv(Sigma)*theta';
w1 = -d/2*log((2*pi)) -1/2*log(det(Sigma)) -d/2*log(s) - s*k/2;
w1 = repmat(w1,N,1);

y = v1 + w1 - rho*(s.^(-1))/2;

y = exp(y);

return
```

We also provide the code to evaluate the characteristic function of the CTS subordinator used in the above Matlab code. As observed above, we implement the code with the alternative parameterization (a, λ, C), where $\alpha = \frac{a}{2}$. We also consider the case in which only a and λ are provided.

```
% characteristic function of the CTS subordinator

function phi = chf_subordinatorCTS(u, param, dt)

if length(param) == 2

    a       = param(1);
    lambda  = param(2);
    lp      = lambda;
    alpha   = a/2;
    C       = lp^(1-alpha)/gamma(1-alpha);

elseif length(param) == 3

    a       = param(1);
    lambda  = param(2);
    C       = param(3);
    alpha   = a/2;
    lp      = lambda;

end

phi = exp(dt*(C.*gamma(-alpha).*((lp - 1i.*u).^alpha - ...
    lp^alpha)));

return;
```

7.4. Simulation Algorithms

Here we describe simple simulation algorithms that can be implemented for all models analyzed in this chapter. From a practical view, the simulation of the four multivariate non-Gaussian models directly follows from (7.41). If one is able to generate random variates from a multivariate normal distribution and from the univariate mixing distribution, then it is simple to implement the code for the simulation of multivariate normal mean–variance mixture discussed in this chapter. Except for the MNTS case, an implementation for the simulation of the mixing random variable is already available (GIG and gamma) or simple to implement (IG). To simulate GIG random numbers, we consider the Matlab function *randraw* which can be downloaded from MathWorks — File Exchange website. Although the simulation of one-sided CTS random numbers is not trivial, an efficient code can be written by following an FFT-based algorithm as shown in Chapter 6 which covers the univariate tempered stable distribution.

In order to have a visual assessment of the simulation algorithm, we run a simulation of the normal and the four non-Gaussian models (MVG, MNIG, MGH, and MNTS) with the parameters estimated using daily log-returns from January 2, 1998 to August 29, 2014. We will not discuss in this chapter how to estimate the parameters of these models. However, we will review a possible estimation approach in Chapter 8. In Figure 7.2, we show the simulated returns from the two randomly selected margins (DAX and FTSEMIB) of the five-dimensional distributions and compare them to real market data. Market log-returns are obtained from Bloomberg daily dividend-adjusted closing prices from January 2, 1998 to August 29, 2014 for five stock indexes: the Deutsche Aktienindex 30 (ticker DAX), the Cotation Assistée en Continu 40 (ticker CAC), the Financial Times Stock Exchange Milano Indice di Borsa (ticker FTSEMIB), Índice

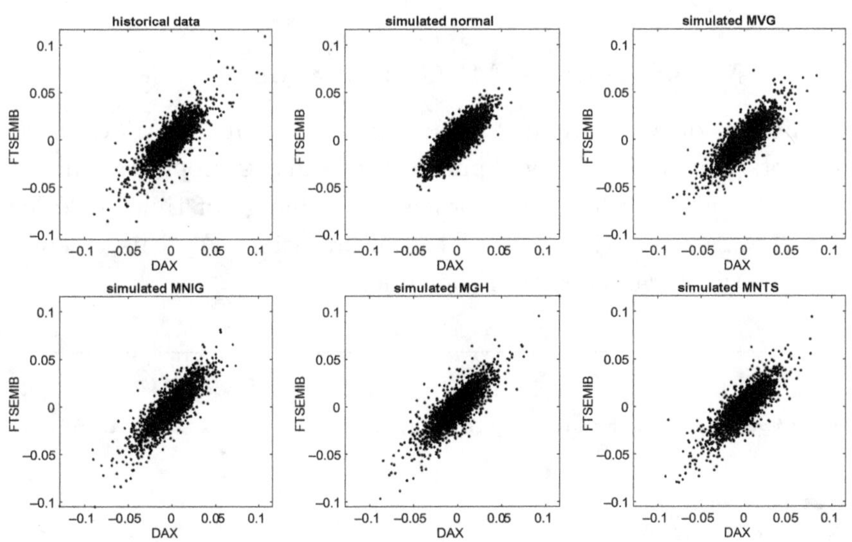

Figure 7.2. Bivariate scatterplot of DAX versus FTSEMIB by considering market data and simulated sample from the multivariate normal, the MVG, the MNIG, the MGH, and the MNTS distribution. Depicted market data are daily log-returns from January 2, 1998 to August 29, 2014.

Bursátil Español (ticker IBEX), and Athens Stock Exchange General Index (ticker ASE) representing five major European indexes. The number of simulations is equal to the number of observed historical data. By looking at the scatter plots, we can assert that the normal model cannot capture tail dependence, while all non-Gaussian models do, at least from a visual perspective. The simulation algorithm directly follows from the normal mean–variance mixture in equation (7.41). The Matlab *gamrnd* function and the Michael–Schucany–Haas algorithm for the simulation of IG random numbers (see Schoutens (2003)), together with *randn* function, are used to simulate MVG and MNIG random vectors. Then, the algorithm for the simulation of GIG random numbers proposed in Dagpunar (1989) and implemented in the *randraw* function of Matlab is considered to simulate MGH random vectors. The inverse transform algorithm for the simulation of one-sided CTS random numbers described in Bianchi *et al.* (2017) and the tempered stable distributions described in the previous chapter have been applied to simulate MNTS random vectors.

7.4.1. Matlab code — MVG simulation

In this section we show how to numerically simulate MVG random numbers given a set of model parameters and a number of simulations *NSim*. Recall that for MVG processes a change of time scale leads to a change of spatial scale of the distribution. For this reason we consider the parameter *dt* in the code.

```
% simulation of MVG random numbers

function ySim = simulateMVG(lambda, psi, mu, theta, Sigma, dt, ...
    NSim)

d = length(Sigma);

Z = randn(NSim,d)*chol(Sigma);

alpha  = lambda*dt;
beta   = psi/2;
```

```
S = gamrnd(alpha,1/beta,1,NSim,1);

S       = S(:);
mu      = mu(:)';
theta   = theta(:)';

ySim = ones(NSim,1)*mu*dt + S*theta + sqrt(S*ones(1,d)).*Z;

return
```

7.4.2. Matlab code — MNIG simulation

In this section we show how to numerically simulate MNIG random numbers given a set of model parameters and a number of simulations *NSim*. As for the MVG case, we consider the parameter *dt* in the code.

```
% simulation of MNIG random numbers

function ySim = simulateMNIG(chi, psi, mu, theta, Sigma, dt, NSim)

d = length(Sigma);

Z = randn(NSim,d)*chol(Sigma);

gamma = sqrt(chi*dt^2);
eta   = sqrt(psi);

S = SimulateIG([gamma eta], NSim);

S       = S(:);
mu      = mu(:)';
theta   = theta(:)';

ySim = ones(NSim,1)*mu*dt + S*theta + sqrt(S*ones(1,d)).*Z;

return

% simulation of IG random numbers

function xSim = SimulateIG(param, NSim, u, v)

if nargin<3
v = randn(1,NSim);
u = rand(1,NSim);
```

```
end

a = param(1);
b = param(2);

y = v.^2;

x = a/b + y/(2*b^2) - sqrt(4*a*b*y + y.^2)/(2*b^2);

xSim = x;

index = [u>a./(a + x*b)];
xSim(index) = a^2./(b^2.*x(index));

return
```

7.4.3. Matlab code — MGH simulation

In this section we show how to numerically simulate MGH random numbers given a set of model parameters and a number of simulations *NSim*. While for MVG (MNIG and MNTS) processes a change of time scale leads to a change of spatial scale of the distribution of the increments, for MGH processes a change of time scale modifies the distributional properties of the increments. For this reason in this case we do not consider the parameter *dt* in the code for the simulation of MGH random variates.

```
% simulation of MGH random numbers

function ySim = simulateMGH(lambda, chi, psi, mu, theta, ...
    Sigma, NSim)

d = length(Sigma);

Z = randn(NSim,d)*chol(Sigma);

S = SimulateGIG([lambda chi psi], NSim);

S       = S(:);
mu      = mu(:)';
theta   = theta(:)';

ySim = ones(NSim,1)*mu + S*theta + sqrt(S*ones(1,d)).*Z;
```

```
return

function xSim = SimulateGIG(param, NSim)

% the "randraw" function can be downloaded from the MathWorks ...
    website
% http://www.mathworks.com/matlabcentral/fileexchange/

xSim = randraw('gig', param, NSim);

return
```

7.4.4. Matlab code — MNTS simulation

In this section we show how to numerically simulate MNTS random numbers given a set of model parameters and a number of simulations *NSim*. As for the MVG and the MNIG cases, we consider the parameter *dt* in the code.

```
% simulation of MNTS random numbers

function ySim = simulateMNTS(a, lambda, C, mu, theta, Sigma, ...
    dt, NSim)

d = length(Sigma);

Z = randn(NSim,d)*chol(Sigma);

S = SimCTSplus_invcdf([a lambda C*dt], 1, NSim, 1, 1e-10, 50, 5);

S       = S(:);
mu      = mu(:)';
theta   = theta(:)';

ySim = ones(NSim,1)*mu*dt + S*theta + sqrt(S*ones(1,d)).*Z;

return
```

We also provide the code to simulate a matrix of dimension $NSim \times MSim$ of random numbers drawn from a CTS subordinator needed in the above Matlab code for simulating MNTS random numbers. Note that the function *cdfEvaluation* is provided in Chapter 6. As observed above, we implement the code with the alternative parameterization (a, λ, C), where $\alpha = \frac{a}{2}$.

```
% simulation of the one-sided CTS random numbers

function xSim = SimCTSplus_invcdf(param, dt, NSim, MSim, xMin, ...
    xMax, dx)

% a       = param(1);
% lambda  = param(2);
% C       = param(3);

chf = @(u) chf_subordinatorCTS(u, param, dt);
[~, cdfi, x] = cdfEvaluation(0, chf, 'chf');

if nargin == 3
    dx = 5;
    xMin = 1e-10;
    xMax = 15;
    MSim = 1;
elseif nargin == 4
    dx = 5;
    xMin = 1e-10;
    xMax = 15;
end

x       = x(1:dx:length(x));
cdfi    = cdfi(1:dx:length(cdfi));

index = cdfi >= 1 | cdfi <= 0 | x<xMin | x > xMax;

cdfi(index) = [];
x(index)    = [];

cdfi = [0; cdfi; 1];
x = [xMin; x; xMax];

figure
plot(x, cdfi)

U = rand(NSim*MSim,1);
xSim = interp1(cdfi,x,U,'pchip');
xSim = reshape(xSim,NSim,MSim);

return
```

Key Points

- In the time-changed Brownian motion, the physical time in the well-known normal model is substituted by an intrinsic time which is a natural way for providing tail effects.
- The increments of a multivariate time-changed Brownian motion have a normal mean–variance mixture law.
- While for MNTS processes a change of time scale leads to a change of spatial scale of the distribution of the increments, for MGH processes a change of time scale modifies the distributional properties of the increments. This means that if the distribution of the increments of a process on a daily basis is MGH, the distribution of the increments of the same process at other time scales is no longer MGH, thereby making it inconvenient for working with data on different time scales.
- The MVG and the MNIG laws are particular cases of the MGH and the MNTS laws.
- In the implementation of the MGH density function, the factors in the multiplications have to be properly arranged to avoid numerical errors.
- The density of the MNTS distribution does not have an analytic solution and therefore a numerical method is needed to evaluate it. In the MNTS case the density function h is computed by means of an FFT because only the characteristic function of the mixing distribution is known in closed form and, conversely, its density function is not. This means in practice that the computational effort is given by two different numerical procedures: (1) an FFT and (2) a numerical integration.
- If one is able to generate random variates from a multivariate normal distribution and from the univariate mixing distribution, then it is simple to implement the code for the simulation of multivariate normal mean–variance mixture.
- The simulation of the mixing distribution of the MNTS law is not trivial since it involves the simulation of a one-sided CTS law.

- The portfolio constructed as a linear combination of the MGH (MNTS) marginal distributions has a univariate GH (NTS) distribution. These models can be used to compute in a relatively simple way widely known portfolio risk measures.

References

K. Aas, I.H. Haff, and X.K. Dimakos. Risk estimation using the multivariate normal inverse Gaussian distribution. *Journal of Risk*, 8:39–60, 2005.

L. Ballotta and E. Bonfiglioli. Multivariate asset models using Lévy processes and applications. *The European Journal of Finance*, 22(13): 1320–1350, 2016.

O.E. Barndorff-Nielsen, P. Blaesild, and V. Seshadri. Multivariate distributions with generalized inverse Gaussian marginals, and associated Poisson mixtures. *Canadian Journal of Statistics*, 20(2):109–120, 1992.

M.L. Bianchi, S.T. Rachev, and F.J. Fabozzi. Tempered stable Ornstein–Uhlenbeck processes: A practical view. *Communications in Statistics — Simulation and Computation*, 46(1):423–445, 2017.

M.L. Bianchi, G.L. Tassinari, and F.J. Fabozzi. Riding with the four horsemen and the multivariate normal tempered stable model. *International Journal of Theoretical and Applied Finance*, 19(4):1650027, 2016.

P.K. Clark. A subordinated stochastic process model with finite variance for speculative prices. *Econometrica*, 41(1):135–155, 1973.

R. Cont and P. Tankov. *Financial Modelling with Jump Processes*. Chapman & Hall/CRC, Boca Raton, FL, 2003.

J.S. Dagpunar. An easily implemented generalised inverse Gaussian generator. *Communications in Statistics — Simulation and Computation*, 18(2):703–710, 1989.

Y. Dominicy, H. Ogata, and D. Veredas. Inference for vast dimensional elliptical distributions. *Computational Statistics*, 28(4):1853–1880, 2013.

E. Eberlein and D.B. Madan. On correlating Lévy processes. *Journal of Risk*, 13(1):3–16, 2010.

G. Frahm. Generalized elliptical distributions: Theory and applications. PhD thesis, Universität zu Köln, 2004.

H. Geman, D.B. Madan, and M. Yor. Time changes for Lévy processes. *Mathematical Finance*, 11(1):79–96, 2001.

H. Geman, D.B. Madan, and M. Yor. Stochastic volatility, jumps and hidden time changes. *Finance and Stochastics*, 6(1):63–90, 2002.

I.S. Gradshteyn and I.M. Ryzhik. *Table of Integrals, Series and Products*, Academic Press, 2007.

J.B. Guerard, S.T. Rachev, and B.P. Shao. Efficient global portfolios: big data and investment universes. *IBM Journal of Research and Development*, 57(5):1–11, 2013.

F. Guillaume. Sato two-factor models for multivariate option pricing. *Journal of Computational Finance*, 15(4):159–192, 2012.

F. Guillaume. The αVG model for multivariate asset pricing: Calibration and extension. *Review of Derivatives Research*, 16(1):25–52, 2013.

M. Hellmich and S. Kassberger. Efficient and robust portfolio optimization in the multivariate generalized hyperbolic framework. *Quantitative Finance*, 11(10):1503–1516, 2011.

A. Hitaj and L. Mercuri. Portfolio allocation using multivariate variance gamma models. *Financial Markets and Portfolio Management*, 27(1): 65–99, 2013.

W. Hu. Calibration of multivariate generalized hyperbolic distributions using the EM algorithm, with applications in risk management, portfolio optimization and portfolio credit risk. PhD thesis, Florida State University, 2005.

W. Hu and A.N. Kercheval. Risk management with generalized hyperbolic distributions. In P. Locke, editor, *Proceedings of the Fourth IASTED International Conference on Financial Engineering and Applications*. ACTA Press, 2007, pp. 19–24.

S.R. Hurst, E. Platen, and S.T. Rachev. Subordinated market index models: A comparison. *Asia-Pacific Financial Markets*, 4(2):97–124, 1997.

S.R. Hurst, E. Platen, and S.T. Rachev. Option pricing for a logstable asset price model. *Mathematical and Computer Modelling*, 29(10):105–119, 1999.

R. Kawai. A multivariate Lévy process model with linear correlation. *Quantitative Finance*, 9(5):597–606, 2009.

R. Kawai and A. Kohatsu-Higa. Computation of Greeks and multidimensional density estimation for asset price models with time-changed Brownian motion. *Applied Mathematical Finance*, 17:201–321, 2010.

Y.S. Kim, J. Lee, S. Mittnik, and J. Park. Quanto option pricing in the presence of fat tails and asymmetric dependence. *Journal of Econometrics*, 187(2):512–520, 2015.

Y.S. Kim, R. Giacometti, S.T. Rachev, F.J. Fabozzi, and D. Mignacca. Measuring financial risk and portfolio optimization with a non-Gaussian multivariate model. *Annals of Operations Research*, 201(1):325–343, 2012.

S. Kring, S.T. Rachev, M. Höchstötter, F.J. Fabozzi, and M.L. Bianchi. Multi-tail generalized elliptical distributions for asset returns. *The Econometrics Journal*, 12(2):272–291, 2009.

P. Leoni and W. Schoutens. Multivariate smiling. *Wilmott Magazine*, March 2008, pp. 82–91.

D. Linders and B. Stassen. The multivariate variance gamma model: Basket option pricing and calibration. *Quantitative Finance*, 16(4):555–572, 2016.

E. Luciano and W. Schoutens. A multivariate jump-driven financial asset model. *Quantitative Finance*, 6(5):385–402, 2006.

E. Luciano and P. Semeraro. Multivariate time changes for Lévy asset models: Characterization and calibration. *Journal of Computational and Applied Mathematics*, 233(8):1937–1953, 2010.

R. Marfé. A generalized variance gamma process for financial applications. *Quantitative Finance*, 12(1):75–87, 2012.

A. McNeil, R. Frey, and P. Embrechts. *Quantitative Risk Management: Concepts, Techniques, and Tools.* Princeton University Press, Princeton, NJ, 2005.

I. Monroe. Processes that can be embedded in Brownian motion. *The Annals of Probability*, 6(1):42–56, 1978.

T.A. Oigard, A. Hanssen, R.E. Hansen, and F. Godtliebsen. EM-estimation and modeling of heavy-tailed processes with the multivariate normal inverse Gaussian distribution. *Signal Processing*, 85(8):1655–1673, 2005.

K. Prause. *The Generalized Hyperbolic Model: Estimation, Financial Derivatives, and Risk Measures.* PhD thesis, Universität Freiburg i. Br., Freiburg i. Br, 1999.

R.S. Protassov. EM-based maximum likelihood parameter estimation for multivariate generalized hyperbolic distributions with fixed λ. *Statistics and Computing*, 14(1):67–77, 2004.

S.T. Rachev, Y.S. Kim, M.L. Bianchi, and F.J. Fabozzi. *Financial Models with Lévy Processes and Volatility Clustering.* Wiley, Hoboken, NJ, 2011.

J. Rosinski. Tempering stable processes. *Stochastic Processes and Their Applications*, 117(6):677–707, 2007.

W. Schoutens. *Lévy Processes in Finance.* Wiley, Chichester, 2003.

P. Semeraro. A multivariate variance gamma model for financial applications. *International Journal of Theoretical and Applied Finance*, 11(1):1–18, 2008.

G.L. Tassinari. *Pricing Equity and Debt Tranches of Collateralized Funds of Hedge Fund Obligations.* PhD thesis, University of Bergamo, 2009.

G.L. Tassinari and M.L. Bianchi. Calibrating the smile with multivariate time-changed Brownian motion and the Esscher transform. *International Journal of Theoretical and Applied Finance*, 17(4), 2014.

G.L. Tassinari and C. Corradi. Pricing equity and debt tranches of collateralized funds of hedge fund obligations: An approach based on stochastic time change and Esscher-transformed martingale measure. *Quantitative Finance*, 13(12):1991–2010, 2013.

G.L. Tassinari and C. Corradi. Valuation of collateralized funds of hedge fund obligations: A basket option pricing approach. In M. Corazza and C. Pizzi, editors, *Mathematical and Statistical Methods for Actuarial Sciences and Finance*. Springer, Berlin, 2014, pp. 277–288.

J. Wang. *The Multivariate Variance Gamma Process and Its Applications in Multi-asset Option Pricing*. PhD thesis, University of Maryland, 2009.

Y.C. Wu, S.L. Liao, and S.D. Shyu. Closed-form valuations of basket options using a multivariate normal inverse Gaussian model. *Insurance: Mathematics and Economics*, 44(1):95–102, 2009.

Chapter 8

Multivariate Time-Changed Brownian Motion: The Expectation–Maximization Estimation Method

The main topics covered in this chapter are:

- a description of a method to estimate the parameters of models based on the multivariate time-changed Brownian motion;
- a review of the expectation–maximization (EM) maximum likelihood estimation (MLE) method to estimate the parameters of multivariate generalized hyperbolic distributions;
- an extension of the EM-based MLE algorithm to normal mean–variance mixture distributions in which only the characteristic function of the mixing distribution is known in closed form, while the density function is not;
- an error analysis of the estimation method applied to the multivariate normal tempered stable case;
- an empirical test showing the model performance on a five- and a 30-dimensional series of index (stock) returns;
- how to evaluate well-known risk measures (i.e., value-at-risk and average value-at-risk) under this framework;
- how to backtest the value-at-risk by taking into account the number of exceedances.

8.1. Introduction

In this chapter, we show a possible estimation method for the multivariate time-changed Brownian motion models defined in Chapter 7. More specifically, we consider four possible parametric examples: the *multivariate variance gamma* (MVG), the *multivariate normal inverse Gaussian* (MNIG), the *multivariate generalized hyperbolic* (MGH), and the *multivariate normal tempered stable* (MNTS) model. By taking into consideration the works of Protassov (2004), Hu (2005), and McNeil *et al.* (2005), we show how it is possible to perform a fast historical estimation on the time series of log-returns through the so-called expectation–maximization (EM) maximum likelihood estimation (MLE) method.

Because the number of parameters is generally large, a direct MLE for these multivariate models is not feasible. The parameters of multivariate normal mean–variance mixture distributions are usually estimated through the EM-based MLE method. By recalling that the distribution of the increments of a multivariate time-changed Brownian motion is a multivariate distribution defined as a normal mean–variance mixture, then it is possible to apply this EM-based estimation method. In practice, this method allows one to separate the estimation of the parameters of the underlying multivariate normal distribution from that of the mixing distribution.

The EM-based MLE algorithm is usually applied in the MGH case (see Protassov (2004), Hu (2005), and McNeil *et al.* (2005)). We show how it is possible to extend the EM algorithm to the MNTS distribution, as proposed by Bianchi *et al.* (2016). Although, in the MGH case, the density function of the MGH and of the generalized inverse Gaussian (GIG) random variables can be expressed in closed form, and in the MNTS case, the density function of the MNTS is not known in closed form. This comes from the fact that for the classical tempered stable (CTS) distribution the density function is also not known in closed form. However, the density of a MNTS random variable can be obtained by a numerical integration that combines the density of a multivariate normal distribution and the density of a univariate tempered stable mixing distribution. By taking into consideration the works of Stoyanov and Racheva-Iotova (2004), Scherer *et al.* (2012), and Bianchi *et al.* (2017), the latter density can be evaluated by means of a fast Fourier transform (FFT). Thus, the computational effort required to evaluate the MNTS density is reduced by the normal mean–variance mixture structure of the MNTS random variable and, as noted above, it is equal to the computation of a univariate density with an FFT algorithm and a numerical integration of a one-dimensional integral.

A review of the EM-based MLE method in both the MGH and the MNTS cases is given in Section 8.2. After discussing the main computational aspects related to the computation of the density function of the MNTS law, an empirical investigation using simulated data is conducted in Section 8.3 to further assess the performance of the estimation algorithm proposed in this chapter. Then, in Section 8.4 we empirically test the estimation method based only on the time series of log-returns. In Section 8.5, we discuss a portfolio market model assuming the MNTS distribution and we empirically illustrate how to compute the portfolio value-at-risk (VaR) and average value-at-risk (AVaR). Finally, we provide a simple algorithm to backtest the VaR.

8.2. EM-based MLE

A direct MLE for these models is not feasible, since the number of parameters is large and it is proportional to the dimension of the multivariate law. For example, the five-dimensional model we consider has 27 parameters to be estimated. For this reason, in Section 8.2.1 we review the EM-based maximum likelihood algorithm proposed in Hu (2005) and McNeil *et al.* (2005) to estimate the parameters of MGH distributions. We point out that the densities of MVG and MNIG are particular cases of the MGH density. Indeed, the proposed MVG and MNIG models are a particular case of the MGH model. This follows from the fact that both the gamma and the inverse Gaussian (IG) random variables are particular cases of the GIG random variable. Then in Section 8.2.2 we show how to extend the EM-based maximum likelihood algorithm to estimate the parameters of MNTS distributions.

For each discrete time step Δt the distribution of the increments of a multivariate time-changed Brownian motion $Y = (Y_t)_{t\geq 0}$ is given by

$$Y_{\Delta t} = \mu \Delta t + \theta S_{\Delta t} + \sqrt{S_{\Delta t}} D_\sigma A Z, \tag{8.1}$$

where

- Z is a multivariate normal distribution with zero mean and unit variance–covariance matrix I of dimension n;
- $S_{\Delta t}$ is the distribution of the increments of the subordinator $S = (S_t)_{t\geq 0}$ with density function h and it is independent of Z;
- A is the lower Cholesky decomposition of a correlation matrix Ω, that is, $\Omega^{1/2} = A$ and $\Sigma^{1/2} = D_\sigma A$, where D_σ is a diagonal matrix with diagonal $\sigma \in \mathbb{R}^n_+$;
- μ and θ are vectors in $\mathbb{R}^n$.

Given a set of N observations $\{Y^k = Y_{t_k} - Y_{t_{k-1}}\}_{k=1,\ldots,N}$, the log-likelihood related to the model (8.1) can be written as

$$\text{LL}(\Theta; Y^1, \ldots, Y^N) = \sum_{k=1}^{N} \log f_Y(Y^k; \Theta). \tag{8.2}$$

We consider the following log-likelihood function instead of the log-likelihood in equation (8.2)

$$\begin{aligned}\text{LL}(\Theta; Y^1, \ldots, Y^N, S^1, \ldots, S^N) &= \sum_{k=1}^{N} \log f_{Y,S}(Y^k, S^k; \Theta)\\ &= \text{LL}_1(\mu, \theta, \Sigma; Y|S) + \text{LL}_2(\Theta_S; S)\end{aligned}$$

$$= \sum_{k=1}^{N} \log f_{Y|S}(Y^k | S^k; \mu, \theta, \Sigma)$$

$$+ \sum_{k=1}^{N} \log h_S(S^k; \Theta_S), \qquad (8.3)$$

where $\Theta = \{\Theta_S, \mu, \theta, \Sigma\}$ is the set of parameters, $\{S^k = S_{t_k} - S_{t_{k-1}}\}_{k=1,\ldots,N}$ is the latent mixing variables coming from the representation (8.1), and Θ_S is the set of parameters of the latent mixing law S. The log-likelihood in equation (8.3) allows one to recursively estimate the parameters of the multivariate normal distribution embedded in the normal mean–variance mixture in equation (8.1) and those of the mixture distribution.

8.2.1. MGH case

Under the notation of Chapter 7 the parameters of a MGH law are λ, χ, ψ, μ, θ, Σ, and the parameters of the latent generalized inverse Gaussian (GIG) random variable S are λ, χ, ψ. Recall that in the MVG case the parameters of the latent gamma random variable S are α, β, and $\Gamma(\alpha, \beta)$ corresponds to $\text{GIG}(\lambda, \chi, \psi)$ with $\lambda = \alpha$, $\chi = 0$, and $\psi = 2\beta$; in the MNIG case the parameters of the latent inverse Gaussian (IG) random variable S are γ, η, and $\text{IG}(\gamma, \eta)$ corresponds to $\text{GIG}(\lambda, \chi, \psi)$ with $\lambda = -1/2$, $\chi = \gamma^2$, and $\psi = \eta^2$ (see Schoutens (2003)).

In the MGH case we have the following log-likelihood function

$$\text{LL}(\Theta; Y^1, \ldots, Y^N, S^1, \ldots, S^N) = \text{LL}_1(\mu, \theta, \Sigma; Y|S)$$

$$+ \text{LL}_2(\lambda, \chi, \psi; S), \qquad (8.4)$$

where $\Theta = \{\lambda, \chi, \psi, \mu, \theta, \Sigma\}$ is the set of parameters. In order to find an MLE based on (8.4), we consider the following iterative algorithm.

Step 1. Set $i = 1$ and select a starting value for $\Theta^{(1)}$, that is $\mu^{(1)} \in \mathbb{R}^n$ is the sample mean, $\theta^{(1)} \in \mathbb{R}^n$ is the zero vector, $V \in \mathbb{R}^n \times \mathbb{R}^n$ is the sample covariance matrix.

Step 2. By considering the relation among joint, marginal and conditional densities, the joint density of (Y, S) can be written as

$$f_{Y,S}(Y^k, s; \Theta) = f_{Y|S}(Y^k|s; \mu, \theta, \Sigma)h(s; \lambda, \chi, \psi)$$

and as

$$f_{Y,S}(Y^k, s; \Theta) = f_{S|Y^k}(s|Y^k, \Theta)f_Y(Y^k; \Theta),$$

it is simple to show that

$$f_{S|Y^k}(s|Y^k, \Theta) = \frac{f_{Y|S}(Y^k|s; \mu, \theta, \Sigma)h(s; \lambda, \chi, \psi)}{f_Y(Y^k; \Theta)}.$$

Furthermore, it is possible to prove that

$$S^k|Y^k \sim \mathrm{GIG}\left(\lambda - \frac{n}{2}, \rho_k + \chi, \psi + \theta'\Sigma^{-1}\theta\right)$$

and defined

$$\begin{aligned}
\delta_k^{(\cdot)} &= E({S^k}^{-1}|Y^k, \Theta^{(\cdot)}),\\
\eta_k^{(\cdot)} &= E(S^k|Y^k, \Theta^{(\cdot)}),\\
\xi_k^{(\cdot)} &= E(\log S^k|Y^k, \Theta^{(\cdot)}),
\end{aligned}$$

compute the following weights:

$$\begin{aligned}
\rho_k^{(i)} &= (Y^k - \mu^{(i)})'(\Sigma^{(i)})^{-1}(Y^k - \mu^{(i)}),\\
\delta_k^{(i)} &= \left(\frac{\rho_k^{(i)} + \chi^{(i)}}{\psi^{(i)} + (\theta^{(i)})'(\Sigma^{(i)})^{-1}\theta^{(i)}}\right)^{-1/2}\\
&\quad \times \frac{K_{\lambda-n/2-1}(\sqrt{(\rho_k^{(i)} + \chi^{(i)})(\psi^{(i)} + (\theta^{(i)})'(\Sigma^{(i)})^{-1}\theta^{(i)})})}{K_{\lambda-n/2}(\sqrt{(\rho_k^{(i)} + \chi^{(i)})(\psi^{(i)} + (\theta^{(i)})'(\Sigma^{(i)})^{-1}\theta^{(i)})})},
\end{aligned}$$

$$\eta_k^{(i)} = \left(\frac{\rho_k^{(i)} + \chi^{(i)}}{\psi^{(i)} + (\theta^{(i)})' \left(\Sigma^{(i)}\right)^{-1} \theta^{(i)}} \right)^{1/2}$$

$$\times \frac{K_{\lambda-n/2+1}(\sqrt{(\rho_k^{(i)} + \chi^{(i)})(\psi^{(i)} + (\theta^{(i)})'(\Sigma^{(i)})^{-1}\theta^{(i)})})}{K_{\lambda-n/2}(\sqrt{(\rho_k^{(i)} + \chi^{(i)})(\psi^{(i)} + (\theta^{(i)})'(\Sigma^{(i)})^{-1}\theta^{(i)})})},$$

$$\xi_k^{(i)} = \log \left(\frac{\rho_k^{(i)} + \chi^{(i)}}{\psi^{(i)} + (\theta^{(i)})'(\Sigma^{(i)})^{-1}\theta^{(i)}} \right)^{1/2}$$

$$+ \frac{g'(0)}{K_{\lambda-n/2}(\sqrt{(\rho_k^{(i)} + \chi^{(i)})(\psi^{(i)} + (\theta^{(i)})'(\Sigma^{(i)})^{-1}\theta^{(i)})})},$$

where the function g is defined as follows

$$g(\alpha) = K_{\lambda-n/2+\alpha}(\sqrt{(\rho_k^{(i)} + \chi^{(i)})(\psi^{(i)} + (\theta^{(i)})'(\Sigma^{(i)})^{-1}\theta^{(i)})}).$$

Matlab code — Step 2

Here we provide the Matlab code to compute the weights in the MGH case.

```
function [delta, delta_bar, eta, eta_bar, xi, xi_bar] ...
    = ... computeWeigthsMGH(lambda, chi, psi, mu, ...
    theta, D, Sigma, data, model)

[N, d] = size(data);

lr = N;
lc = d;

A  = D\chol(Sigma,'lower');
data = data - ones(lr, 1)*mu;

rho = data/(D*A);
rho = sum(rho.^2,2);

C = theta*(Sigma\theta'); % theta*inv(Sigma)*theta';

Z = sqrt((rho + chi) * ( psi + C));
```

```
delta = ( ( rho + chi) / (psi + C) ).^(-1/2) .* ...
        besselk(lambda - d/2 - 1,Z)./besselk(lambda - ...
            d/2,Z);

eta   = ( ( rho + chi) / (psi + C) ).^(1/2) .* ...
        besselk(lambda - d/2 + 1,Z)./besselk(lambda - ...
            d/2,Z);

delta_bar = sum(delta)/N;
eta_bar   = sum(eta)/N;

    % this part does not need for the ECME algorithm
    if strcmp(model,'MNIG') == 0 || ...
        strcmp(model,'mnig') == 0

        g0 =  derivateG(lambda, d, Z, 0);
        xi = 1/2*log(( rho + chi) / (psi + C)) + ...
            g0./besselk(lambda - d/2,Z);
        xi_bar     = sum(xi)/N;

    end

return

function g0 = derivateG(lambda, d, Z, alpha)

  f = @(x) besselk(lambda - d/2 + x, Z);
  x = [alpha-1e-5 alpha+1e-5];
  g0 = (f(x(1)) - f(x(2)))/diff(x);

return
```

Step 3. Evaluate the average values

$$\bar{\delta}^{(i)} = \sum_{k=1}^{N} \delta_k^{(i)}, \qquad \bar{\eta}^{(i)} = \sum_{k=1}^{N} \eta_k^{(i)}, \qquad \bar{\xi}^{(i)} = \sum_{k=1}^{N} \xi_k^{(i)}.$$

Step 4. Get the estimates

$$\theta^{(i+1)} = \frac{N^{-1}\sum_{k=1}^{N} \delta_k^{(i)}(\bar{Y} - Y^k)}{\bar{\delta}^{(i)}\bar{\eta}^{(i)} - 1},$$

$$\mu^{(i+1)} = \frac{N^{-1}\sum_{k=1}^{N} \delta_k^{(i)} Y^k - \theta^{(i+1)}}{\bar{\delta}^{(i)}},$$

$$\Psi = \frac{1}{N}\sum_{k=1}^{N}\delta_k^{(i)}(Y^k - \mu^{(i+1)})(Y^k - \mu^{(i+1)})'$$
$$-\bar{\eta}^{(i)}\theta^{(i+1)}\theta^{(i+1)'},$$
$$\Sigma^{(i+1)} = \frac{|V|^{1/n}\Psi}{|\Psi|^{1/n}}.$$

Matlab code — Step 4

Here we provide the Matlab code to get the estimates of the multivariate normal embedded in the multivariate normal mean–variance mixture law.

```
function [mu, theta, Sigma] = ...
    getMultivariateNormalEstimates(delta, delta_bar, ...
    eta_bar, data, D)

[N, d] = size(data);

lr = N;
lc = d;

data_bar = mean(data);

A = ones(lr, 1)*data_bar - data;
A = (delta*ones(1, lc)).*A;
A = sum(A, 1);

theta = 1/N * A./(delta_bar.*eta_bar - 1);

B = data;
B = (delta*ones(1, lc)).*B;
B = sum(B, 1);

mu = (1/N * B - theta)./delta_bar;

B1 = data - ones(lr, 1)*mu;
B2 = (delta*ones(1, lc)).*B1;

Psi = zeros(d,d);

   for j = 1:N

       Psi = Psi + B2(j,:)'*B1(j,:);
```

```
  end

Psi = 1/N*Psi - eta_bar.*theta'*theta;
Sigma = det(D.^2)^(1/d)/det(Psi)^(1/d)*Psi;

return
```

Step 5. Set

$$\Theta^{(i')} = \{\lambda^{(i)}, \chi^{(i)}, \psi^{(i)}, \mu^{(i+1)}, \theta^{(i+1)}, \Sigma^{(i+1)}\}$$

and calculate the new weights $\bar{\delta}^{(i')}$, $\bar{\eta}^{(i')}$, and $\bar{\xi}^{(i')}$ as done in Steps 2 and 3.

Step 6. To complete the calculation of $\Theta^{(i+1)}$, find λ, χ, and ψ that maximize the log-likelihood function in equation (8.2), that is

$$\mathrm{LL}(\Theta^{(i+1)}; Y^1, \ldots, Y^N) = \sum_{k=1}^{N} \log f_Y(Y^k; \Theta^{(i+1)})$$

where $\Theta^{(i+1)} = \{\lambda, \chi, \psi, \theta^{(i+1)}, \mu^{(i+1)}, \Sigma^{(i+1)}\}$.

Step 7. If $i < 1000$, increment iteration count i and go to step 2, otherwise, stop the algorithm.

As discussed in McNeil *et al.* (2005), the EM algorithm converges to the maximum likelihood estimate because it produces improved parameter estimates at each step, in the sense that the value of the original likelihood is continually increased. We set the maximum number of iterations i equal to 1000.

In this chapter, in Step 6 we consider the expectation conditional maximization either (ECME) approach, that is the original likelihood function in equation (8.2) with respect to λ, χ, and ψ is maximized while all other parameters are kept fixed.

We note that an alternative Step 6 can be considered and a different version of the EM algorithm can be obtained. A possible alternative maximization step is given by finding the maximum with respect to λ, χ, and ψ of the function LL_2 in equation (8.4). As observed in McNeil *et al.* (2005), this is usually referred to as the multicycle expectation conditional maximization (MCECM) algorithm, in which the following maximization step is considered.

Step 6. (Alternative) To complete the calculation of $\Theta^{(i+1)}$, find λ, χ, and ψ that maximize the following function:

$$\begin{aligned} Q(\lambda, \chi, \psi; \mu^{(i+1)}, \theta^{(i+1)}, \Sigma^{(i+1)}) \\ = (\lambda - 1) \sum_{k=1}^{N} \log S^k - \frac{\chi}{2} \sum_{k=1}^{N} S^{k^{-1}} \\ -\frac{\psi}{2} \sum_{k=1}^{N} S^k - \frac{N\lambda}{2} \log \chi \\ +\frac{N\lambda}{2} \log \psi - N \log(2K_\lambda(\sqrt{\chi\psi})), \end{aligned} \tag{8.5}$$

where $S^{k^{-1}}$, S^k, and $\log S^k$ are replaced by their conditional estimates $\delta_k^{(\cdot)} = E(S^{k^{-1}}|Y^k, \theta^{(\cdot)})$, $\eta_k^{(\cdot)} = E(S^k|Y^k, \theta^{(\cdot)})$, and $\xi_k^{(\cdot)} = E(\log S^k|Y^k, \theta^{(\cdot)})$ defined in Step 2.

In both the MNIG and MVG cases, Step 6 of the above algorithm can be simplified. As shown in Hu (2005), in the MNIG case with $\lambda = -\frac{1}{2}$, the parameters χ, and ψ maximizing the function (8.5) are

$$\begin{aligned} \epsilon^{(i+1)} &= \sqrt{\chi^{(i+1)}\psi^{(i+1)}} = \frac{1}{\bar{\eta}^{(i)}\bar{\delta}^{(i)} - 1}, \\ \chi^{(i+1)} &= \frac{\epsilon^{(i+1)}\bar{\eta}^{(i)}K_{-1/2}(\epsilon^{(1+1)})}{K_{1/2}(\epsilon^{(1+1)})}, \\ \psi^{(i+1)} &= \frac{(\epsilon^{(i+1)})^2}{\chi^{(i+1)}}, \end{aligned}$$

while in the MVG case, since $\chi = 0$, one has to find $\lambda^{(i+1)}$ satisfying the following equation:

$$\log \lambda^{(i+1)} - \log \bar{\eta}^{(i)} + \bar{\xi}^{(i)} - \phi(\lambda^{(i+1)}) = 0,$$

where ϕ is the digamma function, and

$$\psi^{(i+1)} = \frac{2\lambda^{(i+1)}}{\bar{\eta}^{(i)}}.$$

The special function ϕ is implemented in the Matlab *psi* function and the parameter $\lambda^{(i+1)}$ can be found by means of the Matlab *fzero* function. Finally, adopting the notation used in Chapter 7 in the MNIG case, $\gamma = \sqrt{\chi}$ and $\eta = \sqrt{\psi}$, and in the MVG case we have $\alpha = \lambda$ and $\beta = \psi/2$.

In this chapter we do not provide the code for the MCECM algorithm that considers the alternative Step 6, but only the code for the ECME algorithm described above. We point out that the evaluation of $\xi^{(i)}$ is needed only for the MCECM algorithm, but not for the ECME algorithm.

Matlab code — MGH estimates

In this section, we show how to implement the optimization algorithm to find an EM-based MLE estimates in the MGH case. In Step 6 we consider the ECME approach, that is we maximize the original log-likelihood in equation (8.2). The code for the MVG and the MNIG cases can be similarly implemented and for this reason we do not provide it in this chapter.

```
% estimation of the MNTS distribution

function [lambda, chi, psi, mu, theta, Sigma, A, D, LL, ...
    numSteps] = ... estimateMGH(data, model, method, x0)

[lr, lc] = size(data);
N = lr;

mu    = mean(data);
Sigma = cov(data);
theta = zeros(1,lc);

std0   = std(data);
D  = diag(std0, 0);

    if strcmp(model,'MGH') == 1 || strcmp(model,'mgh') == 1;
```

```
        if nargin == 5
         lambda = x0(1); chi = x0(2); psi = x0(3);
        else
         lambda = -1/2; chi = 10; psi = 10; x0 = [lambda chi psi];
        end

    end

numberOfIterations = 1000;

    for i = 1:numberOfIterations

        if strcmp(model,'MGH') == 1 || strcmp(model,'mgh') == 1

            if strcmp(method,'ecme') == 1

            [delta, delta_bar, ~, eta_bar]  = ...
                computeWeigthsMGH(lambda, chi, psi, mu, theta, ...
                D, Sigma, data, 'mgh');
            [mu, theta, Sigma] = ...
                getMultivariateNormalEstimates(delta, ...
                delta_bar, eta_bar, data, D);
            [lambda, chi, psi] = ...
                findOptimalParametersMGH_mle(data, mu, theta, ...
                Sigma, [lambda, chi, psi]);

            end

        end

        if  strcmp(model,'MGH') == 1 || strcmp(model,'mgh') == 1

             LL(i) = loglikelihoodMGH(data, [lambda, chi, ...
                 psi], mu, theta, Sigma);

             [i LL(i)/10^4 lambda chi psi]

        end

        if (i>2 && abs(LL(i) - LL(i-1)) <= 1e-5) || i == ...
            numberOfIterations
           break
        end

    end

% to have that A*A' is a correlation matrix
D  = diag(diag(Sigma,0).^(1/2),0);
A  = D\chol(Sigma,'lower');
```

```
LL = LL(end);

mu       = mu(:);
theta    = theta(:);
numSteps = i;

return
```

In this script we show how the *findOptimalParametersMGH_mle* function is implemented. We consider the Matlab optimization function *fmincon*. The upper and lower bounds (*ub* and *lb*) for the parameters are provided in the code and not as input arguments. We show some simple options that one can select in the optimization algorithm. Note that the selection of *ub* and *lb* may depend on the sample to fit. The selected bounds seem to be a proper choice for stock (index) log-returns data.

```
function [lambda, chi, psi] = ...
    findOptimalParametersMGH_mle(data, mu, theta, Sigma, x0)

lb =[ -20 1e-5 1e-5];
ub = [20 20 20];

if nargin == 4
x0 = [-0.5 10 10];
end

options = optimset('Display','iter','Algorithm','active-set',...
                   'UseParallel', 'always');

EstimatedParameters = fmincon(@(x) -loglikelihoodMGH(data, x, ...
    mu, ...
            theta, Sigma), x0,[],[],[],[],lb,ub,[],options);

lambda = EstimatedParameters(1);
chi = EstimatedParameters(2);
psi = EstimatedParameters(3);

return
```

The log-likelihood of the MGH distribution can be easily evaluated as follows.

```
function y = loglikelihoodMGH(data, x, mu, theta, Sigma)

lambda = x(1);
chi = x(2);
psi = x(3);

f = densityMGH_matrix(data, lambda, chi, psi, mu, theta, Sigma);
y = sum(log(f));

end
```

8.2.2. MNTS case

Using the notation of Chapter 7, the parameters of a MNTS law are a, C, λ, μ, θ, Σ, and the parameters of the latent one-sided CTS random variable S are a, C, λ. Recall that the density function of an MNTS distribution can be written as

$$f_Y(y;\Theta) = \int_0^\infty f_{Y|S}(y|s;\mu,\theta,\Sigma)h(s;a,\lambda,C)ds, \tag{8.6}$$

where $Y|S \sim N(\mu+\theta S, S\Sigma)$, h is the density function of the one-sided CTS subordinator, and Θ is the set of model parameters (a, λ, C, θ, μ, Σ). Because it is not possible to find a closed-form solution for f_Y and h, numerical integrations to evaluate these two functions must be employed. We observe that equation (8.6) holds for all multivariate mean–variance mixture distributions where h is the selected mixing distribution (e.g., the GIG distribution in the MGH case).

In the MNTS case we have the following log-likelihood function:

$$\begin{aligned} \mathrm{LL}(\Theta; Y^1,\ldots,Y^N,S^1,\ldots,S^N) &= \mathrm{LL}_1(\mu,\theta,\Sigma;Y|S) \\ &\quad + \mathrm{LL}_2(a,\lambda,C;S), \end{aligned} \tag{8.7}$$

where $\Theta = \{a,\lambda,C,\theta,\mu,\Sigma\}$ is the set of parameters. In order to find an MLE, we slightly modify the above iterative algorithm in Section 8.2.1 as follows.

Step 1. Set $i = 1$ and select a starting value for $\Theta^{(1)}$, that is $\mu^{(1)} \in \mathbb{R}^n$ is the sample mean, $\theta^{(1)} \in \mathbb{R}^n$ is the zero vector, $V \in \mathbb{R}^n \times \mathbb{R}^n$ is the sample covariance matrix.

Step 2. By considering that

$$f_{S|Y^k}(s|Y^k, \Theta) = \frac{f_{Y|S}(Y^k|s; \mu, \theta, \Sigma)h(s; a, \lambda, C)}{f_Y(Y^k; \Theta)}, \quad (8.8)$$

compute the following weights:

$$\begin{aligned}\delta_k^{(\cdot)} &= E(S^{k^{-1}}|Y^k, \Theta^{(\cdot)}),\\ \eta_k^{(\cdot)} &= E(S^k|Y^k, \Theta^{(\cdot)}),\\ \rho_k^{(i)} &= (Y^k - \mu^{(i)})'(\Sigma^{(i)})^{-1}(Y^k - \mu^{(i)}),\end{aligned} \quad (8.9)$$

The expectations in equation (8.9) do not have a closed-form solution, and they are evaluated by numerical integration. In equation (8.8), $f_{Y|S}$ can be written in closed form since $Y|S^k \sim N(\mu + \theta S^k, S^k\Sigma)$ (see Hu (2005)), the density h is computed by means of an FFT procedure, and the denominator which is the density function given in equation (8.6) is evaluated by numerical integration.

Matlab code — Step 2

Note that the function to evaluate the density given the characteristic function (*pdfEvaluation*), the function of the characteristic function of the CTS subordinator (*chf_subordinatorCTS*), and the function of the conditional normal density (*conditionalNormalDensity*) used in this code are provided in previous chapters and for this reason we do not report them again in this chapter.

```
function [delta, delta_bar, eta, eta_bar, xi, xi_bar] ...
    = computeWeigthsMNTS(paramSubordinatorCTS, mu, ...
    theta, Sigma, data, hFFT, NFFT, s0, ds, M)

%a          = paramSubordinatorCTS(1);
%lambda     = paramSubordinatorCTS(2);
%C          = paramSubordinatorCTS(3);

s   = s0:ds:M;
```

```
f1 = conditionalNormalDensity(s, data, mu, theta, Sigma);
chf = @(u) chf_subordinatorCTS(u, ...
    paramSubordinatorCTS, 1);
f2  = pdfEvaluation(s, chf, 'pchip', hFFT, NFFT);

f2 = f2(:);
g = (f1*ds)*f2;

s = s(:);

delta = (f1*ds)*(f2./s)./g;
delta_bar = mean(delta);

eta = (f1*ds)*(f2.*s)./g;
eta_bar = mean(eta);

xi = (f1*ds)*(f2.*log(s))./g;
xi_bar = mean(xi);

return
```

Step 3. Evaluate the average values

$$\bar{\delta}^{(i)} = \sum_{k=1}^{N} \delta_k^{(i)}, \qquad \bar{\eta}^{(i)} = \sum_{k=1}^{N} \eta_k^{(i)}.$$

Step 4. Get the estimates

$$\begin{aligned}
\theta^{(i+1)} &= \frac{N^{-1}\sum_{k=1}^{N}\delta_k^{(i)}(\bar{Y} - Y^k)}{\bar{\delta}^{(i)}\bar{\eta}^{(i)} - 1}, \\
\mu^{(i+1)} &= \frac{N^{-1}\sum_{k=1}^{N}\delta_k^{(i)}Y^k - \theta^{(i+1)}}{\bar{\delta}^{(i)}}, \\
\Psi &= \frac{1}{N}\sum_{k=1}^{N}\delta_k^{(i)}(Y^k - \mu^{(i+1)})(Y^k - \mu^{(i+1)})' \\
&\quad - \bar{\eta}^{(i)}\theta^{(i+1)}\theta^{(i+1)\prime}, \\
\Sigma^{(i+1)} &= \frac{|V|^{1/n}\Psi}{|\Psi|^{1/n}}.
\end{aligned}$$

Step 5. Set

$$\Theta^{(i')} = \{a^{(i)}, \lambda^{(i)}, C^{(i)}, \theta^{(i+1)}, \mu^{(i+1)}, \Sigma^{(i+1)}\}$$

and calculate the new weight $\bar{\eta}^{(i')}$ as done in Steps 2 and 3.

Step 6. To complete the calculation of $\Theta^{(i+1)}$, find a, λ, and C that maximize the likelihood function in equation (8.2), that is

$$\mathrm{LL}(\Theta^{(i+1)}; Y^1, \ldots, Y^N) = \sum_{k=1}^{N} \log f_Y(Y^k; \Theta^{(i+1)}),$$

where $\Theta^{(i+1)} = \{a, \lambda, C, \theta^{(i+1)}, \mu^{(i+1)}, \Sigma^{(i+1)}\}$.

Step 7. If $i < 1000$ and $\mathrm{LL}(i) - \mathrm{LL}(i-1) > 1\mathrm{e}{-5}$, increment iteration count i and go to step 2, otherwise, stop the algorithm.

We point out that in the empirical exercise we conducted in Section 8.4, the algorithm converges after around 10 iterations. As done in Section 8.2.1, we consider the ECME algorithm.

Matlab code — MNTS estimates

In this section, we show how to implement the optimization algorithm to find an EM-based MLE estimates in the MNTS case. In Step 6 we consider the ECME approach, that is, we maximize the original log-likelihood in equation (8.2). The maximization step is similar to the MGH case. One has just to substitute the log-likelihood of the MGH model with that of one of the MNTS model (as shown in the code of the *findOptimalParametersMNTS_mle* function). If in the code for the MNTS estimation one selects *showIterations = 1*, then (1) the histogram of the estimated sample of the latent mixing law and its theoretical probability density function and (2) the time series of the estimated latent observations of the mixing law are shown.

In Figure 8.1, we show the estimated observation of the latent mixing law S obtained through the EM-based maximum likelihood approach applied to five major European stock indexes in the period from January 2, 1998 to August 29, 2014. In periods of high (low) volatility the level of the stochastic time is high (law).

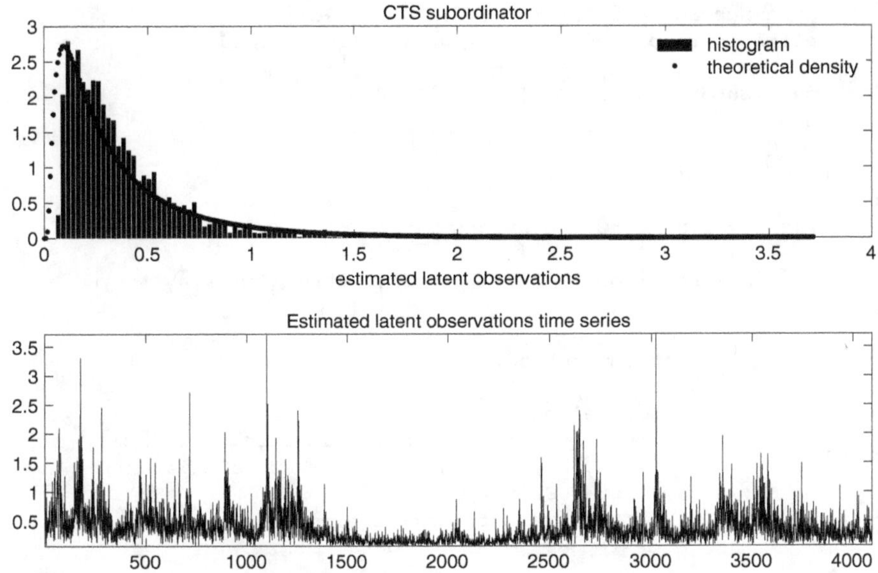

Figure 8.1. Histogram of the estimated sample of the latent mixing law and its theoretical probability density function (top panel) and time series of the estimated latent observations of the mixing law (bottom panel). The estimates are obtained through the EM-based maximum likelihood approach applied to five major European stock indexes in the period from January 2, 1998 to August 29, 2014 (4095 trading days).

```
% estimation of the MNTS distribution

function [a, lambda, C, mu, theta, Sigma, A, D, LL, numSteps] ...
    = estimateMNTS(data, dt, model, method, showIterations, x0)

[~, lc] = size(data);

hFFT = 1e-2;
NFFT = 2^13;
ds   = 1e-2;
M    = 20;
s0   = 1e-10;

mu    = mean(data);
Sigma = cov(data);
theta = zeros(1,lc);

std0   = std(data);
D  = diag(std0, 0);

    if nargin == 6
```

```
    paramSubordinatorCTS = [x0(1:2) x0(3)*dt];
    elseif strcmp(model,'MNTS') == 1 || strcmp(model,'mnts') ...
        == 1
    paramSubordinatorCTS =[1 1 0.5]; % a lambda C
    end

numberOfIterations = 1000;

    for i = 1:numberOfIterations

        if strcmp(model,'MNTS') == 1 || strcmp(model,'mnts') ...
            == 1

            if strcmp(method,'ecme') == 1

            [delta, delta_bar, ~, eta_bar]  = ...
                computeWeigthsMNTS(paramSubordinatorCTS, mu, ...
                theta, Sigma, data, hFFT, NFFT, s0, ds, M);
            [mu, theta, Sigma] = ...
                getMultivariateNormalEstimates(delta, ...
                delta_bar, eta_bar, data, D);

            paramSubordinatorCTS = ...
                findOptimalParametersMNTS_mle(data, ...
                paramSubordinatorCTS, mu, theta, Sigma, 1, ...
                hFFT, NFFT, s0, ds, M);

            end

        end

        if strcmp(model,'MNTS') == 1 || strcmp(model,'mnts') ...
            == 1

             LL(i) = loglikelihoodMNTS(data, ...
                 paramSubordinatorCTS, mu, theta, Sigma, 1, ...
                 hFFT, NFFT, s0, ds, M);

             if showIterations == 1

             [i LL(i)/10^4 paramSubordinatorCTS]

             [~, ~, eta] = ...
                 computeWeigthsMNTS(paramSubordinatorCTS, mu, ...
                 theta, Sigma, data, hFFT, NFFT, s0, ds, M);

                h1 = figure(100);
                subplot 211
                dx = (max(eta) - min(eta))/150;
                binWidth = dx;
```

```
                binCtrs = min(eta):binWidth:max(eta);
                n = length(eta);
                counts = hist(eta,binCtrs);
                prob = counts / (n * binWidth);
                bar(binCtrs,prob,'hist');
                x = linspace(1e-5,max(eta),500);
                hold on
                chf = @(u) chf_subordinatorCTS(u, ...
                    paramSubordinatorCTS, 1);
                pdfCTS  = pdfEvaluation(x, chf, 'pchip', hFFT, ...
                    NFFT);
                plot(x, pdfCTS,'r.')
                title('CTS subordinator')
                hold off
                subplot 212
                plot(eta)
                refreshdata(h1,'caller')
                drawnow; pause(.1)

            end

        end

        if (i>2 && abs(LL(i) - LL(i-1)) <= 1e-5) || i == ...
            numberOfIterations
           break
        end

    end

% to have that A*A' is a correlation matrix
D  = diag(diag(Sigma,0).^(1/2),0);

A  = D\chol(Sigma,'lower');
LL = LL(end);

    % annualized parameters

    if strcmp(model,'MNTS') == 1 || strcmp(model,'mnts') == 1

        a         = paramSubordinatorCTS(1);
            lambda  = paramSubordinatorCTS(2);
            C       = paramSubordinatorCTS(3)/dt;

    end

mu       = mu(:)/dt;
theta    = theta(:);
Sigma    = Sigma;
D        = D;
```

```
numSteps = i;

return
```

In this script we show how the *findOptimalParametersMNTS_mle* function is implemented. In this script we consider the Matlab optimization function *fmincon*. The upper and lower bounds (*ub* and *lb*) for the parameters are provided in the code and not as input arguments. Note that the selection of *ub* and *lb* may depend on the sample to fit. The selected bounds seem to be a proper choice for stock (index) log-returns data.

```
function EstimatedParameters = ...
    findOptimalParametersMNTS_mle(data, x0, mu, theta, Sigma, ...
    dt, hFFT, NFFT, s0, ds, M)

lb =[0.25 1e-2 1e-2];
ub =[1.75 50 20];

options = optimset('Display','iter','Algorithm','active-set',...
                   'UseParallel', 'always');

EstimatedParameters = fmincon(@(x) -loglikelihoodMNTS(data, x, ...
              mu, theta, Sigma, dt, hFFT, NFFT, s0, ds, M), x0, ...
              [],[],[],[],lb,ub,[],options);

return
```

8.3. Some Computational Aspects

In quantitative finance the Fourier transform has been applied for two main reasons: (1) density evaluation (see Stoyanov and Racheva-Iotova (2004), Bianchi *et al.* (2016), and Bianchi *et al.* (2017)) and (2) option pricing (see Levendorskii (2012), Boyarchenko and Levendorskii (2015b), and Boyarchenko and Levendorskii (2015a)). We will consider the first reason.

The integral in equation (8.6) representing the density of the MNTS distribution does not have an analytic solution and therefore a numerical method is needed to evaluate it. As observed above, in the MNTS case the density function h is computed by means of an FFT because only the characteristic function of the mixing distribution is known in closed form and, conversely, its density function is not. This means in practice that the computational effort is given by two different numerical procedures: (1) an FFT and (2) a numerical integration.

In order to test our procedure and to evaluate the error in our numerical approximations, we evaluate the density function of a MGH distribution by considering two different approaches. In the first procedure, we compute it by considering the analytical formula for the density function as given in McNeil *et al.* (2005) and Hu (2005); in the second procedure, we compute the integral (8.6) by considering the characteristic function of a GIG random variable and by applying an FFT together with the numerical integration. We do this by first estimating the MGH parameters on daily log-returns from January 2, 1998 to August 29, 2014 by means of the closed form for the density function. Given the estimates, we evaluate the log-density on each daily log-return with the two different procedures described above. Using the notation introduced in Section 3.2 in Stoyanov and Racheva-Iotova (2004), we consider $\log_{10} h_{\mathrm{FFT}} = -2$ and $q_{\mathrm{FFT}} = 13$, representing the grid spacing and the grid steps in the FFT inversion.

To evaluate the density in equation (8.6), we consider a numerical integration between 1e-10 and 20 with grid spacing $\log_{10} ds = -2$. We find that the mean absolute percentage deviation between the two density evaluations is small (2.4e-5). It would be even smaller (5.8e-7) if one considered $\log_{10} h_{\mathrm{FFT}} = -3$ and $q_{\mathrm{FFT}} = 16$, but the computing time would be more than double. Note that a careful selection of h_{FFT} and q_{FFT} is needed to obtain reliable evaluations (see also Bianchi *et al.* (2017)).

Given the analysis above, we apply the same algorithm to evaluate the MNTS density. Recall that in the MNTS case the density function is not known in closed form, therefore an analysis of the errors as done in the MGH case cannot be conducted. For this reason in Section 8.3.1 we assess through a simulation study whether also in the MNTS case the proposed estimation algorithm is reliable or not.

To perform the estimation using the EM-based MLE method, the computing time for every day is, in median values, 4 seconds in the MGH case, and 65 seconds in the MNTS case. Note that while in the MGH cases we consider the analytical formula for the density function, in the MNTS cases to evaluate the density we apply the procedure described above. As expected, the use of an FFT together with the numerical integration needed to compute the likelihood of MNTS models greatly increases the computing time. However, for all models the number of iterations of the iterative algorithm explained in Section 8.2 is around 10. Note that on each iteration the weights in equation (8.9) are computed twice and an optimization problem is solved. For this reason the numerical evaluation of the density has to be performed many times, particularly during the optimization phase. Even if the use of the MNTS distribution largely increases the computing time with respect to the MGH distribution, it is still acceptable for practical applications.

8.3.1. An error analysis of the estimation method

In addition to estimating the MNTS models on the time series of observed log-returns, we empirically study through a simulation exercise the estimation algorithm presented in Section 8.2. More specifically, we generate MNTS random numbers by considering the set of parameters estimated on the time series from January 2, 1998 to August 29, 2014, and then we estimate the parameter by following the EM-based algorithm. As already highlighted in Chapter 6, to simulate the univariate tempered stable mixing distribution we consider the inverse transform algorithm described by Bianchi *et al.* (2017). We repeat the exercise 1000 times by considering samples with 1500 observations.

We compute the relative error of the parameter estimators and in Figure 8.2, we show the boxplots of the empirical distribution of the difference between the estimated parameters and the true values. On each box, the central mark is the median, the edges of the box are the 25th and 75th percentiles, the whiskers extend to the most extreme data points not considered outliers, and outliers are plotted individually. The relative error of the parameter p is defined by

$$e_{\text{rel}}^{1-\delta}(p) = \max\left\{\left|\frac{q_{\delta/2}(\hat{p}) - p}{p}\right|, \left|\frac{q_{1-\delta/2}(\hat{p}) - p}{p}\right|\right\}, \tag{8.10}$$

and it means that the relative error of the estimator is smaller than $e_{\text{rel}}^{1-\delta}$ with probability $1-\delta$. This error is computed through the empirical distribution of the estimator: $q_\beta(\hat{p})$ represents the βth quantile of the estimator of p.

In Figure 8.2, we observe that medians (the central mark in the box) are close to zero (i.e., the estimated parameters are close to

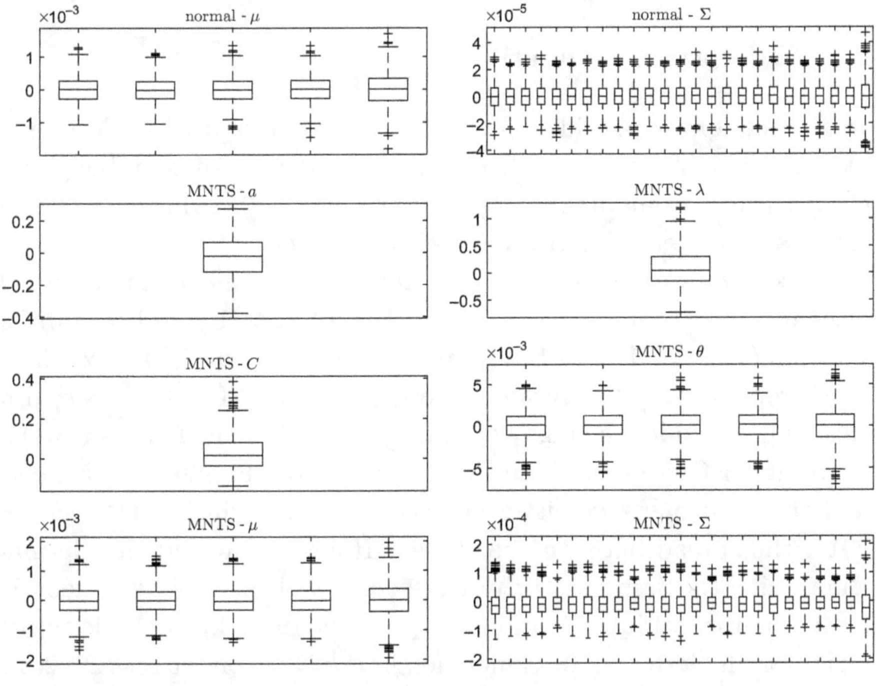

Figure 8.2. Boxplots of the difference between estimates and true parameters of a normal and of an MNTS distribution. Each boxplot consists of 1000 estimates. Normal and MNTS random numbers are generated by considering the set parameters estimated on the time series with 1500 log-return observations ending on August 29, 2014.

the corresponding true values, except for a few outliers). Based on the relative errors, the accuracy of the estimator for a is higher than the one for C and λ, and the one for Σ is higher than the one for μ and θ. The relative error of the estimators of Σ, μ, and θ is generally high. This may be caused by the large number of model parameters and the relative small number of simulations. As shown in Bianchi *et al.* (2017) for tempered stable models, a large number of simulations is needed to obtain small relative errors. We also conducted the same exercise on a multivariate normal distribution. In particular, the accuracy of the estimator $\hat{\Sigma}$ in the normal case is close to that of the $\hat{\Sigma}$ in the MNTS case. Similar results are true also in the MGH case. In the normal case, while the relative error for the elements of Σ ranges between 6.01% and 11.72%, the relative error for the parameter μ exceeds 50%.

8.4. Historical Estimation

In this section we provide a description of the data used in the empirical analysis. We obtained from Bloomberg daily dividend-adjusted closing prices from January 2, 1998 to August 29, 2014 for five stock indexes: the Deutsche Aktienindex 30 (ticker DAX), the Cotation Assistée en Continu 40 (ticker CAC), the Financial Times Stock Exchange Milano Indice di Borsa (ticker FTSEMIB), Índice Bursátil Español (ticker IBEX), Athens Stock Exchange General Index (ticker ASE) representing five major European indexes. The time period in this study includes the high volatility period after the Lehman Brothers filing for Chapter 11 bankruptcy protection (September 15, 2008) and the eurozone sovereign debt crisis, during which, in November 2011, the spread between the 10-year Italian BTP and the German Bund with the same maturity exceeded 500 basis points. Then, for further empirical investigations we used the daily logarithmic return series for all German stocks included in the DAX on August 29, 2014. We obtained from Bloomberg daily dividend-adjusted closing prices from December 31, 2004 to August 29, 2014.

We compare the four non-Gaussian models (MVG, MNIG, MGH, and MNTS) with the multivariate Gaussian model based on the multivariate normal distribution with parameters μ and Σ, where, as usual, μ represents the mean vector and Σ is the variance–covariance matrix. In the empirical analysis on the DAX components, we consider only the Gaussian, the MGH, and the MNTS models, and we do not empirically assess the fitting performance of the MVG and the MNIG models.

Note that the d-dimensional MVG and MNIG (the MGH and MNTS) models have $2 + d$ $(3 + d)$ more parameters compared to the Gaussian model. More precisely, the additional parameters with respect to the Gaussian model are the parameters of the latent mixing law and the parameter θ (i.e., λ, χ, and ψ and the vector θ in the MGH case and a, λ, C, and the vector θ in the MNTS case). Besides estimating the parameters, we apply the Akaike information criterion (AIC) and the Bayesian information criterion (BIC) to identify the superior model. The AIC and the BIC are evaluated as

$$AIC = 2np - 2LL,$$

and

$$BIC = np \log(no) - 2LL,$$

where np is the number of parameters, no is the number of observations, and LL is the model's log-likelihood.

Figures 8.3 and 8.4 show for all calibrated models the boxplot of both the AIC and the BIC computed across the 1736 (933) rolling windows estimations on the time series of indexes (DAX components) returns, respectively. According to both the AIC and the BIC, the MNTS model exhibited the best performance because its AIC and BIC values are the smallest, and, conversely, the Gaussian model is the worst one because its AIC and BIC values are the largest.

In addition, we also compute the Kolmogorov–Smirnov (KS) distance for each marginal distribution of the models analyzed. This distance compares the empirical cumulative distribution function (cdf) with the cdf of the reference distribution. The test statistic of the KS test is defined by the KS distance, and the null hypothesis of the test is rejected if the maximum distance between the two cumulative distribution functions is too large. Recall that the KS distance

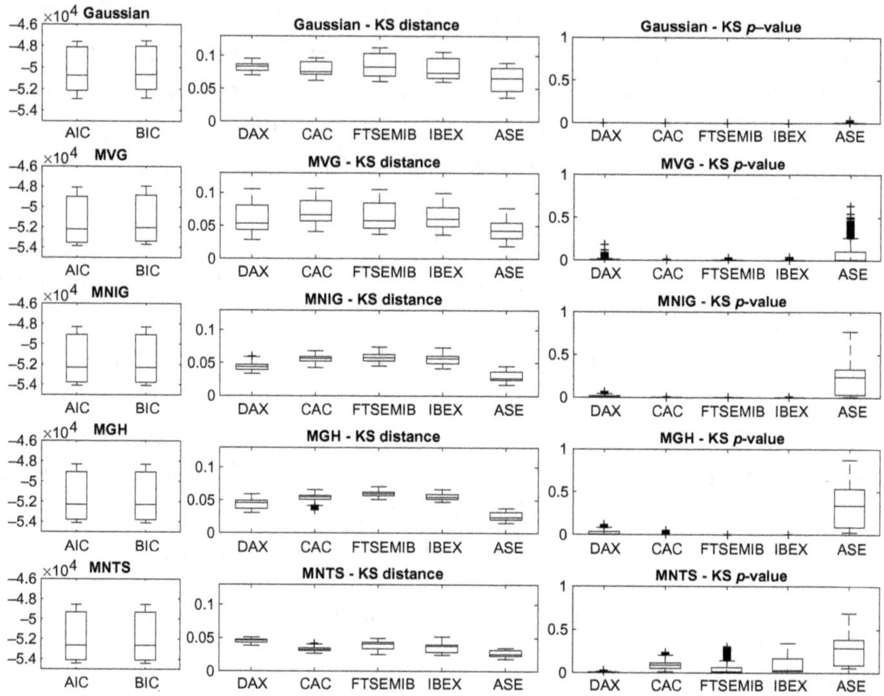

Figure 8.3. (Indexes) Goodness of fit statistics of the MVG, MNIG, MGH, and MNTS models estimated using the EM-based maximum likelihood approach for each trading day from August 29, 2007 to August 29, 2014. For each trading day, a window of fixed size is considered (1500 trading days) for a total of 1736 rolling windows estimations for each model. For each model the time series of the AIC, of the BIC, of the mean univariate KS distance, and of the mean p-value are shown. The non-Gaussian models are compared with the Gaussian one.

is defined as

$$\mathrm{KS}_j(\Theta, \{(Y^1)^j, \ldots, (Y^N)^j\}) = \max_{k=1,\ldots,N} |F((Y^k)^j) - \hat{F}((Y^k)^j)|, \tag{8.11}$$

where F and $\hat{F}$ are the theoretical cdf and the empirical distribution function, respectively, $\{Y^k = Y_{t_k} - Y_{t_{k-1}}\}_{k=1,\ldots,N}$ are the five-dimensional (30-dimensional) observations, and the indicator j represents the stock index (stock) j. This test is implemented in Matlab *kstest* function. Figures 8.3 and 8.4 show the boxplot of the KS statistic computed across the five and 30 marginal distributions

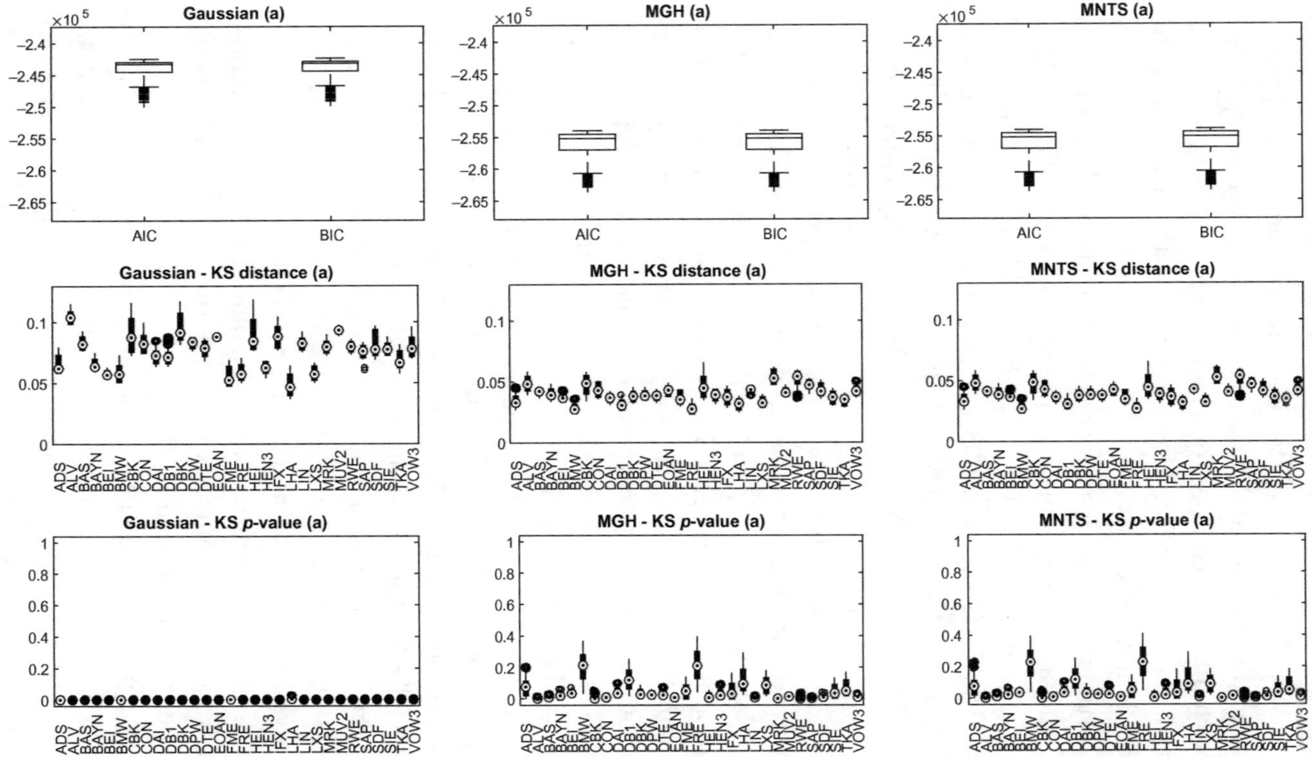

Figure 8.4. (DAX components) Goodness of fit statistics of the MGH and MNTS models estimated using the EM-based maximum likelihood approach for each trading day from December 31, 2010 to August 29, 2014. For each trading day, a window of fixed size is considered (1500 trading days) for a total of 933 rolling windows estimations for each model. For each model the time series of the AIC, of the BIC, of the mean univariate Kolmogorov–Smirnov (KS) distance, and of the mean p-value are shown. The non-Gaussian models are compared with the Gaussian one.

for each model and the boxplot of the p-value. According to the KS statistic, the MNTS model is better because for each index (stock) the KS statistics are smaller and the corresponding p-values are larger compared with all other competitor models. The Gaussian model is the poorest performing model. Moreover, in most of the cases, the MNTS model is not rejected.

In the estimates that consider the five indexes shown in Figure 8.3, the average p-value computed across the 1736 rolling windows ranges from 0.02% (CAC) to 6.34% (ASE) for the MVG model, from 0.03% (CAC) to 23.38% (ASE) for the MNIG model, from 0.01% (FTSEMIB) to 33.66% (ASE) for the MGH model, and from 0.52% (DAX) to 27.68% (ASE) for the MNTS model. The p-values of the Gaussian model are less than 4% in all cases. We point out that if in Step 2 to evaluate the expectations (weights) in equation (8.9) instead with the formulas proposed in the MGH case (i.e., with the Matlab function *computeWeigthsMGH*), one considers the algorithm proposed for the MNTS case (i.e., the Matlab function *computeWeigthsMNTS*); that is, the weights are evaluated by numerical integration, the fitting performance of the three MGH models (MVG, MNIG, and MGH) improves and it becomes similar to the performance of the MNTS model.

In the estimates that consider the DAX components shown in Figure 8.4, the average p-value computed across the 933 rolling windows ranges from 0.07% to 22.55% for the MNTS model, and from 0.07% and 20.98% for the MGH model. In these two cases, in Step 2 the weights are evaluated by numerical integration (i.e., by considering the Matlab function *computeWeigthsMNTS* in the MGH case also). The p-values of the normal model are below 4% in all cases: the average p-value ranges from 5.87e-14% and 0.07%. The average p-value is less than 1% in only six cases for the non-normal models and in all cases for the normal model.

Figure 8.5 reports the time series of the parameters of the MGH and the MNTS models over time (the elements of the matrix A are not reported) estimated on the time series of index returns. In these two cases, in Step 2 the weights are evaluated by numerical

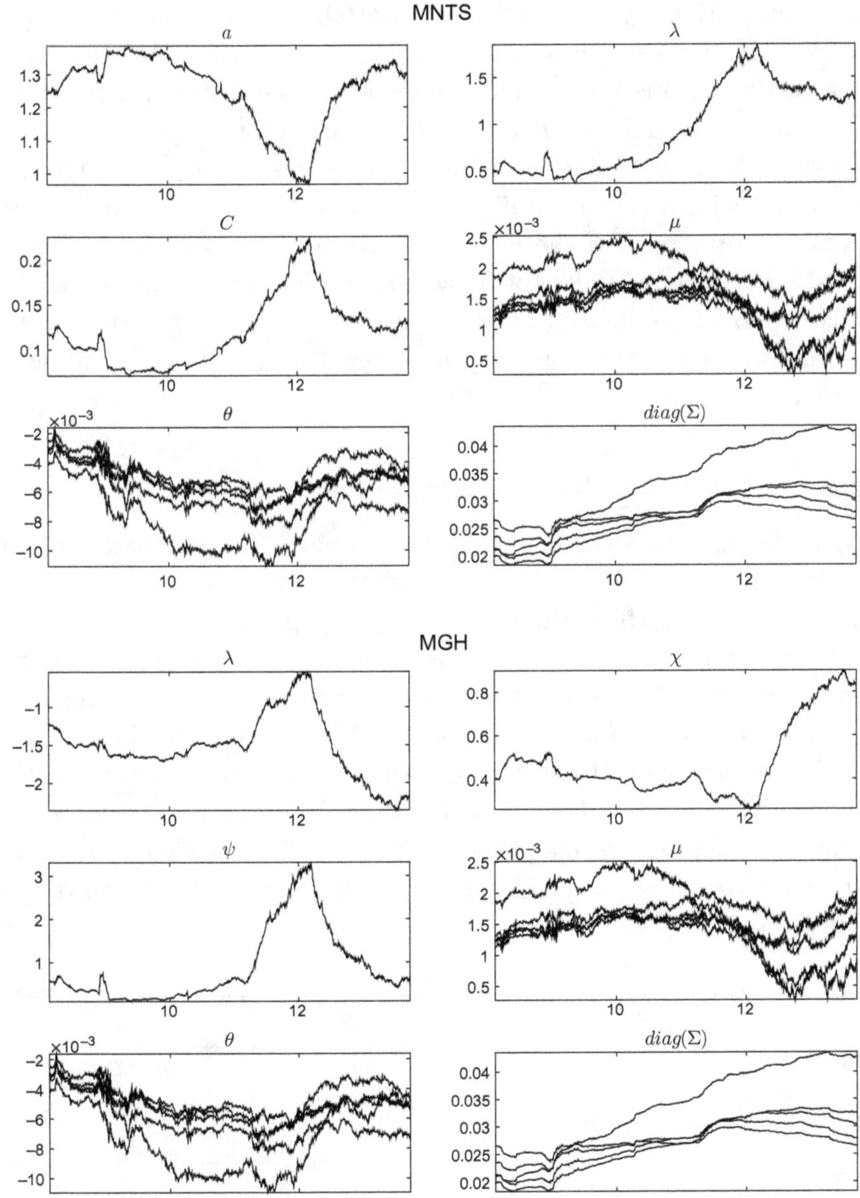

Figure 8.5. MNTS and MGH historical parameter estimates under the EM-based maximum likelihood approach for each trading day from August 29, 2007 to August 29, 2014. For each trading day, a window of fixed size is considered (1500 trading days) for a total of 1736 rolling windows estimations.

integration (i.e., by considering the Matlab function *computeWeigthsMNTS* in the MGH case also), therefore the fitting performance of the two non-Gaussian models are close to each other. We point out that in the MNTS the mean vector is not the parameter vector μ. As expected, the principal diagonal σ of the diagonal matrix D_σ in the MNTS model is similar to the σ in the MGH model. Moreover, the dynamics of the parameter vectors μ and θ are similar for the two models. We do not show the parameters of the MVG and MNIG models estimated on the time series of index returns nor the parameters estimated on the time series of the DAX component returns.

8.4.1. Matlab code — A simple test

Here we show the Matlab script to estimate the parameters of the multivariate normal, the MGH, and the MNTS distribution. The code is based on the estimation method proposed in Section 8.2 and the algorithms shown in Sections 8.2.1 and 8.2.2. The execution of the entire test may take a few seconds. The computing time depends on several factors (e.g., the processor, the Matlab version, the available RAM, the number of cores). Note that some functions used in this code (*cdfEvaluation*, *densityMNTS_matrix*, *densityMGH_matrix*) are provided in previous chapters. Before running this code a matrix of dimension $N \times dim$ of indexes (stocks) log-returns is needed at hand.

```
% The variable 'data' contains multivariate log-returns (N x dim).
% Each stock / index is a column vector of length 'N'.

dt = 1/250;
[N, dim] = size(data);

tic
[mu_normal, Sigma_normal, LL_normal] = ...
    estimateMultivariateNormal(data, dt);
disp(['Normal - time: ', num2str(toc), ' s']);

tic
[a_MNTS, lambda_MNTS, C_MNTS, mu_MNTS, theta_MNTS, Sigma_MNTS, ...
    A_MNTS D_MNTS,  LL_MNTS] =...
```

```
estimateMNTS(data, dt, 'mnts', 'ecme', 1);
disp(['MNTS - time: ', num2str(toc), ' s']);

tic
[lambda_MGH, chi_MGH, psi_MGH, mu_MGH, theta_MGH, Sigma_MGH, ...
    A_MGH, D_MGH,  LL_MGH] =...
estimateMGH(data, 'mgh', 'ecme');
disp(['MGH - time: ', num2str(toc), ' s']);

[N dim] = size(data);

 % KS test

for k = 1:dim

cdfF = @(x) normcdf(x, mu_normal(k)*dt, ...
    sqrt(Sigma_normal(k,k)*dt));
[KS_normal(k),KS_p_normal(k)] = evaluateKS(data(:,k), cdfF);

hFFT = 1e-2;
NFFT = 2^16;
ds   = 1e-2;
M= 20;
s0   = 1e-10;

pdfF = @(x) densityMNTS_matrix(x, [a_MNTS, lambda_MNTS, ...
    C_MNTS],...
       mu_MNTS(k), theta_MNTS(k), Sigma_MNTS(k,k), dt, hFFT, ...
           NFFT, s0, ds, M);
cdfF = @(x) cdfEvaluation(x, pdfF, 'pdf');
[KS_MNTS(k), KS_p_MNTS(k)] = evaluateKS(data(:,k), cdfF);

pdfF = @(x) densityMGH_matrix(x, lambda_MGH, chi_MGH, psi_MGH,...
       mu_MGH(k), theta_MGH(k), Sigma_MGH(k,k));
cdfF = @(x) cdfEvaluation(x, pdfF, 'pdf');
[KS_MGH(k),KS_p_MGH(k)] = evaluateKS(data(:,k), cdfF);

end
```

8.4.2. Matlab code — MLE for the multivariate normal law

In this section, we show how to estimate the parameters of a multivariate normal distribution by MLE. The parameter μ is estimated by considering the sample mean, and the matrix Σ is estimated by considering the sample covariance matrix.

```
function [mu, Sigma, LL] = estimateMultivariateNormal(data, dt)

mu = mean(data);
Sigma = cov(data);

f = mvnpdf(data,mu,Sigma);
LL = sum(log(f));

mu    = mu/dt;
Sigma = Sigma/dt;

return
```

8.4.3. Matlab code — KS estimate

In this section, we show how to compute the KS distance and the p-value for the KS test.

```
function [ksstat,p] = evaluateKS(data, cdfFunction)

[~,p,ksstat] = kstest(data,[data cdfFunction(data)]);

return
```

8.5. A Risk Assessment Study

In this section, we assess the extent to which asymmetric and heavy-tailed multivariate distributions are needed to explain the behavior of stock log-returns and to properly evaluate some well-known portfolio risk measures. The risk measures we use in this study are VaR and AVaR, the average of VaRs greater than the VaR for a given tail probability. As already observed in other chapters, AVaR, also called conditional value-at-risk (CVaR) or expected shortfall, is a superior risk measure to VaR because it satisfies all axioms of a coherent risk measure and it is consistent with preference relations of risk-averse investors (see Rachev *et al.* (2008)). Both measures have been adopted as standard risk measures in the financial industry and by regulators (see Alexander and Sarabia (2012), Emmer *et al.* (2013), Embrechts *et al.* (2014), and the Basel Committee on Banking Supervision (January 2016) document on market risk measurement). We compare the three models (normal, MGH, and MNTS continuous-time models) by considering two equally weighted portfolios composed by the five selected indexes and 30 DAX components, respectively.

Recall that the VaR at tail probability level δ is defined as

$$\text{VaR}_\delta(W) = -\inf\{x|P(W \leq x) > \delta\} = -F_W^{-1}(\delta)$$

and can be computed by inverting the cumulative distribution function F_W, and that the AVaR of a continuous random variable W with finite mean (i.e., $E[W] < \infty$) at tail probability level δ is defined as the average of the VaRs that are greater than the VaR at tail probability δ; that is

$$\text{AVaR}_\delta(W) = \frac{1}{\delta}\int_0^\delta \text{VaR}_p(W)dp = -E[W|W < -VaR_\delta(W)].$$

Therefore, by construction, AVaR is focused on the losses in the tail that are greater than the corresponding VaR level.

In the normal case, the VaR and the AVaR can be easily evaluated. In the MGH case, the VaR and AVaR computation is simpler, since a closed-form expression for the density function can be used. As already observed in Chapter 7 for the MGH case, we have a random variable $W = w'Y_{\Delta t}$ where $w \in \mathbb{R}^n$ which is GH distributed with parameters λ, χ, ψ, $\tilde{\theta}$, $\tilde{\mu}$, $\tilde{\sigma}$, where

$$\tilde{\theta} = w'\theta, \qquad \tilde{\mu} = w'\mu, \qquad \tilde{\sigma} = \sqrt{w'\Sigma w}.$$

Similarly, in the MNTS case $W = w'Y_{\Delta t}$ is NTS distributed with parameters $(a, \lambda, C, \tilde{\theta}, \tilde{\mu}, \tilde{\sigma})$. As already observed in Chapter 6, from Kim *et al.* (2010, 2011) it is possible to obtain a closed formula (up to an integration) to compute the AVaR when only the characteristic function, but not the density function, is known in closed form.

As shown in Table 8.1, although there are not large differences in the case of VaR between the three continuous-time models for all tail probability levels analyzed, the differences in the case of AVaR

Table 8.1. Indexes (DAX components) VaR and AVaR for different tail probability levels δ. The risk measures are evaluated for each trading day from August 29, 2007 (December 31, 2010) to August 29, 2014. For each trading day, a window of fixed size is considered (1500 trading days) for a total of 1736 (933) rolling windows estimations for each model. For each model we report the minimum, the mean, the median, and the maximum values of the risk measures (in percentage points).

			min	mean	median	max	min	mean	median	max
				$\delta = 0.01$				$\delta = 0.025$		
Indexes	VaR	normal	2.53	3.35	3.35	3.89	2.13	2.82	2.82	3.29
		MGH	3.24	4.05	4.12	4.60	2.36	2.95	2.93	3.46
		MNTS	3.22	4.03	4.09	4.53	2.32	2.93	2.91	3.40
	AVaR	normal	2.90	3.83	3.84	4.45	2.54	3.36	3.37	3.91
		MGH	4.44	5.61	5.75	6.12	3.41	4.29	4.41	4.74
		MNTS	4.39	5.44	5.52	5.86	3.37	4.20	4.30	4.67
				$\delta = 0.05$				$\delta = 0.1$		
Indexes	VaR	normal	1.78	2.37	2.37	2.77	1.38	1.85	1.85	2.17
		MGH	1.77	2.23	2.18	2.65	1.23	1.57	1.51	1.93
		MNTS	1.72	2.19	2.15	2.60	1.18	1.52	1.47	1.88
	AVaR	normal	2.24	2.97	2.97	3.46	1.90	2.53	2.53	2.95
		MGH	2.72	3.41	3.44	3.87	2.09	2.64	2.62	3.04
		MNTS	2.68	3.36	3.37	3.81	2.04	2.58	2.56	2.98
				$\delta = 0.01$				$\delta = 0.025$		
DAX	VaR	normal	3.46	3.77	3.83	3.91	2.91	3.18	3.23	3.29
		MGH	3.77	4.26	4.34	4.49	2.82	3.21	3.27	3.39
		MNTS	3.85	4.36	4.45	4.60	2.82	3.21	3.27	3.39
	AVaR	normal	3.97	4.32	4.39	4.47	3.48	3.79	3.85	3.93
		MGH	5.17	5.78	5.88	6.10	4.00	4.50	4.58	4.74
		MNTS	5.25	5.92	6.05	6.26	4.05	4.58	4.70	4.84
				$\delta = 0.05$				$\delta = 0.1$		
DAX	VaR	normal	2.44	2.66	2.71	2.76	1.89	2.07	2.11	2.15
		MGH	2.18	2.50	2.55	2.64	1.58	1.82	1.86	1.93
		MNTS	2.14	2.46	2.51	2.59	1.52	1.77	1.80	1.87
	AVaR	normal	3.06	3.34	3.40	3.46	2.60	2.84	2.89	2.95
		MGH	3.23	3.65	3.72	3.85	2.54	2.89	2.94	3.05
		MNTS	3.24	3.68	3.76	3.89	2.52	2.87	2.93	3.04

are remarkable, at least for the lower tail probability levels (0.01, 0.025). This is particularly evident in some periods, as shown in Figure 8.6. The heavy-tailed nature of both MGH and MNTS distributions clearly affect the estimated AVaR values.

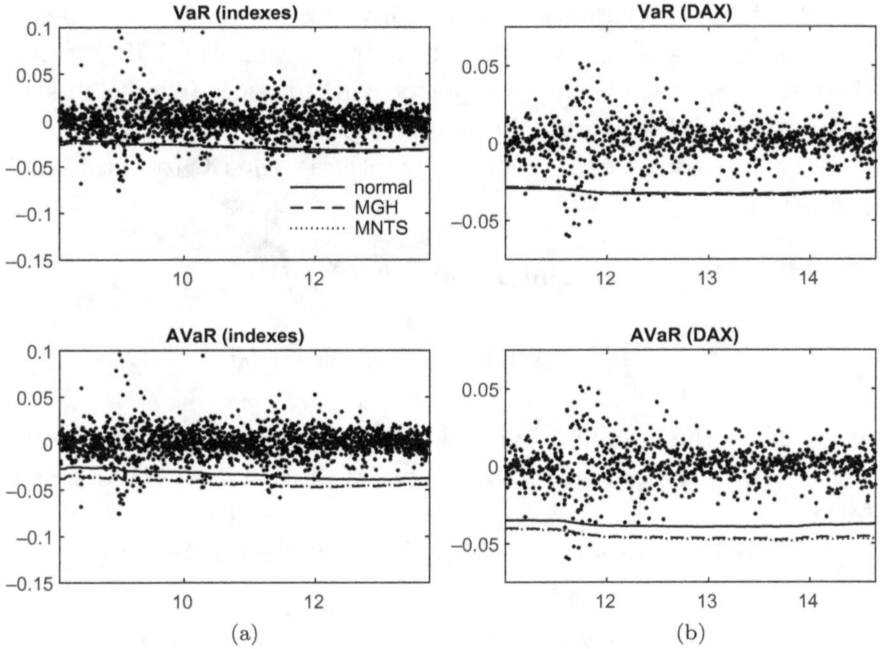

Figure 8.6. VaR and AVaR time series of the equally weighted portfolio for the three considered models: (a) from August 29, 2007 to August 29, 2014, and (b) December 31, 2010 to August 29, 2014. The tail probability level is equal to 0.025. For each trading day, a window of fixed size is considered (1500 trading days) for a total of (a) 1736 and (b) 933 rolling windows estimations for each model. We analyzed three different models (normal, MGH, MNTS). The equally weighed portfolio returns are also reported.

To evaluate the accuracy of forecasting VaR for the models analyzed in this chapter, we perform backtesting using the likelihood ratio (LR) tests proposed by Christoffersen (1998, 2010). The LR tests use the number of violations, where violations occur when the actual loss exceeds the estimated VaR. The LR test consists of three parts: (1) the LR test of unconditional coverage (LR_{uc}), which is the same as the proportion of failures test by Kupiec (1995), (2) the LR test of independence (LR_{ind}), and (3) the joint test of coverage and independence (LR_{cc}).

The LR_{uc} test consists of examining if the realized coverage rate equals the theoretical coverage rate δ of the VaR for a backtesting sample of N observations. This test involves testing if the indicator function $I_t(\delta) = I_{\{w_t<-\mathrm{VaR}_{\delta,t}\}}$, which takes values of 1 if at time t the return w_t is less than the reported VaR measure (change in sign) and 0 otherwise, follows a Bernoulli distribution with parameter δ. Under the unconditional coverage hypothesis, the LR test statistic is defined as

$$\begin{aligned}\mathrm{LR}_{\mathrm{uc}}(\delta) &= -2\ln((1-\delta)^{N-N_v}\delta^{N_v})\\ &\quad +2\ln\left(\left(1-\frac{N_v}{N}\right)^{N-N_v}\left(\frac{N_v}{N}\right)^{N_v}\right),\end{aligned} \tag{8.12}$$

where N_v is the number of violations. As N goes to infinity, the test statistic $\mathrm{LR}_{\mathrm{uc}}$ follows under H_0 a χ^2-distribution with one degree of freedom.

The LR test statistic for independence is defined by

$$\begin{aligned}\mathrm{LR}_{\mathrm{ind}}(\delta) &= -2\ln\left(\left(1-\frac{N_v}{N}\right)^{N-N_v}\left(\frac{N_v}{N}\right)^{N_v}\right)\\ &\quad +2\ln((1-\pi_{01})^{n_{00}}\pi_{01}^{n_{01}}(1-\pi_{11})^{n_{10}}\pi_{11}^{n_{11}}),\end{aligned} \tag{8.13}$$

where n_{ij} is the number of times we have $I_{t+1}(\delta) = j$ and $I_t(\delta) = i$, with both i and j equal to 0 or 1, and with

$$\pi_{01} = \frac{n_{01}}{n_{00}+n_{01}},$$

$$\pi_{11} = \frac{n_{11}}{n_{10}+n_{11}}.$$

The $\mathrm{LR}_{\mathrm{ind}}$ test statistic asymptotically follows under H_0 a χ^2-distribution with one degree of freedom.

The LR test statistic for the conditional coverage is then defined

$$\mathrm{LR}_{\mathrm{cc}}(\delta) = \mathrm{LR}_{\mathrm{uc}}(\delta) + \mathrm{LR}_{\mathrm{ind}}(\delta) \tag{8.14}$$

and as N tends to infinity it follows under H_0 a χ^2-distribution with two degrees of freedom.

Matlab code — Likelihood ratio tests

```
function [LRuc, LRind, LRcc, puc, pind, pcc, n00, n01, n10, ...
    n11] = ...
      BackTestChristoffersen( p, VaRs, nextreturn )

    subret = nextreturn;
    subvars = -VaRs;

    n00 = 0;
    n01 = 0;
    n10 = 0;
    n11 = 0;

    for k = 1:(length(subret)-1)
        if (subret(k)>subvars(k) && subret(k+1)>subvars(k+1))
            n00 = n00+1;
        elseif (subret(k)>subvars(k) && subret(k+1)<=subvars(k+1))
            n01 = n01+1;
        elseif (subret(k)<=subvars(k) && subret(k+1)>subvars(k+1))
            n10 = n10+1;
        else
            n11 = n11+1;
        end
    end

    Nv = length(find(subret<subvars));
    [LRuc,puc] = LR_Kupiec(p,Nv,length(subvars));

    PI1 = [n00/(n00+n01), n01/(n00+n01);n10/(n10+n11), ...
        n11/(n10+n11)];
    pi2 = (n01+n11)/(n00+n10+n01+n11);

    LPI1 = PI1(1,1)^n00*PI1(1,2)^n01*PI1(2,1)^n10*PI1(2,2)^n11;
    LPI2 = (1-pi2)^(n00+n10)*pi2^(n01+n11);

    LRind = -2*log(LPI2/LPI1);
    pind = 1 - chi2cdf(LRind, 1);

    LRcc = LRuc + LRind;
    pcc = 1-chi2cdf(LRcc, 2);

 return

 function [LRuc, puc] = LR_Kupiec(p, Nv, N)

    T = N;
    T0 = N - Nv;
```

```
    T1 = Nv;

    LRuc = - 2*(T0*log(1-p) + T1*log(p) ...
           - T0*log(1 - T1/T) - T1*log(T1/T));
    puc = 1 - chi2cdf(LRuc, 1);

return
```

In Table 8.2 we report the three LR test statistics and the corresponding p-values by considering different δ (0.01, 0.025, 0.05, 0.1). The non-normal models (MGH and MNTS) outperform the competitor models, at least for the tail probability levels that are of interest in banking supervision (0.01, 0.025 that correspond to confidence levels 99% and 97.5%). Conversely, in some cases for larger probability levels the normal model shows a better performance than the MGH and MNTS model. The model's performance depends on the data considered in the empirical study. The normal model shows a good performance at the smallest tail level if one considers the equally weighted portfolio of DAX components.

In Figure 8.6, we show the dynamics of the 1-day VaR and AVaR for each model analyzed and compare them with the observed portfolio return. The portfolio is composed of the five selected indexes (30 DAX components). The risk measures of the three continuous-time models are quite stable over time, and for this reason the LR test does not always show good results. The risk measures of these models are not able to follow the dynamics of portfolio log-returns and for this reason they have not always performed well in terms of the VaR 1-day forecast. Similar dynamics are shown by the equally weighted portfolio of DAX components.

The model proposed in this chapter can explain three stylized facts — heavy tails, negative skew, and asymmetric dependence — but they are not able to capture the volatility clustering effect. As shown in Bianchi *et al.* (2016), one needs the *four horsemen* (see also Allen and Satchell (2014)), that is asymmetric and heavy-tailed multivariate distributions together with volatility clustering to explain the behavior of stock log-returns and to properly evaluate some well-known portfolio risk measures.

Table 8.2. Indexes (DAX components) VaR backtest for each trading day from August 29, 2007 (December 31, 2010) to August 29, 2014. For each trading day, a window of fixed size is considered (1500 trading days) for a total of 1736 (933) rolling windows estimations for each model. For each model we report the number of violations (N_v), the LR test of unconditional coverage (LR_{uc}), the LR test of independence (LR_{ind}), and the joint test of coverage and independence (LR_{cc}). Below the test statistics the corresponding p-value are shown.

		N_v	LR_{uc}	LR_{ind}	LR_{cc}	N_v	LR_{uc}	LR_{ind}	LR_{cc}
			$\delta = 0.01$				$\delta = 0.025$		
Indexes	normal	53	47.82	2.70	50.51	70	14.18	7.14	21.31
			(0.000)	(0.100)	(0.000)		(0.000)	(0.008)	(0.000)
	MGH	33	11.28	0.20	11.47	65	9.61	6.34	15.96
			(0.001)	(0.659)	(0.003)		(0.002)	(0.012)	(0.000)
	MNTS	34	12.61	0.15	12.76	65	9.61	6.34	15.96
			(0.000)	(0.698)	(0.002)		(0.002)	(0.012)	(0.000)
			$\delta = 0.05$				$\delta = 0.1$		
Indexes	normal	100	2.03	4.35	6.39	180	0.27	22.41	22.67
			(0.154)	(0.037)	(0.041)		(0.605)	(0.000)	(0.000)
	MGH	116	9.43	3.47	12.90	239	24.88	17.54	42.43
			(0.002)	(0.063)	(0.002)		(0.000)	(0.000)	(0.000)
	MNTS	122	13.46	3.41	16.87	244	28.64	16.46	45.10
			(0.000)	(0.065)	(0.000)		(0.000)	(0.000)	(0.000)
			$\delta = 0.01$				$\delta = 0.025$		
DAX	normal	8	0.20	0.14	0.34	19	0.87	3.63	4.50
			(0.656)	(0.710)	(0.845)		(0.352)	(0.057)	(0.106)
	MGH	8	0.20	0.14	0.34	19	0.87	3.63	4.50
			(0.656)	(0.710)	(0.845)		(0.352)	(0.057)	(0.106)
	MNTS	7	0.64	0.11	0.74	19	0.87	3.63	4.50
			(0.424)	(0.745)	(0.689)		(0.352)	(0.057)	(0.106)
			$\delta = 0.05$				$\delta = 0.1$		
DAX	normal	25	12.59	4.79	17.38	44	35.18	15.01	50.19
			(0.000)	(0.029)	(0.000)		(0.000)	(0.000)	(0.000)
	MGH	30	7.09	13.73	20.81	65	10.49	15.86	26.35
			(0.008)	(0.000)	(0.000)		(0.001)	(0.000)	(0.000)
	MNTS	30	7.09	13.73	20.81	67	8.98	17.35	26.33
			(0.008)	(0.000)	(0.000)		(0.003)	(0.000)	(0.000)

In this chapter, we will not deal with AVaR backtesting, mainly because of the small number of observations where the VaR is violated (see Christoffersen (2010)). However, as shown in Broda and Paolella (2011) and given the estimation algorithm proposed in this chapter, it is not difficult to obtain the bootstrap distribution for

AVaR and an approximative approach to the backtesting can be based on a representation of AVaR as an integrated VaR (see Acerbi and Tasche (2002) and Emmer *et al.* (2013)).

Key Points

- In the EM-based MLE algorithm, in the maximization step, one can choose to find the optimal values of the parameters of the mixing law by finding the maximum value of the likelihood of the original multivariate law (i.e., the ECME approach) or only a part of the alternative likelihood function on which the recursive algorithm is based (i.e., the MCECM approach).
- The EM-based MLE algorithm can be extended to the normal mean–variance mixture distributions in which only the characteristic function of the mixing distribution is known in closed form and, conversely, its density function is not.
- In quantitative finance the Fourier transform has been applied for two main purposes: (1) density evaluation and (2) option pricing. For the estimation of the MNTS parameters, one needs to cast the FFT algorithm into the EM-based MLE approach.
- The computational effort needed to evaluate the likelihood function of an MNTS distribution is equal to the computation of a univariate density with an FFT algorithm and a numerical integration of a one-dimensional integral.
- Both the MGH and the MNTS laws can be used in portfolio allocation problems since well-known risk measures can be easily computed. A linear combination of MGH (MNTS) margins is GH (NTS) distributed.
- Multivariate continuous-time MGH and MNTS models can explain three stylized facts — heavy tails, negative skew, and asymmetric dependence — but they are not able to capture the volatility clustering effect.
- While there are numerous techniques for backtesting the VaR measure, the AVaR backtesting is not trivial mainly because of the small number of observations exceeding VaR.

References

C. Acerbi and D. Tasche. On the coherence of expected shortfall. *Journal of Banking & Finance*, 26(7):1487–1503, 2002.

C. Alexander and J.M. Sarabia. Quantile uncertainty and value-at-risk model risk. *Risk Analysis*, 32(8):1293–1308, 2012.

D. Allen and S. Satchell. The four horsemen: heavy-tails, negative skew, volatility clustering, asymmetric dependence. Discussion paper 4, University of Sydney, 2014.

Basel Committee on Banking Supervision. Minimum capital requirements for market risk, January 2016. https://www.bis.org/bcbs/publ/d352.pdf.

M.L. Bianchi, S.T. Rachev, and F.J. Fabozzi. Tempered stable Ornstein–Uhlenbeck processes: A practical view. *Communications in Statistics — Simulation and Computation*, 46(1):423–445, 2017.

M.L. Bianchi, G.L. Tassinari, and F.J. Fabozzi. Riding with the four horsemen and the multivariate normal tempered stable model. *International Journal of Theoretical and Applied Finance*, 19(4):1650027, 2016.

S.I. Boyarchenko and S.Z. Levendorskii. Efficient variations of Fourier transform in applications to option pricing. *Journal of Computational Finance*, 18(2):57–90, 2015a.

S.I. Boyarchenko and S.Z. Levendorskii. Ghost calibration and the pricing of barrier options and CDS in spectrally one-sided Lévy models: the parabolic Laplace inversion method. *Quantitative Finance*, 15(3): 421–441, 2015b.

S.A. Broda and M.S. Paolella. Expected shortfall for distributions in finance. In P. Cizek, W.K. Härdle, and R. Weron, editors, *Statistical Tools for Finance and Insurance*. Springer, Berlin, 2011, pp. 57–99.

P.F. Christoffersen. Evaluating interval forecasts. *International Economic Review*, 39(4):841–862, 1998.

P.F. Christoffersen. Backtesting. In *Encyclopedia of Quantitative Finance*. Wiley, Chichester, 2010.

P. Embrechts, G. Puccetti, L. Rüschendorf, R. Wang, and A. Beleraj. An academic response to Basel 3.5. *Risks*, 2(1):25–48, 2014.

S. Emmer, M. Kratz, and D. Tasche. What is the best risk measure in practice? A comparison of standard measures. Working Paper 1322, ESSEC Business School, 2013.

W. Hu. *Calibration of Multivariate Generalized Hyperbolic Distributions Using the EM Algorithm, with Applications in Risk Management, Portfolio Optimization and Portfolio Credit Risk*. PhD thesis, Florida State University, 2005.

Y.S. Kim, S.T. Rachev, M.L. Bianchi, and F.J. Fabozzi. Computing VaR and AVaR in infinitely divisible distributions. *Probability and Mathematical Statistics*, 30(2):223–245, 2010.

Y.S. Kim, S.T. Rachev, M.L. Bianchi, I. Mitov, and F.J. Fabozzi. Time series analysis for financial market meltdowns. *Journal of Banking & Finance*, 35(8):1879–1891, 2011.

P.H. Kupiec. Techniques for verifying the accuracy of risk measurement models. *Journal of Derivatives*, 3(2): 73–84, 1995.

S.Z. Levendorskii. Efficient pricing and reliable calibration in the Heston model. *International Journal of Theoretical and Applied Finance*, 15(7), 2012.

A. McNeil, R. Frey, and P. Embrechts. *Quantitative Risk Management: Concepts, Techniques, and Tools.* Princeton University Press, Princeton, NJ, 2005.

R.S. Protassov. EM-based maximum likelihood parameter estimation for multivariate generalized hyperbolic distributions with fixed λ. *Statistics and Computing*, 14(1):67–77, 2004.

S.T. Rachev, S.V. Stoyanov, and F.J. Fabozzi. *Advanced Stochastic Models, Risk Assessment, and Portfolio Optimization: The Ideal Risk, Uncertainty, and Performance Measures.* Wiley, Hoboken, NJ, 2008.

M. Scherer, S.T. Rachev, Y.S. Kim, and F.J. Fabozzi. Approximation of skewed and leptokurtic return distributions. *Applied Financial Economics*, 22:1305–1316, 2012.

W. Schoutens. *Lévy Processes in Finance.* Wiley, Chichester, 2003.

S.V. Stoyanov and B. Racheva-Iotova. Numerical methods for stable modeling in financial risk management. In S.T. Rachev, editor, *Handbook of Computational and Numerical Methods in Finance.* Birkhäuser, Boston, MA, 2004, pp. 299–329.

Chapter 9

Extreme Value Theory

The main topics covered in this chapter are:

- what extreme value theory is and how it differs from classical statistics;
- the two pillars of extreme value theory: Fisher–Tippett–Gnedenko theorem and Pickands–Balkema–de Haan theorem;
- the three classes that the limit distribution of maxima will fall into: the Fréchet, Weibull, or Gumbel distribution;
- the generalized Pareto distribution;
- the maximum domain of attraction of an extreme value distribution and the concept of tail equivalence;
- the theory of maxima for stationary processes;
- extreme value theory for multivariate distributions;
- the role of copula in multivariate extreme value theory;
- the three types of copulas;
- three estimation methods for distributions: maximum likelihood estimation method, method of moments, and special estimators;

- the Hill estimator and the Pickands estimator for estimating the shape parameter of a distribution;
- use and limitations of the quantile plot (QQ-plot) for verifying statistical hypotheses by examining the degree of deviations of the linearity plot of a hypothesized distribution;
- three different approaches to compute widely-known risk measures (VaR and AVaR).

9.1. Introduction

In this chapter, we describe extreme value theory (EVT), the key EVT theorems, and the problem of estimation for EVT which is fundamentally the problem of estimating the tail of a distribution from a finite sample. We postpone applications of EVT until to Chapter 12.

EVT differs from classical statistics. In classical statistics, the focus is on modeling the average behavior of a distribution. In EVT, the focus is on modeling the extreme behavior of a distribution (i.e., the tail of a distribution). In contrast to classical statistics, EVT typically involves datasets that are very small.

In finance, the use of classical statistics is to see how financial markets tend to behave under routine or near normal conditions. In contrast, the use of EVT is important in looking at the major consequences of extreme values on some financial measure such as portfolio returns or credit losses despite the fact that extreme values are rare. That is, EVT is concerned with both the occurrence and sizes of rare events. EVT allows a portfolio manager to statistically model extreme events, extreme value measures such as VaR or AVaR can be estimated so as to improve risk management. The measures are therefore based on a well-founded theory. Readers are referred to Rocco (2014) for a complete overview of applications of EVT to finance and to Chernobai *et al.* (2008) for possible applications of EVT to the modeling of operational risk.

The crucial element of EVT is that the type of asymptotic distribution of extreme values does not depend on the exact cumulative distribution function F of returns. The precise form of F can thus be ignored and a non-parametric or a semi-parametric method can be used to estimate VaR. This is important, given that the whole tail of the distribution of returns is unknown and that, although financial time series usually exhibit skewed and fat-tailed distributions, there is usually no complete agreement on what distribution would fit them best.

As we will explain, to measure extreme values, three methods are used.

(1) The first is the block maxima method. With this method, extreme values are extreme observations from successive periods that are called blocks.
(2) The second method for measuring extreme values is the peak-over-threshold method. With this method, the focus is on observations that exceed a specified threshold value.
(3) The third is a compromise method that is based on order statistics, referred to as the r-largest order statistic.

Furthermore, EVT is based on the following two pillars: (1) the Fisher–Tippett–Gnedenko theorem on the distribution of rescaled and re-centered block maxima, and (2) the Pickands–Balkema–de Haan theorem on the distribution of exceedances over a threshold.

Estimation methods for distributions include maximum likelihood estimation (MLE) method, method of moments, and special estimators. EVT focuses on the tails and doing so involves studying block maxima and peaks-over-threshold estimators.

The principal result of EVT is that, for large classes of distributions, regardless of the shape of the distribution from which the extreme values are derived, which can be referred to as the *parent distribution*, the limit distribution of maxima will fall into one of three classes: the Fréchet, Weibull, or Gumbel distribution. In addition, the tails of the parent distribution will be approximately a generalized Pareto distribution. The main source for this chapter is

Embrechts *et al.* (1997) plus specific references (e.g., Marinelli *et al.* (2007) and Rocco (2014) for Section 9.10).

9.2. Probabilistic Concepts Underlying EVT

In this section we introduce a number of important probabilistic concepts that form the conceptual basis of EVT. The estimation of a distribution tail from a finite sample is the objective of EVT. In the case of heavy-tailed distributions, no finite sample can be considered exhaustive. Therefore, we have to estimate a tail using data that represent the tail only partially. Estimation is accomplished in a very specific way: we use theoretical properties shared by all tails.

There are two concepts that play a crucial role in EVT that relate to the behavior of the sample. The first concept relates to the behavior of the upper-order statistics (i.e., the sample largest k values) and the sample maxima. The second concept relates to the behavior of the points exceeding a given threshold, that is, the tail. Below we discuss these two concepts: the limit distributions of maxima and the distribution of the points of exceedances of a high threshold. Given these two behavioral concepts, we describe a number of estimators of the tail index in sequences of independent and identically distributed (iid) random variables.

9.2.1. Maxima

In the theory of sums, the principal result is that, regardless of the shape of individual summands, the behavior of sums simplifies in the limit of properly scaled and centered infinite sums. Moreover, sums of finite-mean, finite-variance variables, normalized by dividing by appropriate constants, always converge to a normal variable (see Embrechts *et al.* (1997)). The two main results follow.

Consider a sequence of finite-mean iid variables with mean $\mu = E[X]$. Consider the sum of the first n elements. The first form of the *weak law of large numbers* holds, that is

$$\lim_{n\to\infty} \bar{X}_n = \lim_{n\to\infty} \frac{\sum_{i=1}^{n} X_i}{n} = \mu,$$

where limit is in probability. If the scaling factor is the $\sqrt{n}$ instead of n, then averages do not converge. However, if the variables are iid with finite mean μ and finite variance σ^2, then the central limit theorem (CLT) holds, that is

$$\frac{S_n - n\mu}{\sigma\sqrt{n}} \xrightarrow{d} N(0,1),$$

where

$$S_n = \sum_{i=1}^{n} X_i,$$

$\sigma^2 = \mathrm{Var}[X]$, and $N(0,1)$ is a normal random variable with zero mean and unit variance.

In summary:

- *law of large number*: sums of finite-variance, finite-mean variables normalized by the factor n tend to a constant;
- *central limit theory*: centered sums of finite-variance, finite-mean variables normalized by the square root of n tend to a normal random variable.

If we drop the finite variance conditions then standard CLT does not hold. However, the variables belong to the domain of attraction of an α-stable random variable. That is, the following relationship holds:

$$\frac{S_n - a_n}{b_n} \xrightarrow{d} \mathcal{S}_\alpha,$$

where $a_n \in \mathbb{R}$ and $b_n > 0$, and $\mathcal{S}_\alpha$ is an α-stable random variable.

For α-stable random variable, the following identity in distribution holds for any number $n \geq 2$

$$\sum_{i=1}^{n} X_i \stackrel{d}{=} c_n X + d_n,$$

where X_i are identical independent copies of X, c_n and d_n are constants.

A similar theory can be formulated for maxima, which can be informally defined as the sample largest value. If it exists, the maxima limit distribution belongs to one of three possible distributions: Fréchet, Weibull, or Gumbel. These three distributions are described later in this section. This result, known as the Fisher–Tippett–Gnedenko theorem, is the basis of what can be described as classical EVT. Each maxima limit distribution possesses its own *maximum domain of attraction*. Moreover, limit laws are closed with respect to maxima, a property referred to as *max-stable*. The behavior of sums is more robust than the behavior of maxima: maxima do not converge to limit distributions for important classes of distributions, such as Poisson or geometric distributions.

Consider a sequence of independent variables X_i with a common, nondegenerate distribution F. Assume that the maxima of samples extracted from this sequence is as follows:

$$M_1 = X_1$$

and

$$M_n = \max(X_1, \ldots, X_n).$$

Although the maxima M_n form a new sequence of random variables, they are not independent. Similarly, we can define the minima of a sample

$$m_n = \min(X_1, \ldots, X_n).$$

Because it is assumed that the variables of the sequence X_i are independent, the distribution F_n of the maxima M_n can be expressed as

$$F_n(x) = P(X_1 \leq x \wedge \cdots \wedge X_n \leq x) = F^n(x),$$

where $\wedge$ is the logical symbol for "and". If the distribution F, which is a non-decreasing function, reaches 1 at a finite point x_F, that is, if

$$x_F = \sup\{x : F(x) < 1\} < \infty,$$

then, when n tends to infinity,

$$P(M_n < x) = \lim_{n \to \infty} F^n(x) = 0$$

for $x < x_F$. If x_F is finite, then

$$P(M_n < c) = F^n(x) = 1$$

for $x \geq x_F$. The point x_F is called the *right endpoint* of the distribution F.

For example, consider a variable X with uniform distribution in the interval $(0, 1)$, that is, its probability density function (pdf) is

$$f(x) = \begin{cases} 0 & \text{for } x < 0, \\ 1 & \text{for } 0 \leq x \leq 1, \\ 0 & \text{for } x > 1. \end{cases}$$

Let us compute the distribution of the maxima of samples of size n and compute the limit of the distribution of the maxima when n tends to infinity. The cumulative distribution function (cdf) of the uniform distribution is

$$F(x) = \begin{cases} 0 & \text{for } x < 0, \\ x & \text{for } 0 \leq x \leq 1, \\ 1 & \text{for } x > 1, \end{cases}$$

and the cdf of the maxima is therefore

$$F^n(x) = \begin{cases} 0 & \text{for } x < 0, \\ x^n & \text{for } 0 \leq x \leq 1, \\ 1 & \text{for } x > 1. \end{cases}$$

This expression tends to 0 when m tends to infinity for $x < 1$, and is equal to 1 for $x \geq 1$. The pdf of the maxima is

$$f^n(x) = \frac{dF^n(x)}{dx} \begin{cases} 0 & \text{for } x < 0, \\ nx^{n-1} & \text{for } 0 \leq x \leq 1, \\ 1 & \text{for } x > 1. \end{cases}$$

We can also compute the expectation and its limit, that is

$$E[M_n] = \int_0^1 nx^{n-1}x dx = \left[\frac{n}{n+1}x^{n+1}\right] = \frac{n}{n+1}$$

and therefore

$$\lim_{n\to\infty} E[M_n] = \lim_{n\to\infty} \frac{n}{n+1} = 1.$$

We can also easily compute the distribution of the minima. In fact, because it is assumed that the variables of the sequence X_i are independent, the distribution G_n of the random variable m_n can be expressed as

$$G_n(x) = 1 - P(X_1 \geq x \wedge \cdots \wedge X_n \geq x) = 1 - (1 - F(x))^n.$$

We can illustrate the behavior of maxima in the case of a normal distribution. Assuming a normal distribution with mean zero and unit variance, 100,000 samples of 20 elements each are selected. The maximum (minimum) for each sample is then chosen. The distribution of the maxima (minima) and the empirical distribution of independent draws from the same normal are illustrated in Figure 9.1.

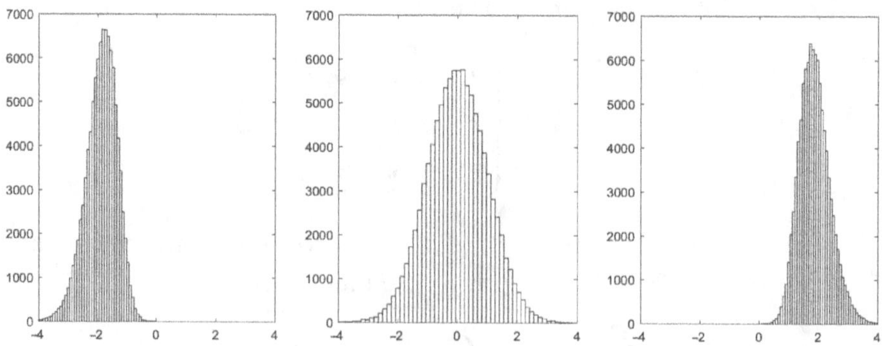

Figure 9.1. The left figure is the histogram of the minima of 100,000 samples formed by 20 elements each. The figure in the center is the histogram corresponding to 100,000 values extracted from a standard normal distribution. The right figure is the histogram of the maxima of 100,000 samples formed by 20 elements each.

We show the Matlab code needed to draw Figure 9.1.

```
N = 10^5;

% simulate a normal random variable
X1 = randn(N,1);

% simulate N samples of 20 independent normal random variables
mu = zeros(1,20);
I = eye(20);
M = mvnrnd(mu,I,N);

% take the maximum of each sample
max20 = max(M,[],2);

% take the minimum of each sample
min20 = min(M,[],2);

nbins = 60;

subplot 131
histogram(min20,nbins)
ylim([0 7000])
xlim([-4 4])
subplot 132
histogram(X1,nbins)
ylim([0 7000])
xlim([-4 4])
subplot 133
histogram(max20,nbins)
ylim([0 7000])
xlim([-4 4])
```

Given that for any cdf $F(x) \leq 1$, the limit distribution $\lim_{n\to\infty} F^n(x)$ is always zero except where the cdf assumes the value of 1. However, if we rescale appropriately, we find non-degenerate distributions. By considering sequences of normalized and centered maxima, greater insight as to how maxima behave can be gained. Consider the following sequence

$$\frac{M_n - d_n}{c_n},$$

where c_n and d_n are constants in $\mathbb{R}$ and $c_n > 0$. There is a fundamental result on the behavior of maxima, which was first established by Fisher and Tippett (1928) and in full mathematical form by Gnedenko (1943). The Fisher–Tippett–Gnedenko theorem can be stated as follows.

Theorem 9.1 (Fisher–Tippett–Gnedenko theorem). *Consider a sequence of iid variables X_i and the relative sequence of maxima M_n. If there exist two sequences of constants $c_n > 0$ and $d_n \in \mathbb{R}$ and a nondegenerate distribution function H such that*

$$\frac{M_n - d_n}{c_n} \xrightarrow{D} H, \tag{9.1}$$

then H is one of the following random variables:

(1) *a Fréchet random variable, that is a random variable with cdf*

$$F(x) = \begin{cases} 0 & \text{for } x \leq 0, \\ \exp(-x^{-\alpha}) & \text{for } x > 0, \end{cases}$$

where $\alpha > 0$;

(2) *a Weibull random variable, that is a random variable with cdf*

$$F(x) = \begin{cases} \exp(-(-x)^{\alpha}) & \text{for } x < 0, \\ 1 & \text{for } x \geq 0, \end{cases}$$

where $\alpha > 0$;

(3) *a Gumbel random variable, that is a random variable with cdf*

$$F(x) = \exp(-e^{-x})$$

with $x \in \mathbb{R}$.

The proof of this theorem can be found in Fisher and Tippett (1928) and Gnedenko (1943). In the sense that different sequences of normalizing constants determine the same distribution, the limit distribution H is unique.

The Fréchet, Weibull, and Gumbel distributions are referred to as *standard extreme value distributions.* All three distributions are continuous functions for every real x. A random variable is referred to as an *extremal random variable* if it is distributed according to one of the standard extreme value distributions.

To illustrate, consider a standard exponential random variable X. Because the cdf of X is

$$F(x) = P(X \leq x) = 1 - e^{-x},$$

the distribution of maxima is

$$P(M_n \leq x) = F^n(x) = (1 - e^{-x})^n$$

with $x \geq 0$. Selecting $d_n = \log(n)$, we can write

$$P(M_n - d_n \leq x) = P(M_n \leq \log(n) + x) = (1 - n^{-1}e^{-x})^n,$$

with $x \geq 0$. For any given x,

$$(1 - n^{-1}e^{-x})^n \to \exp(-e^{-x}),$$

demonstrating that the maxima of standard exponential variables centered with $d_n = \log(n)$ tend to a Gumbel distribution.

Figures 9.2, 9.3, and 9.4 show the pdf and cdf of the Fréchet, Gumbel, and Weibull random variables, respectively. From Figure 9.2 it can be seen that for the Fréchet distribution the tails decline slowly and are known as fat or heavy tails. The Student's t-distribution and the Pareto distributions are examples. Therefore, this distribution

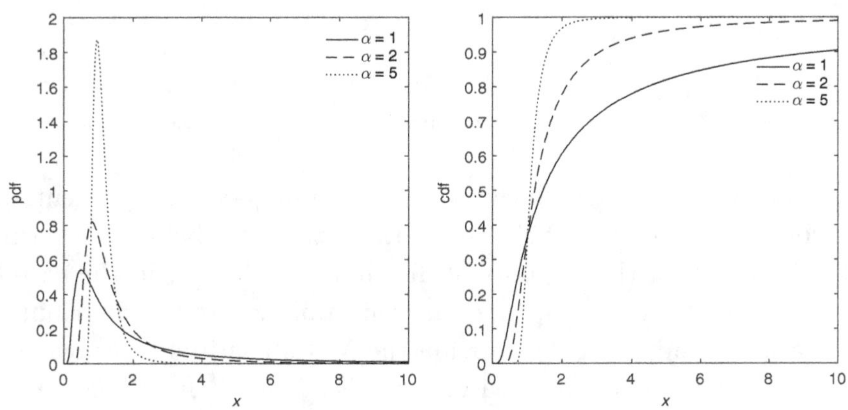

Figure 9.2. Illustration of the pdf and cdf of the Fréchet random variable.

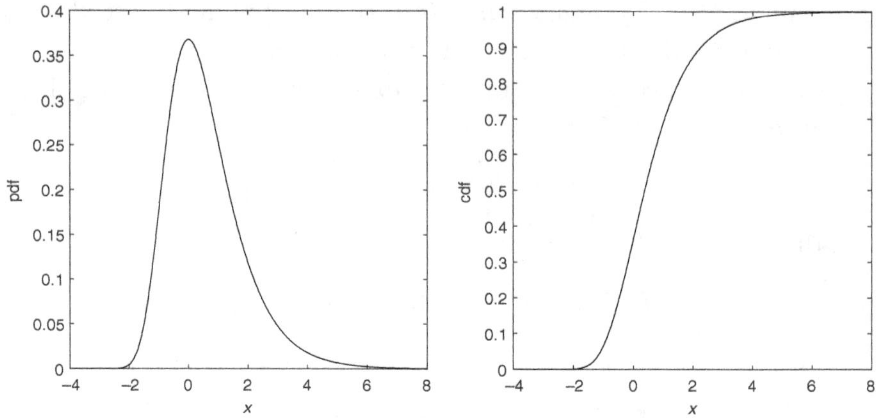

Figure 9.3. Illustration of the pdf and cdf of the Gumbel random variable.

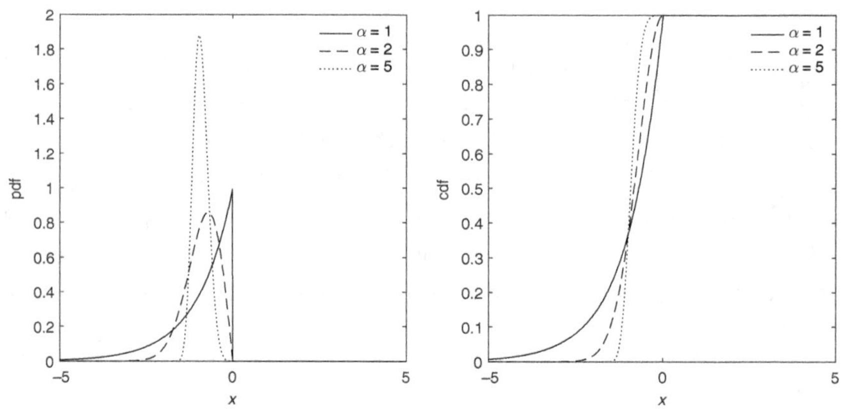

Figure 9.4. Illustration of the pdf and cdf of the Weibull random variable.

is well-suited for fat or heavy tails. The tails decline exponentially in the case of the Gumbel distribution, examples being the normal distribution and the log-normal distribution. Finally, in the case of the Weibull distribution, there are thin tails where the distribution has a finite endpoint. We provide the Matlab scripts to draw these figures. The scripts are based on the Matlab functions *gevpdf* (*wblpdf*) and *gevcdf* (*wblcdf*), that is the Matlab proprietary implementation of the pdf and cdf of the generalized extreme value

(Weibull) random variable. The generalized extreme value will be described in Section 9.4.

We show the Matlab code needed to draw Figure 9.2.

```
% draw Fréchet pdf and cdf
x = 0:0.01:10;

[f1pdf, f1cdf] = getFrechet(x,1,1,0);
[f2pdf, f2cdf] = getFrechet(x,2,1,0);
[f3pdf, f3cdf] = getFrechet(x,5,1,0);

% set figure dimension
scrsz = get(0,'ScreenSize');
h4 = figure('Position',[50 50 scrsz(3)/2 scrsz(4)/3]);

subplot 121
plot(x, f1pdf, 'k',...
     x, f2pdf, 'k--',...
     x, f3pdf, 'k:')
xlabel('x')
ylabel('pdf')
legend('\alpha=1','\alpha=2','\alpha=5','Interpreter','latex')
legend boxoff

subplot 122
plot(x, f1cdf, 'k',...
     x, f2cdf, 'k--',...
     x, f3cdf, 'k:')
xlabel('x')
ylabel('cdf')
legend('\alpha=1','\alpha=2','\alpha=5','Interpreter','latex')
legend boxoff
```

Recall that if Z is a continuous random variable with pdf $f_Z(z)$ and $x = g(z)$ with g a strictly increasing continuously differentiable function with inverse $z = h(x)$, then $X = g(Z)$ is a continuous random variable with pdf given by

$$f_X(x) = f_Z(h(x))h'(x).$$

We used this result to compute the pdf of a Fréchet random variable starting from the pdf of a generalized extreme value random variable implemented in Matlab (i.e., *gevpdf*). More in details, in the following script we consider the change of variable

$$x = g(z) = 1 + \xi \frac{z - \mu}{\sigma}$$

with inverse

$$z = h(x) = \frac{(x-1)\sigma}{\xi} + \mu,$$

and because the equality

$$h'(x) = \frac{\sigma}{\xi}$$

holds, the pdf is multiplied by the factor $\sigma\xi^{-1}$. A similar argument holds in the Weibull case: we consider the change of variable

$$x = g(z) = -1 - \xi \frac{z - \mu}{\sigma}$$

with inverse

$$z = h(x) = -\frac{(x+1)\sigma}{\xi} + \mu,$$

and because the equality

$$h'(x) = -\frac{\sigma}{\xi}$$

holds, the pdf is multiplied by the factor $-\sigma\xi^{-1}$.

```
function [pdf, cdf] = getFrechet(x,alpha,sigma,mu)

% get the pdf and cdf of a Fréchet random variable from a
% generalized extreme value random variable parameterization
xi = 1/alpha;
z = (x - 1)*sigma/xi + mu;
pdf = sigma/xi*gevpdf(z,xi,sigma,mu);
cdf = gevcdf(z,xi,sigma,mu);

return
```

We show the Matlab code needed to draw Figure 9.3.

```
% draw Gumbel pdf and cdf
x = -4:0.01:8;

g1pdf = gevpdf(x,0,1,0);
g1cdf = gevcdf(x,0,1,0);

% set figure dimension
scrsz = get(0,'ScreenSize');
h4 = figure('Position',[50 50 scrsz(3)/2 scrsz(4)/3]);

subplot 121
plot(x, g1pdf, 'k')
xlabel('x')
ylabel('pdf')

subplot 122
plot(x, g1cdf, 'k')
xlabel('x')
ylabel('cdf')
```

We show the Matlab code needed to draw Figure 9.4. The Matlab functions, *wblpdf*, correspond to the mirror image of the pdf of the Weibull random variable defined in Theorem 9.1.

```
% draw Weibull pdf and cdf
x = -5:0.01:5;

[w1pdf, w1cdf] = getWeibull(x,1,1,0);
[w2pdf, w2cdf] = getWeibull(x,2,1,0);
[w3pdf, w3cdf] = getWeibull(x,5,1,0);

% set figure dimension
scrsz = get(0,'ScreenSize');
h4 = figure('Position',[50 50 scrsz(3)/2 scrsz(4)/3]);

subplot 121
plot(x, w1pdf, 'k',...
     x, w2pdf, 'k--',...
     x, w3pdf, 'k:')
xlabel('x')
ylabel('pdf')
legend('\alpha=1','\alpha=2','\alpha=5','Interpreter','latex')
```

```
legend boxoff

subplot 122
plot(x, w1cdf, 'k',...
     x, w2cdf, 'k--',...
     x, w3cdf, 'k:')
xlabel('x')
ylabel('cdf')
legend('\alpha=1','\alpha=2','\alpha=5','Interpreter','latex')
legend boxoff
```

```
function [pdf, cdf] = getWeibull(x,alpha,sigma,mu)

% get the pdf and cdf of a Weibull random variable from a
% generalized extreme value random variable parameterization
xi = -1/alpha;
z = -(x +  1)*sigma/xi + mu;
pdf = -sigma/xi*gevpdf(z,xi,sigma,mu);
cdf = gevcdf(z,xi,sigma,mu);

return
```

9.3. Maximum Domain of Attraction

Are there conditions on the distribution F that ensure the existence of centering and scaling constants and the convergence to an extreme value distribution? To answer this question, it is necessary to introduce the concept of the *maximum domain of attraction* (MDA) of an extreme value distribution H or MDA(H). For details see Embrechts *et al.* (1997) and Leadbetter *et al.* (2012).

A random variable X is said to belong to the MDA(H) of the extreme value distribution H if there exist constants $c_n > 0$ and $d_n \in \mathbb{R}$ such that

$$\frac{M_n - d_n}{c_n} \xrightarrow{D} H.$$

Consider two distribution functions F and G. If they have the same right endpoints and the following condition holds:

$$\lim_{x\to\infty} \frac{\bar{F}(x)}{\bar{G}(x)} = c$$

with $0 < c < \infty$ and

$$\bar{F}(x) = 1 - F(x),$$

then the two distributions are said to be *tail equivalent.*

The concept of tail equivalence is important for characterizing MDAs. In fact, it can be shown that every MDA(H) is closed with respect to tail equivalence. That is, if two distribution functions F and G are tail equivalent, then $F \in$ MDA(H) if and only if $G \in$ MDA(H). Tail equivalence allows for a powerful characterization of the three MDAs.

Let us first define the *quantile function* that will be useful in the following sections. Given a distribution function F, the quantile function of F is

$$F^{\leftarrow}(x) = \inf\{s \in \mathbb{R} : F(s) \geq x\}, \tag{9.2}$$

where $0 < x < 1$.

9.3.1. The MDA of the Fréchet distribution

The cdf of a Fréchet random variable is written as

$$F(x) = \exp(-x^{-\alpha}).$$

Observe that the tail of the Fréchet distribution decays as an inverse power law and can be expressed as

$$1 - F(x) = 1 - \exp(-x^{-\alpha}) \approx x^{-\alpha}$$

for $x \to \infty$.

A distribution function F belongs to the MDA of a Fréchet random variable X if and only if there is a slowly-varying function L such that

$$F\hat{(x)} = x^{-\alpha}L(x).$$

In this special case, where the constants assume the values,

$$c_n = F^{\leftarrow}(1 - n^{-1}) = \left(\frac{1}{\bar{F}}\right)^{\leftarrow}(n)$$

and $d_n = 0$. This condition can be expressed more compactly as follows:

$$F \in \text{MDA}(X) \Longleftrightarrow \hat{F} \in \mathcal{R}_{-\alpha}.$$

If the distribution function F is absolutely continuous and admits a density f and if the following condition is satisfied

$$\lim_{x \to \infty} \frac{x f(x)}{\bar{F}(x)} = \alpha > 0, \tag{9.3}$$

then $F \in \text{MDA}(X)$. The proof of this was provided by von Mises (1964). It can be shown that $\text{MDA}(X)$ consists of all the distribution functions that satisfy the von Mises condition and their tail-equivalent distribution functions.

From the above definitions and theorems it can be shown that the Pareto, Cauchy, Burr, alpha-stable laws with exponent $\alpha < 2$, and log-gamma distributions belong to the MDA of the Fréchet law.

9.3.2. The MDA of the Weibull distribution

The cdf of a Weibull random variable is written as

$$F(x) = \exp(-(-x)^{\alpha}).$$

It is clear from the definition that the following relationship holds:

$$\exp(-(-x)^{\alpha}) = \exp(-(-1/x)^{-\alpha})$$

for $x < 0$. Consequently, the MDA of the Weibull and Fréchet distributions are expected to be closely related. In fact, it can be demonstrated that a distribution function F belongs to the MDA of a Weibull distribution if and only if

$$x_F < \infty$$

and

$$\hat{F}(x_F - x^{-1}) = x^{-\alpha} L(x),$$

where L is a slowly varying function. If $F \in \mathrm{MDA}(X)$, then

$$\frac{M_n - x_F}{c_n} \xrightarrow{D} X.$$

The Weibull distribution MDA admits a characterization in terms of tail equivalence that is similar to that of the Fréchet distribution MDA. Suppose that a distribution function F is absolutely continuous, admitting a density f that is positive on some finite interval (z, x_F). Also suppose that the von Mises condition given by equation (9.3) is satisfied. If those two assumptions hold, it can be shown that $\mathrm{MDA}(X)$ consists of all the distribution functions that satisfy the von Mises condition and their tail-equivalent distribution functions.

The MDA of the Weibull distribution includes important distributions such as the uniform distribution in $[0, 1]$, power laws truncated to the right, and beta distributions.

9.3.3. The MDA of the Gumbel distribution

The cdf of a Gumbel random variable is written as

$$F(x) = \exp(-e^{-x}).$$

Using a Taylor expansion, it can be shown that the Gumbel distribution has exponential tails. The closure properties of tail equivalence hold: two distributions F and G belong to the MDA of the Gumbel distribution if and only if they are tail equivalent. This closure property, however, does not imply that F and G distributions must be tail equivalent to the Gumbel distribution. That is, the two distributions are not necessarily tail equivalent to an exponential distribution.

However, it can be shown that a Gumbel distribution MDA consists of von Mises functions and of their tail-equivalent distributions. Suppose that F is a distribution function with right endpoint $x_F \leq \infty$ and that there is a real number z such that the following equality holds:

$$\bar{F}(x) = c \exp\left\{-\int_z^x \frac{1}{a(t)} dt\right\}$$

for $z < x < x_F$, where c is a positive constant and a is a positive and absolutely continuous function such that $\lim_{x \to x_F^-} a'(x) = 0$. Then, F is referred to as *von Mises function.*

Gumbel distribution's MDA includes a large class of distributions: the exponential distribution, the normal distribution, and the lognormal distribution. Although the Gumbel distribution has exponential tails, its MDA includes subexponential distributions such as the Berktander distribution (see Goldie and Resnick (1988)).

9.3.4. Max-stable distributions

While α-stable distributions remain unchanged after summation, *max-stable distributions* remain unchanged after taking maxima. A non-degenerate random variable X and the relative distribution are called max-stable if there are constants $c_n > 0$ and $d_n \in \mathbb{R}$ such that the following conditions are satisfied:

$$\max(X_1, \ldots, X_n) \stackrel{D}{=} c_n X + d_n,$$

where X_i are iid random variables.

It can be demonstrated that the class of max-stable distributions coincides with the class of possible limit laws for normalized and centered maxima. In view of the previous discussions, the max-stable laws are the three possible limit laws: Fréchet, Weibull, and Gumbel.

9.4. Generalized Extreme Value Random Variable

The three extreme value distributions (i.e., Fréchet, Weibull, and Gumbel distributions) can be represented as a one-parameter family of distributions through the *standard generalized extreme value distribution* (GEV) of von Mises (1964) and Jenkinson (1955). Define the distribution function

$$F(x) = \begin{cases} \exp(-(1+\xi x)^{-\frac{1}{\xi}}) & \text{for } \xi \neq 0, \\ \exp(-\exp(-x)) & \text{for } \xi = 0, \end{cases} \tag{9.4}$$

where $1 + \xi x > 0$. From this definition it can be seen that

- $\xi = \alpha^{-1} > 0$ and $x = (y-1)\xi^{-1}$ corresponds to the Fréchet distribution;
- $\xi = 0$ corresponds to the Gumbel distribution;
- $\xi = -\alpha^{-1} < 0$ and $x = -(y+1)\xi^{-1}$ corresponds to the Weibull distribution.

We can now introduce the related location-scale dependent family H by replacing x with $(x-\mu)/\sigma$. The key results of EVT established in the previous sections can be summarized in the following propositions and the definition of the quantile function $F^{\leftarrow}(x)$ given by equation (9.2). Define

$$U(x) = F^{\leftarrow}(1 - x^{-1})$$

for $x > 0$. It can be confirmed that the following statements are equivalent:

(1) $F \in \text{MDA}(H)$;
(2)

$$\lim_{u \to x_F} \frac{\bar{F}(u + xa(u))}{\bar{F}(u)} = \begin{cases} \exp(-(1+\xi x)^{-\frac{1}{\xi}}) & \text{for } \xi \neq 0, \\ \exp(-\exp(-x)) & \text{for } \xi = 0; \end{cases}$$

(3) for $x, y > 0$ and $y \neq 1$

$$\lim_{u \to \infty} \frac{U(ux) - U(s)}{U(uy) - U(s)} = \begin{cases} \dfrac{x^{\xi} - 1}{y^{\xi} - 1} & \text{for } \xi \neq 0, \\ \dfrac{\log(x)}{\log(y)} & \text{for } \xi = 0. \end{cases}$$

Figures 9.5 shows the pdf and cdf of the GEV random variable.

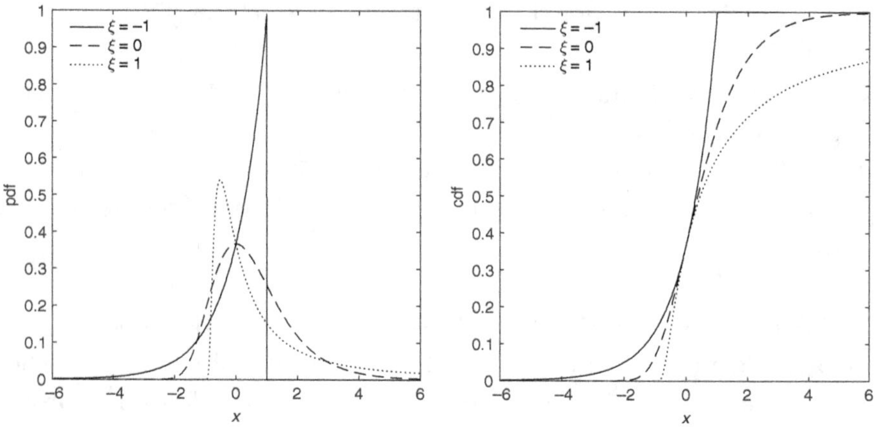

Figure 9.5. Illustration of the pdf and cdf of the GEV random variable.

We show the Matlab code needed to draw Figure 9.5.

```
% draw GEV pdf and cdf
x = -6:0.01:6;

gev1pdf = gevpdf(x,-1,1,0);
gev2pdf = gevpdf(x,0,1,0);
gev3pdf = gevpdf(x,1,1,0);

gev1cdf = gevcdf(x,-1,1,0);
gev2cdf = gevcdf(x,0,1,0);
gev3cdf = gevcdf(x,1,1,0);

% set figure dimension
scrsz = get(0,'ScreenSize');
h4 = figure('Position',[50 50 scrsz(3)/2 scrsz(4)/3]);

subplot 121
plot(x, gev1pdf, 'k',...
     x, gev2pdf, 'k--',...
     x, gev3pdf, 'k:')
xlabel('x')
ylabel('pdf')
legend('\xi=-1','\xi=0','\xi=1','Interpreter','latex')
legend boxoff

subplot 122
plot(x, gev1cdf, 'k',...
```

```
    x, gev2cdf, 'k--',...
    x, gev3cdf, 'k:')
xlabel('x')
ylabel('cdf')
legend('\xi=-1','\xi=','\xi=1','Interpreter','latex')
legend boxoff
```

9.5. Point of Exceedances or Peak-Over-Threshold

We have now reviewed the behavior of sums, maxima, and upper order statistics of continuous random variables. Still another approach to EVT is based on the point of exceedances of some high threshold. This approach exists in two versions: (1) given a threshold, one might simply consider the distribution of the points that exceed the threshold or (2) one might represent those points of exceedances as point processes. The latter is called peaks-over-threshold (POT).

Consider a sample $(x_1, \ldots, x_n)$ from a random variable X with cdf F. For example, suppose X are the losses of an insurance or financial portfolio. For any given level l we define the excess distribution over the threshold l as the conditional distribution of $X - l$ given $X > l$.

The cdf of the exceedances is

$$F_l(x) = P(X - l \leq x | X > l) = \frac{F(x+l) - F(x)}{1 - F(l)},$$

where $x > 0$. The mean of $F_l(x)$ as a function of l is called the *mean excess function*

$$e(l) = E(X - l | X > l).$$

If F is an exponential function then, due to the memoryless property of exponential distributions, $F_l(x) = F(x)$.

A random variable with cdf

$$G_\xi(x) = \begin{cases} 1 - (1 + \xi x)^{-\frac{1}{\xi}} & \text{for } \xi \neq 0, \\ 1 - e^{-x} & \text{for } \xi = 0, \end{cases}$$

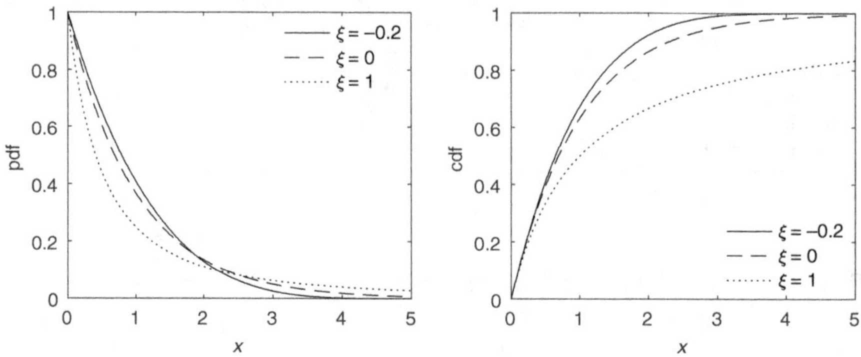

Figure 9.6. Illustration of the pdf and cdf of the GP random variable.

where

$$\begin{cases} x \geq 0 & \text{for } \xi \geq 0, \\ 0 \leq x \leq -\frac{1}{\xi} & \text{for } \xi < 0 \end{cases}$$

is referred to as generalized Pareto (GP). We will call ξ the shape parameter, as in the GEV distribution. One can introduce the related location-scale family $G_{\xi,\nu,\beta}$ by replacing the argument x above by $(x-\nu)/\beta$, for $\nu \in \mathbb{R}$ and $\beta > 0$. The support has to be adjusted accordingly. In Theorem 9.2, we will consider the case $\nu = 0$, to which we refer to as two-parameter generalized Pareto.

Figures 9.6 shows the pdf and cdf of the GP random variable.

We show the Matlab code needed to draw Figure 9.6.

```
% draw generalized Pareto pdf and cdf
x = 0:0.01:5;

gp1pdf = gppdf(x,-0.2,1,0);
gp2pdf = gppdf(x,0,1,0);
gp3pdf = gppdf(x,1,1,0);

gp1cdf = gpcdf(x,-0.2,1,0);
gp2cdf = gpcdf(x,0,1,0);
gp3cdf = gpcdf(x,1,1,0);

% set figure dimension
scrsz = get(0,'ScreenSize');
```

```
h4 = figure('Position',[50 50 scrsz(3)/2 scrsz(4)/3]);

subplot 121
plot(x, gp1pdf, 'k',...
    x, gp2pdf, 'k--',...
    x, gp3pdf, 'k:')
xlabel('x')
ylabel('pdf')
legend('\xi=-0.2','\xi=0','\xi=1','Interpreter','latex')
legend boxoff

subplot 122
plot(x, gp1cdf, 'k',...
    x, gp2cdf, 'k--',...
    x, gp3cdf, 'k:')
xlabel('x')
ylabel('cdf')
legend('\xi=-0,2','\xi=','\xi=1','Interpreter','latex')
legend boxoff
```

If F is a GPD with parameters (ξ, ν, β), then $F_l(x)$ is also a GPD. A GPD has finite mean if and only if $\xi < 1$. The mean excess function of a two-parameter generalized Pareto with parameters $\xi < 1$ and β can be written as

$$e(l) = E[X - l|X > l] = \frac{\beta + l\xi}{1 - \xi}.$$

The Pickands–Balkema–de Haan theorem states that the distribution of the block maxima M_n converges toward a GEV with shape parameter ξ if and only if the excess distribution $F_l(x)$ over a level l converges uniformly in x as l increases toward a GPD with shape parameter ξ and a scale parameter possibly varying with l.

Theorem 9.2 (Pickands–Balkema–de Haan theorem). *It is possible to find a positive-measurable function $\beta(u)$ such that*

$$\lim_{u \to x_F} \sup_{0 \le x < x_F - u} |F_u(x) - G_{\xi,\beta(u)}| = 0,$$

if and only if $F \in \mathrm{MDA}(H_\xi)$, where $\xi \in \mathbb{R}$.

For a proof see Pickands (1975). That is, if the distribution of block maxima is a GEV, then the distribution of the exceedances of a large threshold is approximately a GPD. In other words, those cumulative distribution functions, whose excess distribution over the threshold l converges (as l grows to the right endpoint x_F) to a GPD with shape parameter ξ, are exactly those lying in the maximum domain of attraction of a GEV distribution with the shape parameter ξ.

9.5.1. Point processes

Thus far we have considered the points of exceedances without a structure related to their arrivals. However, we can look at them as point processes. Point processes can be defined in several ways. To illustrate the mathematics of point processes, let us first introduce the homogeneous Poisson process.

A homogeneous Poisson process $N = (N_t)_{t \geq 0}$, with parameter $\lambda > 0$, is a non-decreasing, integer-valued process starting at $N_0 = 0$, with independent and stationary increments which are Poisson distributed, that is

$$P(N_t - N_s = k) = (\lambda(t-s))^k \frac{e^{-\lambda(t-s)}}{k!}$$

for all $0 \leq s \leq t$ and $k \in \mathbb{N} \cup \{0\}$. The Poisson process is typically used as a model for the occurrence of rare independent events. In particular, N_t counts the number of events that occurred up to time t, and $N(t)-N(s)$ the number of events occurred in the time interval $(t, s]$. The parameter λ is called intensity of the process because it counts the expected number of times a given event occurs per unit of time. In other terms, for all $t > 0$,

$$E[N_t - N_{t-1}] = \lambda,$$

and thus, the expected number of times the event occurs in the interval $(t, s]$ is proportional to its length, that is

$$E[N_t - N_s] = \lambda(t-s).$$

As shown in Chapter 3, a Poisson process belongs to the family of Lévy processes.

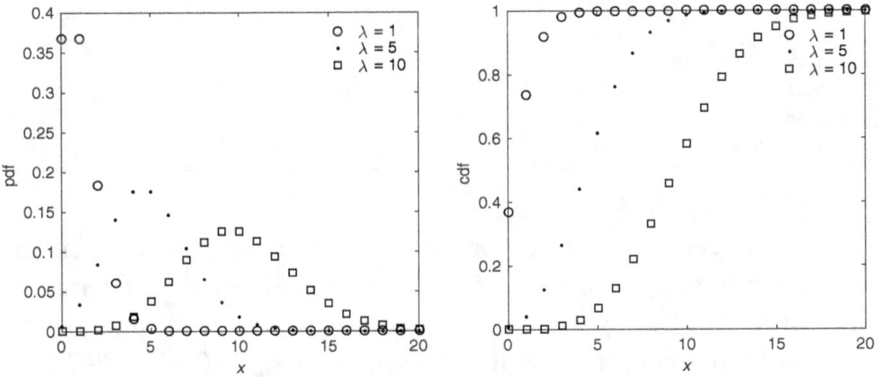

Figure 9.7. Illustration of the pdf and cdf of the Poisson random variable.

We show the Matlab code needed to draw Figure 9.7 in which the pdf and cdf of three Poisson random variables with different intensities are illustrated.

```
% draw Poisson pdf and cdf
x = 0:1:20;

p1pdf = poisspdf(x,1);
p2pdf = poisspdf(x,5);
p3pdf = poisspdf(x,10);

p1cdf = poisscdf(x,1);
p2cdf = poisscdf(x,5);
p3cdf = poisscdf(x,10);

% set figure dimension
scrsz = get(0,'ScreenSize');
h4 = figure('Position',[50 50 scrsz(3)/2 scrsz(4)/3]);

subplot 121
plot(x, p1pdf, 'ko',...
     x, p2pdf, 'k.',...
     x, p3pdf, 'ks')
ylabel('pdf')
legend('\lambda=1','\lambda=5','\lambda=10','Interpreter','latex')
legend boxoff

subplot 122
plot(x, p1cdf, 'ko',...
```

```
    x, p2cdf, 'k.',...
    x, p3cdf, 'ks')
xlabel('x')
ylabel('cdf')
legend('\lambda=1','\lambda=5','\lambda=10','Interpreter','latex')
legend boxoff
```

A homogeneous Poisson process can be constructed by modeling the interarrival times of the events as iid exponential random variables. Specifically, let $T_1, \ldots, T_n$ be iid exponential random variables with parameter λ. Define the sequence $\tau_n = T_1 + \cdots + T_n$ and $N_t = \sup\{n : \tau_n \leq t\}$, then it can be shown that $N = (N_t)_{t \geq 0}$ is a Poisson process with intensity λ. In other words, a homogeneous Poisson process can be seen as a random sequence of points in the real line separated by iid random intervals exponentially distributed.

This construction can be generalized to define point processes. Intuitively, a generic point process is a random collection of discrete points in some space. It is convenient from a mathematical perspective to express a point process through the distribution of the number of points that fall in an arbitrary set (see Cox and Isham (1980)). Let us look at the special case of a homogeneous Poisson process. Consider the number of points that fall in a given interval. It is convenient to consider a wider class of sets for a generic point process.

Consider (1) a subspace E of a finite-dimensional Euclidean space of dimension n and (2) the σ-algebra $\mathcal{E}$ of the Borel sets generated by open sets in E. We refer to the space E as the state space. The Dirac measure, denoted by ε_x, can be defined for each point x in E and for each set $A \in \mathcal{E}$, by

$$\varepsilon_x(A) = \begin{cases} 1 & \text{if } x \in A, \\ 0 & \text{if } x \notin A. \end{cases}$$

Define the following set function for any given sequence x_j of points in E as

$$m(A) = \sum_{i=1}^{\infty} \varepsilon_{x_i}(A) = \#\{i : x_i \in A\},$$

where $A \in \mathcal{E}$. It can be demonstrated that $m(A)$ is a measure on $\mathcal{E}$, which is referred to as a *counting measure.* A counting measure is referred to as a point measure if it is finite on each compact set. That is, any given countable sequence in E results in a counting measure on $\mathcal{E}$.

The joint probability distribution for a point process obtained by associating to each family of sets $A_i \in \mathcal{E}$ is given by

$$P(m(A_i) = n_i),$$

where $i \in \{1, \ldots, k\}$ and $k \in \mathbb{N}$.

Let us make this definition for a point process more mathematically rigorous. Armed with an appropriate σ-algebra, a point process can be defined as a measurable map from some probability space to the set of all point measures. It should be evident that point processes are defined by the probability distribution of the number of points that fall in each set A of some σ-algebra. There are two key features of point processes. The first feature is that there are counting measures associated with each set A the number of points of each discrete sequence that fall in A with the additivity restrictions of measures. The second feature of pointing processes relates to probability distributions. Specifically, for point processes, probability distributions are defined over the space of counting measures.

By understanding the general concept of point processes, the concept of point process of exceedances can be defined. Suppose that there are (1) a threshold which is formed by any real number u and (2) a sequence of random variables X_i. The point process of exceedances with state space $E = (0, 1]$ counts the number of occurrences where the random variables X_i exceed the threshold u; that is,

$$N_n(A) = \sum_{i=1}^{\infty} \varepsilon_{x_i}(A) = \#\{i \leq n \wedge X_i > u\}.$$

The state space stipulates the sample size in this case.

9.6. Order Statistics

As we will see in Section 9.9, maxima of blocks of data are used to estimate the tail of the distribution of data. However, the block maxima method makes a rather inefficient use of data. In fact, only one data per block, the maximum, is used. In order to mitigate this inefficiency a method based on upper order statistics is used. The r-upper order statistic method uses the r-largest values of each block instead of the maxima. The basic motivation for this method is that the joint distributions of the upper order statistics admit limit distributions similar to those of maxima.

The ordered values from a sample of observations are referred to as *order statistics of the sample*; they are non-parametric statistics. The smallest and largest observations are called the *extreme values.* Both order statistics of a sample and the extreme values of a sample are of interest in finance since they focus on the lows and highs of a distribution. That is, the behavior of order statistics is a useful tool for characterizing fat-tailed distributions.

Suppose that we have a random sample $(X_1, \ldots, X_n)$ made of n independent draws from the same distribution F. We can order the sample in decreasing order $(X_{n,n}, \ldots, X_{1,n})$. The random variable $X_{k,n}$ is referred to as the *kth upper order statistic.* Let us call $F_{k,n}$ the distribution of $X_{k,n}$. As we have seen in the previous section, we know the distribution of $F_{1,n}$ and of which are the distributions of the minimum and the maximum respectively

$$F_{1,n} = 1 - (1 - F)^n = 1 - \bar{F}^n$$

and

$$F_{n,n} = F^n.$$

It can be easily demonstrated that the distribution of the kth upper order statistic is

$$F_{k,n} = P(X_{k,n} \leq x) = \sum_{i=k}^{n} \binom{n}{i} F^i(x) \bar{F}^{n-i}(x). \tag{9.5}$$

In fact, the statement that the kth upper order statistic is less than or equal to a real number x is equivalent to the statement that at least k elements of the sample are less than or equal to x

$$F_{k,n} = P(X_{k,n} \le x) = \sum_{i=k}^{n} P(\Omega_F),$$

where Ω_F is the event

$$\Omega_F = \{\text{exactly } i \text{ elements are less or equal to } x\}.$$

The probability $P(\Omega_F)$ is equal to the probability that a binomial variable has i successes and $n-i$ failures where the probability of success is $F(x)$ and the probability of failure is $\bar{F}(x)$. Therefore, equation (9.5) follows.

If the distribution F is continuous and it admits a density f, then the equality

$$F_{k,n} = \int_{-\infty}^{x} f_{k,n}(z)dz$$

holds. Then the distribution $F_{k,n}$ admits a density that can be easily computed taking its derivative, that is

$$F'_{k,n}(x) = f_{k,n}(x) = n\binom{n-1}{k-1} F^{k-1}(x)\bar{F}^{n-k} f(x).$$

Thus far we have discussed the distribution of individual upper statistics. However, from the point of view of applications of EVT, the joint distribution of upper order statistics is more useful. Unfortunately, expressions become rapidly very complex if one wants to consider even a small number of upper statistics.

The differences between two consecutive variables in a sample

$$X_{k,n} - X_{k+1,n}$$

are random variables called *spacings*. The zeroth spacing is defined as

$$X_{0,n} - X_{1,n} = x_F - X_{1,n}$$

for variables with a finite right endpoint. The distribution of spacings depends on the distribution F. It can be demonstrated that the spacings of an exponential random variables are independent exponential random variables with mean $1/n$ for an n-sample. Spacings are a key concept for the definition of the Hill estimator, as explained later in this section.

Another way to visualize order statistics, which is related to spacings, is that of quantile function (or inverse distribution function). Suppose $(X_1, \ldots, X_n)$ is a vector of iid variables with distribution function F and let $(U_1, \ldots, U_n)$ be a vector of iid variables uniformly distributed on the interval $[0, 1]$. Given a distribution function F, recall that the quantile function of F is

$$F^{\leftarrow}(x) = \inf\{s \in \mathbb{R} : F(s) \geq x\}.$$

It can be demonstrated that the following results hold:

- $F^{\leftarrow}(U_i) \stackrel{d}{=} X_i$;
- $(X_{1,n}, \ldots, X_{n,n}) \stackrel{d}{=} (F^{\leftarrow}(U_{1,n}), \ldots, F^{\leftarrow}(U_{n,n}))$;
- the random variable $F(X_i)$ has a uniform distribution on $[0, 1]$ if and only if F is a continuous function.

To appreciate the importance of the quantile function, we introduce the concept of empirical distribution function and then the Glivenko–Cantelli theorem. The empirical distribution function F_n of a sample $(X_1, \ldots, X_n)$ is defined as

$$\hat{F}_{(n)}(x) = \frac{1}{n} \sum_{i=1}^{n} I_{\{X_i \leq x\}},$$

where I is the indicator function. That is, for each x, the empirical distribution function counts the number of samples that have a value that is less than or equal to x.

The theoretical underpinning of non-parametric statistics, such as order statistics, is the Glivenko–Cantelli theorem. This theorem states that, if the samples are independent draws from the distribution F, the empirical distribution function tends to F for large n in the sense that

$$\Delta_n = \sup_{x \in \mathbb{R}} |\hat{F}_{(n)}(x) - F(x)| \xrightarrow{a.s.} 0$$

for $n \to \infty$. In cases where F is a Pareto distribution, the quantile function indicates that if we approximate n random draws from a uniformly distributed variable as the sequence $(1, 2, \ldots, n)$, then the corresponding values of the sample $(X_1, \ldots, X_n)$ will be $(1, \frac{1}{2}, \ldots, \frac{1}{n})$. This is a statement of Zipf's law formulated in the 1930s by the American linguist George Kingsley Zipf who found that in the English language, the word most often used occurs about twice as often as the word that is second most often used and about three times as the word that is the third most often used, and so on.

From the quantile function, the limit law of the ratio between two successive order statistics can also be inferred. Suppose that (1) an (infinite) population is distributed according to a distribution $F \in \mathcal{R}_\alpha$ with regularly varying tails and (2) n samples are randomly and independently drawn from this distribution and ordered in function of size

$$X_{n,n} \geq \cdots \geq X_{1,n}.$$

The following property can be shown to hold:

$$\frac{X_{k,n}}{X_{k+1,n}} = 1$$

when

$$\frac{k}{n} \longrightarrow 0.$$

9.7. Stationary Processes

Thus far we have discussed the distribution of the extremes of iid sequences. The assumption that sequences are iid is not appropriate in finance given that volatility exhibits significant autocorrelations. However, the fundamental result that the distributions of block maxima belong only to extreme value distributions holds.

Consider a strictly stationary sequence (X_i), where i is an integer, with stationary distribution F. The sequence (X_i) is assumed to be stationary and therefore it might exhibit autocorrelations. Autoregressive moving average (ARMA), autoregressive conditional heteroscedasticity (ARCH), and generalized autoregressive conditional heteroscedasticity (GARCH) sequences are examples of stationary, autocorrelated series. Let M_n denote the maximum on n-samples, that is

$$M_n = \max(X_1, \ldots, X_n).$$

Consider now the associated iid process formed by a sequence $(\tilde{X}_i)$ that has the same distribution F but it is iid. Analogously, call $\tilde{M}_n$ the maximum on n-samples, that is

$$\tilde{M}_n = \max(\tilde{X}_1, \ldots, \tilde{X}_n).$$

It can be demonstrated that, for many processes, a real number θ in the interval $(0, 1)$ exists such that the following condition holds:

$$\lim_{n\to\infty} P\left(\frac{\tilde{M}_n - d_n}{c_n} \leq x\right) = H(x)$$

if and only if

$$\lim_{n\to\infty} P\left(\frac{M_n - d_n}{c_n} \leq x\right) = H^{\theta}(x).$$

The parameter θ is called the extremal index. ARMA processes with Gaussian innovation have $\theta = 1$ (see Embrechts *et al.* (1997) for details) while ARCH and GARCH processes have $\theta < 1$ (see de Haan *et al.* (1989) for ARCH and Mikosch and Starica (2000) for GARCH).

9.8. Multivariate EVT

Thus far we have discussed EVT applied to univariate distributions. In this section we will discuss EVT for multivariate distributions. Concepts of multivariate EVT are subject to some arbitrariness. This is because there is no natural way to define the maximum of a set of vectors. However, using the concept of copula it is possible to establish the multivariate equivalent of the theory of maxima and of points of exceedances.

Let us consider a sample of n iid random vectors $X_i = (X_{i,1}, \ldots, X_{i,d})$. Call $F(x_1, \ldots, x_d)$ the multivariate cdf of X_i

$$F(x_1, \ldots, x_d) = P(X_{i,1} \leq x_1, \ldots, X_{i,d} \leq x_d)$$

and $F_1, \ldots, F_d$ their marginal distributions.

Consider the vector $M_n = (M_{n,1}, \ldots, M_{n,d})$ formed taking the maximum of each component, that is

$$M_{n,j} = \max(X_{1,j}, \ldots, X_{n,j}),$$

where $j \in \{1, \ldots, d\}$. The vector generally is not one of the vectors in the sample.

To discuss the limit distribution, we need to introduce the concept of copula.

9.8.1. Copula functions

Every joint distribution function for a random vector contains both a description of their marginal behavior and of their dependence structure.[1] Copulas provide a way to separate the two elements. Copulas allow a deeper understanding of dependence than correlation and allow a bottom-up approach to multivariate analysis.

A d-dimensional copula is a distribution function on $[0,1]^d$ with standard uniform marginal distributions. We denote copulas as functions $C(u_1, \ldots, u_d)$. Copulas are mappings $[0,1]^d \longrightarrow [0,1]$ characterized by the following properties:

- the function $C(u_1, \ldots, u_d)$ is increasing in each component;
- $C(1, \ldots, 1, u_i, 1, \ldots, 1) = u_i$ for each $i \in \{1, \ldots, d\}$ and $u_i \in [0,1]$;

[1] See Cherubini *et al.* (2004), Nelsen (2007), and Cherubini *et al.* (2011).

- the function $C(u_1, \ldots, u_d)$ is d-increasing, that is for all $(a_1, \ldots, a_d)$ and $(b_1, \ldots, b_d) \in [0,1]^d$ with $a_i \leq b_i$,

$$\sum_{u \in \{a_1,b_1\} \times \cdots \times \{a_d,b_d\}} (-1)^{N(c)} C(u_1, \ldots, u_d) \geq 0,$$

where $u = (u_1, \ldots, u_d)$ and $N(c) = \#\{k : u_k = a_k\}$.

The first property is clearly required by any multivariate distribution function. The second property is the condition of uniform marginal distributions. The third property ensures that probabilities are never negative. For example, consider the two-dimensional case. The following equalities hold:

$$0 \leq P(U_1 \leq u_1, U_2 \leq u_2) = C(u_1, u_2),$$

and

$$\begin{aligned} 0 &\leq P(u_{11} \leq U_1 \leq u_{12}, u_{21} \leq U_2 \leq u_{22}) \\ &= P(U_1 \leq u_{12}, U_2 \leq u_{22}) - P(U_1 \leq u_{11}, U_2 \leq u_{22}) \\ &\quad - P(U_1 \leq u_{12}, U_2 \leq u_{21}) + P(U_1 \leq u_{11}, U_2 \leq u_{21}) \\ &= C(u_{12}, u_{22}) - C(u_{11}, u_{22}) - C(u_{12}, u_{21}) + C(u_{11}, u_{21}). \end{aligned}$$

The theory of copulas is based on Sklar's theorem which can be stated as follows.

Theorem 9.3 (Sklar). *Let $F(x_1, \ldots, x_d)$ be a joint distribution function with margins $F_1, \ldots, F_d$. Then there exists a copula function $C(u_1, \ldots, u_d)$ such that*

$$F(x_1, \ldots, x_d) = C(F_1(x_1), \ldots, F_d(x_d)).$$

If the margins are continuous, then the copula is unique. Conversely, if $F(x_1, \ldots, x_d) = C(F_1(x_1), \ldots, F_d(x_d))$, then F is a joint distribution function.

Theorem 9.4. *Let $X_1, \ldots, X_d$ be a random vector with continuous margins and copula C and let $T_1, \ldots, T_d$ be strictly increasing functions. Then the random vector*

$$(T_1(X_1), \ldots, T_d(X_d))$$

also has copula C.

For every copula C the following inequalities hold:

$$\max\left\{\sum_{i=1}^{d} u_i + 1 - d\right\} \leq C(u) \leq \min(u_1, \ldots, u_d).$$

We refer to these inequalities as *Fréchet bounds.*

9.8.2. Some common copula functions

There are three types of copulas:

(1) *fundamental copulas*: that are copulas of important dependence structure;
(2) *implicit copulas*: extracted from well-known distribution functions but no closed-form expression exists;
(3) *explicit copulas*: that have explicit closed-form expressions.

Comonotonicity and countermonotonicity copulas

The comonotonicity copula is defined through the Fréchet upper bound as follows:

$$C(u_1, \ldots, u_d) = \min(u_1, \ldots, u_d).$$

The comonotonicity copula represents perfect dependence. The countermonotonicity copula is defined in two variables as

$$C(u_1, u_2) = \max(u_1 + u_2 - 1, 0).$$

It represents perfect negative dependence.

Independence copula

The independence copula is the copula of a vector of independent variables. Their joint distribution is the product of marginal distributions and therefore the independence copula is

$$C(u_1, \ldots, u_d) = \prod_{i=1}^{d} u_i.$$

Implicit copulas

The Gaussian copula is an implicit copula. Suppose X is a multivariate Gaussian random vector distributed as $N_d(\mu, \Sigma)$. By Theorem 9.4, its copula is the same as that of $N_d(0, \Omega)$ where Ω is the correlation matrix of X. Therefore, the Gaussian copula is

$$C(u_1, \ldots, u_d) = \Phi_\Omega(\Phi^{-1}(u_1), \ldots, \Phi^{-1}(u_d)),$$

where Φ is the univariate normal cdf and Φ_Ω is the multivariate normal distribution with zero mean and correlation matrix Ω. No explicit expression is known. If Ω is equal to the identity matrix I, we obtain the independence copula; if Ω is a matrix of ones, the comonotonicity copula.

We can use the same scheme to obtain implicit copulas from other multivariate distribution functions. For example, the d-dimensional t-copula is defined as

$$C(u_1, \ldots, u_d) = t_{\nu,\Omega}(t_\nu^{-1}(u_1), \ldots, t_\nu^{-1}(u_d)),$$

where t_ν is the univariate Student-t cdf and $t_{\nu,\Omega}$ is the multivariate Student's t-distribution with correlation matrix Ω. Both the univariate and multivariate random variables have tail parameter ν.

We show the Matlab code needed to generate Figure 9.8 in which 1000 simulations from the Gaussian and t-copula are drawn. We set

$$\Omega = \begin{pmatrix} 1 & 0.75 \\ 0.75 & 1 \end{pmatrix}.$$

as correlation matrix. In the t copula case we have $\nu = 2$.

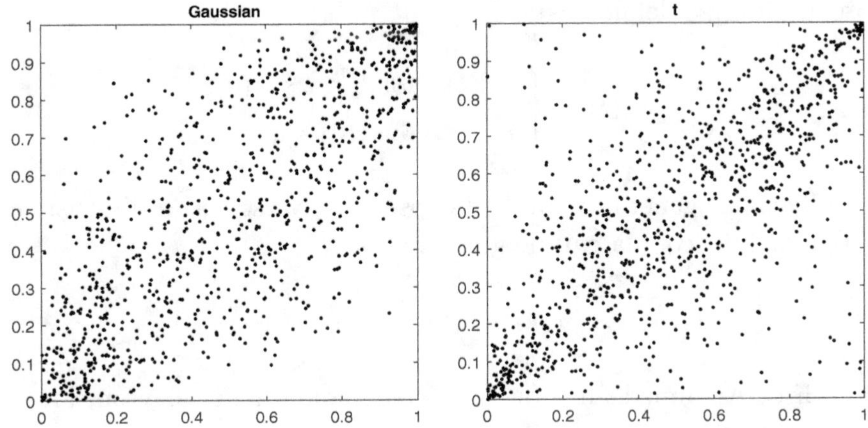

Figure 9.8. Random draws (1000 simulations) from the Gaussian and t-copula.

```
Omega = [1 0.75; 0.75 1];
nu = 2;
n = 1000;

u1 = copularnd('Gaussian',Omega,n);
u2 = copularnd('t',Omega,nu,n);

% set figure dimension
scrsz = get(0,'ScreenSize');
h4 = figure('Position',[50 50 scrsz(3)/2 scrsz(4)/3]);

subplot 121
scatter(u1(:,1),u1(:,2),'k.')
title('Gaussian')
box on

subplot 122
scatter(u2(:,1),u2(:,2),'k.')
title('t')
box on
```

Archimedean copulas

Archimedean copulas can be built using a continuous, strictly decreasing, and convex function ς defined on $[0,1]$ with values belonging to $[0,+\infty]$, such that $\varsigma(1) = 0$, called generator, and its pseudo-inverse defined as

$$\varsigma^{[-1]}(t) = \begin{cases} \varsigma^{-1}(t) & \text{for } 0 \leq t \leq \varsigma(0), \\ 0 & \text{for } \varsigma(0) \leq t \leq \infty. \end{cases}$$

Given a generator and its pseudo-inverse, a d-dimensional Archimedean copula can be generated as

$$C(u_1, \ldots, u_d) = \varsigma^{[-1]}(\varsigma(u_1) + \cdots + \varsigma(u_2)).$$

Three Archimedean copulas are commonly used: Clayton, Frank, and Gumbel.

(1) The Gumbel copula is defined as

$$C(u_1, \ldots, u_d) = \exp(-((-\log u_1)^\theta + \cdots + (-\log u_d)^\theta)^{\frac{1}{\theta}})$$

with $1 \leq \theta < \infty$, which is the independence copula for $\theta = 1$ and tends to the comonotonicity copula when $\theta \to \infty$.

(2) The Clayton copula is defined as

$$C(u_1, \ldots, u_d) = \max((u_1^{-\theta} + \cdots + u_d^{-\theta} - 1)^{-\frac{1}{\theta}}, 0)$$

with $-1 \leq \theta < \infty$ and $\theta \neq 0$, which is the independence copula for $\theta \to 0$ and tends to the comonotonicity copula when $\theta \to \infty$.

(3) The Frank copula is defined as

$$C(u_1, \ldots, u_d) = -\frac{1}{\theta} \log \left(1 + \frac{(e^{-u_1\theta} - 1) \cdots (e^{-u_d\theta} - 1)}{e^{-\theta} - 1} \right)$$

with $\theta \neq 0$, which is the independence copula for $\theta \to 0$ and tends to the comonotonicity copula when $\theta \to \infty$.

We show the Matlab code needed to generate Figure 9.9 in which 1000 simulations from the three Archimedean copulas with $\theta = 5$ are drawn.

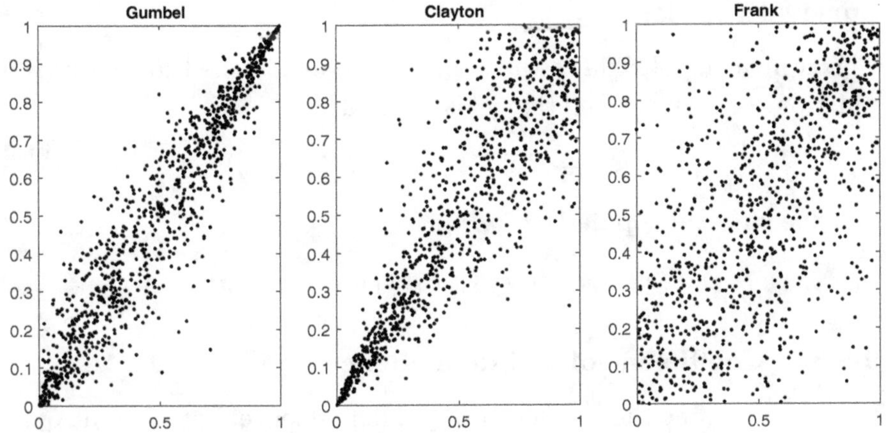

Figure 9.9. Random draws (1000 simulations) from the three Archimedean copulas with $\theta = 5$.

```
theta = 5;
n = 1000;

u1 = copularnd('Gumbel',theta,n);
u2 = copularnd('Clayton',theta,n);
u3 = copularnd('Frank',theta,n);

% set figure dimension
scrsz = get(0,'ScreenSize');
h4 = figure('Position',[50 50 scrsz(3)/2 scrsz(4)/3]);

subplot 131
scatter(u1(:,1),u1(:,2),'k.')
title('Gumbel')
box on

subplot 132
scatter(u2(:,1),u2(:,2),'k.')
title('Clayton')
box on

subplot 133
scatter(u3(:,1),u3(:,2),'k.')
title('Frank')
box on
```

Survival copulas

The equivalent of Sklar's theorem can be established for a survival function. A survival function $\bar{F}$ is defined as

$$\bar{F}(x) = P(X_1 > x_1, \dots, X_d > x_d)$$

and the survival copula is

$$\bar{F}(x_1, \dots, x_d) = \bar{C}(\bar{F}_1(x_1), \dots, \bar{F}_d(x_d)).$$

9.8.3. Coefficient of tail dependence

Let us now define the coefficient of tail dependence. Let X_1 and X_2 be random variables with distribution functions F_1 and F_2, respectively. The coefficients of upper and lower tail dependence are defined as

$$\lambda_u = \lambda_u(X_1, X_2) = \lim_{q \to 1^-} P(X_2 > F_2^{\leftarrow}(q) | X_1 > F_1^{\leftarrow}(q))$$

and

$$\lambda_l = \lambda_l(X_1, X_2) = \lim_{q \to 0+} P(X_2 \leq F_2^{\leftarrow}(q) | X_1 \leq F_1^{\leftarrow}(q)).$$

They can be expressed in terms survival copulas

$$\lambda_u = \lim_{q \to 1^-} \frac{\bar{C}(q, q)}{q}$$

and copulas

$$\lambda_l = \lim_{q \to 0+} \frac{C(q, q)}{q}.$$

9.8.4. The multivariate theory of maxima

The distribution of maxima is equal to the following multivariate cdf:

$$P(M_{n,1} \leq x_1, \dots, M_{n,d} \leq x_d) = F^n(x_1, \dots, x^d).$$

The limit for $n \to \infty$ is degenerate. However, suppose exist two sequences of constant vectors in $\mathbb{R}^d$

$$c_n = (c_{n,1}, \dots, c_{n,d}) > 0$$

and

$$d_n = (d_{n,1}, \ldots, d_{n,d}) \in \mathbb{R}^d$$

so that the vectors

$$\frac{M_n - d_n}{c_n} = \left(\frac{M_{n,1} - d_{n,1}}{c_{n,1}}, \ldots, \frac{M_{n,d} - d_{n,d}}{c_{n,d}} \right)$$

converge in distribution to a random vector with joint cdf H, that is

$$\lim_{n\to\infty} P\left(\frac{M_n - d_n}{c_n} \leq x \right) = \lim_{n\to\infty} F^n(c_n x + d_n) = H(x).$$

If this happens, we say that F is in the maximum domain of attraction of H written as $F \in \mathrm{MDA}(H)$ and H is a multivariate extreme value (MEV) distribution.

It can be proven that, the margins of H, if non-degenerate, can only be extreme value distributions of the Fréchet, Gumbel, or Weibull type. As these distributions are continuous, the copula function of H is unique and must satisfy the equality

$$C(u^t) = C^t(u),$$

for all $t > 0$.

Any copula with this property is called *extreme value copula.* It can be demonstrated that the Gumbel copula, defined in Section 9.8.2, and the Galambos copula defined as

$$C(u_1, u_2) = u_1 u_2 \exp(((\log u_1)^{-\theta} + (\log u_2)^{-\theta})^{-\frac{1}{\theta}})$$

with $0 \leq \theta < \infty$, are extreme value copulas.

9.9. Estimation

The key topics presented up to this point deal with the probability structure of a distribution tails. Specifically, we have described how they can be light-tailed or fat-tailed. For the remainder of this chapter, we explain the problem of estimation. The problem of estimation for EVT is basically the problem of estimating from a finite sample the tail of a distribution. The key in this estimation as it relates to EVT is to include in the sample only observations that populate the distribution tail and not the entire sample.

To construct the sample data for the estimation of the tail distribution, it is necessary to identify criteria for separating the tail from the bulk of the distribution. Consequently, there are three tasks associated with the estimation of an EVT distribution. The first task is identifying where the tail begins. The second task is to identify the shape of the tail. More specifically, it is important to ascertain if the tail can be characterized as a power law tail. Given the identification of where the tail begins and the shape of the tail, the final task is to estimate the tail index.

Unfortunately, these three tasks cannot be easily accomplished independently. There is no dependable way for handling all these tasks automatically. Specifically, statistical model selection, that is the selection of the distribution that can best describe the data is a classical problem in statistics for formulating and validating a scientific hypothesis in a probabilistic context. Fortunately, there are many tools and tests that are available for this purpose: quantile plot and the MLE method, the latter includes the Pickands estimator and the Hill estimator. We describe each of these in this section.

9.9.1. The quantile plot

The quantile plot, also referred to as the QQ-plot, is a graphical representation of data. It is in particular defined as the following set:

$$\left\{X_{k,n}, F^{\leftarrow}\left(\frac{n-k+1}{n+1}\right)\right\},$$

for $k = 1, \ldots, n$.

The QQ-plot must be approximately linear based on the quantile transformation and the Glivenko–Cantelli theorem. In the case where F is Pareto distribution, the QQ-plot linearity is another statement of Zipf's law.

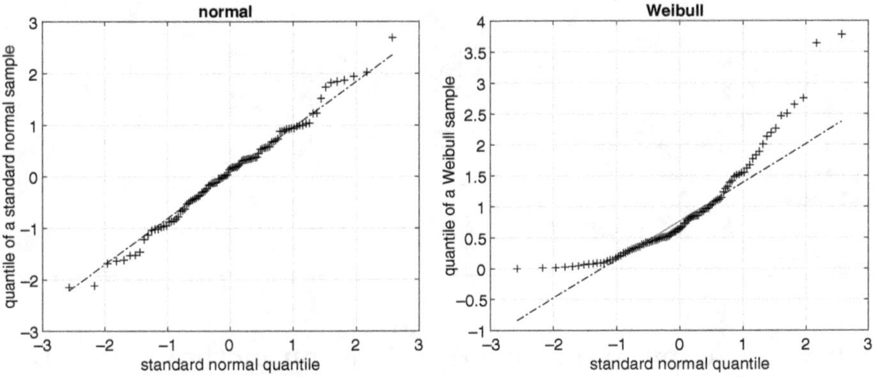

Figure 9.10. Illustration of the QQ-plot.

The QQ-plot is a valuable tool for two reasons. First, it offers a quick tool for verifying statistical hypotheses by examining the degree of deviations of the linearity plot of a hypothesized distribution. Second, it aids in the first task by allowing the formation of a preliminary view as to where the tail begins and whether the distribution fails at the far end of the tail. In Figure 9.10 we show the QQ-plot for two different samples: (1) a standard normal sample and (2) a Weibull sample with parameter $\alpha = 1$. In both cases the hypothesized distribution is a standard normal one.

We show the Matlab code needed to draw Figure 9.10.

```
% draw normal random numbers
sample1 = normrnd(0,1,[100,1]);

% draw Weibull random numbers
sample2 = wblrnd(1,1,[100,1]);

% set figure dimension
scrsz = get(0,'ScreenSize');
h4 = figure('Position',[50 50 scrsz(3)/2 scrsz(4)/3]);

subplot 121
qqplot(sample1)
title('normal')
ylabel('quantile of a standard normal sample')
xlabel('standard normal quantile')
```

```
box on
grid on

subplot 122
qqplot(sample2)
title('Weibull')
ylabel('quantile of a Weibull sample')
xlabel('standard normal quantile')
box on
grid on
```

9.9.2. Maximum likelihood estimation method

Despite the QQ-plot being a useful exploratory tool, it suffers from the usually criticism that graphical presentation of data depend on both human judgment and intuition. Rather than relying on an exploratory tool, it is necessary to use rigorous statistical tests. A starting point is parameter estimation for the generalized extreme value (GEV) distribution that is given by equation (9.4), that is

$$F(x) = \exp\left(-\left(1+\xi\frac{x-\mu}{\sigma}\right)^{-\frac{1}{\xi}}\right),$$

where

$$1+\xi\frac{x-\mu}{\sigma} > 0.$$

If $\xi = 0$, then it corresponds to the Gumbel distribution

$$F(x) = \exp\left(-\exp\left(-\frac{x-\mu}{\sigma}\right)\right),$$

for all $x \in \mathbb{R}$.

Recall that in Section 9.2.1, assuming that they exist, these distributions are the limit distribution of the normalized maxima of iid sequences. Assuming that the data to be used in the estimation are independent draws from some GEV is, admittedly, a rather strong assumption. However, as we progress, we will be relaxing this assumption. This assumption is plausible in domains where there is a long series of data available such that when the sample data are the maxima of blocks of consecutive data. However, it is probably too strong in applying EVT distributions to finance. However, it will be insightful to discuss its consequences.

In this case there are standard methodologies that exist for parameter estimation. In particular, given the sample data, the usual maximum likelihood estimation (MLE) method can be used for fitting the best GEV distribution. Note that if the above distributions fit maxima we have to divide data into blocks and consider the maxima of each block. In order to apply the MLE method, the likelihood function on the data must be computed to select the parameters that maximize the likelihood function. To do so, numerical methods can be employed. As an alternative, the Matlab functions *gevfit* and *gpfit* can be considered. This functions return maximum likelihood estimates of the parameters for the GEV and the two-parameter generalized Pareto.

It should be noted that the method of moments (MoM) is an alternative estimation method to the MLE method. The basic idea behind the MoM is equating empirical moments with their corresponding theoretical moments. There are a good number of references describing various versions of the MoM (for a discussion of the different methods see Smith (1990)). For example, Hosking *et al.* (1985) describe the method of probability-weighted moments.

Let us relax the assumption that the sequence of empirical data is represented by independent draws from an exact GEV. We will replace this strong assumption with the following weaker assumption: the empirical data are independent draws from $F \in \mathrm{MDA}(H)$. Assume that the limit distribution is a Fréchet distribution. Then in this case the data must be independent draws from some distribution F whose tail has the form $\bar{F} = x^{-\alpha}L(x)$, where, as described in Section 9.3.2, L is a slowly varying function. When this weaker assumption is made, estimation is semi-parametric in nature.

Next we describe estimators of the shape parameter ξ: the Pickands estimator and the Hill estimator. The estimation methods reviewed below are based on the behavior of maxima and upper order statistics. There is another methodology that employs the points of exceedances of high thresholds. To employ estimation methodologies that are based on the points of exceedances an appropriate model is needed for the point process of exceedances that was defined in general terms earlier in this chapter.

The Pickands estimator

For an n-sample of independent draws from a distribution $F \in \text{MDA}(H)$, the Pickands estimator $\hat{\xi}_{k,n}^{(P)}$, introduced by Pickands (1975), is defined as

$$\hat{\xi}_{k,n}^{(P)} = \frac{1}{\log 2} \log \frac{X_{k,n} - X_{2k,n}}{X_{2k,n} - X_{4k,n}},$$

where $X_{k,n}$ are upper-order statistics.

In Figure 9.11, we show the estimates for $\hat{\xi}_{k,n}^{(P)}$ for different values of k on a GEV random sample (100,000 draws) with parameter $\xi = 1$.

The theory of the Pickands estimator is based on the property, established in Section 9.4, that for $x, y > 0$ and $y \neq 1$ the following equality holds:

$$\lim_{u \to \infty} \frac{U(ux) - U(s)}{U(uy) - U(s)} = \begin{cases} \dfrac{x^{\xi} - 1}{y^{\xi} - 1} & \text{for } \xi \neq 0, \\ \dfrac{\log(x)}{\log(y)} & \text{for } \xi = 0. \end{cases}$$

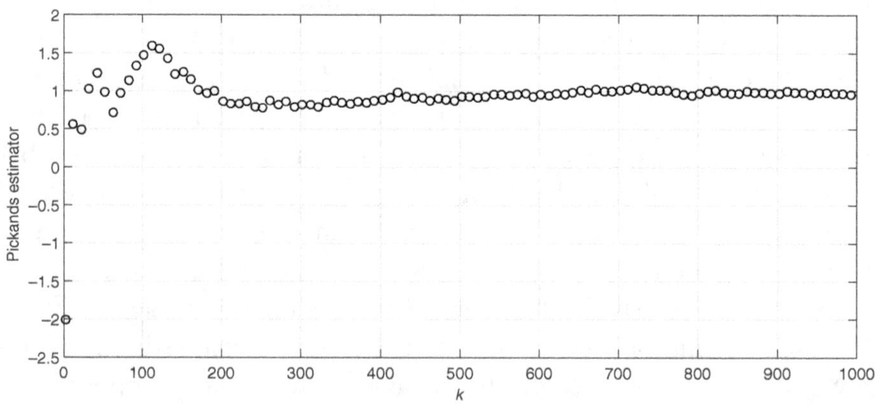

Figure 9.11. Pickands estimates for different values of k on a GEV random sample (100,000 draws) with parameter $\xi = 1$.

The following properties for the Pickands estimator can be proven:

- *weak consistency*, that is
$$\hat{\xi}_{k,n}^{(P)} \xrightarrow{P} \xi$$
with $n \to \infty$, $k \to \infty$, and $k/n \to 0$;
- *strong consistency*, that is
$$\hat{\xi}_{k,n}^{(P)} \xrightarrow{a.s.} \xi$$
with $n \to \infty$, $k/\log\log n \to \infty$, and $k/n \to 0$;
- *asymptotic normality*, that is, under additional hypotheses,
$$\sqrt{k}(\hat{\xi}_{k,n}^{(P)} - \alpha) \xrightarrow{D} N(0, \alpha^2).$$

Finally, in order to estimate the parameter ξ, there is no assumption on the type of limit distribution for the Pickands estimator. A major flaw of the Pickands estimator is its high volatility.

We show the Matlab code needed to draw Figure 9.11.

```
% draw GEV random numbers
sample1 = gevrnd(1,1,0,[100000,1]);
X = sort(sample1, 'descend');
k = 2:10:1000;

% preallocate
alphaP = zeros(1,length(k));

% compute the Hill estimator
for i = 1:length(k)
alphaP(i) = 1/log(2)*log((X(k(i)) - X(2*k(i)))/(X(2*k(i)) - ...
    X(4*k(i))));
end

% set figure dimension
scrsz = get(0,'ScreenSize');
h4 = figure('Position',[50 50 scrsz(3)/2 scrsz(4)/3]);

% draw the figure
plot(k,alphaP,'ko')
ylabel('Pickands estimator')
xlabel('k')
grid on
```

The Hill estimator

We now turn to the estimator proposed by Hill (1975). Unlike the Pickands estimator, the Hill estimator requires prior knowledge that the sample data are independent draws from a Fréchet distribution.

Assume $(x_1, \ldots, x_n)$ are independent draws from a distribution $F \in \mathrm{MDA}(H)$ with $\alpha > 0$ so that $\hat{F} = x^{-\alpha}L(x)$, where L is a slowly varying function. The Hill estimator is an MLE based on the k upper order statistics. The Hill estimation can also be calculated using a regular variation approach which is similar to the approach used for the Pickands estimator that we describe next.

The Hill estimator takes the following form:

$$\hat{\alpha}_{k,n}^{(H)} = \left(\frac{1}{k} \sum_{j=1}^{k} \log X_{j,n} - \log X_{k,n} \right)^{-1}.$$

In Figure 9.12, we show the estimates for $\bar{\alpha}_{k,n}^{(H)}$ for different values of k on a GEV random sample (100,000 draws) with parameter $\xi = 1$. We refer to Figure 9.12 as *Hill plot.*

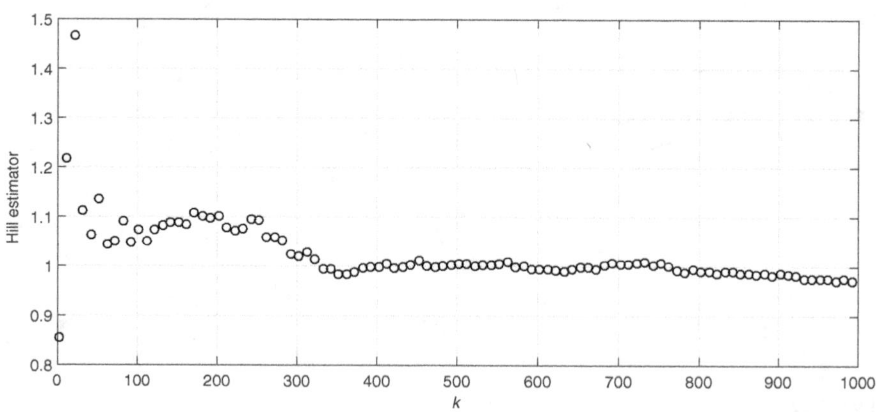

Figure 9.12. Hill estimates for different values of k on a GEV random sample (100,000 draws) with parameter $\xi = 1$.

We show the Matlab code needed to draw Figure 9.12.

```
% draw GEV random numbers
sample1 = gevrnd(1,1,0,[100000,1]);
X = sort(sample1, 'descend');
k = 2:10:1000;

% preallocate
alphaH = zeros(1,length(k));

% compute the Hill estimator
for i = 1:length(k)
alphaH(i) = 1/mean(log(X(1:k(i))/X(k(i))));
end

% set figure dimension
scrsz = get(0,'ScreenSize');
h4 = figure('Position',[50 50 scrsz(3)/2 scrsz(4)/3]);

% draw the figure
plot(k,alphaH,'ko')
ylabel('Hill estimator')
xlabel('k')
grid on
```

Both the Hill estimator and the Pickands estimator have the same weak and strong consistency property as well as asymptotic normality. Of the two approaches for estimating the tail index, the Hill estimator is by far the more popular estimator. The Hill estimator has the advantage of being robust to some dependency in the data but can perform poorly in instances where there are deviations from strict Pareto behavior. Its properties have been extensively studied and therefore the Hill estimator is well known.

The estimator thus depends on the parameter k which represents the cut-off between the observations considered as belonging to the centre of the distribution and those pertaining to the upper tail. It is subject to a bias-variance trade-off in the following sense: the variance of the Hill estimator depends on the ratio k/n, decreasing for increasing k. However, using a large fraction of the data will introduce bias in the estimator.

As explained earlier in this chapter, a critical principle of EVT is the fitting of the tail rather than the fitting of the entire distribution. There have been a number of studies proposing the automatic determination of the optimal subset of samples that should be included in the tail. Drees and Kaufmann (1998) propose an approach to the automatic determination of the tail sample using the variance-bias trade-off. Dacorogna *et al.* (1995) and Danielsson and de Vries (1997) suggest an alternative approach based on a bootstrap approach.

There is a generalization of the Hill estimator, the moment ratio estimator. Consider the following estimator of the second-order moments of the k-th upper order statistic

$$\hat{M}_{k,n} = \left(\frac{1}{k} \sum_{j=1}^{k} \log X_{j,n} - \log X_{k+1,n} \right)^2 .$$

The moment ratio estimator is defined as follows:

$$\hat{\alpha}_{k,n}^{(m)} = \frac{\hat{M}_{k,n}}{\hat{\alpha}_{k,n}^{(H)}}.$$

Extensive simulation analysis of various estimators has been performed by Wagner and Marsh (2002). They report that the moment ratio estimator outperforms the Hill estimator in sequences with a dependence structure.

The Hill estimator is an estimator of the tail index α, since this is the parameter of interest in most applications to finance. Anyway, since $\xi = \alpha^{-1}$, it can be used as an estimator of the shape parameter ξ.

Dekkers and de Haan (1989) provide an extension of the Hill estimator that covers the entire range of shape parameters ξ. A number of other estimators have been proposed. In particular, estimation procedures based on regression analysis have been proposed assuming that financial data follow a stable process. For example, Nicolau and Rodrigues (2015) propose a regression-based tail index estimator that does not involve order statistics. Moreover, their approach can be applied in more general cases than just the Pareto case. In fact, the assumption of stable behavior, or at least of exact Pareto tail, naturally leads to fitting a linear model in a logarithmic scale. There is an ample literature on this topic with a number of useful discussions, though empirical studies based on Monte Carlo simulations are still limited (see Diebold *et al.* (2000)).

9.10. Risk Measures

A great variety of approaches to VaR computation have been proposed in the literature. The main alternatives can be divided roughly into three families:(1) non-parametric methods, the most common being historical simulation (see Barone Adesi *et al.* (2008)); (2) semi-parametric methods, such as EVT; (3) parametric methods, such as the approach proposed in Chapter 6.

The historical simulation approach is based on the empirical distribution obtained from the data: the δ tail level VaR is identified with the δ empirical quantile. While this approach avoids the problem of choosing a distribution for the time series considered, it does have two important shortcomings: it cannot predict the occurrence of future extreme events, unless they have some ancestor in the time series, and it cannot provide reliable estimates of high confidence level VaR, which would require too many data.

By contrast, fully parametric methods use an explicit distributional assumption to model the time series. The model is fitted to the data and, by means of the estimated parameters, allows the VaR to be calculated at any specified confidence level. The main problem with this approach is that the actual distribution of the data at hand is unknown (often a major problem that needs to be tackled). Parametric methods can therefore lead to inaccurate estimates of VaR, potentially yielding either severe losses (if the actual risk level is underestimated) or conservative positions (in the case of overestimation).

Semi-parametric methods try to address this issue by combining the advantages of non-parametric and parametric methods.

In this sense, EVT can be regarded as a semi-parametric approach to VaR as it avoids imposing a given distribution on the whole set of data (thus reducing model risk) but focuses on one tail, trying to model its asymptotic behavior (and therefore not running into the main difficulties of wholly non-parametric methods either).

Irrespective of the approach chosen to calculate VaR, it is generally impossible to avoid the two flaws mentioned above, i.e., the VaR inability to control what happens far away in the tail and its lack of convexity. Further risk measures have been proposed as a solution, the best-known being AVaR, which by construction takes into account what happens in the tail, beyond the corresponding VaR.

In the following we describe three possible approaches to estimate VaR and AVaR based on EVT: (1) peak-over-threshold method, (2) block maxima method, and (3) tail estimation by means of the Hill estimator. In all three examples log-returns are simulated by considering a t-location-scale random variable (see Chapter 4) with parameters (μ, σ, ν) estimated on the time series of Deutsche Bank daily log-returns from June 30, 2002 to April 30, 2017 (see Chapter 11). Instead of considering returns we estimate the model on losses: if X is the random variable of log-returns, we compute VaR and AVaR of $-X$.

Peak-over-threshold approach

The GPD-based approach to model threshold exceedances can be considered to estimate VaR and AVaR as follows:

- fix a high threshold l and fit the two-parameters GPD to the exceedances over l, obtaining estimates of the shape parameter ξ and the scale parameter β (e.g., by means of a maximum likelihood estimation algorithm);
- estimate the tail probability $\bar{F}(x)$ by means of the following formula:

$$\hat{\bar{F}}(x) = \frac{n_l}{n}\left(1 + \hat{\xi}\frac{x-l}{\hat{\beta}}\right)^{-\frac{1}{\hat{\xi}}}$$

for $x \geq l$, where n_l is the number of exceedances observed in a given sample, and n is the size of the sample;

- invert the previous formula to obtain an estimate of the δ quantile

$$\mathrm{VaR}_\delta(X) = l + \frac{\hat{\beta}}{\hat{\xi}}\left(\left(\frac{n}{n_l}(1-\delta)\right)^{-\hat{\xi}} - 1\right);$$

- compute the AVaR by means of the following formula:

$$\mathrm{AVaR}_\delta(X) = \frac{\mathrm{VaR}_\delta(X)}{1-\hat{\xi}} + \frac{\hat{\beta} - \hat{\xi} l}{1-\hat{\xi}}.$$

As observed in Marinelli *et al.* (2007), there is no general rule to optimally select the threshold l. This choice is nonetheless very important, as for l too high the estimator has high variance, and for l too small the estimator becomes biased. McNeil and Frey (2000) suggested to select a random threshold that considers the top 10% of the observed negative returns (positive losses).

We show a possible Matlab implementation in which also the empirical and the estimated GPD pdf are drawn (see Figure 9.13).

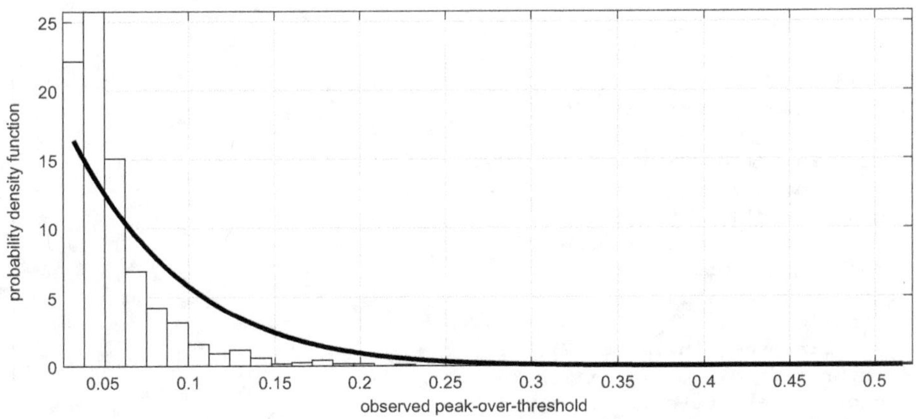

Figure 9.13. Empirical and estimated pdf based on the peak-over-threshold of simulated log-returns.

```
% simulate t location-scale random numbers
mut = 0;
sigmat = 0.02;
nut = 3;
logReturns = random('tlocationscale', mut, sigmat, nut,[10000,1]);

% set the tail level
delta = 0.95;

X = logReturns;
n = length(X);
X = -X;

% fix a threshold
l        = quantile(X,0.90);
index    = X>l;
Y        = X(index);
nl       = sum(index);

parmhat = gpfit(Y);
xi       = parmhat(1);
beta     = parmhat(2);

VaR  = l + beta/xi*((n/nl*(1 - delta))^(-xi) - 1);
AVaR = (VaR + beta - xi*l)/(1 - xi);

disp(['VaR: ', num2str(VaR), ...
      '  AVaR: ', num2str(AVaR),...
      ' - tail level: ', num2str(delta)]);

close all
% draw the empirical and estimated (GPD) pdf
hold on
dx = (max(Y) - min(Y))/40;
binWidth = dx;
binCtrs = min(Y):binWidth:max(Y);
n = length(Y);
counts = hist(Y,binCtrs);
prob = counts / (n * binWidth);
bar(binCtrs,prob,'hist');
h = get(gca,'child');
xlabel('observed peak-over-threshold');
x = linspace(min(Y),max(Y),500);
returnsPdf = gppdf(x, xi, beta, l);
plot(x, returnsPdf,'k','LineWidth', 3)
ylabel('probability density function')
grid on
box on
axis tight
axis tight
hold off
```

Block maxima approach

Another possible parametric model to estimate the VaR is by means of the δ^n tail level VaR for the maxima M_n, taking into account that

$$P(X \leq \text{VaR}_\delta(X)) = 1 - \delta$$

if and only if

$$P(M_n \leq \text{VaR}_\delta(X)) = (1 - \delta)^n,$$

where n is the total number of block maxima. By assuming that the distribution of the maxima M_n is approximately GEV, one can estimate the parameter (ξ, μ, σ) of the GEV (e.g., by means of a maximum likelihood estimation algorithm) and, using simple properties of the distribution function of the maximum of a finite family of iid random variables, obtains an estimate of VaR.

$$\text{VaR}_{\delta^n}(M_n) = \hat{\mu} - \frac{\hat{\sigma}}{\hat{\xi}}(1 - (-\log \delta^n)^{-\hat{\xi}}).$$

In this case the AVaR has not been a closed form solution. It can be obtained numerically by integrating the density function f of M_n, that is

$$\text{AVaR}_{\delta^n}(M_n) = \frac{1}{\delta} \int_{VaR_{\delta^n}(M_n)}^{\infty} x f(x) dx.$$

where f is the derivative with respect to x of the function F (see Nadarajah *et al.* (2014)).

We show a possible Matlab implementation in which both the empirical and the estimated GEV pdf are drawn (see Figure 9.14).

```
% set the tail level
delta = 0.95;

X = logReturns;
X = -X;

nMaxima = 20;
Mn = movmax(X,nMaxima);

parmhat = gevfit(Mn);
xi      = parmhat(1);
```

```
sigma    = parmhat(2);
mu       = parmhat(3);

VaR  = mu - sigma/xi*(1 - (- log(delta^nMaxima))^(-xi));
AVaR = integral(@(x) x.*gevpdf(x,xi,sigma,mu), VaR, 10);

disp(['VaR: ', num2str(VaR), ...
      '  AVaR: ', num2str(AVaR),...
      ' - tail level: ', num2str(delta)]);

close all
% draw the empirical and estimated (GEV) pdf
hold on
dx = (max(Mn) - min(Mn))/40;
binWidth = dx;
binCtrs = min(Mn):binWidth:max(Mn);
n = length(Mn);
counts = hist(Mn,binCtrs);
prob = counts / (n * binWidth);
bar(binCtrs,prob,'hist');
h = get(gca,'child');
xlabel('observed block maxima');
x = linspace(min(Mn),max(Mn),500);
returnsPdf = gevpdf(x, xi, sigma, mu);
plot(x, returnsPdf,'k','LineWidth', 3)
ylabel('probability density function')
grid on
box on
axis tight
axis tight
hold off
```

Tail estimation approach

Non-parametric estimates of VaR and AVaR can be obtained by estimating the tail of the distribution by

$$P(X \leq x) = 1 - \frac{k}{n}\left(\frac{x}{X_{k+1,n}}\right)^{-\frac{1}{\hat{\xi}^{(H)}_{k,n}}},$$

where $\hat{\xi}^{(H)}_{k,n}$ is the Hill estimator. Specifically, VaR and AVaR can be computed as follows:

$$\text{VaR}_\delta(X) = X_{k+1,n}\left(\frac{n}{k}(1-\delta)\right)^{-\hat{\xi}^{(H)}_{k,n}}$$

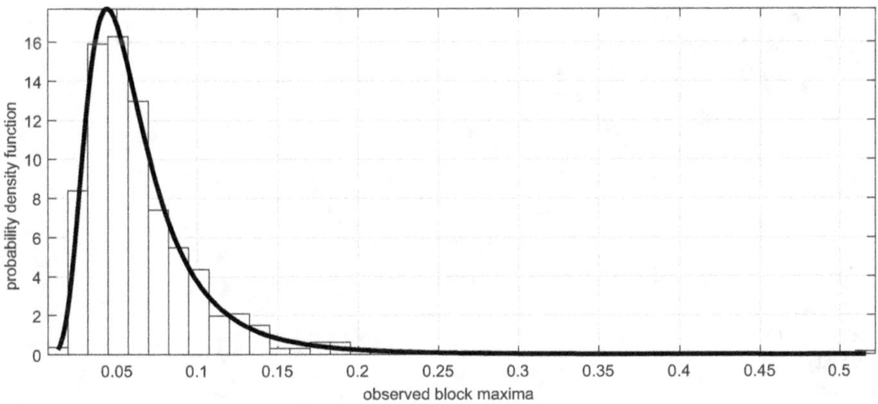

Figure 9.14. Empirical and estimated pdf based on the maxima of simulated log-returns.

and

$$\mathrm{AVaR}_\delta(X) = \frac{k}{n(1-\delta)(1-\hat{\xi}^{(H)}_{k,n})} X_{k+1,n}^{\frac{1}{\hat{\xi}^{(H)}_{k,n}}} (\mathrm{VaR}_\delta(X))^{1-\frac{1}{\hat{\xi}^{(H)}_{k,n}}},$$

where $X_{k,n}$ denotes the kth-order statistic (the same one chosen as a cut-off when estimating the shape parameter ξ with the Hill estimator). The choice of k in the above estimator is a very delicate issue, and there is an extensive literature on the subject (see Embrechts *et al.* (1997) and references therein). Usually k is chosen by visual inspection of a Hill plot, and as such is not amenable to an automated computer implementation.

We show a possible Matlab implementation in which also the Hill plot is drawn (see also in Figure 9.12). For the selected value of k ($k(end)$ in the below script), the Hill estimator reliable is observed by the visual inspection of the Hill plot.

```
% set the tail level
delta = 0.95;

X = logReturns;
X = -X;
```

```
X = sort(X, 'descend');
n = length(X);
k = 2:250;

% compute the Hill estimator
for i = 1:length(k)
alphaH(i) = 1/mean(log(X(1:k(i))/X(k(i))));
end

% draw the Hill plot
plot(k,alphaH,'ko')
ylabel('Hill estimator')
xlabel('k')
title('Hill plot')
grid on

xiH = 1/alphaH(end);

VaR  = X(k(end)+1)*(n/k(end)*(1 - delta))^(-xiH);
AVaR = k(end)/((n*(1-delta)*(1 - ...
    xiH)))*X(k(end)+1)^(1/xiH)*VaR^(1-1/xiH);

disp(['VaR: ', num2str(VaR), ...
      ' AVaR: ', num2str(AVaR),...
      ' - tail level: ', num2str(delta)]);
```

Key Points

- The focus of classical statistics is on modeling the average behavior of a distribution while extreme value theory focuses on modeling the extreme behavior of a distribution (i.e., the tail of a distribution).
- EVT deals with extreme values of a distribution, that is observations that are large deviations from some location measure of a probability distributions.
- The three methods to measure extreme values are block maxima method, peak-over-threshold method, and r-largest order statistic method.
- The ordered values from a sample of observations are referred to as order statistics of the sample; the smallest and largest observations are called the extreme values.

- Extreme value distributions are the limiting distributions for the maximum of large collections of independent random variables from the same parent distribution.
- The two pillars of EVT are the Fisher–Tippett–Gnedenko theorem and the Pickands–Balkema–de Haan theorem.
- The Fisher–Tippett–Gnedenko theorem deals with the distribution of rescaled and re-centered block maxima.
- The Pickands–Balkema–de Haan theorem deals with the distribution of point of threshold exceedances.
- The principal result of EVT is that regardless of the shape of the distribution from which the extreme values are derived, the limit distribution of maxima will fall into one of three classes: the Fréchet, Weibull, or Gumbel distribution.
- The Fréchet, Weibull, and Gumbel distributions are referred to as standard extreme value distributions.
- A random variable is referred to as an extremal random variable if it is distributed according to one of the standard extreme value distributions.
- The three extreme value distributions can be represented as a one-parameter family of distributions through the standard generalized extreme value distribution.
- Each maxima's limit distribution possesses its own maximum domain of attraction.
- For any given level l we define the excess distribution over the threshold l as the conditional distribution of $X - l$ given $X > l$.
- Cumulative distribution functions, whose excess distribution over the threshold l converges to a GPD with shape parameter ξ, are exactly those lying in the maximum domain of attraction of a GEV distribution with shape parameter ξ.
- Concepts of multivariate EVT are subject to some arbitrariness because there is no natural way to define the maximum of a set of vectors.
- In multivariate EVT by using the concept of copula it is possible to establish the multivariate equivalent of the theory of maxima and of points of exceedances.

- Because every joint distribution function for a random vector contains both a description of their marginal behavior and of their dependence structure, a copula provides a way to separate the two elements.
- Copulas allow a deeper understanding of dependence than correlation and allow a bottom-up approach to multivariate analysis.
- There are three types of copulas: (1) fundamental copulas that are important dependence structure; (2) implicit copulas that are extracted from well-known distribution functions but no closed-form expression exists; (3) explicit copulas that have explicit closed-form expressions.
- The quantile plot (QQ-plot) for verifying statistical hypotheses is by examining the degree of deviations of the linearity plot of a hypothesized distribution but there are limitations to its use.
- The three estimation methods for distributions include maximum likelihood estimation method, method of moments, and special estimators.
- The Hill estimator and the Pickands estimator are used to estimate the shape parameter of a distribution.
- It is possible to compute VaR and AVaR under the three different methods to measure extreme values.

References

G. Barone Adesi, R.F. Engle, and L. Mancini. A GARCH option pricing model with filtered historical simulation. *Review of Financial Studies*, 21(3):1223–1258, 2008.

A.S. Chernobai, S.T. Rachev, and F.J. Fabozzi. *Operational Risk: A Guide to Basel II Capital Requirements, Models, and Analysis*. Wiley, Hoboken, NJ, 2008.

U. Cherubini, E. Luciano, and W. Vecchiato. *Copula Methods in Finance*. Wiley, Chichester, 2004.

U. Cherubini, S. Mulinacci, F. Gobbi, and S. Romagnoli. *Dynamic Copula Methods in Finance*. Wiley, Chichester, 2011.

D.R. Cox and V. Isham. *Point Processes*. Chapman & Hall, London, 1980.

M.M. Dacorogna, U.A. Müller, O.V. Pictet, and C.G. de Vries. The distribution of extremal foreign exchange rate returns in extremely large data sets. *O&A Research Group*, 1995.

J. Danielsson and C.G. de Vries. Tail index and quantile estimation with very high frequency data. *Journal of Empirical Finance*, 4(2–3): 241–257, 1997.

L. de Haan, S.I. Resnick, H. Rootzén, and C.G. de Vries. Extremal behaviour of solutions to a stochastic difference equation with applications to ARCH processes. *Stochastic Processes and Their Applications*, 32(2):213–224, 1989.

A.L.M. Dekkers and L. De Haan. On the estimation of the extreme-value index and large quantile estimation. *The Annals of Statistics*, 17(4): 1795–1832, 1989.

F.X. Diebold, T. Schuermann, and J.D. Stroughair. Pitfalls and opportunities in the use of extreme value theory in risk management. *The Journal of Risk Finance*, 1(2):30–35, 2000.

H. Drees and E. Kaufmann. Selecting the optimal sample fraction in univariate extreme value estimation. *Stochastic Processes and Their Applications*, 75(2):149–172, 1998.

P. Embrechts, C. Klüppelberg, and T. Mikosch. *Modelling Extremal Events for Insurance and Finance*. Springer, Berlin, 1997.

R.A. Fisher and L.H.C. Tippett. Limiting forms of the frequency distribution of the largest or smallest member of a sample. *Mathematical Proceedings of the Cambridge Philosophical Society*, 24(2):180–190, 1928.

B. Gnedenko. Sur la distribution limite du terme maximum d'une serie aleatoire. *Annals of Mathematics*, 44(3):423–453, 1943.

C.M. Goldie and S. Resnick. Distributions that are both subexponential and in the domain of attraction of an extreme-value distribution. *Advances in Applied Probability*, 20(4):706–718, 1988.

B.M. Hill. A simple general approach to inference about the tail of a distribution. *The Annals of Statistics*, 3(5):1163–1174, 1975.

J.R.M. Hosking, J.R. Wallis, and E.F. Wood. Estimation of the generalized extreme-value distribution by the method of probability-weighted moments. *Technometrics*, 27(3):251–261, 1985.

A.F. Jenkinson. The frequency distribution of the annual maximum (or minimum) values of meteorological elements. *Quarterly Journal of the Royal Meteorological Society*, 81(348):158–171, 1955.

M.R. Leadbetter, G. Lindgren, and H. Rootzén. *Extremes and Related Properties of Random Sequences and Processes*. Springer, New York, 2012.

C. Marinelli, S. d'Addona, and S.T. Rachev. A comparison of some univariate models for value-at-risk and expected shortfall. *International Journal of Theoretical and Applied Finance*, 10(6):1043–1075, 2007.

A.J. McNeil and R. Frey. Estimation of tail-related risk measures for heteroscedastic financial time series: an extreme value approach. *Journal of Empirical Finance*, 7(3–4):271–300, 2000.

T. Mikosch and C. Starica. Limit theory for the sample autocorrelations and extremes of a GARCH (1,1) process. *Annals of Statistics*, 28(5): 1427–1451, 2000.

S. Nadarajah, B. Zhang, and S. Chan. Estimation methods for expected shortfall. *Quantitative Finance*, 14(2):271–291, 2014.

R.B. Nelsen. *An Introduction to Copulas.* Springer, New York, 2007.

J. Nicolau and P.M.M. Rodrigues. A new regression-based tail index estimator: An application to exchange rates. *Banco de Portugal*, 14, 2015. Working Papers.

J. Pickands III. Statistical inference using extreme order statistics. *The Annals of Statistics*, 3(1):119–131, 1975.

M. Rocco. Extreme value theory in finance: A survey. *Journal of Economic Surveys*, 28(1):82–108, 2014.

R.L. Smith. Extreme value theory. In W. Ledermann, E. Lloyd, S. Vajda, and C. Alexander, editors, *Handbook of Applicable Mathematics Supplement,* Volume 7, Chapter 14. Wiley, Chichester, 1990, pp. 437–472.

R. von Mises. *Selected Papers of Richard von Mises: Vol. 2, Probability and Statistics.* American Mathematical Society, Providence, RI, 1964.

N. Wagner and T. Marsh. On adaptive tail index estimation for financial return models. Research Program in Finance Working Papers, 295, 2002. University of California at Berkeley.

Part III
Applications

Chapter 10

A Portfolio Selection Analysis with Non-Gaussian Models

The main topics covered in this chapter are:

- a description of a method to perform portfolio selection based on normal, multivariate generalized hyperbolic, and multivariate normal tempered stable distributions;
- a review of the main properties of the average value-at-risk measure;
- a review of the mean–variance and mean-risk portfolio optimization strategies;
- an implementation of the minimum-variance and equally weighted portfolio criteria;
- a review of portfolio performance measures to evaluate and backtest different portfolio strategies;
- an empirical test showing the portfolio selection performance on a 50-dimensional series of stock returns;
- an evaluation of different portfolio rebalancing frequencies.

10.1. Introduction

The two main conventional approaches to modeling asset returns for the evaluation of portfolio risk measures are based either on a historical or a normal (Gaussian) distribution of returns. Neither approach adequately captures unusual behavior of asset prices and returns. The historical model is bounded by the extent of the available observations and the normal distribution model inherently cannot produce extreme returns. There are many studies exploring the non-normality of assets returns and suggesting alternative approaches. Specifying properly the distribution of assets returns is vital for risk management and optimal asset allocation. A failure may lead to significant underestimation of portfolio risk and, consequently, to wrong decisions.

The distributional modeling of financial variables has several dimensions. First, there should be a realistic model for the returns of each financial variable considered separately. That is, we should employ realistic one-dimensional models. Second, the model should capture properly the dependence between the one-dimensional variables. Therefore, we need a true multivariate model with the above two building blocks correctly specified. For the purpose of portfolio risk estimation, constructing one-dimensional models for financial instruments is incomplete. For the return dependencies between financial instruments may be fatal for the analysis. There are different ways to build a complete multivariate model. Here we explore the more natural one; that is, we hypothesize a multivariate distribution directly. Assumptions of this type include the multivariate normal and the multivariate Student's t-distributions.

Sometimes, especially among practitioners, in analyzing dependence, an explicit assumption is not made and the covariance matrix is very often relied on to describe the dependence of asset returns. However, while an explicit multivariate model is not specified, it should be kept in mind that the use of the correlation matrix is equivalent to the multivariate normal hypothesis. Furthermore, the covariance matrix can describe only linear dependencies and this is a huge limitation. Here we compare the multivariate normal distribution the multivariate normal distribution to the multivariate generalized hyperbolic (MGH) and the multivariate normal tempered stable (MNTS) models reviewed in Chapters 7 and 8. These distributions allow for a more complex dependency structure.

In order to have a reliable risk model, not only do we need a realistic assumption for return distributions but also a true downside risk measure. For this reason, we employ the average value-at-risk (AVaR), as defined in Rachev *et al.* (2008). After having reviewed the main properties of this risk measure and the classical Markowitz approach for portfolio construction, we show how to implement a portfolio selection analysis and then how to evaluate and backtest different portfolio strategies. The minimum-AVaR (MA) strategy is compared to the minimum-variance (MV) and the equally weighted (EW) portfolio. Furthermore, different portfolio rebalancing frequencies are investigated.

10.2. The Average Value-At-Risk

Recall from Chapters 5, 6, and 8 that the value-at-risk (VaR) of a continuous random variable X at tail level δ is

$$\mathrm{VaR}_\delta(X) = -\inf\{x|P(X \leq x) > \delta\} = -F_X^{-1}(\delta)$$

and it can be computed by inverting the cumulative distribution function F_X or, equivalently, by finding the δ-quantile of the random variable X and changing its sign. The VaR measure has been adopted as a standard risk measure in the financial industry. Nonetheless, it has a number of deficiencies recognized by financial professionals and bank regulators (Basel Committee on Banking Supervision, January 2016).

First, VaR is not a coherent risk measure (Artzner *et al.* (1999)) because it does not respect the sub-additivity property. Thus, VaR cannot always account for diversification. There are cases in which the portfolio VaR is larger than the sum of the VaRs of the portfolio constituents. This shows that VaR cannot be used as a true risk measure.

A second deficiency of VaR is that it is not very informative. We will use the following example to illustrate this deficiency with

VaR. Suppose that X and Y are two random variables describing the return distribution of two financial instruments. If at a given tail probability level $\text{VaR}_\delta(X) = \text{VaR}_\delta(Y) = -q_\delta$, can we state that the two financial instruments are equally risky? The answer is no because while we know that losses larger than $-q_\delta$ for both financial instruments will occur with the same probability δ, we are not sure about the magnitude of these losses.

AVaR is a risk measure which is a superior alternative to VaR. Not only does it lack the deficiencies of VaR, but it also has an intuitive interpretation: it measures the riskiness of a position by considering both the size and the likelihood of losses above a certain tail level. As reviewed in Chapters 5 and 6, the AVaR at tail probability δ is defined as the average of the VaRs which are larger than the VaR at level δ. Recall that for a continuous random variable X the AVaR at tail probability δ is defined as

$$\text{AVaR}_\delta(X) = \frac{1}{\delta}\int_0^\delta \text{VaR}_p(X)dp = -E\big[X \big| X < -\text{VaR}_\delta(X)\big]. \quad (10.1)$$

Thus, by definition, the AVaR is focused on the losses in the tail which are larger than the corresponding VaR level.

The AVaR is a convex and smooth function of portfolio weights, and it is therefore attractive for optimizing portfolios as shown by Rockafellar and Uryasev (2002). Moreover, it satisfies all axioms of coherent risk measures, as defined by Artzner *et al.* (1999), and it is consistent with the preference relations of risk-averse investors, as proven by Rachev *et al.* (2008).

Unfortunately, due to the definition of AVaR, a closed formula for the valuation of (10.1) is available only for a few distributions (e.g., in the normal case). However, there are convenient ways for computing and estimating AVaR which allows its application in optimal portfolio problems. It can be computed through a minimization formula, as described in Rockafellar and Uryasev (2002) (see also Stoyanov *et al.* (2010)), which is particularly useful in scenario-based portfolio optimization, or by means of a numerical integration, as reviewed in Chapter 6. In this chapter we will consider this latter approach.

10.3. A Portfolio Selection Problem

In this chapter, we briefly review the mean–variance and the mean-risk approach to portfolio selection, and then we study a minimum-risk strategy.

10.3.1. Mean–variance approach

Since its introduction by Markowitz (1952), the mean–variance approach has become the industry standard for asset allocation and portfolio construction. Let

$$X = (X_1, \dots, X_n)'$$

be the column vector of stock returns X_i, with $i = 1, \dots, n$, on the market we are considering, for instance the EuroStoxx 50 universe,

$$\mu = (\mu_1, \dots, \mu_n)'$$

the column vector representing the expected returns

$$\mu_i = E[X_i],$$

and

$$w = (w_1, \dots, w_n)'$$

the column vector of weights of our portfolio, with

$$\sum_{i=1}^{n} w_i = 1.$$

Thus, r_p is the random variable representing our portfolio return, i.e.,

$$r_p = \sum_{i=1}^{n} w_i X_i = w'X.$$

The expected portfolio return can be expressed by the vector of weights and expected assets returns,

$$E[r_p] = \sum_{i=1}^{n} w_i \mu_i = w'\mu,$$

and the variance of portfolio returns can be expressed by means of portfolio weights and the covariances σ_{ij} between the assets returns,

$$\sigma_{r_p}^2 = \sum_{i=1}^{n}\sum_{j=1}^{n} w_i w_j \sigma_{ij} = w'\Sigma w,$$

where Σ is the covariance matrix related to the multivariate distribution of asset returns.

The optimization problem behind mean–variance analysis is

$$\begin{aligned} &\min_{w} \ w'\Sigma w \\ &\text{subject to} \ \ w'e = 1 \\ &\qquad\qquad\quad w'\mu \geq R_* \\ &\qquad\qquad\quad w \geq 0, \end{aligned} \tag{10.2}$$

where $e \in \mathbb{R}^n$ is the column vector with all elements equal to 1 and $w \geq 0$ means that all components of the vector are non-negative, that is $w_i \geq 0$, for each i from 1 to n (this means that no short selling is permitted). The objective function is the variance of portfolio returns and R_* is the lower bound on the expected performance. In mean–variance analysis, the variance is employed to measure the degree of diversification. In fact, the main idea of Markowitz portfolio theory is that the optimal trade-off between risk and return should be the basis of financial decision-making.

The popularity of the portfolio allocation problem (10.2) also brought up several technical issues in practical applications, and there has been a large amount of further development addressing them. One main direction of related research is dedicated to the impact of parameter uncertainty on the investment performance. The practical pitfalls of the mean–variance analysis are mainly related to the extreme weights that often arise when the sample efficient portfolio is constructed. This typically happens in asset allocation, that is in portfolio strategies involving different asset classes. Furthermore, in portfolio construction, that is, in portfolio strategies in which one has to select assets in the same asset class, there is the issue of the sensitivity. The high sensitivity of the estimated mean–variance efficient portfolio to estimation errors in the underlying distribution parameters (expectations and covariances of asset returns) may lead to highly non-robust results. Barry (1974) showed the high sensitivity in particular when estimating the expected returns. This phenomenon was studied by Merton (1980), who among others argued that the estimates of the variances and the covariances of the asset returns are more accurate than the estimates of the means. Best and Grauer (1991) showed that the sample efficient portfolio is extremely sensitive to changes in the asset means. Chopra and Ziemba (1993) concluded for a real data set that errors in means are over 10 times as damaging as errors in variances and over 20 times as errors in covariances. For that reason many authors assume equal means for the portfolio asset returns or, in other words, the MV portfolio, that is

$$\begin{aligned} &\min_{w} w'\Sigma w \\ &\text{subject to } \; w'e = 1 \\ &\qquad\qquad\; w \geq 0. \end{aligned} \tag{10.3}$$

Some authors found that the pure MV portfolio may outperform the mean–variance efficient portfolio (see Gupta *et al.* (2013)).

Another direction of related research instead focuses on the enhancement of the objective function in the optimization problem (10.2). In the portfolio selection problem in equations (10.2) and (10.3), the variance of portfolio return is not a true risk measure but a measure of dispersion. An extension of the mean–variance approach can be obtained by employing a true risk measure and then studying the optimal trade-off between risk and return. We refer to this approach as *mean-risk* portfolio selection strategy. The corresponding optimization problem can be formulated as

$$\begin{aligned} &\min_{w} \rho(r_p) \\ &\text{subject to } w'e = 1 \\ &\qquad\qquad w'\mu \geq R_* \\ &\qquad\qquad w \geq 0, \end{aligned} \tag{10.4}$$

where the objective function of (10.4) is the risk of portfolio return $r_p = w'X$ as computed by the selected risk measure ρ and R_* is the lower bound on the expected portfolio return.

10.3.2. Mean-risk approach

Mean-risk optimization problems are different from the Markowitz counterpart. In order to calculate the risk of the portfolio return $\rho(r_p)$, we need to know the multivariate distribution of the asset returns. This is true also in the mean–variance case, even if one only needs the covariance matrix in the optimization problem (10.2): the approach leads to reasonable decision-making only under certain distributional hypotheses such as the multivariate normal distribution. The principal difference between mean-risk and mean-variance optimization problems is that the risk measure ρ may capture completely different characteristics of the portfolio return distribution.

In the spirit of the mean-risk analysis and by taking into consideration that the portfolio expectation may be difficult to estimate, we want to solve a simpler portfolio problem, where there is no need to estimate any expectation, that is

$$\begin{aligned} &\min_{w} \mathrm{AVaR}_{\delta}(r_p) \\ &\text{subject to } \; w'e = 1 \\ &\omega \geq 0, \end{aligned} \tag{10.5}$$

to which we refer to as *minimum-AVaR* (MA) strategy (see Stoyanov *et al.* (2010)). With the last inequality, we consider long-only portfolios. Even if the structure of the problem (10.5) closely resembles the MV problem, the idea behind it is quite different. While the variance is a dispersion measure around the expectation, the AVaR is the expected loss given that the loss has exceeded the VaR at a tail probability δ.

Now what we will demonstrate how to apply equation (10.5) in our setting by proceeding in the following way:

Step 1: Select a statistical model capable of explaining a number of observed phenomena in stock returns, such as heavy-tails and asymmetries.

Step 2: Estimate the model parameters by taking into account a sample of observed stock returns from a predetermined time window of 1500 trading days.

Step 3: On the basis of the estimated parameters from the previous step, cast the portfolio risk measure into a minimization algorithm to find optimal weights of the problem described in equation (10.5).

10.4. The Empirical Analysis

Before starting with the empirical analysis we provide a description of the data used in this chapter. In the empirical investigation we used the daily logarithmic return series for all Euro denominated stocks included in the EuroStoxx 50 on April 30, 2017. We obtained from Datastream daily dividend-adjusted closing prices from June 30, 2002 to April 30, 2017.[1]

In this section we provide the necessary definitions needed to implement an MA portfolio selection criterion and show the backtest of a portfolio selection strategy applied to a 50-dimensional case.

By considering the result of Proposition 1 in Kim *et al.* (2012) (recall Chapter 7), it follows that if Y is a MNTS random variable with parameters a, λ, C, θ, μ, σ, and $w \in \mathbb{R}^n$ is the vector of portfolio weights, then the random variable $w'Y$ is a NTS distribution with parameters a, λ, C, $\tilde{\theta}$, $\tilde{\mu}$, $\tilde{\sigma}$, where

$$\begin{aligned}
\tilde{\theta} &= w'\theta,\\
\tilde{\mu} &= w'\mu,\\
\tilde{\sigma} &= \sqrt{w'\Sigma w}.
\end{aligned}$$

[1]CRH and Engie stock price data are not available for the entire period. For this reason we selected Unicredit Bank and Assicurazioni Generali instead.

This means that a portfolio of NTS margins with joint MNTS distribution is an NTS random variable. The same result holds in the MGH case; that is, if Y is a MGH random variable with parameters λ, χ, ψ, θ, μ, σ, and $w \in \mathbb{R}^n$ is the vector of portfolio weights, then the random variable $w'Y$ has a GH distribution with parameters λ, χ, ψ, $\tilde{\theta}$, $\tilde{\mu}$, $\tilde{\sigma}$, where

$$\begin{aligned}\tilde{\theta} &= w'\theta,\\ \tilde{\mu} &= w'\mu,\\ \tilde{\sigma} &= \sqrt{w'\Sigma w}.\end{aligned}$$

This property is very useful for computing portfolio risk measures, since it reduces the dimension of the problem from n to 1 and the portfolio distribution belongs to the same parametric family of the margins. Furthermore, from Kim *et al.* (2010, 2011), it is possible to obtain a closed formula (up to an integration) to compute the AVaR in the MNTS case. In the MGH case, the VaR and AVaR computation is simpler, since a portfolio of GH margins with joint MGH distribution is a GH random variable and a closed-form expression for the density function can be used. In the normal case, VaR and the AVaR can be easily evaluated.

The portfolio selection strategy proposed here is based on the MA approach. Given a distributional assumption, we find the weights minimizing the AVaR at a given tail level. We suppose that short selling is not allowed ($w_j \geq 0$) and that it is not permissible to invest more than 10% of the wealth in a specific stock ($w_j \leq 0.1$). We consider an AVaR tail level equal to 5%. Due to the convexity property of AVaR, the problem has a unique minimum which can be obtained through the standard first-order optimality conditions for constrained optimization problems. We implement the MA strategy under the normal, the MGH, and the MNTS hypothesis (MA normal, MA MGH, and MA MNTS, respectively).

The performance of the proposed long-only strategies is benchmarked against the MV portfolio and the EW portfolio with rebalancing. These fundamental strategies are important benchmarks for large-scale applications. We implement the MV strategy by considering the historical variance; that is, we implicitly assume a multivariate normal distribution for stock returns (MV normal). As done in Mainik *et al.* (2015), the comparison includes annualized portfolio returns, maximum drawdowns, transaction costs, portfolio concentration, and asset diversity in the portfolio. Note that none of the three methods being compared (MV, EW, and MA) looks at expected returns.

The computation of portfolio weights utilizes the estimates based on the time series of stock daily log-returns from the six years prior to each Wednesday. For example, the optimal portfolio for July 1, 2009 is estimated from the stock price data for the period from July 2, 2003 to July 1, 2009. While the estimates are based on daily data, the rebalancing is performed on a weekly basis.

In Figure 10.1, we report the behavior of the total wealth for all strategies and for the EuroStoxx 50 index. The performance of the MGH and the MNTS model are almost indistinguishable. The values of the estimated optimal weights are shown in Figure 10.2. At a first look, it seems that the optimal portfolios are concentrated on less than ten stocks. The estimated optimal AVaR of the MA strategy

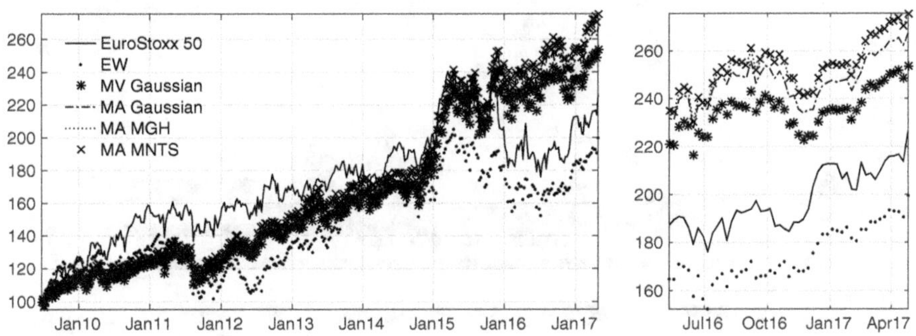

Figure 10.1. Portfolio selection backtest with weekly rebalancing for each Wednesday from June 30, 2009 to April 31, 2017 (for a total of 409 rebalancing days) for the MV, EW, and MA strategies. The resulting portfolio values are scaled to 100 for the first date of the backtest period.

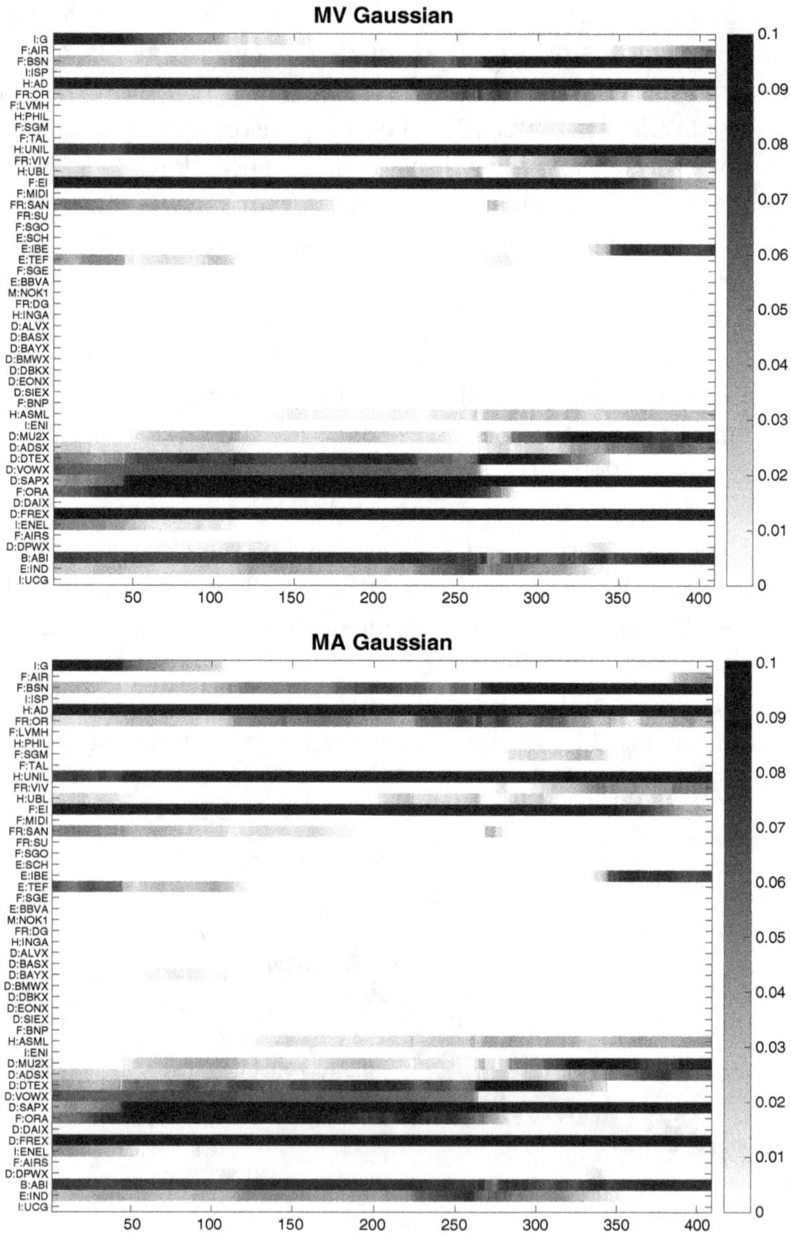

Figure 10.2. Portfolio weights with weekly rebalancing for each Wednesday from June 30, 2009 to April 31, 2017 (for a total of 409 rebalancing days) for the MV and MA strategies.

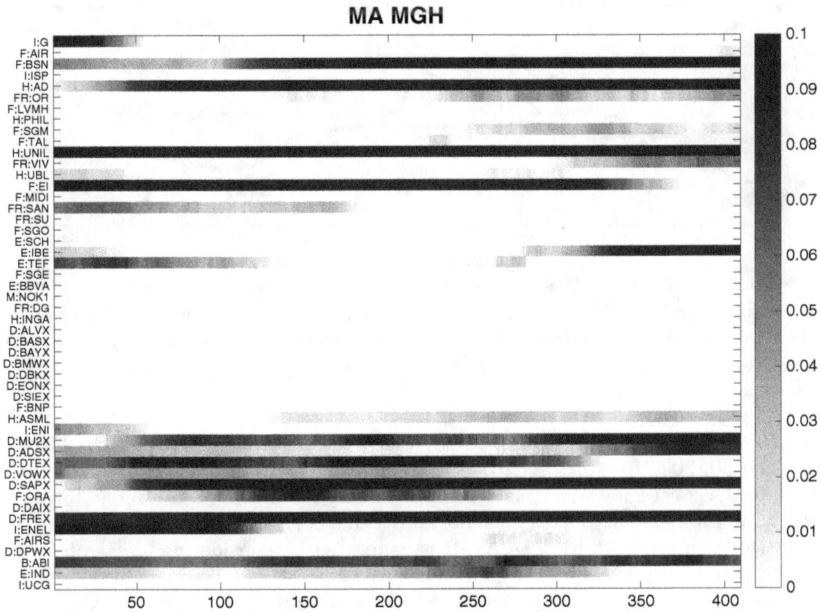

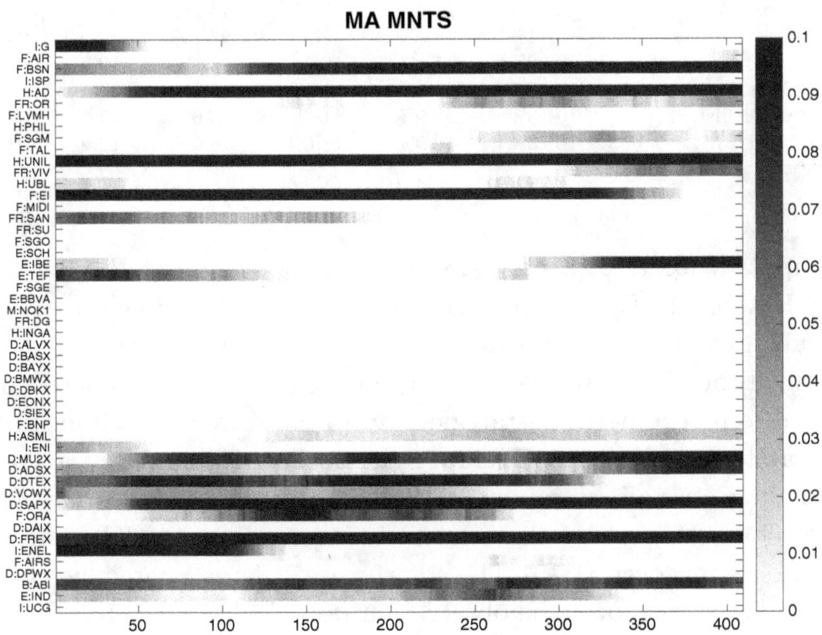

Figure 10.2. (*Continued*)

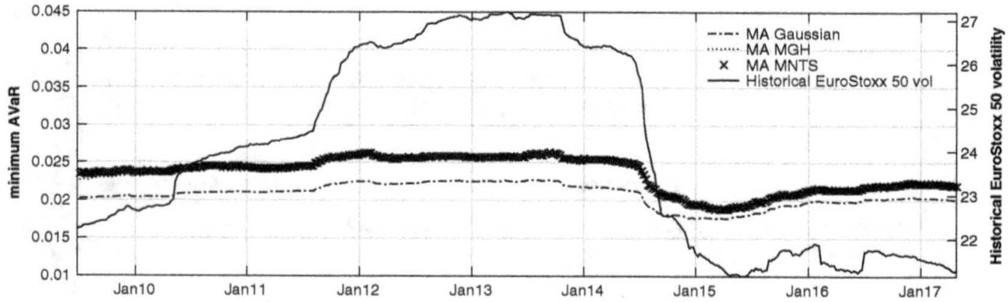

Figure 10.3. AVaR estimates for the MA portfolio strategy with weekly rebalancing for each Wednesday from June 30, 2009 to April 31, 2017 (for a total of 409 rebalancing days). For the MGH and the MNTS, both the historical and the double calibration approaches are considered to find the MA weights.

Table 10.1. Portfolio selection backtest with weekly rebalancing for each Wednesday from June 30, 2009 to April 31, 2017 (for a total of 409 rebalancing days) for the MV, EW, and MA strategies. We consider the following performance measures: total return (TR), annualized return (AR), Sharpe ratio (Sharpe), concentration coefficient (CC), and portfolio turnover (PT). The AVaR is computed by considering a 5%-tail level.

	TR	AR	Sharpe	MaxDD	CC	PT
EuroStoxx 50	81.6%	10.4%	7.2%	27.1%	—	—
EW	69.1%	8.8%	5.9%	27.7%	50.0	0.0202
MV Gaussian	93.0%	11.8%	10.7%	15.8%	12.3	0.0161
MA Gaussian	98.3%	12.5%	11.3%	15.2%	12.4	0.0160
MA MGH historical	101.3%	12.9%	11.6%	16.6%	12.1	0.0157
MA MNTS historical	101.3%	12.9%	11.6%	16.6%	12.2	0.0157

computed under different distributional assumptions (Gaussian, MGH, and MNTS) is reported in Figure 10.3. The dynamics of the AVaR are compared with the historical volatility.

In Table 10.1, we show the results of the MA approach compared to its competitor strategies and to the EuroStoxx 50 index.We report the total return (TR) over the observation period, the corresponding annualized return (AR), the Sharpe ratio,the maximum drawdown (MaxDD), the concentration coefficient (CC), and the portfolio turnover (PT). The Sharpe ratio is estimated using the portfolio returns over the observation period (409 weeks)

$$\text{Sharpe}_p = \frac{E[r_p] - r_f}{\sigma_{r_p}}$$

and r_f is the risk-free rate. We considered a zero risk-free rate in this application.

The MaxDD is the maximum decline of a series, measured as return, from a peak to a nadir over a period of time. It can be evaluated with the Matlab *maxdrawdown* function.

The results show that the MA strategy indeed outperforms MV and EW portfolios in many respects. In particular, the MA optimal portfolio gives higher returns (12.9% on an annual basis) and a higher Sharpe ratio (11.6%), at least in the MGH and MNTS cases. The overperformance on a yearly basis with respect to the EuroStoxx 50 index is 2.5% in the MA MGH and MA MNTS cases, 2.1% in the MA Gaussian case, and 1.4% in the MV Gaussian case. The EW strategy underperforms the EuroStoxx 50 index. The lowest maximal drawdown is obtained for the MA Gaussian strategy (15.2%), even if the value obtained for the MV strategy is quite close (15.8%). It is slightly higher in the MA MGH and MA MNTS cases (16.6%).

We measure the portfolio stock concentration through the concentration coefficient (CC), also known as the Herfindahl–Hirschman index, defined as

$$\mathrm{CC}_t = \left(\sum_{i=1}^{N} w_{i,t}^2\right)^{-1},$$

where $w_{i,t}$ is the portfolio weight of asset i after rebalancing. The CC of an EW portfolio is the number of assets n. As the portfolio becomes concentrated in fewer assets, the CC decreases. In Table 10.1, we report the average CC computed over the different rebalancing weeks. The numbers in that table indicate that the MA and the MV strategies are quite selective, whereas the number of stocks in the MA portfolio under the Gaussian distributional assumption is slightly higher.

As a proxy for transaction costs, we consider the portfolio turnover (PT) defined as

$$\mathrm{PT}_t = \sum_{j=1}^{n} |w_t^j - w_{t\_}^j|,$$

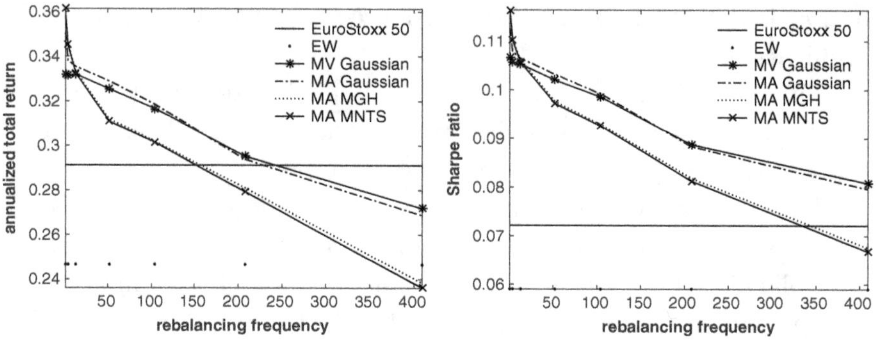

Figure 10.4. For each model we compare the portfolio strategy performance by switching from weekly to monthly (four weeks), to quarterly (13 weeks), to semi-annually (26 weeks), to annually (52 weeks), to biennially (104 weeks), to quadrennially (208 weeks) rebalancing, and to the buy-and-hold (no rebalancing over the entire observation period). Total annualized returns and Sharpe ratios are reported.

where w_t^j is the portfolio weight of asset j after rebalancing (according to the portfolio selection strategy) at time t, and $w_{t_-}^j$ is the portfolio weight of asset j just before rebalancing. The average turnover of MA optimal portfolios ranges from 0.0157 to 0.0160 and it is close to that of the MV portfolio (0.0161). Surprisingly, the EW strategy shows the highest PT (0.0202).

As a further empirical investigation, we switch from weekly to monthly (four weeks), to quarterly (13 weeks), to semi-annually (26 weeks), to annually (52 weeks), to biennially (104 weeks), to quadrennially (208 weeks) rebalancing, and to the buy-and-hold (no rebalancing over the entire observation period). The calculation of portfolio weights is still based on the same estimates. This allows the using of all observations in the historical window, and not only a subset. The Sharpe ratio is estimated using the portfolio returns over the observation period (409 weeks). Both the AR and the Sharpe ratios are reported in Figure 10.4. The performance decreases if one decreases the rebalancing frequency.

10.4.1. Matlab code — VaR and AVaR

In the normal case, the VaR and AVaR can be easily computed. Here we recall the formulas and provide the Matlab scripts. If X is a normal random variable with mean μ and standard deviation σ, then VaR can be computed as follows:

$$\mathrm{VaR}_\delta(X) = -\sigma F_Z^{-1}(\delta) - \mu,$$

where Z is a normal random variable with zero mean and unit variance, and AVaR as (see Rachev *et al.* (2008))

$$\mathrm{AVaR}_\delta(X) = \frac{\sigma}{\delta\sqrt{2\pi}} e^{-\frac{-\mathrm{VaR}_\delta(Z)^2}{2}} - \mu.$$

```
function VaR = VaR_normal(epsilon, paramNormal, dt)

mu    = paramNormal(1)*dt;
sigma = paramNormal(2)*sqrt(dt);

VaR = - sigma*norminv(epsilon) - mu;

return
```

```
function AVaR = AVaR_normal(epsilon, paramNormal, dt)

mu    = paramNormal(1)*dt;
sigma = paramNormal(2)*sqrt(dt);

VaR = VaR_normal(epsilon, [0,1], 1);

AVaR = sigma / (epsilon*sqrt(2*pi))*exp(VaR^2/2) - mu;

return
```

10.4.2. Matlab code — Margins

In the following box we show the Matlab script to compute the mean and variance of the MGH and MNTS margins. The formulas are provided in Chapter 7.

```
% mean and variance of MGH margins

function [meanMarginMGH, varianceMarginMGH] = ...
    getMomentsMarginsMGH(lambda, chi, psi, mu, theta, Sigma)
```

```
[meanGIG, varianceGIG] = getMomentsGIG(lambda, chi, psi);

mu    = mu(:);
theta = theta(:);

meanMarginMGH = mu + meanGIG*theta;

sigma_MGH = sqrt(diag(Sigma));

varianceMarginMGH = meanGIG*sigma_MGH.^2 + varianceGIG*theta.^2;

return

% mean and variance of a GIG random variable

function [meanGIG, varianceGIG] = getMomentsGIG(lambda, chi, psi)

z = sqrt(chi*psi);

meanGIG = (chi/psi)^(1/2)*besselk(lambda + 1, ...
    z)/besselk(lambda, z);

varianceGIG = (chi/psi)*(besselk(lambda + 2, ...
    z)/besselk(lambda, z) ...
                        - (besselk(lambda + 1, ...
                            z)/besselk(lambda, z))^2);

return
```

```
% mean and variance of MNTS margins

function [meanMarginMNTS, varianceMarginMNTS] = ...
             getMomentsMarginsMNTS(paramSubordinatorCTS, mu, ...
                 theta, Sigma)

a_CTS         = paramSubordinatorCTS(1);
lambda_CTS    = paramSubordinatorCTS(2);
C_CTS         = paramSubordinatorCTS(3);

mu = mu(:);
theta = theta(:);

omega_CTS = a_CTS/2;

meanMarginMNTS = mu - ...
    theta*omega_CTS*C_CTS*gamma(-omega_CTS)*lambda_CTS^(omega_CTS ...
    - 1);
```

```
sigma_CTS = sqrt(diag(Sigma));

k_CTS = omega_CTS*(omega_CTS - ...
    1)*C_CTS*gamma(-omega_CTS)*lambda_CTS^(omega_CTS - 2);

varianceMarginMNTS = k_CTS*(theta.^2 + ...
    sigma_CTS.^2*lambda_CTS/(1- omega_CTS));

end
```

In the following box, we provide the code related to the characteristic function of both MGH and MNTS margins. A portfolio of MNTS margins is a NTS random variable, and the margins of a MNTS random variable are NTS distributed. A similar argument holds for the MGH random variable.

```
% characteristic function of a MGH margin

function y = chf_MGH_margin(u, param, ~)

% We do not consider dt in this function

lambda = param(1);
chi    = param(2);
psi    = param(3);
mu     = param(4);
theta  = param(5);
Sigma  = param(6);

epsilon = sqrt(chi*psi);

C = u.^2*Sigma^2;

g = 1i*u*theta' - 1/2*C;
Z = sqrt(chi*(psi - 2*g));

c1 = exp(1i*u*mu');
c2 = (1 - 2/psi*g).^(-lambda/2);
c3 = besselk(lambda, Z);

y = c1.*c2.*c3/besselk(lambda, epsilon);

return
```

```
% characteristic function of a MNTS margin

function y = chf_MNTS_margin(u, param, dt)

a      = param(1);
lambda = param(2);
C      = param(3);
mu     = param(4);
theta  = param(5);
sigma  = param(6);

k = (lambda - 1i*theta*u + sigma^2*u.^2/2).^(a/2) - lambda^(a/2);

y = exp(dt*(1i*mu.*u + C*gamma(-a/2)* k ) );

return;
```

10.4.3. Matlab code — Computing portfolio risk measures

In the next box, we show the Matlab script to compute the portfolio risk measures for the multivariate normal, the MGH, and the MNTS distribution. The code is based on the estimation method proposed in Chapter 6. The execution of the entire test lasts some seconds. The computing time varies for different reasons (e.g., the processor, the Matlab version, the available RAM, the number of cores). Note that some functions used in this code (*cdfEvaluation*, *densityMGH_matrix*, *densityMNTS_matrix*, *VaR_given_cdf*, *AVaR_quad*) are provided in previous chapters. Before running this code, a matrix of dimension $N \times dim$ of indexes (stocks) log-returns is needed.

```
% find portfolio risk measure in the normal, MNTS and MGH case

function riskMeasure = getPortfolioRiskMeasure(w, model, ...
    param, dt, riskMeasureType, epsilon)

w = w(:);

hFFT = 1e-2;
nFFT = 2^16;
ds   = 1e-2;
M    = 20;
```

```
s0   = 1e-10;

if strcmp(model,'Normal') == 1 || strcmp(model,'normal') == 1

    mu = param{1,1};
    Sigma = param{1,2};

    paramNormalPortfolio = [w'*mu, sqrt(w'*Sigma*w)];

    if strcmp(riskMeasureType,'VaR') == 1
    VaR = VaR_normal(epsilon, paramNormalPortfolio, dt);
    riskMeasure = VaR;
    elseif strcmp(riskMeasureType,'AVaR') == 1
    VaR = VaR_normal(epsilon, paramNormalPortfolio, dt);
    AVaR = AVaR_normal(epsilon, paramNormalPortfolio, dt);
    riskMeasure = AVaR;
    elseif strcmp(riskMeasureType,'variance') == 1
    variance = w'*Sigma*w;
    riskMeasure = variance;
    end

elseif strcmp(model,'MNTS') == 1 || strcmp(model,'mnts') == 1

    a = param{1,1};
    lambda = param{1,2};
    C = param{1,3};
    mu = param{1,4};
    theta = param{1,5};
    Sigma = param{1,6};


    pdfF = @(x) densityMNTS_matrix(x, [a lambda C], ...
                w'*mu, w'*theta, w'*Sigma*w, dt, hFFT, NFFT, ...
                    s0, ds, M);
    cdfF = @(x) cdfEvaluation(x, pdfF, 'AVaR', 10);

    paramMNTSPortfolio = [a, lambda, C, w'*mu, w'*theta, ...
        sqrt(w'*Sigma*w)];

    if strcmp(riskMeasureType,'VaR') == 1
    VaR = VaR_given_cdf(epsilon, cdfF);
    riskMeasure = VaR;
    elseif strcmp( riskMeasureType,'AVaR') == 1
    VaR = VaR_given_cdf(epsilon, cdfF);
    AVaR = AVaR_quad(epsilon, VaR, paramMNTSPortfolio, dt,...
                    @chf_MNTS_margin, nFFT, hFFT);
    riskMeasure = AVaR;
    elseif strcmp(riskMeasureType,'variance') == 1
    [~, variance] = getMomentsMarginsMNTS([a, lambda, C],...
                    w'*mu, w'*theta, w'*Sigma*w);
```

```
    riskMeasure = variance;
    end

elseif strcmp(model,'MGH') == 1 || strcmp(model,'mgh') == 1

    lambda = param{1,1};
    chi = param{1,2};
    psi = param{1,3};
    mu = param{1,4};
    theta = param{1,5};
    Sigma = param{1,6};

    pdfF = @(x) densityMGH_matrix(x, lambda, chi, psi,...
                                 w'*mu, w'*theta, w'*Sigma*w,...
                                 dt);
    cdfF = @(x) cdfEvaluation(x, pdfF, 'AVaR', 4.5);
    paramMGHPortfolio = [lambda, chi, psi, w'*mu, w'*theta, ...
        sqrt(w'*Sigma*w)];

    if strcmp(riskMeasureType,'VaR') == 1
    VaR = VaR_given_cdf(epsilon, cdfF);
    riskMeasure = VaR;
    elseif strcmp( riskMeasureType,'AVaR') == 1
    VaR = VaR_given_cdf(epsilon, cdfF);
    AVaR = AVaR_quad(epsilon, VaR, paramMGHPortfolio, dt,...
                    @chf_MGH_margin, nFFT, hFFT);
    riskMeasure = AVaR;
    elseif strcmp( riskMeasureType,'variance') == 1
    [~, variance] = getMomentsMarginsMGH(lambda, chi, psi,...
                                        w'*mu, w'*theta, ...
                                           w'*Sigma*w);
    riskMeasure = variance;
    end
end

return
```

10.4.4. Matlab code — Finding the optimal weights

In the following box we provide the script to find optimal weights obtained by minimizing a given risk measure (i.e., variance or AVaR). Even if by a numerical perspective it is possible to minimize the portfolio VaR, for the reasons discussed above, this does not seem to be correct from a theoretical perspective. The script is implemented by considering the Matlab *fmincon*.

```
% find optimal portfolio weights

function [optWeights, fval] = ...
                getOptimalWeights(w0, riskMeasureFunction, ...
                    dim, maxWeight)

w0 = w0(:);

lb = zeros(dim,1);
ub = ones(dim,1)*maxWeight;

Aeq = ones(1,dim);
beq = 1;

MaxFunEvals = 1000;

options = optimset('MaxFunEvals', MaxFunEvals,...
          'TolFun',1e-10,'TolCon',1e-10,'TolX',1e-10,...
          'Display','off','Algorithm','active-set','UseParallel',
          'always');

[optWeights, fval] = fmincon(@(w) riskMeasureFunction(w),...
                            w0,[],[],Aeq,beq,lb,ub,[],options);

return
```

10.4.5. Matlab code — Portfolio optimization

The script in the next box merges together the functions described above. The function *riskMeasureFunction* is an input for the *getOptimalWeights*. The former function is passed to the latter through a function handle. A function handle is a data type that stores an association to a function. In the *DatesStocks* variable dates are stored, and *dataStocks* represents the matrix of stock prices. It is clear that before running this code, for each date a parameter estimation has to be performed. The function *tick2ret* converts price series to return series and, with the option '*Continuous*', returns are continuously compounded.

```
% run a portfolio optimization example

epsilon = 0.05;
w0 = 1/dim*ones(dim,1);

optWeightsEW = repmat(w0, 1, lVars);

dataPortfolio = zeros(lVars, dim);
datesPortfolio = zeros(lVars,1);
time_optimization = zeros(lVars,1);

i = 0;

 for j =  initialDateImpVol:finalDateImpVol

    clc

    if weekday(DatesStocks(j)) == 4 % 4 = Wednesday

    i = i + 1;

    disp([num2str(i), ' - ', datestr(DatesStocks(j))])

    index = find(dataForStocks(:,1)<=DatesStocks(j),1,'last');

    startDate = index - 1500;
    endDate   = index;

    data = zeros(endDate - startDate, dim);
    dataPrices = zeros(endDate - startDate, dim);

    for k = 1:dim
    data(:,k)         = ...
        tick2ret(dataStocks(startDate:endDate,k),[], ...
        'Continuous');
    dataPortfolio(i,k)   = dataStocks(endDate,k);
    end

    datesPortfolio(i) = dataForStocks(endDate,1);

    % delete zeros vectors in the data
    indexData = sum(data==0,2) == dim;
    data(indexData,:) = [];

    [N, dim] = size(data);
    tic

    % normal
```

```
display(' Normal ')

paramNormal{1,1} = mu_normal(:,i);
paramNormal{1,2} = Sigma_normal(:,:,i);

if i > 1
   w0 =  optWeightsNormal_variance(:,i-1);
end

riskMeasureFunction = @(w) getPortfolioRiskMeasure(w,...
      'Normal', paramNormal, dt, 'variance', epsilon);
[optWeightsNormal_variance(:,i), optNormal_variance(i)] = ...
                        getOptimalWeights(w0, ...
                            riskMeasureFunction, dim);

if i > 1
   w0 =  optWeightsNormal_AVaR(:,i-1);
end

riskMeasureFunction = @(w) getPortfolioRiskMeasure(w, ...
      'Normal', paramNormal, dt, 'AVaR', epsilon);
[optWeightsNormal_AVaR(:,i), optNormal_AVaR(i)] = ...
                        getOptimalWeights(w0, riskMeasure ...
                            Function, dim);

% MNTS

display(' MNTS historical ')

paramMNTS{1,1} = a_MNTS(i);
paramMNTS{1,2} = lambda_MNTS(i);
paramMNTS{1,3} = C_MNTS(i);
paramMNTS{1,4} = mu_MNTS(:,i);
paramMNTS{1,5} = theta_MNTS(:,i);
paramMNTS{1,6} = Sigma_MNTS(:,:,i);

if i > 1
   w0 =  optWeightsMNTS_AVaR(:,i-1);
end

riskMeasureFunction = @(w) getPortfolioRiskMeasure(w, ...
     'MNTS',  paramMNTS, dt, 'AVaR', epsilon);
[optWeightsMNTS_AVaR(:,i), optMNTS_AVaR(i)] = ...
                        getOptimalWeights(w0, ...
                            riskMeasureFunction, dim);

% MGH

display(' MGH historical ')
```

```
    paramMGH{1,1} = lambda_MGH(i);
    paramMGH{1,2} = chi_MGH(i);
    paramMGH{1,3} = psi_MGH(i);
    paramMGH{1,4} = mu_MGH(:,i);
    paramMGH{1,5} = theta_MGH(:,i);
    paramMGH{1,6} = Sigma_MGH(:,:,i);

    if i > 1
       w0 =  optWeightsMGH_AVaR(:,i-1);
    end

    riskMeasureFunction = @(w) getPortfolioRiskMeasure(w, ...
         'MGH', paramMGH, dt, 'AVaR', epsilon);
    [optWeightsMGH_AVaR(:,i), optMGH_AVaR(i)] = ...
             getOptimalWeights(w0, riskMeasure ...
                Function, dim);

    end

 end
```

10.4.6. Matlab code — Computing performance measures

Finally, the following script may be useful to evaluate the performance of different portfolio strategies. The variable *dataPortfolio* is a matrix containing the stock prices on the day in which the optimization is performed.

```
function [totalReturn, SharpeRatio, concentrationIndex, ...
    turnoverIndex, ... MaxDD]
       =  getPerformanceMeasures(dataPortfolio, weights, ...
           numOfWeeks)

[dim, N] = size(weights);

if nargin == 2
   numOfWeeks = 1;
end

for k = 1:dim
returnPortfolio(1:N,k) = tick2ret(dataPortfolio(:,k),[], ...
    'Continuous');
end
```

```
if numOfWeeks == 1

   for i = 1:N

   totalReturn(i) = returnPortfolio(i,:)*weights(:,i);
   c1 = weights(:,i).*exp(returnPortfolio(i,:)');
   c2 = exp(returnPortfolio(i,:)) * weights(:,i);
   weights_old(:,i) = c1/c2;

   end

   SharpeRatio = mean(totalReturn)/std(totalReturn);
   concentrationIndex = mean(1./sum(weights.^2,1));
   turnoverIndex = mean(sum(abs(weights_old - weights)),2);
   MaxDD = maxdrawdown(exp(cumsum(totalReturn)));

elseif numOfWeeks > 1 && nargout == 2


   for i = 1:N

   if mod(i,numOfWeeks) == 1
       weights_new(:,i) = weights(:,i);
   else
       weights_new(:,i) = weights_new(:,i-1);
   end

   totalReturn(i) = returnPortfolio(i,:)*weights_new(:,i);

   end

   SharpeRatio = mean(totalReturn)/std(totalReturn);

end

return;
```

Key Points

- The mean–variance approach is the industry standard for asset selection. This approach has two main drawbacks: (1) the optimal portfolio is extremely sensitive to changes in asset means and (2) the variance of portfolio returns is not a true risk measure but a measure of dispersion.

- Possible alternatives to the mean–variance approach are the minimum-variance, the mean-risk, and the minimum-risk approaches.
- To measure the risk of a portfolio, one needs to know the multivariate distribution of asset returns.
- The average value-at-risk (AVaR) can be applied for mean-AVaR and min-AVaR portfolio strategies.
- AVaR gives an informed view of losses beyond VaR. It is a convex function of portfolio weights, and is therefore attractive to optimize portfolios. It is sub-additive and satisfies a set of intuitively appealing coherent risk measure properties. AVaR is a form of expected loss (i.e., a conditional expected loss) and it is a very convenient form for use in scenario-based portfolio optimization.

References

P. Artzner, F. Delbaen, J.M. Eber, and D. Heath. Coherent measures of risk. *Mathematical Finance*, 9(3):203–228, 1999.

C.B. Barry, Portfolio analysis under uncertain means, variences, and covariences. *Journal of Finance*, 29(2):515–522, 1974.

Basel Committee on Banking Supervision. Minimum capital requirements for market risk, January 2016. https://www.bis.org/bcbs/publ/d352.pdf.

M.J. Best and R.R. Grauer. On the sensitivity of mean–variance-efficient portfolios to changes in asset means: Some analytical and computational results. *Review of Financial Studies*, 4(2):315–342, 1991.

V.K. Chopra and W.T. Ziemba. The effect of errors in means, variances, and covariances on optimal portfolio choice. *Journal of Portfolio Management*, 19(2):6–11, 1993.

A.K. Gupta, T. Varga, and T. Bodnar. *Elliptically Contoured Models in Statistics and Portfolio Theory*. Springer, 2013.

Y.S. Kim, S.T. Rachev, M.L. Bianchi, and F.J. Fabozzi. Computing VaR and AVaR in infinitely divisible distributions. *Probability and Mathematical Statistics*, 30(2):223–245, 2010.

Y.S. Kim, R. Giacometti, S.T. Rachev, F.J. Fabozzi, and D. Mignacca. Measuring financial risk and portfolio optimization with a non-Gaussian multivariate model. *Annals of Operations Research*, 201(1): 325–343, 2012.

Y.S. Kim, S.T. Rachev, M.L. Bianchi, I. Mitov, and F.J. Fabozzi. Time series analysis for financial market meltdowns. *Journal of Banking & Finance*, 35(8):1879–1891, 2011.

G. Mainik, G. Mitov, and L. Rüschendorf. Portfolio optimization for heavy-tailed assets: Extreme risk index vs. Markowitz. *Journal of Empirical Finance*, 32:115–134, 2015.

H. Markowitz. Portfolio selection. *Journal of Finance*, 7(1):77–91, 1952.

R.C. Merton. On estimating the expected return on the market: An exploratory investigation. *Journal of Financial Economics*, 8(4): 323–361, 1980.

S.T. Rachev, S.V. Stoyanov, and F.J. Fabozzi. *Advanced Stochastic Models, Risk Assessment, and Portfolio Optimization: The Ideal Risk, Uncertainty, and Performance Measures.* Wiley, Hoboken, NJ, 2008.

R.T. Rockafellar and S. Uryasev. Conditional value-at-risk for general loss distributions. *Journal of Banking & Finance*, 26(7):1443–1471, 2002.

S.V. Stoyanov, B. Racheva-Iotova, S.T. Rachev, and F.J. Fabozzi. Stochastic models for risk estimation in volatile markets: A survey. *Annals of Operations Research*, 176(1):293–309, 2010.

Chapter 11

Implied Volatility Smile with Non-Gaussian Processes

The main topics covered in this chapter are:

- basic definitions of option contracts, from plain-vanilla to more complex structures;
- a review of the Black–Scholes option pricing framework with the standard formulas to price call and put options;
- a review of the Lévy option pricing framework together with the implementation of the quasi-closed-form solution for the pricing of European contracts;
- a discussion on the implied volatility smile;
- an implementation of the algorithm to extract risk-neutral parameters from observed market quotes;
- an implementation of the joint calibration to simultaneously extract risk-neutral parameters from observed market quotes and historical parameters from the time series of stock returns;
- a detailed description of the change of measure for Lévy processes with a focus on the CTS model;
- a deep dive into the Esscher transform under a univariate framework for the CTS, the NTS, and the GH models;
- an implementation of a Monte Carlo pricing algorithm with non-Gaussian processes.

11.1. Introduction

Since the seminal work of Black and Scholes (1973) and Merton (1974) who derived the arbitrage-free option price solely from the stochastic dynamics of the underlying stock and the risk-free interest rate, researchers have focused on improving the fit of the theoretical stock price dynamics and option prices to market data. Since the mid-1970s, an enormous number of papers identifying the problems with Black-Scholes option pricing models have been published. One issue that researchers have dealt with is the inconsistency between the homoscedasticity and the lognormality assumptions made in the Black–Scholes framework and stylized facts about volatility clustering and leptokurtosis that have been observed for asset prices in both historical log-returns and option prices (see Cont (2001)).

As Derman (2004), a well-known market practitioner, wrote:

> The history of quants on Wall Street is the history of the ways in which practitioners and academics have refined and extended the Black–Scholes model. (p. 8)

As described by Patterson (2010), in 1987 soon after the Black Monday crash, options traders started to notice a strange pattern on charts of stock-option prices. Graphs of prices for deep out-of-the-money puts compared with prices for puts closer to the current price of the stock displayed a curvy kink around such options that, according to the Black–Scholes model, should not exist. Traders named this phenomenon *volatility smile*,irrespective of the actual shape, even if for some asset classes (e.g., equity) the shape is better described as a skew or a smirk. Although in this book we are dealing with stock (index) price models only and we do not intend to discuss model risk issues, we share the following paragraph extracted from Derman and Miller (2016):

> Traders and quants in every product area have had to model the smile in their own market. At many firms, not only does each front-office trading desk have its own particular smile models, but the firm-wide risk management group is likely to have its own models as well. The modeling of the volatility smile is likely one of the largest sources of model risk within finance. (p. 5)

The volatility smile disobeys the orderly world postulated by Black–Scholes and modern portfolio theory. Additionally, the volatility smile appeared and it is still there. This smile is nothing more than a graph of the so-called implied volatility, which is the value of the volatility such that, for each strike and for each maturity, and given a value for the risk-free rate, dividend yield and stock price, the price obtained from the Black–Scholes formula matches the observed price of the traded option. This means that for each strike and for each maturity, there is potentially a different value for the implied volatility. The graph of this implied volatility across strikes and maturities is known as the volatility surface, which has become a market standard reference for derivatives trading, pricing, and hedging because typically options are quoted in terms of Black–Scholes implied volatility. However, as observed by Corcuera *et al.* (2009) the volatility surface depends on the model and the well-known volatility smile might disappear if one considers the volatility obtained by considering a Lévy-based model for asset returns. Specifically, the authors showed that any smiling or smirking Black–Scholes volatility curve can be transformed into a flatter Lévy volatility curve under a well-chosen set of parameters.

The underlying assumption made in most models is that the uncertainty in financial markets can be explained by a Brownian motion. However, there is an extensive body of empirical evidence showing that the Brownian motion, and thus the normality assumption, is not rich and flexible enough to explain the dynamics of complex financial products. The drawbacks of the normal model are by no means recent. The first fundamental criticism is due to Mandelbrot (1963). He strongly rejected normality as a distributional model for asset returns based on his study of commodity returns and interest rates. For this reason, academic researchers as well as practitioners have increasingly applied more complex non-normal distribution models in finance, particularly since the turn of the century. The introduction of jumps and heavy-tails in the dynamics of stock returns (see Rachev and Mittnik (2000), Schoutens (2003), and Cont and Tankov (2003)) has been followed by the introduction of jumps in volatility dynamics (Barndorff-Nielsen and Shephard (2001), Nicolato and Venardos (2003)). In particular, jumps in stock price returns models can be introduced in three ways: (1) by modeling the price dynamics as a jump–diffusion or as a more general Lévy process; (2) by including them in the volatility dynamics; or (3) by using them as a time scale of a Lévy process (i.e., a Brownian motion or a *pure jump* Lévy process) driving jumps in the price, increasing or decreasing the level of uncertainty by speeding up or slowing down the rate at which time passes (see Schoutens (2003)). There is also the possibility of assuming a stochastic interest rate and adding jumps to the instantaneous interest rate process. However, these more complex models are usually applied only to the pricing of hybrid products.

In both the continuous-time and discrete-time frameworks, the asset return model is specified under the historical measure P and cannot be directly used to price options. One possibility is to specify a change of measure between P and a possible risk-neutral measure Q. This approach is particularly attractive because the parameters can be easily estimated using historical asset returns and used for pricing purposes. Unfortunately, the failure to explain observed option prices by considering only time series information is well known. This approach leads to rather poor pricing performance and it is largely dominated by option pricing models calibrated to option prices (see Chernov and Ghysels (2000), Christoffersen *et al.* (2010), and Barone Adesi *et al.* (2008)). Additionally, in most of the continuous-time models discussed in the literature on option pricing, the change of measure from the market measure P to the risk-neutral measure Q (needed to evaluate the prices of the options) is not unique. This means that to find a proper change of measure one has to estimate the model by also considering the prices of options traded in the market. The connection between the historical measure P and the risk-neutral measure Q in Lévy models is given by a proper change of measure as defined in the work of Sato (1999) (see also Gerber and Shiu (1994), Cont and Tankov (2003), Kim and Lee (2007), and Bianchi and Tassinari (2018)). Based on these results. We then discuss in this chapter a possible approach to jointly estimate historical asset returns (under the historical measure) and calibrate option implied volatilities (under the risk-neutral measure).

The remainder of this chapter is organized as follows. Section 11.2 briefly recalls basic properties of option contracts. After reviewing the Black–Scholes pricing framework in Section 11.3, we describe the pricing model in which the underlying stock price is driven by an exponential Lévy process that is more flexible than the geometric Brownian motion.We then discuss the change of measure problem and derive the option pricing formula under this framework in Section 11.4. In Section 11.5, we show how to compute the implied volatility smile under a Lévy framework. In Section 11.6, we describe how to calibrate model parameters to the smile observed in the market. In Section 11.7, we describe the double calibration algorithm in which we simultaneously calibrate the implied volatility surface, by minimizing (1) the average relative percentage error which is a measure of the distance between model and observed implied volatilities, and (2) the Kolmogorov–Smirnov distance between the theoretical and empirical historical distributions. Finally, we show how to use these estimates to simulate trajectories of the stock price process for the pricing of derivatives under a Monte Carlo approach.

11.2. Option Contracts

Two parties are involved in an option contract: the *buyer* and the *seller*. The seller offers the buyer the right, but not the obligation, to purchase or to sell the underlying asset at a specified price within a specified period of time (or at a specified date). The seller offers this right to the buyer in exchange for a certain sum of money called the *option price* or *option premium*. The price at which the underlying asset may be bought or sold is called the *exercise price* or *strike price*. The date at which an option expires is referred to as the *maturity date*.

When an option grants the buyer the right to purchase the underlying asset from the seller, it is referred to as *call option.* When the option buyer has the right to sell the underlying asset to the writer, the option is called a *put option.*

The timing of the possible exercise of an option is an important characteristic of the contract. There are options that may be exercised at any time up to and including the maturity date. Such options are referred to as *American options.* Other options may be exercised only at the expiration date. These contracts are called *European options.* Some options instead can be exercised at several specified dates as well as at the expiration. These options are referred to as *Bermudan options.*

The cash flow of an option at the exercise date is referred to as its *payoff.* The payoff of a European call option with maturity T is

$$\max(S_T - K, 0),$$

and that one of a plain-vanilla put option is

$$\max\,(K - S_T, 0),$$

where S_T is the value of the underlying asset at time T, and K is the strike price. In the following, we refer to a plain-vanilla call option and a plain-vanilla put option as a call option (or simply call) and a put option (or simply put), respectively.

We can define various types of payoff. In the design of structured products, practitioners use dozens of possible payoffs (see Blümke (2009) and De Weert (2011)). Some simple examples are *Asian options* and *barrier options.* If the payoff depends on the difference between the strike price and the average price of the underlying asset in a given time interval, then we refer to the option as an *Asian option.* We refer to an option as a *barrier option*, if its payoff depends on whether or not the underlying asset has reached a predetermined level in a given time period. In most of the cases, the payoffs of a complex option can be replicated by a combination of simpler ones. Building a replicating portfolio is sometimes useful to speed up the computing time of the evaluation algorithm.

Typically when we describe the evolution of an asset price by a stochastic process and analyze its distributional properties we are considering implicitly the *real-world* probability measure P. Since the price process parameters can be estimated using a time series of return data, this probability measure is also known as the *market measure*, *physical measure*, or *historical measure*. A probability measure Q equivalent to P is called an *equivalent martingale measure* (EMM) if the discounted price process $\tilde{S} = (\tilde{S}_t)_{t\in[0,T]}$ of an underlying asset is a Q-martingale, where $\tilde{S}_t = e^{-rt}S_t$ and r is the risk-free rate of return. If the underlying asset is a stock that pays a continuous dividend at a constant rate d, then Q is a probability measure equivalent to P such that the process $\tilde{S} = (\tilde{S}_t)_{t\in[0,T]}$ is a Q-martingale, where $\tilde{S}_t = e^{-(r-d)t}S_t$. We refer to this measure Q as *risk-neutral* measure.

The no-arbitrage price of a European option can be obtained as the expectation of the present value of its payoff under the risk-neutral measure Q (see Harrison and Pliska (1981)). Thus, the no-arbitrage value V_t at time $t \leq T$ of a European option with maturity T and payoff $\Pi(T)$ is given by

$$V_t = e^{-r(T-t)}E_Q[\Pi(T)|\mathcal{F}_t], \tag{11.1}$$

where $\mathcal{F}_t$ is the filtration originated by the price process.

It follows that the price of a European call option at $t = 0$ can be calculated by the discounted value of its expected payoff under Q, that is

$$C_0 = e^{-rT}E_Q[(S_T - K)^+],$$

where T is the time to maturity, r is the risk-free rate of return, S_T is the underlying asset price at time T, and K is the strike price (we are assuming $d = 0$). In order to calculate the above expectation, we need to define a dynamics for the stock price S_t. As we will see in Section 11.3, the price of plain-vanilla options can be obtained in closed form under the assumption that stock log-returns are normally distributed. However, if one moves from the normality assumption, or if one considers more complex payoffs, closed-form solutions to find the price of the option are usually not available. In these cases one has to turn to (1) quasi-closed-form solutions (e.g., numerical integration); (2) Monte Carlo simulations, or (3) more complex numerical methods (i.e., numerical methods for partial differential equations). Monte Carlo simulation is usually the last resort method to evaluate an option in case other methods are too complex or not available. In some cases, simulation is the only way to find the price of a derivative.

11.3. Option Pricing Under the Black–Scholes Model

The Black–Scholes model assumes that the riskless continuously compounded interest rate is constant over time and that the stock price process $S = (S_t)_{t\in[0,T]}$ is a geometric Brownian motion under the historical measure P, that is

$$dS_t = \mu S_t dt + \sigma S_t dW_t,$$

where $W = (W_t)_{t\in[0,T]}$ is a standard Brownian motion. By solving the previous stochastic differential equation (SDE), it is possible to express the stock price as an exponential arithmetic Brownian motion, that is

$$S_t = S_0 e^{\mu t - \frac{\sigma^2}{2}t + \sigma W_t}.$$

The parameters μ and σ can be estimated from historical data.

To compute the price of an option on the stock it is necessary to find the dynamics of the underlying asset under an equivalent martingale measure Q. Since the Black–Scholes model is arbitrage free and complete, there exists a unique risk-neutral measure. In this case, the measure Q can be found by the Girsanov theorem. Thus, the stock price dynamics under the measure Q is given by the following SDE:

$$dS_t = (r - d)S_t dt + \sigma S_t dW_t^Q,$$

where as defined earlier r and d are the risk-free rate of return and the dividend rate, respectively, and $W^Q = (W_t^Q)_{t\in[0,T]}$ is a standard

Brownian motion under the risk-neutral measure Q with

$$W_t^Q = \frac{\mu - r + d}{\sigma} t + W_t.$$

By solving this SDE we get the stock price under the probability measure Q

$$S_t = S_0 e^{(r-d)t - \frac{\sigma^2}{2}t + \sigma W_t^Q}.$$

Then by substituting $\Pi(T) = (S_T - K)^+$ and $\Pi(T) = (K - S_T)^+$ into equation (11.1), we obtain the Black–Scholes call option price (C_t) and the put option price (P_t) at time t, respectively:

$$C_t = S_t e^{-d(T-t)} N(x) - K e^{-r(T-t)} N(x - \sigma\sqrt{T-t}), \tag{11.2}$$

$$P_t = K e^{-r(T-t)} N(-x + \sigma\sqrt{T-t}) - S_t e^{-d(T-t)} N(-x), \tag{11.3}$$

where

$$x = \frac{\ln(S_t/K) + (r - d + \sigma^2/2)(T-t)}{\sigma\sqrt{T-t}},$$

and $N(x)$ denotes the cumulative distribution function of the standard normal random variable evaluated at x.

11.4. Option Pricing Under the Lévy Model

In the Black–Scholes option pricing model the stock price follows a geometric (or exponential) Brownian motion. In this section we consider a richer dynamics for the underlying by introducing jumps in the price process. In particular, the stock price process is modeled as an exponential Lévy process more general than the geometric Brownian motion, that is

$$S_t = S_0 \exp(X_t), \tag{11.4}$$

where $X = (X_t)_{t\in[0,T]}$ is a Lévy process. Specifically, we define the return process $X = (X_t)_{t\in[0,T]}$ as

$$X_t = (\mu - \psi_J(-i))t + J_t, \tag{11.5}$$

where $\mu - \psi_J(-i)$ represents the deterministic drift, $J = (J_t)_{t\in[0,T]}$ is a Lévy process, and ψ_J is its characteristic exponent. This parameterization is useful because it allows one to get the very intuitive expression

$$\begin{aligned} E[S_t] &= S_0 \exp((\mu - \psi_J(-i))t) E[\exp(J_t)] \\ &= S_0 \exp((\mu - \psi_J(-i))t) \exp(\psi_J(-i)t) \\ &= S_0 \exp(\mu t). \end{aligned} \tag{11.6}$$

Note that in the Black and Scholes model, the only exponential Lévy model with continuous trajectories for the price process, we have $\mu - \psi_J(-i) = \mu - \sigma^2/2$ and $J_t = \sigma W_t$. In this section we will consider three different geometric Lévy models: classical tempered stable (CTS), normal tempered stable (NTS), and generalized hyperbolic (GH). Since we are interested in calibrating both the market measure and the risk-neutral one, we will describe the relation between the two measures and the dynamics of stock price log-returns under P, which is used to conduct an analysis from a time series perspective, and under Q, which is used to calibrate implied volatilities. The two measures can be calibrated separately, as done in Section 11.6. Alternatively, it is possible to perform a joint calibration of the model by fitting at the same time both the time series of index log-returns and the cross-section of implied volatilities, as proposed in Section 11.7.

Assuming the existence of a riskless asset providing a continuously compounded rate of return, it is possible to show that an exponential Lévy model is arbitrage free (see Cont and Tankov (2003)) and this ensures the existence of an equivalent martingale measure. However, since the risk due to jumps cannot be hedged, geometric Lévy option pricing models with jumps are not complete. Thus, the equivalent martingale measure is not unique. Under a possible risk-neutral measure we must have

$$X_t^Q = (r - d - \psi_{J^Q}(-i))t + J_t^Q, \tag{11.7}$$

where r and d represent the deterministic risk-free rate and the dividend yield, respectively, and $J^Q = (J_t^Q)_{t\in[0,T]}$ is a Lévy process under

Q and ψ_{JQ} is its characteristic exponent. Under this setting, while $X = (X_t)_{t\in[0,T]}$ is a Lévy process under the measure P with Lévy triplet $(a_c, 0, \nu)$, $X^Q = (X_t^Q)_{t\in[0,T]}$ is a Lévy process under the measure Q with Lévy triplet $(a_c^Q, 0, \nu^Q)$, and the following equality has to be fulfilled:

$$a_c^Q - a_c = \int_{\mathbb{R}} c(x)(\nu^Q - \nu)(dx), \tag{11.8}$$

where a_c^Q and a_c are the risk-neutral and physical drifts with truncation function c, and ν^Q and ν are the risk-neutral and market Lévy measures of the return process. This result directly follows from Theorem 33.1 in Sato (1999) (this theorem is discussed in Chapter 3).

While in the CTS case the Lévy measure has a simple form and the integral in equation (11.8) can be easily evaluated, in both the NTS and the GH cases the integral in equation (11.8) do not have a simple expression. In the NTS case, the Lévy measure is the product of an exponential, a power, and a Bessel function, and in the GH case the Lévy measure does not have a closed-form expression (it is expressed as an integral with the integrand involving exponential and Bessel functions). For this reason, in Section 11.4.1 we will describe the Esscher transform.

11.4.1. The Esscher transform

Among possible candidates for a suitable risk-neutral probability measure in this section we select the Esscher equivalent martingale measure proposed in Gerber and Shiu (1994). The Q_h Esscher measure associated with the log-return process $X = (X_t)_{t\in[0,T]}$ is defined by the following Radon–Nikodym derivative:

$$\frac{dQ_h}{dP}|\mathcal{F}_t = \frac{\exp(hX_t)}{E\left[\exp(hX_t)\right]}, \tag{11.9}$$

where $\mathcal{F}_t$ is the filtration originated by the price process. To get the Esscher risk-neutral dynamics of $X = (X_t)_{t\in[0,T]}$ we need to find a parameter h such that the discounted price process of the asset is a martingale under the new probability measure Q_h equivalent to P, that is

$$E_{Q_h}[S_t e^{-(r-d)t}] = S_0. \tag{11.10}$$

It follows that

$$\begin{aligned}
E_{Q_h}[S_t e^{-(r-d)t}] &= E\left[S_t e^{-(r-d)t}\frac{dQ_h}{dP}|\mathcal{F}_t\right] \\
&= e^{-(r-d)t}E\left[S_0 \exp(X_t)\frac{\exp(hX_t)}{E\left[\exp(hX_t)\right]}\right] \\
&= S_0 e^{-(r-d)t}\frac{E(\exp((h+1)X_t))}{E\left[\exp(hX_t)\right]}
\end{aligned}$$

and

$$\frac{E[\exp((h+1)X_t)]}{E[\exp(hX_t)]} = e^{(r-d)t}.$$

Since X_t is infinitely divisible, the last condition can be written in terms of the P characteristic function of X_1

$$\frac{\phi_{X_1}(-i(h+1))}{\phi_{X_1}(-ih)} = \exp(r-d). \tag{11.11}$$

We cannot be sure that an equivalent martingale measure exists. However, if the previous equation has a solution, then the existence of the Q_h measure is ensured. Assuming that a solution exists, it is possible to compute the characteristic function of X_t under the Esscher risk-neutral probability measure as follows:

$$\phi_{X_t}^{Q_h}(u) = \frac{E[\exp((iu+h)X_t)]}{E[\exp(hX_t)]} = \frac{\phi_{X_t}(u-ih)}{\phi_{X_t}(-ih)}. \tag{11.12}$$

Given a set of parameters under the historical measure P and h, the corresponding parameters under the risk-neutral measure Q_h can be found. Similarly, given the parameters under the risk-neutral measure Q, the inverse procedure can be implemented to find the parameters under the historical measure P_h. We refer to this measure change as the inverse Esscher transform. The inverse Esscher transform will allow us to apply the double calibration procedure.

More specifically, in order to find the inverse Esscher transform, that is the value of h that allows one to find the historical parameters (under the historical measure P_h) starting from the risk-neutral ones (under the risk-neutral measure Q), we have to solve the following equation:

$$\psi_{X^Q}(-i) = r - d$$

under the constraints allowing one to express historical parameters as a function of the risk-neutral ones. By using the procedure described above it is easy to show that in all three models considered, the Esscher change of measure does not modify the nature of the stock price process.

In the following we will consider the same parameterization for the CTS, the NTS, and the GH models; that is, we consider the truncation function

$$c(x) = xI_{|x|<1}.$$

Under this parameterization, the characteristic function of the return process in the CTS case under the historical measure is

$$\begin{aligned}\phi_{X_t}(u) = \exp(iukt + Ct\Gamma(-\alpha)((\lambda_+ - iu)^\alpha - \lambda_+^\alpha \\ + (\lambda_- + iu)^\alpha - \lambda_-^\alpha)),\end{aligned} \tag{11.13}$$

where $k = \mu - C\Gamma(-\alpha)((\lambda_+ - 1)^\alpha - \lambda_+^\alpha + (\lambda_- + 1)^\alpha - \lambda_-^\alpha)$. Note that under this parameterization, as previously shown, we have

$$E[\exp(X_t)] = \exp(\mu t)$$

but, as for both NTS and GH models,

$$E[X_t] \neq kt.$$

By equation (11.11), the equation to solve to find the parameter h may be written as

$$\begin{aligned}&(\lambda_+ - h - 1)^\alpha - (\lambda_+ - h)^\alpha + (\lambda_- + h + 1)^\alpha - (\lambda_- + h)^\alpha \\ &\quad = \frac{r - d - k}{C\Gamma(-\alpha)},\end{aligned} \tag{11.14}$$

and the relations among risk-neutral and historical parameters are

$$\begin{aligned}
&k_{Q_h} = k, \quad \alpha^{Q_h} = \alpha, \quad C^{Q_h} = C,\\
&\lambda_+^{Q_h} = \lambda_+ - h,\\
&\lambda_-^{Q_h} = \lambda_- + h.
\end{aligned}$$

In the NTS case, the characteristic function under the historical measure P is

$$\phi_{X_t}(u) = \exp\left\{t\left[iuk + C\Gamma(-\alpha)\left(\left(\lambda - iu\theta + \frac{1}{2}u^2\sigma^2\right)^\alpha - \lambda^\alpha\right)\right]\right\},$$

where $k = \mu - C\Gamma(-\alpha)\left(\left(\lambda - \theta - \frac{1}{2}\sigma^2\right)^\alpha - \lambda^\alpha\right)$.

The equation to solve to find the parameter h may be written as

$$\left(\lambda - l(h) - \theta - \frac{1}{2}\sigma^2 - h\sigma^2\right)^\alpha - (\lambda - l(h))^\alpha = \frac{r-d-k}{C\Gamma(-\alpha)}, \tag{11.15}$$

where

$$l(h) = h\theta + \frac{1}{2}h^2\sigma^2.$$

The relations among risk-neutral and historical parameters are

$$\begin{aligned}
&k^{Q_h} = k, \quad \sigma^{Q_h} = \sigma, \quad C^{Q_h} = C, \quad \alpha^{Q_h} = \alpha,\\
&\theta^{Q_h} = \theta + h\sigma^2,\\
&\lambda^{Q_h} = \lambda - h\theta - \frac{1}{2}h^2\sigma^2.
\end{aligned}$$

In the GH option pricing model the market characteristic function is

$$\begin{aligned}
\phi_{X_{\Delta t}}(u) = \exp(iu\mu)&\left[1 - \frac{2}{\psi}\left(iu\theta - \frac{1}{2}u^2\sigma^2\right)\right]^{-\frac{\lambda}{2}}\\
&\times\frac{K_\lambda\left(\sqrt{\chi\left(\psi - 2\left(iu\theta - \frac{1}{2}u^2\sigma^2\right)\right)}\right)}{K_\lambda\left(\sqrt{\chi\psi}\right)},
\end{aligned}$$

where

$$k = \mu + \frac{\lambda}{2}\ln\left[1 - \frac{2\left(\theta + \frac{1}{2}\sigma^2\right)}{\psi}\right]$$
$$- \ln K_\lambda\left(\sqrt{\chi\left(\psi - 2\left(\theta + \frac{1}{2}\sigma^2\right)\right)}\right) + \ln K_\lambda\left(\sqrt{\chi\psi}\right). \tag{11.16}$$

The equation to find the parameter h may be written as

$$-\frac{\lambda}{2}\ln\left[1 - \frac{2\left(\theta + \frac{1}{2}\sigma^2 + h\sigma^2\right)}{\psi - 2l(h)}\right]$$
$$+ \ln K_\lambda\left(\sqrt{\chi\left(\psi - 2\left(l(h) + \theta + \frac{1}{2}\sigma^2 + h\sigma^2\right)\right)}\right)$$
$$- \ln K_\lambda(\sqrt{\chi\,(\psi - 2l(h))}) = r - d - k, \tag{11.17}$$

where again

$$l(h) = h\theta + \frac{1}{2}h^2\sigma^2.$$

The relations among risk-neutral and historical parameters are

$$k^{Q_h} = k, \quad \sigma^{Q_h} = \sigma, \quad \chi^{Q_h} = \chi, \quad \lambda^{Q_h} = \lambda,$$
$$\theta^{Q_h} = \theta + h\sigma^2,$$
$$\psi^{Q_h} = \psi - 2\left(h\theta + \frac{1}{2}h^2\sigma^2\right).$$

11.4.2. An alternative parameterization

To be consistent with the CTS parameterization introduced in Chapter 6, in this section we slightly modify (11.5). Specifically, we define the return process $X = (X_t)_{t\in[0,T]}$ as

$$X_t = (m - E[J])t + J_t, \tag{11.18}$$

where m is a constant, $J = (J_t)_{t\in[0,T]}$ is a pure jump Lévy process, and $E[J] = E[J_1]$, where

$$E[J_1] = C\Gamma(1-\alpha)(\lambda_+^{\alpha-1} - \lambda_-^{\alpha-1}).$$

By setting

$$\tilde{J}_t = J_t - E[J]t,$$

and

$$m = \mu - \psi_{\tilde{J}}(-i)$$

in equation (11.18) we can write the return process as

$$X_t = (\mu - \psi_{\tilde{J}}(-i))t + \tilde{J}_t, \tag{11.19}$$

where $\tilde{J} = (\tilde{J}_t)_{t\in[0,T]}$, by construction, is a Lévy process with zero mean, and $\psi_{\tilde{J}} = \psi_J - E[J]$ is its characteristic exponent. This parameterization allows one to write

$$E[X_t] = mt$$

and

$$E[\exp(X_t)] = \exp(\mu t).$$

Observe that this parameterization is equivalent to assuming that the truncation function considered to obtain the characteristic function is

$$c(x) = x.$$

Under a possible risk-neutral measure we must have

$$X_t^Q = (r - d - \psi_{\tilde{J}^Q}(-i))t + \tilde{J}_t^Q, \tag{11.20}$$

where $\tilde{J}^Q = (\tilde{J}_t^Q)_{t\in[0,T]}$ is a Lévy process under Q with zero mean and $\psi_{\tilde{J}^Q}$ is its characteristic exponent. Even if all the models described in this chapter can be reparameterized in this fashion, in the illustrations in this chapter, we are going to do it only in the CTS case. In particular, we will show how to change measure by considering equation (11.8), having as a particular case the Esscher transform.

In the CTS case, recall that the Lévy measure is

$$\nu(z) = \frac{Ce^{-\lambda_+ z}}{z^{1+\alpha}} 1_{z>0} + \frac{Ce^{-\lambda_- |z|}}{|z|^{1+\alpha}} 1_{z<0},$$

and by considering that

$$\int_{\mathbb{R}} x(\nu^Q - \nu)(dx) = C\Gamma(1-\alpha)(\lambda_+^{Q^{\alpha-1}} - \lambda_-^{Q^{\alpha-1}} - \lambda_+^{\alpha-1} + \lambda_-^{\alpha-1})$$

and truncation function

$$c(x) = x,$$

then equation (11.8) becomes

$$\begin{aligned} r - d - \psi_{J^Q}(-i) - m \\ = C\Gamma(1-\alpha)(\lambda_+^{Q^{\alpha-1}} - \lambda_-^{Q^{\alpha-1}} - \lambda_+^{\alpha-1} + \lambda_-^{\alpha-1}) \end{aligned} \quad (11.21)$$

with $\alpha^Q = \alpha$ and $C^Q = C$. Given a set of historical parameters, all possible risk-neutral parameters λ_+^Q and λ_-^Q satisfying equation (11.21) define a proper change of measure between P and Q. Similarly, given a set of risk-neutral parameters, all possible historical parameters λ_+ and λ_- satisfying equation (11.21) define a proper change of measure between Q and P. The Esscher risk-neutral measure, or simply the Q_h measure, is a particular case of this general change of measure. Under this setting, the physical characteristic function of the return process is

$$\begin{aligned} \phi_{X_t}(u) = \exp(iut(m - C\Gamma(1-\alpha)(\lambda_+^{\alpha-1} - \lambda_-^{\alpha-1})) \\ + tC\Gamma(-\alpha)((\lambda_+ - iu)^\alpha - \lambda_+^\alpha + (\lambda_- + iu)^\alpha - \lambda_-^\alpha)), \end{aligned} \quad (11.22)$$

where

$$\begin{aligned} m = \mu - \psi_{\tilde{J}}(-i) = \mu + C\Gamma(1-\alpha)(\lambda_+^{\alpha-1} - \lambda_-^{\alpha-1}) \\ - C\Gamma(-\alpha)((\lambda_+ - 1)^\alpha - \lambda_+^\alpha + (\lambda_- + 1)^\alpha - \lambda_-^\alpha)) \end{aligned}$$

and

$$\psi_{\tilde{J}}(-i) = \psi_J(-i) - E[J].$$

By considering that

$$\begin{aligned} m^Q &= r - d - \psi_{\tilde{J}^Q}(-i) \\ &= m + C\Gamma(1-\alpha)(\lambda_+^{Q\alpha-1} - \lambda_-^{Q\alpha-1} - \lambda_+^{\alpha-1} + \lambda_-^{\alpha-1}) \end{aligned}$$

we obtain

$$\begin{aligned} m^Q - C\Gamma(1-\alpha)(\lambda_+^{Q\alpha-1} - \lambda_-^{Q\alpha-1}) \\ = m - C\Gamma(1-\alpha)\left(\lambda_+^{\alpha-1} - \lambda_-^{\alpha-1}\right). \end{aligned}$$

Thus, under a structure-preserving equivalent martingale measure, the risk-neutral characteristic function can be written as

$$\begin{aligned} \phi_{X_t^Q}(u) &= \exp(iut(m - C\Gamma(1-\alpha)(\lambda_+^{\alpha-1} - \lambda_-^{\alpha-1}))) \\ &\quad \times \exp(tC\Gamma(-\alpha)((\lambda_+^Q - iu)^\alpha - \lambda_+^{Q\alpha} + (\lambda_-^Q + iu)^\alpha - \lambda_-^{Q\alpha})). \end{aligned} \tag{11.23}$$

Now, we want to show that the Esscher risk-neutral measure is a special case of a suitable equivalent martingale measure. From equation (11.11) it follows that the equation to find the parameter h may be written as

$$\begin{aligned} \alpha(\lambda_+^{\alpha-1} - \lambda_-^{\alpha-1}) + (\lambda_+ - h - 1)^\alpha - (\lambda_+ - h)^\alpha \\ +(\lambda_- + h + 1)^\alpha - (\lambda_- + h)^\alpha = \frac{r-d-m}{C\Gamma(-\alpha)}. \end{aligned} \tag{11.24}$$

By equation (11.12) we get the Q_h risk-neutral characteristic function of the return process

$$\begin{aligned} \phi_{X_t^{Q_h}}(u) = \exp(iut(m - C\Gamma(1-\alpha)(\lambda_+^{\alpha-1} - \lambda_-^{\alpha-1})) \\ + tC\Gamma(-\alpha)((\lambda_+ - h - iu)^\alpha - (\lambda_+ - h)^\alpha \\ + (\lambda_- + h + iu)^\alpha - (\lambda_- + h)^\alpha)). \end{aligned} \tag{11.25}$$

By comparing (11.25) with (11.22) one can see that the Q_h return process is still a CTS process with

$$\begin{aligned} \alpha^{Q_h} &= \alpha, C^{Q_h} = C, \\ \lambda_+^{Q_h} &= \lambda_+ - h, \\ \lambda_-^{Q_h} &= \lambda_- + h, \end{aligned}$$

and the risk-neutral drift is equal to the physical one; that is,

$$\begin{aligned} m^{Q_h} &- C\Gamma(1-\alpha)({\lambda_+^{Q_h}}^{\alpha-1} - {\lambda_-^{Q_h}}^{\alpha-1}) \\ &= m - C\Gamma(1-\alpha)(\lambda_+^{\alpha-1} - \lambda_-^{\alpha-1}). \end{aligned}$$

Clearly equation (11.25) is a particular case of equation (11.23).

11.4.3. Matlab code — Characteristic exponent

The characteristic function ϕ is built by following equation (11.7); that is, the parameter k in the function implementing the characteristic exponent ψ in the below script takes into consideration the parameters r and d, as well as the convexity correction $\psi(-i)$. The characteristic exponent is given in the following scripts. In the CTS case, the *chExp_CTS_Optimization* function calls *chExp_CTS*. This latter function is reused in *MCprice* and *getARPEandKS*. While in the function *chExp_CTS* the truncation function $c(x) = x$ is chosen, in both *chExp_NTS* and *chExp_GH* the truncation function $c(x) = xI_{|x|<1}$ is selected. For this reason a slightly modification is needed in the CTS case if the Esscher transform is considered.

```
% risk-neutral characteristic exponent for the CTS model

function y = chExp_CTS_Optimization(u, param, dt, r, d, method)

C     = param(1);
lm    = param(2);
lp    = param(3);
alpha = param(4);

if strcmp(method,'Esscher')
    c1 = C*gamma(1-alpha)*(lp^(alpha - 1) - lm^(alpha - 1));
    k = r - d  - chExp_CTS(-1i, [C lm lp alpha c1], 1);
    m = k + c1;
    y = chExp_CTS(u, [C lm lp alpha m], dt);
else
    m = r - d - chExp_CTS(-1i, [C lm lp alpha 0], 1);
    y = chExp_CTS(u, [C lm lp alpha m], dt);
end

return;
```

```
% characteristic exponent for the CTS model

function phi = chExp_CTS(u, param, dt)

C = param(1);
lm = param(2);
lp = param(3);
alpha = param(4);
m = param(5);

c1 = C*gamma(1-alpha)*(lp^(alpha - 1) - lm^(alpha - 1));

phi = dt*( 1i*u*(m - c1) ...
      + C*gamma(-alpha)*((lp - 1i*u).^alpha - lp^alpha ...
                 + (lm + 1i*u).^alpha  - lm^alpha ));
return;
```

Similar arguments hold in the NTS and the GH cases. The corresponding scripts are given below. The script *chExp_GH_Optimization* is not reported since it is very similar to *chExp_NTS_Optimization.*

```
% characteristic exponent for the NTS model

function y = chExp_NTS(u, param, dt)

a      = param(1);
lambda = param(2);
C      = param(3);
sigma  = param(4);
theta  = param(5);
k     = param(6);

y = dt*(1i*k.*u + C*gamma(-a/2)*...
    ((lambda - 1i*theta*u + sigma^2*u.^2/2).^(a/2) - ...
        lambda^(a/2)));

 return;
```

```
% risk-neutral characteristic exponent for the NTS model

function y = chExp_NTS_Optimization(u, param, dt, r, d, ~)

a      = param(1);
lambda = param(2);
C      = param(3);
```

```
sigma  = param(4);
theta  = param(5);
k = r - d - chExp_NTS(-1i, [a lambda C sigma theta 0], 1);

y = chExp_NTS(u,  [a lambda C sigma theta k], dt);

 return;
```

```
% characteristic exponent for the GH model

function phi = chExp_GH(u, param, dt)

lambda = param(1);
chi    = param(2);
psi    = param(3);
sigma  = param(4);
theta  = param(5);
k      = param(6);

xau = chi*(psi - 2*(1i*u*theta - (1/2)*u.^2*sigma.^2));
xbu = chi*psi;

phi =    dt*(1i*u*k ...
          - lambda/2*log(1 - 2*(1i*u*theta - ...
              (1/2)*u.^2*sigma.^2)/psi) ...
          + log(besselk(lambda, sqrt(xau))) ...
          - log(besselk(lambda, sqrt(xbu))));

return;
```

11.5. The Implied Volatility Smile

In contrast to the classical Black–Scholes case, continuous-time Lévy models are different from Brownian motion because it is very often not possible to find an explicit formula for European call and put option prices. However, it is not necessary to resort to Monte Carlo simulations since, due to the closed-form solution for the characteristic function of the driving processes we analyze in this chapter, we can follow the widely-known analytical (up to an integration) pricing method for standard vanilla options proposed by Carr and Madan (1999) (see also Schoutens (2003) and Cherubini *et al.* (2010)). Let ρ

be a positive constant such that the ρth moment of the price exists and ϕ the characteristic function of the random variable $\log S_T$.

As observed by Schoutens (2003), a value of $\rho = 0.75$ will typically do fine. The price of a call option at inception $t = 0$ is given by the following equality:

$$C_0 = \frac{\exp\left(-\rho \log K\right)}{\pi} \int_0^\infty \exp(-iv \log K)\varrho(v)dv, \tag{11.26}$$

where

$$\varrho(v) = \frac{\exp(-rT)\phi(v - (\rho+1)i)}{\rho^2 + \rho - v^2 + i(2\rho+1)v}.$$

Equation (11.26) holds if the integrand function is analytical on the horizontal strip of the complex plane, on which the line $L_\rho = \{x+i\rho \in \mathbb{C} | -\infty < x < \infty\}$ lies (see Lukacs (1970) for a detailed explanation of the properties of characteristic functions, and Lewis (2001) for a detailed description of this option pricing formula).

11.5.1. Matlab code — Implied volatility

Here we show the Matlab script to compute both call option prices and corresponding implied volatilities starting from the characteristic function of the underlying distribution for log-returns.

Therefore, given the characteristic exponent, one can compute the call option prices and derive the corresponding smile by using the following code. We use a function found on the Paul Wilmott website (see http://www.wilmott.com) to compute the implied volatilities from option prices. We refer to it as *impvolWilmott.* This function is much faster than the standard Matlab *blsimpv* function. By selecting *ShowFigure* = 1, the script allows one to draw the smile or the surface, depending on how many different maturities T are given as input.

```
% evaluation of call option prices and implied volatilities

function [EstimatedImpVol, Call] = RiskNeutralCallImpVol ...
    (param, CallMatrix, chExp, model)
```

```
S0       = CallMatrix(:,1);  % stock price at time 0
K        = CallMatrix(:,2);  % strike prices
T        = CallMatrix(:,3);  % maturity in year
r        = CallMatrix(:,4);  % constant interest rate annualized
d        = CallMatrix(:,5);  % constant dividend rate annualized
ImpVol   = CallMatrix(:,6);  % market implied vol

[MarketCall, ~] = blsprice(S0, K, r, T, ImpVol/100, d);

x   = log(K);
M   = 5000;
n   = 1:M;
du  = 1/5;

% Carr and Madan(1998) option pricing formula

kk  = exp( -1i.*du.*n.'*x.');

fty = rho( du.*n, param, K, S0, T, r, d, chExp, model);
a0  = rho( 0, param, K, S0, T, r, d, chExp, model);

Call = 1/(2.*pi).*2.*real(diag(du.*(fty*kk))+du.*a0/2);

Call(Call<0) = 0.01;

CP = 0;

if CP == 0

Put = Call - S0.*exp(-d.*T) + K.*exp(-r.*T);
Put(Put<0) = 0;

[nrow, ~] = size(CallMatrix);
EstimatedImpVol = zeros(nrow,1);

indexPut    = K<S0;
indexCall   = K>=S0;

EstimatedImpVol(indexPut) = blsimpv(S0(indexPut), K(indexPut), ...
    r(indexPut), T(indexPut), Put(indexPut), 1, d(indexPut), ...
    [], {'Put'})*100;
EstimatedImpVol(indexCall) = blsimpv(S0(indexCall), ...
    K(indexCall), r(indexCall), T(indexCall), Call(indexCall), ...
    1, d(indexCall), [], {'Call'})*100;

elseif CP == 1

    EstimatedImpVol = blsimpv(S0, K, r, T, Call, 1, d, [], ...
        {'Call'})*100;
```

```
elseif CP == 2

    EstimatedImpVol = impvolWilmott(Call,S0,K,r,T,d)*100;

end

     ShowFigure = 0;

     if ShowFigure == 1

     lx = length(unique(K));
     ly = length(unique(T));

         if ly > 1
             % plot the implied volatility surface
             h  = figure(100);
             subplot(2,2,1)
             plot(K./S0, Call,'rx', K./S0, MarketCall,'bo');
             subplot(2,2,2)
             axis tight
             plot(1:length(EstimatedImpVol), ...
                 EstimatedImpVol,'rx',...
                  1:length(ImpVol), ImpVol,'bo');
             title(['Prices - ',model])
             subplot(2,2,[3 4])
             X = sort(unique(K))/S0(1);
             Y = sort(unique(T));
             surf(X, Y, reshape(ImpVol - EstimatedImpVol,lx,ly)')
             colormap('autumn')
             xlim([min(X) max(X)])
             ylim([min(Y) max(Y)])
             title(['Implied volatility - ',model])
             refreshdata(h,'caller')
             drawnow; pause(0.01)
         elseif ly == 1
             % plot the implied volatility smile
             h  = figure(100);
             subplot 121
             plot(K./S0, Call,'kx', K./S0, MarketCall,'ko');
             xlabel('moneyness')
             ylabel('price')
             grid on
             subplot 122
             axis tight
             plot(K./S0, EstimatedImpVol,'kx',...
                  K./S0, ImpVol,'ko');
             xlabel('moneyness')
             ylabel('implied volatility')
             legend('model','market')
```

```
                grid on
                refreshdata(h,'caller')
                drawnow; pause(0.01)
           end
      end
return;

function Q = rho( v, param, K, S0, T, r, d, chExp, model)

alpha    = 0.75;
x        = log(K);
xi       = (v - (alpha+1)*1i);

if strcmp(model,'GH') || strcmp(model,'NTS') || ...
     strcmp(model,'CTS')
% in case r, d and the convexity correction are already in the chf
     omega    = 0;
     r        = 0;
     d        = 0;
else
% in case they are not considered
     omega    = feval(chExp, -1i, param, 1);
end

if strcmp(model,'NTS') || strcmp(model,'CTS')
B        = feval(chExp, xi, param, 1);
elseif strcmp(model,'GH')
B        = feval(chExp, xi, param);
end

psi      = exp( 1i.*xi.'*log(S0).' + ...
     1i.*xi.'*(T.*((r-d)-omega))'+ B.'*T.');

k = (((alpha.^2 + alpha - v.^2 + 1i*(2.*alpha+1).*v)).^(-1));

Q = k.'*(exp(-alpha.*x - r.*T))'.*psi;
Q = Q.';

return;
```

11.6. Risk-Neutral Calibration

There are two possible methodologies to estimate a continuous-time option pricing model: (1) one can fit the model to the daily implied volatilities calculated from market prices and check both the model capabilities and the stability of the parameters, or (2) one can calibrate both daily returns and implied volatilities by using a filter. The Kalman filter and its extensions can be taken into consideration when the model is Gaussian, otherwise a more sophisticated framework has to be applied (e.g., Christoffersen *et al.* (2010)).

The first approach can be viewed as a short-term static estimation, for the purpose of pricing and hedging on a daily basis, and for this reason we refer to it as the *market maker approach.* Under this approach, at a given point in time, model prices have to be as close as possible to the observed market prices and even if the market is quoting unreasonable prices, market makers have to find the parameters that replicate those prices. The market maker needs to achieve static consistency in order to provide at each given point in time two-sided quotes.

The second approach is a long-term dynamic estimation where the stock price process is estimated by filtering the information coming from the stock market by looking at the underlying under the historical measure and from the derivatives market by looking at the implied volatilities under the risk-neutral measure for the purpose of assessing the long-term behavior. Estimated model parameters can be used to make projections on price movements and then used to trade the difference between the model predictions and the market quotes. These investment strategies (known as *statistical arbitrage strategies*) are commonly used by hedge funds and proprietary trading desks and, for this reason, we refer to this second estimation methodology as the *long-term convergence trader approach.* Under this second approach, one assumes that the model can beat the market, or, more precisely, one trusts the model prices and tries to find possible arbitrage opportunities by looking at the differences between model and market prices. Dynamic consistency is important for long-term convergence trading and pricing errors represent trading opportunities (see Wu (2008)).

Here we implement the market maker approach. From a practical perspective, on each trading day we solve the following minimization problem:

$$\widehat{\Theta}^Q = \min_{\Theta^Q} \mathrm{ARPE}(\Theta^Q), \tag{11.27}$$

where the average relative percentage error (ARPE) is given by

$$\begin{aligned}\mathrm{ARPE}(\Theta^Q) &= \frac{1}{\text{number of observations}} \\ &\quad \times \sum_{T_n}\sum_{K_m} \frac{|\mathrm{iVol}^{\text{market}}_{T_n K_m} - \mathrm{iVol}^{\text{model}}_{T_n K_m}(\Theta^Q)|}{\mathrm{iVol}^{\text{market}}_{T_n K_m}},\end{aligned} \tag{11.28}$$

where T_n and K_m are the different maturities and strikes, and Θ^Q is the parameter vector according to a given model under the risk-neutral measure Q.

In practice, we want to find Θ^Q such that the model implied volatility ($\mathrm{iVol}^{\text{model}}$) is as close as possible to the market implied volatility ($\mathrm{iVol}^{\text{market}}$). Since the minimization of equation (11.27) with respect to the parameters vector Θ^Q has neither a closed-form solution nor a global minimum, a numerical optimization routine is needed to find a local minimum. We use the Matlab r2016b function *fmincon* for the optimization routine.

11.6.1. Matlab code — Smile calibration

In this section, we show how to implement the optimization algorithm to calibrate the risk-neutral parameters of the GH, NTS, and CTS models. We consider the 1-month smile of Deutsche Bank options on April 30, 2017. Since we considered dividend-adjusted closing prices, we assume that $d = 0$. Option data are given in the variable *CallMatrix*. The optimization problem is given in equation (11.27); that is, we find the model parameters minimizing the distance between the theoretical and the market observed smile. During the optimization, the script provides the graph of both prices and implied volatilities as in Figure 11.1. The estimated risk-neutral parameters together with the calibration errors are reported in Table 11.1. The estimated parameters strictly depend on the starting point x_0, as well as on the bounds *lb* and *ub*.

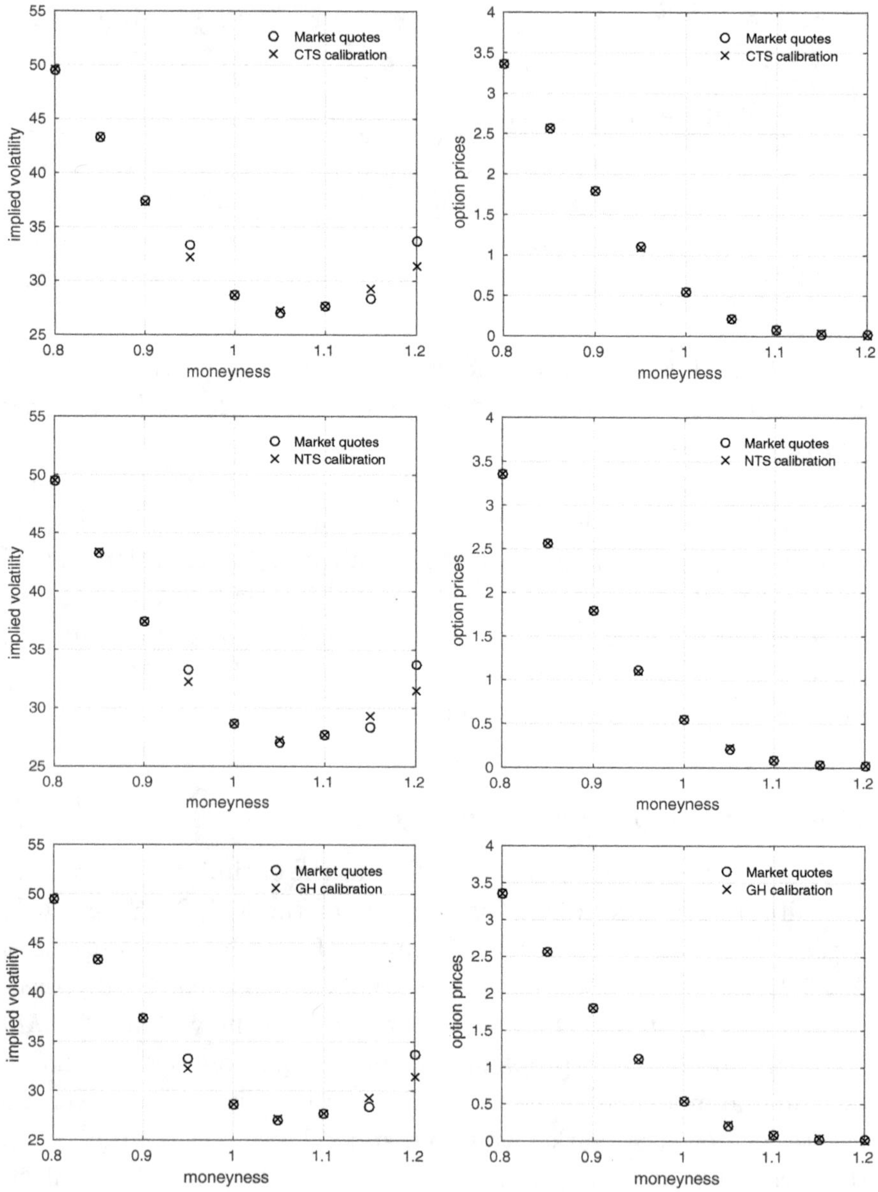

Figure 11.1. Calibration results for the CTS, NTS, and GH models: model and market implied volatilities are shown on the left, model and market prices are shown on the right.

Table 11.1. Estimated parameters and corresponding estimation errors. Three (two) different calibration methods are analyzed for the CTS (NTS and GH) model: (a1) and (a2) risk-neutral calibration (in the CTS case we consider the two alternative parameterizations), (b1) joint calibration with the general change of measure, and (b2) joint calibration with the Esscher transform. The ARPE and the KS statistic with the corresponding p-value are reported. While in the first case ($w = 1$) the algorithm is more focused on minimizing the calibration error of the smile fitting, in the second case ($w = 2$) the algorithm is more focused on minimizing the estimation error of the log-returns.

				C	λ_-	λ_-	α	m	k	ARPE	KS
CTS		(a1)	Q	0.100	1.216	9.097	1.310			1.63%	
		(a2)	Q	0.100	1.218	9.099	1.310			1.63%	
	($w = 1$)	(b1)	Q	0.024	1.002	41.561	1.751			9.54%	
			P	0.024	0.378	41.796	1.751	−0.131			0.029(0.379)
		(b2)	Q	0.035	1.002	32.373	1.674		−0.022	7.58%	
			P_h	0.035	1.336	32.039	1.674		−0.022		0.055(0.006)
	($w = 2$)	(b1)	Q	0.025	1.004	26.054	1.745			10.35%	
			P	0.025	0.422	26.297	1.745	−0.131			0.025(0.572)
		(b2)	Q	0.022	1.018	22.267	1.762		−0.022	10.55%	
			P_h	0.022	1.352	21.933	1.762		−0.022		0.028(0.440)
				a	λ_-	C	σ	θ	k	ARPE	KS
NTS		(a2)	Q	1.424	0.147	1.470	0.111	−0.049		1.55%	
	($w = 1$)	(b2)	Q	1.750	0.060	1.291	0.082	−0.048	0.589	3.62%	
			P_h	1.750	0.108	1.291	0.082	−0.041	0.589		0.085(0.000)
	($w = 2$)	(b2)	Q	1.714	1.019	1.989	0.092	−0.041	0.469	13.90%	
			P_h	1.714	1.057	1.989	0.092	−0.032	0.469		0.026(0.548)
				λ	χ	ψ	σ	θ	k	ARPE	KS
GH		(a2)	Q	−0.818	0.074	0.015	0.027	−0.003		1.54%	
	($w = 1$)	(b2)	Q	−1.492	0.296	0.010	0.035	−0.014	0.004	5.29%	
			P_h	−1.492	0.296	0.037	0.035	−0.013	0.004		0.091(0.000)
	($w = 2$)	(b2)	Q	−2.604	0.267	0.047	0.072	−0.030	0.002	14.75%	
			P_h	−2.604	0.267	0.101	0.072	−0.025	0.002		0.017(0.943)

Note that in the CTS case we consider two different parameterizations given by (11.13) and (11.22). We empirically show that the calibration error does not depend on the parameterizations.

```
% calibration of risk-neutral parameters and draw the results

CallMatrix = [16.53 13.224  0.0833 -0.00329 0 49.44;
              16.53 14.0505 0.0833 -0.00329 0 43.29;
```

```
              16.53 14.877  0.0833 -0.00329 0 37.40;
              16.53 15.7035 0.0833 -0.00329 0 33.26;
              16.53 16.53   0.0833 -0.00329 0 28.62;
              16.53 17.3565 0.0833 -0.00329 0 26.99;
              16.53 18.183  0.0833 -0.00329 0 27.65;
              16.53 19.0095 0.0833 -0.00329 0 28.33;
              16.53 19.836  0.0833 -0.00329 0 33.71];

dt = 1/250;

S0      = CallMatrix(1,1);
K       = CallMatrix(:,2);
T       = CallMatrix(1,3);
r       = CallMatrix(1,4);
d       = CallMatrix(1,5);
ImpVol  = CallMatrix(:,6);

[MarketCall, ~] = blsprice(S0, K, r, T, ImpVol/100, d);

model = 'CTS';
method = 'Esscher';

if strcmp(model,'CTS')
   lb =[1e-2 1 1 0.75];
   ub =[0.1   100   100   1.9];
   x0 = [0.05 10 10 1.7];
elseif strcmp(model,'NTS')
   lb =[0.75  1e-2 1e-2 0.01 -0.15];
   ub =[1.75  5    2  0.25   0.01];
   x0 = [1.25 1 10 0.1 -0.02];
elseif strcmp(model,'GH')
   lb =[-4.5 1e-2 1e-2 0.01  -0.15];
   ub =[-0.5 5    2    0.25  0.01];
   x0 =[-4.5   0.2    3    0.1    0];
end

options = optimset('TolFun',1e-15,'TolCon',1e-15,'TolX',1e-15,...
                   'MaxFunEvals', 1500,'Display','iter',...
                   'Algorithm','interior-point','UseParallel', ...
                       'always');

[RiskNeutralParameters, EstimatedARPE] = ...
 fmincon(@(x) getARPE(x,CallMatrix,model,dt, ...
     method),x0,[],[],[],[],lb,ub,[],options);

if strcmp(model,'NTS')
chExp = @(u, param, dt) chExp_NTS_Optimization(u, param, dt, ...
    r, d, method);
elseif strcmp(model,'GH')
```

```
chExp = @(u, param) chExp_GH_Optimization(u, param, 1/dt, ...
    r*dt, d*dt, method);
elseif strcmp(model,'CTS')
chExp = @(u, param, dt) chExp_CTS_Optimization(u, param, dt, ...
    r, d, method);
end

[ChfImpVol, ChfCall] = ...
    RiskNeutralCallImpVol(RiskNeutralParameters, CallMatrix, ...
        chExp, model);

getPlotCalibration
```

Note that in the GH case, one has to slightly modify this by a factor of dt, the input of the characteristic exponent. As discussed in Bianchi *et al.* (2016) and in Chapters 6 and 7, from a theoretical standpoint, while for CTS and NTS processes a change of time scale leads to a change of spatial scale of the distribution of the increments, for GH processes a change of time scale modifies the distributional properties of the increments (see Cont and Tankov (2003)). This means that if the distribution of the increments of a process on a daily basis is GH, the distribution of the increments of the same process at other time scales is no longer GH, thereby making it inconvenient when working with data on different time scales. This inconvenience comes from generalized inverse Gaussian processes and it may occur in practical applications such as in option pricing models that combine the information provided by daily log-returns observations and implied volatilities, usually computed on a yearly basis as done in Section 11.7 (see also Tassinari and Bianchi (2014) and Bianchi and Tassinari (2018)).

As an alternative it would be possible to write a smarter code by using a function handle to pass the *chExp* to the function implemented to compute the implied volatility error under the risk-neutral measure (i.e., *getARPE* and *getARPEandKS*), but for clarity purposes we write the code by using *if* and *elseif* statements in the scripts provided here and in Section 11.7.

```
% find ARPE for NTS, GH and CTS models

function y = getARPE(param, CallMatrix, model, dt, method)

r            = CallMatrix(1,4);  % constant interest rate
d            = CallMatrix(1,5);  % constant dividend rate
MarketImpVol = CallMatrix(:,6);  % market implied vol

n = length(MarketImpVol);

if strcmp(model,'NTS')
chExp = @(u, param, dt) chExp_NTS_Optimization(u, param, dt, ...
    r, d, method);
elseif strcmp(model,'GH')
chExp = @(u, param) chExp_GH_Optimization(u, param, 1/dt, ...
    r*dt, d*dt, method);
elseif strcmp(model,'CTS')
chExp = @(u, param, dt) chExp_CTS_Optimization(u, param, dt, ...
    r, d, method);
end

EstimatedImpVol = RiskNeutralCallImpVol(param, CallMatrix, ...
    chExp, model);

if sum(isnan(EstimatedImpVol)) || ...
    sum(isfinite(EstimatedImpVol) == 0)
   disp('NaN or Inf in implied volatility')
   y = 10^7;
else
   y = 1/n * sum(abs(MarketImpVol - ...
       EstimatedImpVol)./MarketImpVol)*10000;
end

return
```

Finally, in Figure 11.1 we show the results of the calibration. On the left (right), we report the comparison between market quoted implied volatilities (prices), and model estimated ones. The figures are drawn with the script *getPlotCalibration*.

```
% plot risk-neutral joint calibration
% option prices and implied volatilities


scrsz = get(0,'ScreenSize');
h4 = figure('Position',[50 50 scrsz(3)/2 scrsz(4)/3]);
subplot 121
```

```
plot(K/S0, ImpVol, 'ko',...
     K/S0, ChfImpVol,'kx')
xlabel('moneyness');
ylabel('implied volatility')
legend('Market quotes',[model, ' calibration'])
legend boxoff
grid on
box on
axis tight
ylim([25 55])
subplot 122
plot(K/S0, MarketCall, 'ko',...
     K/S0, ChfCall,'kx')
xlabel('moneyness');
ylabel('option prices')
legend('Market quotes',[model, ' calibration'])
legend boxoff
grid on
box on
axis tight
ylim([0 4])
```

11.7. Historical and Risk-Neutral Joint Calibration

The failure to explain observed option prices by considering only time series information is well known (see Chernov and Ghysels (2000)). For this reason, in this section we consider a calibration framework in which we jointly estimate the model parameters on the time series of log-returns, by minimizing (1) the Kolmogorov–Smirnov distance (under the historical measure) and (2) the ARPE in the calibration of the implied volatility surface (under the risk-neutral measure). We refer to this method as *double calibration* (Tassinari and Bianchi (2014) and Bianchi and Tassinari (2018)). By following this approach, both the implied volatility and log-return calibration errors are jointly minimized.

From a practical perspective, on each trading day we solve the following minimization problem:

$$\widehat{\Theta}^Q = \min_{\Theta^Q}\left(\frac{\mathrm{ARPE}(\Theta^Q) + w\mathrm{KS}(\Theta^P)}{w+1}\right), \tag{11.29}$$

where the ARPE is given in equation (11.28), KS is the Kolmogorov–Smirnov distance between the theoretical and the empirical historical distribution functions estimated through an appropriate change of measure allowing one to derive the set of parameters Θ^P from the risk-neutral one Θ^Q. In particular, given a set of risk-neutral parameters, we find the historical parameters using the change of measure described in equation (11.8). Finally $w \in \mathbb{R}_+$ is a positive number representing the contribution of the historical estimation to the overall calibration. For example, if we select $w = 1$ more emphasis is given in fitting the implied volatility smile. Conversely, if we select $w = 2$ more emphasis is given in fitting the log-returns. We will show the results in both cases.

The KS distance measures the absolute value of the maximum distance between the empirical distribution function $\hat{F}$ and the theoretical distribution function F, putting equal weight on each observation,

$$\mathrm{KS} = \sup_{x_i} |F(x_i) - \hat{F}_n(x_i)|, \tag{11.30}$$

where $\{x_i\}_{1 \leq i \leq n}$ is a given set of observations. Based on this distance, a non-parametric test can be defined where the null hypothesis is that the population distribution of the data is equal to the hypothesized distribution. This test is implemented in the Matlab function *kstest*.

Note that a similar approach can be considered in a multivariate setting. For example, in Tassinari and Bianchi (2014) by considering the EM-based maximum likelihood estimation method together with the multivariate Esscher transform needed to find the connection between historical and risk-neutral parameters, a joint calibration/estimation of the univariate option surfaces, and of the time series of log-returns has been proposed for both the multivariate variance gamma (MVG) and multivariate normal inverse Gaussian (MNIG) model. Bianchi and Tassinari (2018) extend the framework to multivariate generalized hyperbolic (MGH) and multivariate normal tempered stable (MNTS) models to perform a portfolio optimization exercise. A similar calibration procedure which rests on a joint calibration of univariate option surfaces and pairwise correlations has been introduced by Guillaume (2012, 2013). In all cases briefly described above, the calibrated parameters then can be used to price multiunderlying derivatives as well as to perform portfolio optimization problems.

11.7.1. Matlab code — Joint calibration

In this section, we show how to implement the optimization algorithm to calibrate both the risk-neutral and historical parameters of the CTS model. We consider the 1-month smile of Deutsche Bank on April 30, 2017 together with the time series of log-returns (i.e., 1000 observations ending on April 30, 2017). Since we considered dividend-adjusted closing prices, we assumed that $d = 0$. Option data are given in the variable *CallMatrix* and log-returns in the variable *logReturns*. The optimization problem is given in equation (11.29); that is, we consider a calibration framework in which we jointly calibrate the implied volatility surface (under the risk-neutral measure) by minimizing the ARPE, and estimate the model parameters on the time series of log-returns (under the historical measure) by minimizing the Kolmogorov–Smirnov distance. Thus, by following this approach both the implied volatility and log-return calibration errors are estimated simultaneously. As starting point in the optimization algorithm we consider the parameters estimated in Section 11.6.1.

```
% perform a joint calibration and draw the results

% load Deutsche Bank stock prices
load('historicalDataDB', 'stockPrices')
logReturns   = tick2ret(stockPrices,[], 'Continuous');
logReturns(logReturns==0) = [];

model = 'GH';
method = 'Esscher';
weight = 2;

if strcmp(model,'CTS')
   lb =[1e-2 1 1 0.75];
```

```
   ub =[0.1   100   100   1.9];
   x0 =[0.09976 1.2165 9.0970 1.3105];
elseif strcmp(model,'NTS')
   lb =[0.75  1e-2 1e-2 0.01 -0.25];
   ub =[1.75  5  2 0.25 0.01];
   x0 =[1.4241 0.1466 1.4699 0.1112 -0.0516];
elseif strcmp(model,'GH')
   lb =[-4.5 1e-2 1e-2 0.01  -0.15] ;
   ub =[-0.5 5    2     0.25  0.01];
   x0 =[-0.8177 0.0736 0.0149 0.0273 -0.0030];
end

options = optimset('TolFun',1e-20,'TolCon',1e-20,'TolX',1e-20,...
                  'MaxFunEvals', 1500,'Display','iter',...
                  'Algorithm','interior-point','UseParallel', ...
                      'always');

[RiskNeutralParameters, EstimatedARPE] = ...
 fmincon(@(x) getARPEandKS(x,CallMatrix,logReturns,model, ...
     dt,method, weight),...
         x0,[],[],[],[],lb,ub,[],options);

[y, ARPE, KS, KS_p, fval, HistoricalParameters, paramCM] = ...
getARPEandKS(RiskNeutralParameters,CallMatrix,logReturns,model,dt,
  method,weight);

S0       = CallMatrix(1,1);
K        = CallMatrix(:,2);
T        = CallMatrix(1,3);
r        = CallMatrix(1,4);
d        = CallMatrix(1,5);

if strcmp(model,'NTS')
   chExp = @(u, param, dt) chExp_NTS_Optimization(u, param, ...
       dt, r, d, method);
elseif strcmp(model,'GH')
   chExp = @(u, param) chExp_GH_Optimization(u, param, 1/dt, ...
       r*dt, d*dt, method);
elseif strcmp(model,'CTS')
   chExp = @(u, param, dt) chExp_CTS_Optimization(u, param, ...
       dt, r, d, method);
end

[ChfImpVol, ChfCall] = ...
RiskNeutralCallImpVol(RiskNeutralParameters, CallMatrix, ...
    chExp, model);

getPlotJointCalibration
```

In the *jointCalibration* script, the optimization problem (11.29) is solved. The objective function is implemented in a separate script *getARPEandKS*. The connection between the two measures is computed by finding the historical parameters satisfying equation (11.24) and (11.14) in the CTS case, equation (11.15) in the NTS case, and equation (11.17) in the GH case. This is implemented by using the Matlab *fsolve* function. The objective function of this root-finding algorithm is provided in the function *getHistoricalParameters*. The functions *cdfEvaluation* and *evaluateKS* are called in the script. While the latter is given here, the former has been already provided in Chapter 6. In the GH case, to speed up the calibration it is possible to compute the probability density function in closed form, by leveraging on the function *densityMGH_matrix* provided in Chapter 7 and used here under a univariate framework. In all other cases, the probability density function is computed by means of the fast Fourier transform (FFT) methods. Note that in the GH case, one has to slightly modify by a factor of *dt* the input of the characteristic exponent (see the assignment of the function handler *chExp*). The estimated risk-neutral and historical parameters together with the calibration errors (ARPE and KS) are reported in Table 11.1. For the reason explained above, while in the GH case the parameters are on a daily basis, in both the CTS and the NTS case they are annualized. We consider two different values for the parameter *weight* in the script *jointCalibration* (w in the script *getARPEandKS*): $weight = 1$ and $weight = 2$.

```
% find ARPE and KS distance for the NTS, GH and CTS models
% (for the CTS model we consider two different changes of measure)

function [y, ARPE, KS, KS_p, fval, paramH, paramCM] = ...
   getARPEandKS(paramRN, CallMatrix, logReturns, model, dt, ...
        method, weight)

r             = CallMatrix(1,4);   % constant interest rate
d             = CallMatrix(1,5);   % constant dividend rate
MarketImpVol  = CallMatrix(:,6);   % market implied vol

n = length(MarketImpVol);
```

```
meanH = mean(logReturns)/dt;
if strcmp(model,'NTS')
chExp = @(u, param, dt) chExp_NTS_Optimization(u, param, dt, ...
    r, d, method);
elseif strcmp(model,'GH')
meanH = meanH*dt;
r = r*dt;
d = d*dt;
chExp = @(u, param) chExp_GH_Optimization(u, param, 1/dt, r, ...
    d, method);
elseif strcmp(model,'CTS')
chExp = @(u, param, dt) chExp_CTS_Optimization(u, param, dt, ...
    r, d, method);
end

EstimatedImpVol = RiskNeutralCallImpVol(paramRN, CallMatrix, ...
    chExp, model);

if sum(isnan(EstimatedImpVol)) || ...
     sum(isfinite(EstimatedImpVol) == 0)
   disp('NaN or Inf in implied volatility')
   ARPE = 10;
else
   ARPE = 1/n * sum(abs(MarketImpVol - ...
       EstimatedImpVol)./MarketImpVol);
end

if strcmp(method,'Esscher')
x0 = [-1 paramRN(2:3)];
else
x0 = paramRN(2:3);
end

% find the historical parameters, given the risk-neutral ones

options = optimset('MaxFunEvals',1500,'TolFun',1e-20,'TolX', ...
    1e-20,...
                    'Display','off','Algorithm', ...
                        'levenberg-marquardt');

[paramCM, fval]  = fsolve(@(x) getHistoricalParameters(x, ...
    paramRN,...
                                r, d, meanH, model, method), ...
                                    x0, options);

if sum(imag([paramCM, fval])) > 1
   disp('error in the change of measure')
end

if strcmp(method,'Esscher')
```

```
[~, kH]   = getHistoricalParameters(paramCM, ...
                              paramRN, r, d, meanH, model, method);
kH = real(kH);
end


if strcmp(model,'CTS')

    if strcmp(method,'Esscher')
    % h = real(paramCM(1));
    C = paramRN(1);
    lmH = real(paramCM(2));
    lpH = real(paramCM(3));
    alpha = paramRN (4);
    c1 = C*gamma(1-alpha)*(lpH^(alpha - 1) - lmH^(alpha - 1));
    paramH = [C lmH lpH alpha kH + c1];
    else
    C = paramRN(1);
    lmH = real(paramCM(1));
    lpH = real(paramCM(2));
    alpha = paramRN (4);
    paramH = [C lmH lpH alpha meanH];
    end

    chf = @(z) chf_CTS(z, paramH, dt);
    cdf = @(x) cdfEvaluation(x, chf, 'chf');

elseif strcmp(model,'NTS')

    if strcmp(method,'Esscher')
    h         = real(paramCM(1));
    a         = paramRN(1);
    lambdaRN  = paramRN(2);
    C         = paramRN(3);
    sigma     = paramRN(4);
    thetaRN   = paramRN(5);

    lambdaH = real(paramCM(2));
    thetaH = real(paramCM(3));

    paramH = [a lambdaH C sigma thetaH kH];
    end

    chf = @(z) exp(chExp_NTS(z, paramH, dt));
    cdf = @(x) cdfEvaluation(x, chf, 'chf');

elseif strcmp(model,'GH')

    if strcmp(method,'Esscher')
    % h         = real(paramCM(1));
```

```
    lambda    = paramRN(1);
    chi       = paramRN(2);
    % psiRN     = paramRN(3);
    sigma     = paramRN(4);
    % thetaRN   = paramRN(5);

    psiH   = real(paramCM(2));
    thetaH = real(paramCM(3));

    paramH = [lambda chi psiH sigma thetaH kH];
    end

    cdfEvaluationMethod = 'pdf';
    if strcmp(cdfEvaluationMethod,'chf')
    chf = @(z) chf_GH(z, paramH, 1);
    cdf = @(x) cdfEvaluation(x, chf, 'chf');
    elseif strcmp(cdfEvaluationMethod,'pdf')
    pdf = @(x) densityMGH_matrix(x, lambda, chi, psiH, kH, ...
        thetaH, sigma^2);
    cdf = @(x) cdfEvaluation(x, pdf,'pdf');
    end
end

[KS,KS_p] = evaluateKS(logReturns, cdf);

y = (ARPE + weight*KS)/(weight + 1)*10^4;

return
```

Here we implement the change of measure conditions provided in Sections 11.4.1 and 11.4.2; that is, given a set of parameter under the risk-neutral measure, we find the corresponding parameters under the historical one.

```
% change of measure conditions

function [y, kH] = getHistoricalParameters(x, param, r, d, ...
    meanH, model, method)

if strcmp(model,'CTS')

    if strcmp(method,'Esscher')

    h = x(1);
    lmH = x(2);
    lpH = x(3);
    C = param(1);
    lmRN = param(2);
```

```
    lpRN = param(3);
    alpha = param(4);

    c1 = C*gamma(1-alpha)*(lpRN^(alpha - 1) - lmRN^(alpha - 1));
    kRN = r - d  - chExp_CTS(-1i, [C lmRN lpRN alpha c1], 1);
    kH = kRN;

    lmQ = lmH + h;
    lpQ = lpH - h;

    y(1) = r - d - kRN ...
        - C*gamma(-alpha)*((lpQ - 1)^alpha - lpQ^alpha ...
                           +(lmQ + 1)^alpha - lmQ^alpha);

    y(2) = lmQ - lmRN;
    y(3) = lpQ - lpRN;

    else

    lmH = x(1);
    lpH = x(2);

    C = param(1);
    lmRN = param(2);
    lpRN = param(3);
    alpha = param(4);

    y = meanH  - r + d + chExp_CTS(-1i, [C lmRN lpRN alpha 0], ...
        1) +...
        C*gamma(1-alpha)*(lpRN^(alpha - 1) - lmRN^(alpha - 1) ...
        - lpH^(alpha - 1) + lmH^(alpha - 1));

    end

elseif strcmp(model,'NTS')

    if strcmp(method,'Esscher')

    h = x(1);
    lambdaH = x(2);
    thetaH = x(3);
    a        = param(1);
    lambdaRN = param(2);
    C        = param(3);
    sigma    = param(4);
    thetaRN  = param(5);

    kRN = r - d - chExp_NTS(-1i, [a lambdaRN C sigma thetaRN ...
        0], 1);
    kH = kRN;
```

```
    thetaQ = thetaH + h*sigma^2;
    lambdaQ = lambdaH - h*thetaH - (1/2)*h^2*sigma^2;

    xa = (lambdaQ - thetaQ - (1/2)*sigma^2)^(a/2);
    xb = (lambdaQ)^(a/2);

    y(1) = r - d - kRN - C*gamma(-a/2)*(xa - xb);
    y(2) = lambdaQ - lambdaRN;
    y(3) = thetaQ - thetaRN;

    end

elseif strcmp(model,'GH')

    if strcmp(method,'Esscher')

    h = x(1);
    psiH = x(2);
    thetaH = x(3);
    lambda  = param(1);
    chi     = param(2);
    psiRN   = param(3);
    sigma   = param(4);
    thetaRN = param(5);

    kRN = r - d - chExp_GH(-1i, [lambda chi psiRN sigma ...
        thetaRN 0], 1);
    kH = kRN;

    thetaQ = thetaH + h*sigma^2;
    lh = h*thetaH + (1/2)*h^2*sigma^2;
    psiQ = psiH - 2*lh;

    k1 = thetaQ + (1/2)*sigma^2;
    xa = sqrt(chi*(psiQ - 2*k1));
    xb = sqrt(chi*psiQ);

    y(1) = r - d - kRN + lambda/2*log(1 - 2*k1/psiQ) ...
        - log(besselk(lambda, xa)) ...
        + log(besselk(lambda, xb));

    y(2) = psiQ - psiRN;
    y(3) = thetaQ - thetaRN;

    end
end

return
```

In the following script the implementation of the Kolmogorov–Smirnov distance is provided. This script makes use of the Matlab function *kstest* in which both the test static (i.e., the Kolmogorov–Smirnov distance) and the corresponding p-value are computed.

```
% evaluate the Kolmogorov-Smirnov distance

function [ksstat,p] = evaluateKS(data, cdfFunction)

data = sort(data);

cdf = cdfFunction(data);

checkIncreasing = all(diff(cdf) >= 0);

if checkIncreasing == 0
    disp('Non increasing cdf')
    p = 0;
    ksstat = 1;
elseif checkIncreasing == 1
    [~,p,ksstat] = kstest(data,[data cdf]);
end

return
```

Finally, in Figures 11.2 and 11.3 we show the results of the joint calibration with $w = 1$ and with $w = 2$. While in the first case ($w = 1$), the algorithm is more focused on minimizing the calibration error of the smile fitting; in the second case ($w = 2$), the algorithm is more focused on minimizing the estimation error of the log-returns. On the left-hand side of the figures, we report the comparison between market quoted implied volatilities, and model estimated ones. On the right-hand side, we show the histogram of the observed log-returns together with the estimated density function. The figures are drawn with the script *getPlotJointCalibration*.

```
% getPlotJointCalibration.m
% plot joint calibration
% historical and theoretical pdfs

ImpVol   = CallMatrix(:,6);
```

```
hFFT = 0.001;
nFFT = 2^17;

scrsz = get(0,'ScreenSize');
h4 = figure('Position',[50 50 scrsz(3)/2 scrsz(4)/3]);
subplot 121
plot(K/S0, ImpVol, 'ko',...
 K/S0, ChfImpVol,'kx')
xlabel('moneyness');
ylabel('implied volatility')
legend('Market quotes',[model, ' calibration ', method])
legend boxoff
axis tight
ylim([25 55])
% implied volatility
subplot 122
hold on
dx = (max(logReturns) - min(logReturns))/50;
binWidth = dx;
binCtrs = min(logReturns):binWidth:max(logReturns);
n = length(logReturns);
counts = hist(logReturns,binCtrs);
prob = counts / (n * binWidth);
bar(binCtrs,prob,'hist');
h = get(gca,'child');
xlabel('observed returns');
x = linspace(min(logReturns),max(logReturns),500);
if strcmp(model,'CTS')
    chf = @(z) chf_CTS(z, HistoricalParameters, dt);
elseif strcmp(model,'NTS')
    chf = @(z) chf_NTS(z, HistoricalParameters, dt);
elseif strcmp(model,'GH')
    chf = @(z) chf_GH(z, HistoricalParameters, 1);
end
returnsPdf = pdfEvaluation(x, chf, 'pchip', hFFT, nFFT);
plot(x, returnsPdf,'k','LineWidth', 3)
ylabel('probability density function')
grid on
box on
axis tight
axis tight
ylim([0 30])
hold off
```

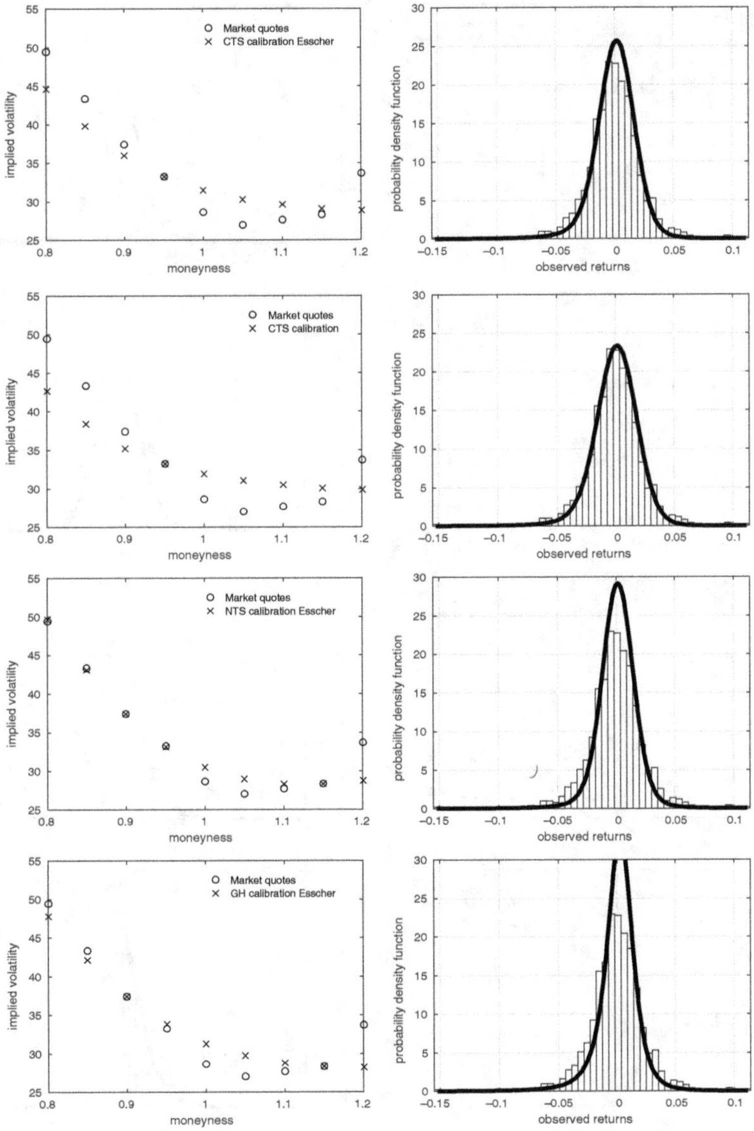

Figure 11.2. Joint calibration results for the CTS, NTS, and GH models: model and market implied volatilities are shown on the left, the histogram of observed log-returns and the estimated density function is shown on the right. For the CTS model we report the two different changes of measure (the Esscher transform and the more general one). This figure is generated at the end of the *jointCalibration* script by *getPlotJointCalibration*. Here $w = 1$ is selected: the algorithm is more focused in minimizing the calibration error of the smile fitting.

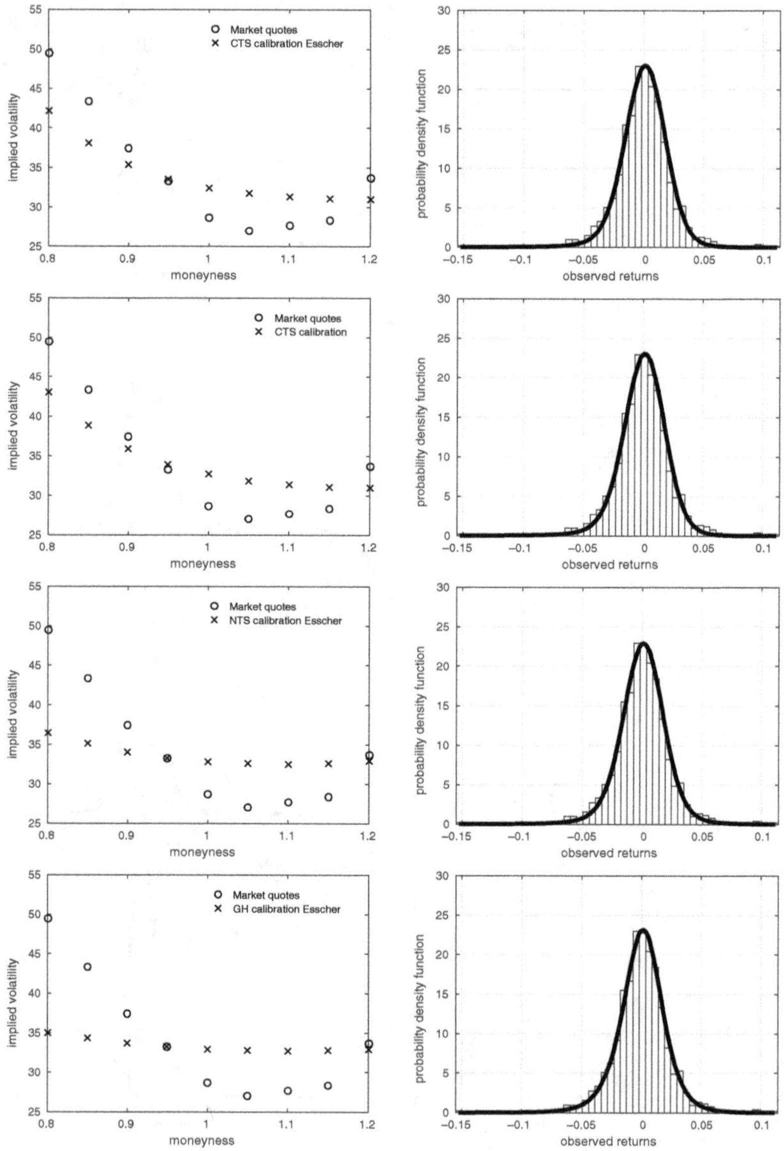

Figure 11.3. Joint calibration results for the CTS, NTS, and GH models: model and market implied volatilities are shown on the left, the histogram of observed log-returns and the estimated density function is shown on the right. For the CTS model we report the two different changes of measure (the Esscher transform and the more general one). This figure is generated at the end of the *jointCalibration* script by *getPlotJointCalibration*. Here $w = 2$ is selected: the algorithm is more focused in minimizing the estimation error of the log-returns.

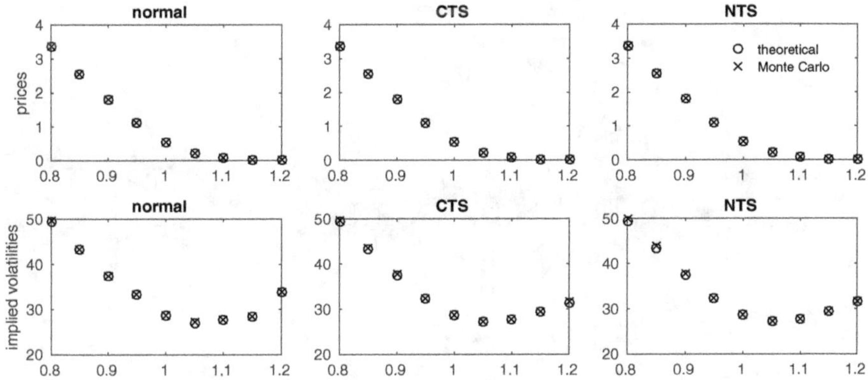

Figure 11.4. Monte Carlo simulations for the normal, CTS, and NTS models for log-returns. While theoretical and simulated prices are shown above, theoretical and simulated implied volatilities are shown below.

11.8. A Monte Carlo Exercise

In this section, we assess the goodness of fit of random number generators proposed in Chapter 6. A brief Monte Carlo study is performed and prices of European call options with different strikes are calculated. By taking into account equation (11.7), the stock price process under a risk-neutral martingale measure Q is of the form

$$S_t = S_0 e^{(r-d-\psi_{J^Q}(-i))t+J_t^Q}, \tag{11.31}$$

where r and d represent the deterministic risk-free rate and the dividend yield and $J^Q = (J_t^Q)_{t\in[0,T]}$ is a Lévy process under Q and ψ_{J^Q} is its characteristic exponent. The price of a European call (see Figure 11.4) can be calculated by discounting with the risk-free rate its expected risk-neutral payoff, that is

$$C_0 = e^{-rT} E_Q[(S_T - K)^+]$$

under the assumption of a constant interest rate. The expectation above is calculated using Monte Carlo simulation, where a given number *MC_paths* of paths is generated.

We analyze three different distributional assumptions: normal, CTS, and NTS (see Figure 11.5). In the normal case, for each moneyness we consider the corresponding market implied volatility. In the

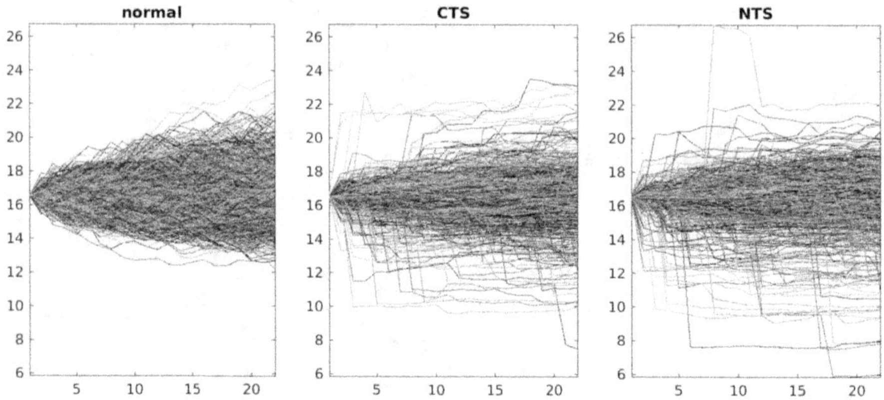

Figure 11.5. Monte Carlo stock prices for the normal, CTS, and NTS models for log-returns.

CTS and the NTS cases, we use the risk-neutral parameters estimated by following the approach defined in Section 11.6. In the CTS case, we set $C = 0.0265$, $\lambda_- = 0.0109$, $\lambda_+ = 38.7949$, $\alpha = 1.7110$, and in the NTS case, $a = 1.4140$, $\lambda = 0.0908$, $C = 2.1508$, $\sigma = 0.0860$, and $\theta = -0.0292$. Furthermore, in all cases we set interest rate $r = -0.00329$, dividend yield $d = 0$, initial stock price $S_0 = 16.53$, and annualized maturity $T = 0.0833$.

11.8.1. Matlab code — Monte Carlo

In this section, we show how to implement a Monte Carlo simulation for three different distributional assumptions for log-returns (normal, CTS, and NTS). The risk-neutral dynamics are simulated to price call options and find the corresponding implied volatilities. The number of paths is one million. However, the computing time is few seconds. In the code we simulate the entire path of the stock and save it in the variable S. By writing the command *plot*(S'), it is possible to see the different simulated trajectories of the price process.

```
% test a Monte Carlo simulation for normal, CTS and NTS ...
    distributions

[Call, ~] = blsprice(S0, K, r, T, ImpVol/100, d);
```

```
dt = 1/250;
MC_paths = 10^6;

model = 'normal';
RiskNeutralParameters = ImpVol/100;
[McImpVolNormal, McCallNormal] = MCprice(S0, K, r, d, T, ...
    RiskNeutralParameters, dt, model, MC_paths);

model = 'CTS';
RiskNeutralParameters = [0,0998 1,2165  9,0970  1,3105];
[McImpVolCTS, McCallCTS] = MCprice(S0, K, r, d, T, ...
    RiskNeutralParameters, dt, model, MC_paths);

chExp = @(u, param, dt) chExp_CTS_Optimization(u, param, dt, ...
    r, d);
[ChfImpVolCTS, ChfCallCTS] = ...
    RiskNeutralCallImpVol(RiskNeutralParameters, CallMatrix, ...
    chExp, model);

model = 'NTS';
RiskNeutralParameters = [1,4241 0,1466  1,4699  0,1112  -0,0485];
[McImpVolNTS, McCallNTS] = MCprice(S0, K, r, d, T, ...
    RiskNeutralParameters, dt, model, MC_paths);

chExp = @(u, param, dt) chExp_NTS_Optimization(u, param, dt, ...
    r, d);
[ChfImpVolNTS, ChfCallNTS] = ...
    RiskNeutralCallImpVol(RiskNeutralParameters, CallMatrix, ...
    chExp, model);
```

In the Monte Carlo pricing it should be noted that the CTS and NTS random numbers simulation is conducted by the inverse transform algorithm. In both cases, the cumulative distribution function F_X is not available in closed form. To find values of F_X, first we have to invert the characteristic function in order to find both the density (f_X) and the cumulative distribution function (F_X), and then find the value x_i satisfying the equality $F_X(x_i) - u_i = 0$, where u_i is the ith draw in a sequence $u_1, \ldots, u_n$ of independent draws from a uniform random variable on $[0, 1]$. The density (cumulative distribution) function is computed by means of *pdfEvaluation* (*cdfEvaluation*).

As described in Chapter 6, in the script *pdfEvaluation* the parameters of the FFT method N (the number of summands in the Riemann sum) and h (the grid spacing) should be carefully chosen in order to have a proper accuracy. Wrong choices may led to a strange shape for the smile obtained through simulation. In this example we select $N = 2^{25}$ and $h = 1\text{e} - 5$, in the CTS case, and $N = 2^{20}$ and $h = 1\text{e} - 4$, in the NTS case. In the script *pdfEvaluation* we select the option *pchip* in the *interp1* Matlab function.

```
% evaluate call option prices and implied volatilities via ...
    Monte Carlo
% simulations

function [ImpVol, Call, S] = MCprice(S0, K, r, q, T, param, ...
    dt, model, MC_paths)

days = round(T/dt);
lK   = length(K);

if strcmp(model,'normal')
    for j = 1:lK
    sigma = param(j);
    mu = r - q - 1/2*sigma^2;
    Z = mu*dt + sigma*sqrt(dt)*randn(MC_paths, days);
    Z = cumsum(Z,2);
    S = S0*exp([zeros(MC_paths,1) Z]);
    ST = S(:,end);
    Call(j,1) = mean(max(ST-K(j),0))*(exp(-r*T));
    Put(j,1)  = mean(max(K(j)-ST,0))*(exp(-r*T));
    end
elseif strcmp(model,'CTS')
    C = param(1);
    lm = param(2);
    lp = param(3);
    alpha = param(4);
    m = r - q - log(chf_CTS(-1i, [C lm lp alpha 0], 1));

    Z = SimCTS_invcdf([C lm lp alpha m], dt, MC_paths, days, ...
        0.5, 1);
    Z = cumsum(Z,2);
    S = S0*exp([zeros(MC_paths,1) Z]);
    ST = S(:,end);
    for j = 1:lK
    Call(j,1) = mean(max(ST-K(j),0))*(exp(-r*T));
    Put(j,1)  = mean(max(K(j)-ST,0))*(exp(-r*T));
    end
elseif strcmp(model,'NTS')
    a = param(1);
    lambda = param(2);
    C  = param(3);
```

```
    sigma = param(4);
    theta  = param(5);
    k      = r - q - C*gamma(-a/2)*...
        ((lambda - theta - 1/2*sigma^2).^(a/2) - lambda^(a/2));
    NSim = MC_paths*days;
    Z = simulateMNTS(a, lambda, C, k, theta, sigma^2, dt, NSim);
    Z = reshape(Z ,MC_paths, days);
    Z = cumsum(Z,2);
    S = S0*exp([zeros(MC_paths,1) Z]);
    ST = S(:,end);
    for j = 1:lK
    Call(j,1) = mean(max(ST-K(j),0))*(exp(-r*T));
    Put(j,1)  = mean(max(K(j)-ST,0))*(exp(-r*T));
    end

end

ImpVol(:,1) = blsimpv(S0, K, r, T, Call, 1, q, 1e-6, ...
    {'Call'})*100;

return;
```

Key Points

- In 1987 soon after the Black Monday crash, options traders started to notice a strange pattern on charts of stock-option prices. They named this phenomenon *volatility smile.*
- Traders and quants in every product area have had to model volatility smile in their own market. At many firms, not only does each front-office trading desk have its own particular smile models, but the firm-wide risk management group is likely to have its own models as well. The modeling of the volatility smile is likely one of the largest sources of model risk within finance.
- If one moves from the normality assumptions or if one considers more complex payoffs, closed-form solutions to evaluate the price of an option are usually not available. In these cases the evaluation alternatives are (1) quasi-closed form solutions (e.g., numerical integration); (2) Monte Carlo simulations, and (3) more complex numerical methods (i.e., numerical methods for partial differential equations).

- The failure to explain observed option prices by considering only log-returns time series information is well known. There are two possible approaches to calibrate a continuous-time option pricing model: (1) one can fit the model by considering only implied volatilities quotes, or (2) one can jointly calibrate the model parameters on the time series of log-returns and implied volatilities quotes, by considering a proper change of measure. This can be applied to a univariate or a multivariate framework.
- In Monte Carlo simulations conducted through the inverse transform combined with a fast Fourier transform algorithm to find the density function, the parameters of the FFT method N (the number of summands in the Riemann sum) and h (the grid spacing) should be carefully chosen in order to have a proper accuracy. Wrong choices may led to a strange shape for the smile obtained through simulation.

References

O.E. Barndorff-Nielsen and N. Shephard. Non-Gaussian Ornstein–Uhlenbeck-based models and some of their uses in financial economics. *Journal of the Royal Statistical Society: Series B (Statistical Methodology)*, 63(2):167–241, 2001.

G. Barone Adesi, R.F. Engle, and L. Mancini. A GARCH option pricing model with filtered historical simulation. *Review of Financial Studies*, 21(3):1223–1258, 2008.

M.L. Bianchi and G.L. Tassinari. Forward-looking portfolio selection with multivariate non-Gaussian models and the Esscher transform. *Preprint*, 2018. https://arxiv.org/abs/1805.05584.

M.L. Bianchi, G.L. Tassinari, and F.J. Fabozzi. Riding with the four horsemen and the multivariate normal tempered stable model. *International Journal of Theoretical and Applied Finance*, 19(4):1650027, 2016.

F. Black and M. Scholes. The pricing of options and corporate liabilities. *Journal of Political Economy*, 81(3):637–654, 1973.

A. Blümke. *How to Invest in Structured Products: A Guide for Investors and Asset Managers.* Wiley, Chichester, 2009.

P. Carr and D. Madan. Option valuation using the fast Fourier transform. *Journal of Computational Finance*, 2(4):61–73, 1999.

M. Chernov and E. Ghysels. A study towards a unified approach to the joint estimation of objective and risk neutral measures for the purpose of options valuation. *Journal of Financial Economics*, 56(3):407–458, 2000.

U. Cherubini, G. Della Lunga, S. Mulinacci, and P. Rossi. *Fourier Transform Methods in Finance.* Wiley, Chichester, 2010.

P. Christoffersen, K. Jacobs, and K. Mimouni. Volatility dynamics for the S&P500: Evidence from realized volatility, daily returns, and option prices. *Review of Financial Studies*, 23(8):3141–3189, 2010.

R. Cont. Empirical properties of asset returns: Stylized facts and statistical issues. *Quantitative Finance*, 1:223–236, 2001.

R. Cont and P. Tankov. *Financial Modelling with Jump Processes.* Chapman & Hall/CRC, Boca Raton, FL, 2003.

J.M. Corcuera, F. Guillaume, P. Leoni, and W. Schoutens. Implied Lévy volatility. *Quantitative Finance*, 9(4):383–393, 2009.

F. De Weert. *Exotic Options Trading.* Wiley, Chichester, 2011.

E. Derman. *My Life as a Quant: Reflections on Physics and Finance.* Wiley, Hoboken, NJ, 2004.

E. Derman and M.B. Miller. *The Volatility Smile.* Wiley, Hoboken, NJ, 2016.

H.U. Gerber and E.S. Shiu. Option pricing by Esscher transform. *Transactions of the Society of Actuaries*, 46:99–144, 1994.

F. Guillaume. Sato two-factor models for multivariate option pricing. *Journal of Computational Finance*, 15(4):159–192, 2012.

F. Guillaume. The αVG model for multivariate asset pricing: Calibration and extension. *Review of Derivatives Research*, 16(1):25–52, 2013.

J.M. Harrison and S.R. Pliska. Martingales and stochastic integrals in the theory of continuous trading. *Stochastic Processes and Their Applications*, 11(3):215–260, 1981.

Y.S. Kim and J.H. Lee. The relative entropy in CGMY processes and its applications to finance. *Mathematical Methods of Operations Research*, 66(2):327–338, 2007.

A.L. Lewis. A simple option formula for general jump–diffusion and other exponential Lévy processes. Technical report, Envision Financial Systems and OptionCity.net, 2001.

E. Lukacs. *Characteristic Functions.* Griffin, 1970.

B.B. Mandelbrot. The variation of certain speculative prices. *Journal of Business*, 36:394–419, 1963.

R.C. Merton. On the pricing of corporate debt: The risk structure of interest rates. *Journal of Finance*, 29(2):449–470, 1974.

E. Nicolato and E. Venardos. Option pricing in stochastic volatility models of the Ornstein–Uhlenbeck type. *Mathematical Finance*, 13(4): 445–466, 2003.

S. Patterson. *The Quants: How a New Breed of Math Whizzes Conquered Wall Street and Nearly Destroyed It.* Crown Business, New York, 2010.

S.T. Rachev and S. Mittnik. *Stable Paretian Models in Finance.* Wiley, Chichester, 2000.

K.I. Sato. *Lévy Processes and Infinitely Divisible Distributions.* Cambridge University Press, Cambridge, 1999.

W. Schoutens. *Lévy Processes in Finance.* Wiley, Chichester, 2003.

G.L. Tassinari and M.L. Bianchi. Calibrating the smile with multivariate time-changed Brownian motion and the Esscher transform. *International Journal of Theoretical and Applied Finance*, 17(4):1450023, 2014.

L. Wu. Modeling financial security returns using Lévy processes. In J.R. Birge and V. Linetsky, editors, *Handbooks in Operations Research and Management Science. Volume 15: Financial Engineering.* Elsevier, 2008, pp. 117–162.

Chapter 12

Application of Extreme Value Theory to Estimate Tail Thickness for Asset Return Distributions

The main topics covered in this chapter are:

- assessment of value-at-risk (VaR) and average value-at-risk (AVaR) through extreme value theory (EVT) using the peaks-over-threshold method;
- how to combine the standard GARCH model which takes into account the clustering of volatility with EVT for risk estimation;
- extensive in-sample and out-of-sample analysis of the upper and the lower tail thickness of the return distributions of market indices;
- AVaR-based and standard VaR-based statistical tests to assess the out-of-sample behavior of the risk model.

12.1. Introduction

The properties of asset return distributions have been studied extensively since the fundamental work of Mandelbrot (1963) and Fama (1965). Empirical research has established the following four stylized facts about asset returns[1]: (i) clustering of volatility, (ii) autoregressive behavior, (iii) skewness, and (iv) fat tails. The degree to which these properties are present in return data depends heavily on the frequency of returns and asset class. The higher the frequency is, the more pronounced is the clustering of volatility and fat tailedness. With respect to asset class, returns of large-cap stocks are known to exhibit lighter tails, reflecting better liquidity and lower probability of default.

Although the functional form of return distributions is unknown, there have been significant efforts to determine the functional form of the tail decay. This is theoretically possible because of a fundamental result in extreme value theory (EVT) discussed in Chapter 9 asserts that the extreme tail can be approximately described through a simple parametric family of distributions known as the generalized Pareto distribution (GPD). In finance, the rate of tail decay has important implications for portfolio risk measurement and stress-testing where a distributional model is combined with a risk measure sensitive to tail events, as well as option pricing where the so-called smile effect discussed in Chapter 11 can be explained by the presence of fat tails.

EVT has played an increasing role in the estimation of the frequency of extreme events and risk forecasting using value-at-risk (VaR) or average value-at-risk (AVaR) as risk measures. Studies on predictive performance of various VaR methods have found the EVT-based method to be particularly accurate.[2] In studies of the distribution of extreme stock returns, the principal finding is that market index returns exhibit Pareto-like tails.[3] However, in these studies, EVT is applied unconditionally without explaining away the clustering of volatility which is known to cause fat tails in the unconditional return distribution.[4]

[1]See Stoyanov *et al.* (2011) for a discussion and literature review.

[2]See Danielsson and de Vries (1997), Pownall and Koedij (1999), McNeil and Frey (2000), Bekiros and Georgoutsos (2005), Fernandez (2005), Tolikas *et al.* (2007).

[3]See, for example, Jondeau and Rockinger (2003), Gettinby *et al.* (2004), Longin (2005), and Tolikas and Gettinby (2009).

[4]The ARCH(1) example in Section 8.4 of Embrechts *et al.* (1997) illustrates that a distribution of the error term which is in the maximum domain of attraction (MDA) of

12.2. Estimation Methodology

In finance, EVT has been traditionally applied to estimate probabilities of extreme losses or loss thresholds such that losses beyond them occur with a predefined small probability. In fact, EVT provides a model for the extreme tail of the distribution which turns out to have a relatively simple structure described through the GPD. Here we begin by briefly reviewing the peak-over-threshold method described in Chapter 9. Then, we briefly describe with the backtesting methodology for the out-of-sample tests.

12.2.1. The peak-over-threshold method

As explained in Chapter 9, the most common approach for testing for extreme values is the peak-over-threshold (POT) method. Suppose that we have selected a high loss threshold u and we are interested in the conditional probability distribution of the excess losses beyond u. We denote this distribution by $F_u(x)$ which is expressed through the unconditional distribution in the following way:

$$F_u(x) = P(X - u \leq x | X > u) = \frac{F(x+u) - F(u)}{1 - F(u)}, \tag{12.1}$$

where $x > 0$. The expression in (12.1) is restated in terms of the tail $\overline{F}_u(x) = P(X - u > x | X > u) = 1 - F_u(x)$,

$$\overline{F}_u(x) = \frac{\overline{F}(x+u)}{\overline{F}(u)}. \tag{12.2}$$

A celebrated result in EVT states that as the threshold increases, the conditional tail probability converges to the GPD,[5]

$$\lim_{u \uparrow x_F} \sup_{0<x<x_F - u} |\overline{F}_u(x) - \overline{G}_{\xi,\beta(u)}(x)| = 0, \tag{12.3}$$

the Gumbel distribution generates Pareto-type tails for the ARCH process. See Chapter 9 for a discussion of the MDA and the Gumbel distribution.

[5]See Chapter 3 in Embrechts *et al.* (1997).

where

$$\overline{G}_{\xi,\beta}(x) = \begin{cases} (1+\xi x/\beta)^{-\frac{1}{\xi}}, & \xi \neq 0, \\ \exp(-x/\beta), & \xi = 0, \end{cases} \tag{12.4}$$

and $\beta(u)$ is a scaling depending on the selected threshold u.

The limit result in (12.3) can be used to construct an approximation for the tail of the losses exceeding a high threshold u. If we denote by $y = u + x$ and express x in terms of y in (12.2), we obtain

$$\overline{F}(y) \approx \overline{F}(u) \times \overline{G}_{\xi,\beta}(y-u) \tag{12.5}$$

having substituted the limit law $\overline{G}_{\xi,\beta}$ for $\overline{F}_u$. For a fixed threshold u, note that $\overline{F}(u)$ is a constant and the tail $\overline{F}(y)$ for $y > u$ is determined entirely by the GPD tail $\overline{G}_{\xi,\beta}$.

It is possible to define sets of portfolio loss distributions also known as maximum domains of attraction (MDA) such that the limit relation in (12.3) leads to a GPD with one and the same tail parameter ξ. Since EVT is used to study rare events, the tail behavior of the portfolio loss distribution turns out to be the important feature determining the MDA; other features of F are irrelevant. The theory distinguishes between three different classes of distributions: the Fréchet MDA ($\xi > 0$), the Gumbel MDA ($\xi = 0$), and the Weibull MDA ($\xi < 0$). Each of these distributions is described and graphically illustrated in Chapter 9. The Fréchet MDA contains fat-tailed distributions with unbounded moments of order $1/\xi$ (e.g., Student's t). The Gumbel MDA is much more diverse — it contains distributions with an asymptotic exponential tail decay, which can be heavy-tailed (e.g., the log-normal), light-tailed (e.g., the normal, the exponential), or even distributions with bounded support with exponential behavior near x_F. The Weibull MDA contains only distributions with bounded support (e.g., the uniform).[6]

To apply (12.5) in practice, we need to choose a high threshold u and also estimate the probability $\overline{F}(u)$. In addition, we need

[6]For further information, see, for example, Chapter 3 in Embrechts *et al.* (1997).

estimates of ξ and $\beta(u)$. Regarding the choice of u, different strategies have been adopted in the literature. One general recommendation is to set u so that a given percentage of the sample is excess. Chavez-Demoulin and Embrechts (2004) report that a 10% threshold provides a good trade-off between the bias and variability of the estimator of the important shape parameter ξ when the sample size is about 1000 observations. A similar guideline is provided by McNeil and Frey (2000).[7] If the threshold is allowed to vary, then the probability $\overline{F}(u)$ can be estimated through the empirical cumulative distribution function (cdf) as suggested, for example, in McNeil and Frey (2000). For instance, suppose that $(x_1, \ldots, x_n)$ is an observed sample of iid portfolio losses. If u is chosen such that exactly m observations are excesses, then the approximation in (12.5) becomes

$$\hat{\overline{F}}(y) \approx \frac{m}{n} \times \overline{G}_{\hat{\xi},\hat{\beta}}(y - x_{s,n}), \tag{12.6}$$

where $s = 1 - m/n$ and $x_{s,n}$ is the sth observation in the sample sorted in increasing order and $\hat{\xi}$ and $\hat{\beta}$ are estimates of ξ and β, respectively.

Regarding estimation, a variety of estimators can be employed to estimate ξ and β. We use the maximum likelihood estimator (MLE) which is rationalized by the uniform convergence in (12.3). Under the assumption that data are distributed exactly according to the GPD, then given a sample of iid observations $y = (y_1, \ldots, y_n)$ the log-likelihood function equals

$$LL(\xi, \beta; y) = -n_1 \log \beta - \left(\frac{1}{\xi} + 1\right) \sum_{i=1}^{n_1} \log\left(1 + \frac{\xi}{\beta} y_i\right)$$

and can be maximized numerically. The MLE $(\hat{\xi}_{n_1}, \hat{\beta}_{n_1}) = \max_{(\xi,\beta)\in D} LL(\xi, \beta; y)$, where $D = (-1/2, \infty) \times (0, \infty)$, satisfies the

[7]An approach based on adaptive calibration of the threshold is adopted by some authors. Gonzalo and Olmo (2004) describe a method that calibrates the quantile level based on minimizing the distance between the empirical $\overline{F}_u(x)$ and the tail of the GPD with parameters estimated through the maximum likelihood method. The suggested distance is the Kolmogorov–Smirnov statistic. For our dataset, the optimal quantile levels turned out to be close to 10% for all estimation time-windows in our backtests.

following asymptotic property:

$$n_1^{1/2}\left(\hat{\xi}_{n_1}-\xi,\frac{\hat{\beta}_{n_1}}{\beta}-1\right)\xrightarrow{d}\mathcal{N}(0,\Sigma), \tag{12.7}$$

where

$$\Sigma=\begin{pmatrix}(1+\xi)^2 & -(1+\xi)\\ -(1+\xi) & 2(1+\xi)\end{pmatrix}$$

and $\mathcal{N}(0,\Sigma)$ denotes a bivariate normal distribution.[8] Since data do not exactly follow the GPD law but are in its MDA, we use the GPD log-likelihood and the result in (12.7) only as an approximation. In practice, the GPD is estimated from the sample $y_i = x_{s+i,n} - x_{s,n}$, where $i = 1,\ldots,n_1 = n - s$ and s is defined as $s = 1 - m/n$.[9]

```
function [par, stat] = gpmlefit(extr, x0)

% function [par, stat] = gpmlefit(extr, x0)
%
% Estimates the GPD parameters using MLE and provides 95% ...
    asymptotic confidence  bounds. The input parameter extr ...
    contains the extremes of the original sample and x0 is a ...
    2x1 vector with initial parameter estimates.

par = fminsearch(@(x)loglikelihood(x, extr), x0, ...
    optimset('Display',...
    'Off', 'TolX', 10^-8, 'TolFun', 10^-8));

par(1) = -0.5 + exp(par(1));

% calculate asymptotic confidence bounds
if nargout == 2
    I = [(1 + par(1))^2, -(1 + par(1)); -(1 + par(1)), 1 + (1 ...
        + par(1))^2];
    stat.I = I;

    confLev = 0.95;
    q = norminv((1 - confLev)/2);
    ci(:,1) = par' + q*sqrt(diag(I))/sqrt(length(extr));
```

[8]For additional details, see Section 6.5 in Embrechts *et al.* (1997).

[9]Information about other estimators, such as the Hill and the Pickands estimator, and further detail on the relevance of the MLE are described in Chapter 9.

```
    ci(:,2) = par' - q*sqrt(diag(I))/sqrt(length(extr));
    stat.ci = ci;
end

% The log-likelihood function relies on the Matlab built-in ...
    function gppdf
function l = loglikelihood(par, data)

l = -sum(log(gppdf(data, -0.5 + exp(par(1)), par(2), 0)))
```

The Matlab implementation of the MLE method provided here (*gpmlefit*) uses the sample of extremes $y_i = x_{s+i,n} - x_{s,n}$ and needs a vector of initial parameter estimates. However, it is more convenient to work directly with the initial sample of returns. The following Matlab function estimates the GPD parameters of the left tail of the returns provided in the input sample by first identifying the extreme losses defined as the extreme observations beyond a given quantile and then by running ML estimator implemented in the function *gpmlefit*.

```
function [xi, beta, q, stat] = evtfit(sample, qlevel)

% function [xi, beta, q, stat] = evtfit(sample, qlevel)
%
% Fits the parameters of GPD of the left tail of sample.
% qlevel is the quantile level defining the extremes (e.g. 0.1)

[extr, q] = getExtremes(sample, qlevel);

if nargout == 4
    [par, stat] = gpmlefit(-extr, [0.2, 0.5]);
else
    par = gpmlefit(-extr, [0.2, 0.5]);
end

xi = par(1);
beta = par(2);

function [extr, q] = getExtremes(sample, qlevel)

q = quantile(sample, qlevel);
extr = sample(sample < q) - q;
```

12.2.2. A GARCH-EVT model for tail risk estimation

To deal with the clustering of volatility, a GARCH model must be used. The resulting model is referred to as the GARCH-EVT model for tail risk estimation. The specific GARCH model used is the GARCH (1,1) model. We apply this model to portfolio losses.

Denoting the time series of portfolio losses by X_t, the GARCH(1, 1) model is given by

$$\begin{aligned} X_t &= \mu + \sigma_t Z_t, \\ \sigma_t^2 &= K + a\epsilon_{t-1}^2 + b\sigma_{t-1}^2, \end{aligned} \tag{12.8}$$

where $\epsilon_t = \sigma_t Z_t$. The innovations Z_t are iid random variables with zero mean, unit variance, and marginal distribution function $F_Z(x)$, and K, a, and b are the positive parameters with $a+b < 1$. The model in (12.8) is fitted to the data and then the standardized residuals are derived. If we assume that the data are generated by the model in (12.8), then the standardized residual is a sample from the distribution F_Z. EVT is then applied by fitting the GPD to the residual using MLE.

In combination with the probabilistic model, two measures of risk can be used to perform out-of-sample tests on the predictive properties of the model: VaR and AVaR at three different tail probabilities (1%, 2.5%, and 5%). Here we provide definitions and explicitly state the risk forecasts built through the probabilistic model. Notice, however, that in setting the GARCH-EVT, any risk measure necessarily becomes a function of σ_t, ξ, and β. As a result, the behavior of these parameters have implications not only for VaR and AVaR but for any other downside risk measure defined on the tail of the return distribution.

We continue to assume that the random variable X describes portfolio losses; that is, positive numbers are interpreted as percentage losses rather than gains. Therefore, the risk measures VaR and AVaR are defined on the right tail of the loss distribution rather than the left tail of the portfolio return distribution. This is not restrictive because the quantiles in the right tail of the return distribution (left tail of the loss distribution) are obtained from the definitions below by considering $-X$ instead of X; that is, the downside of a short position is the upside of the corresponding long position. The risk functionals are, however, multiplied by -1 to preserve the interpretation that negative risk means a potential for profit.

Value-at-risk

The VaR of a random variable Y describing portfolio losses at a tail probability p, $\text{VaR}_p(Y)$, is implicitly defined as a loss threshold such that over a given time horizon losses higher than it occur with a probability p. By construction, VaR is the negative of the pth quantile of the portfolio return distribution or the $(1-p)$th quantile of the portfolio loss distribution. In the industry, VaR is often defined in terms of a confidence level but we prefer to reserve the term confidence level for the context of statistical testing. Thus, to map the terms properly, in practice one refers to VaR at the 95% and 99% confidence levels, which respectively correspond to VaR at 5% and 1% tail probability.

Formally, if we assume that Y describes portfolio losses, then VaR at tail probability p is defined as

$$\text{VaR}_p(Y) = F_Y^{-1}(1-p), \tag{12.9}$$

where F_Y^{-1} denotes the inverse of the cdf $F_Y(x) = P(Y \leq x)$ which is also known as the quantile function of Y.

As explained earlier, we employ EVT to estimate high quantiles of the loss distribution. To this end, we adopt the approximation of the tail in (12.5). Solving for the value of y yielding a tail probability of p, we get

$$\mathrm{VaR}_p(Y) = \begin{cases} u + \dfrac{\beta}{\xi}\left(\left(\dfrac{p}{\overline{F}(u)}\right)^{-\xi} - 1\right), & \xi \neq 0, \\ u - \beta \log\left(\dfrac{p}{\overline{F}(u)}\right), & \xi = 0. \end{cases} \tag{12.10}$$

The estimator is derived from (12.6) in the same way. Suppose that $y_{1,n} < y_{2,n} < \cdots < y_{n,n}$ denote the observed value of the order statistics, then using (12.6) we get

$$\widehat{\mathrm{VaR}}_p(Y) = \begin{cases} x_{s,n} + \dfrac{\hat{\beta}}{\hat{\xi}}\left(\left(\dfrac{pn}{m}\right)^{-\hat{\xi}} - 1\right), & \hat{\xi} \neq 0, \\ x_{s,n} - \hat{\beta} \log\left(\dfrac{pn}{m}\right), & \hat{\xi} = 0, \end{cases} \tag{12.11}$$

where $s = 1 - m/n$ and m denotes the number of observations that are considered excesses.

The approximation in (12.11) is usually interpreted in the following way: the estimate of VaR equals the empirical quantile $y_{s,n}$, which is such that $p < m/n$, plus a correction term obtained through the GPD. In practice, typically we set $m/n = 0.1$ and, thus, in terms of quantiles the 99% quantile ($\widehat{\mathrm{VaR}}_{0.01}(Y)$) equals the 90% quantile ($X_{(0.9 \times n)}$) plus the corresponding correction term.

Under this assumption of a GARCH(1, 1) process, the conditional VaR model is given by

$$\widehat{\mathrm{VaR}}_p(X_{t+1}|I_t) = \hat{\mu} + \hat{\sigma}_{t+1|t}\widehat{\mathrm{VaR}}_p(Z), \tag{12.12}$$

where I_t denotes the information available at time t and $\widehat{\mathrm{VaR}}_p(Z)$ is given in (12.11) and calculated from the sample of the standardized residuals.

Average value-at-risk

An important criticism of VaR is that it is uninformative about the extreme losses beyond it. Indeed, the only information provided is the probability of losing more than VaR which is equal to the tail probability level p. However, should any such loss occur, there is no information about its possible magnitude. AVaR is constructed to overcome this deficiency: AVaR at tail probability p, $\mathrm{AVaR}_p(Y)$, equals the average loss provided that the loss exceeds $\mathrm{VaR}_p(Y)$.

AVaR is formally defined as an average of VaRs,

$$\mathrm{AVaR}_p(Y) = \frac{1}{p}\int_{1-p}^{1} \mathrm{VaR}_h(Y)dh \tag{12.13}$$

and if we assume that the portfolio loss distribution has a continuous cdf, then AVaR can be expressed as a conditional expectation,

$$\mathrm{AVaR}_p(Y) = E(X|X > \mathrm{VaR}_p(Y)). \tag{12.14}$$

Assuming that $\xi < 1$, the expectation in (12.14) can be calculated explicitly through the GPD,

$$\mathrm{AVaR}_p(Y) = \begin{cases} \dfrac{\mathrm{VaR}_p(Y)}{1-\xi} + \dfrac{\beta - \xi u}{1-\xi}, & \xi \neq 0, \\ u - \beta\left(1 + \log\left(\frac{p}{\overline{F}(u)}\right)\right), & \xi = 0, \end{cases} \tag{12.15}$$

where

$$\mathrm{VaR}_p(Y) = u + \frac{\beta}{\xi}\left(\left(\frac{p}{\overline{F}(u)}\right)^{-\xi} - 1\right).$$

Plugging in $\widehat{\mathrm{VaR}}_p(Y)$ from (12.11) and the corresponding estimates, we get[10]

$$\widehat{\mathrm{AVaR}}_p(Y) = \begin{cases} \dfrac{\widehat{\mathrm{VaR}}_p(Y)}{1-\hat{\xi}} + \dfrac{\hat{\beta} - \hat{\xi} y_{s,n}}{1-\hat{\xi}}, & \hat{\xi} \neq 0, \\ y_{s,n} - \hat{\beta}\left(1 + \log\left(\frac{pn}{m}\right)\right), & \hat{\xi} = 0. \end{cases} \tag{12.16}$$

Under the assumption of a GARCH(1, 1) process for the portfolio loss distribution, the counterpart of (12.12) for AVaR equals

$$\widehat{\mathrm{AVaR}}_p(X_{t+1}|I_t) = \hat{\mu} + \hat{\sigma}_{t+1|t}\widehat{\mathrm{AVaR}}_p(Z), \tag{12.17}$$

[10]For derivations and further detail, see Section 7.2.3 in McNeil *et al.* (2005).

where $\widehat{\text{AVaR}}_p(Z)$ is given in (12.16) and is estimated from the sample of the standardized residuals.

Regarding the choice of the standard GARCH formulation, Furió and Climent (2013) find no evidence of any impact when the standard GARCH(1, 1) formulation is replaced by an asymmetric GARCH specification for VaR estimation.[11] Jalal and Rockinger (2008) perform a Monte Carlo study to analyze the impact of an incorrectly specified GARCH model for VaR modeling. The Monte Carlo study includes GARCH(1, 1) with Gaussian and Student's t-residuals, EGARCH(1, 1) with Gaussian innovations, a switching regime volatility model, stochastic volatility with jumps, and a pure jump process for the data generating process. Jalal and Rockinger (2008) conclude that the two-step procedure of GARCH(1, 1) and EVT is remarkably robust compared to a two-step parametric approach with the Gaussian or Student's t-distribution. In line with the empirical literature, for the estimation of the GARCH(1, 1) model in (12.8) we use the pseudo-maximum likelihood method assuming F_Z follows the Gaussian distribution.[12]

Although very difficult to formally compare across empirical studies because different authors focus on different estimation methods, different market indices, and different time periods, studies applying EVT unconditionally in the estimation of equity market tail risk generally report higher estimated ξ values than studies with a conditional EVT model. Jondeau and Rockinger (2003) apply an unconditional EVT model through the block-of-maxima method and cover 20 equity market indices.[13] For the S&P 500, FTSE 100, and Nikkei 225, for example, they report $\hat{\xi}_L = 0.282, 0.273, 0.265$ and $\hat{\xi}_R = 0.132, 0.33, 0.268$ for the lower and the upper tails, respectively. They find that both the left and the right tails of the return

[11]The particular specifications are EGARCH(1, 1) and TGARCH(1, 1).

[12]The pseudo-maximum likelihood method provides consistent and asymptotically normal estimator, see Chapter 4 in Gouriéroux (1997).

[13]The block-of-maxima method is used to estimate the parameters of the other important limit distribution in EVT, the Generalized Extreme Value distribution. Its tail index parameter is the same as the tail index of the GPD. See Chapter 9.

distributions belong to the domain of attraction of the Fréchet law because $\hat{\xi} = 0$ is rejected for all markets and for both tails. In contrast, Furió and Climent (2013) use a conditional EVT model under three different GARCH specifications. For the same three markets, they report $\hat{\xi}_L = 0.211,\ 0.083,\ 0.536$ and $\hat{\xi}_R = -0.1, 0.082, 0.006$, respectively, under the GARCH(1, 1) and $\hat{\xi}_L = 0.272, -0.037, 0.213$ and $\hat{\xi}_R = -0.17, -0.047, 0.024$ under the EGARCH(1, 1) specifications. Based on the reported standard errors, which are higher than those in Jondeau and Rockinger (2003) because of a smaller sample size, it is not possible to reject $\hat{\xi} = 0$ for the upper tails of the three markets and also for the lower tail of the Nikkei 225 index.[14]

```
function [var, avar] = evtrisk(xi, beta, tailProb, q, qlevel)

% function [var, avar] = evtrisk(xi, beta, tailProb, q, qlevel)
%
% Computes the VaR and AVaR approximated through the GPD
% xi and beta are the GPD parameters, tailProb is the tail ...
    probability parameter of the VaR/AVaR risk measure, q is ...
    the quantile at probability qlevel of the original sample.

if xi == 0
    xi = 0.00001;
end

var = -(-q + beta*((tailProb/qlevel)^(-xi) - 1)/xi);
avar = -(-q + beta*((tailProb/qlevel)^(-xi)/(1 - xi) - 1)/xi);
```

The Matlab implementation takes advantage of the continuity of the GPD density function in the parameter space and does not implement a special case for $\xi = 0$. The input parameters are easily mapped to the corresponding formulae, see for example (12.10). Finally, the AVaR implementation does not rely on VaR as in (12.16) and is done by simplifying the formula provided in (12.15).

[14]Jondeau and Rockinger (2003) consider samples of about 9000 observations for the three markets while Furió and Climent (2013) use samples of about 6000 observations.

12.2.3. Backtesting and statistical tests

Assume that the dataset includes daily returns. The risk model can be backtested using a rolling-time window a certain number of days[15] for parameter estimation and by building forecasts for the VaR and AVaR at different tail probabilities for both tails through the estimated model on a daily basis. Each day it can be verified if the realized return violates any of the forecasted quantile levels in the upper and the lower tails and an exceedance is counted if any such violation occurs. At the end of the backtesting period, we then obtain a sequence of indicators for each tail and each tail probability level. Using the sequence of indicators and the realized returns conditioned on a violation, three well-known statistical tests for VaR and one for AVaR can be employed. Confidence bounds for the estimated ξ using (12.7) can also be calculated.

The VaR-based tests include the well-known test on the average number of exceedances (Kupiec (1995)), the test on independence of the exceedances (Christoffersen (1998)), and the combination of the two (Christoffersen (1998)). A statistical test on AVaR can be based on the differences between the realized losses and the forecasted AVaR conditioned on the events of VaR exceedances.[16] Consider the differences

$$D_{t+1} = (X_{t+1} - \text{AVaR}_p(X_{t+1}|I_t)) \times \mathbf{1}_{\{\text{VaR}_p(X_{t+1}|I_t)<X_{t+1}\}} \tag{12.18}$$

in which $\mathbf{1}_{\{A\}}$ denotes the indicator of the event A. Because AVaR is the expected shortfall of the continuous loss distribution, $E[D_{t+1}] = 0$ and, therefore, D_{t+1} forms a martingale difference series. Under the assumption of a GARCH model, the normalized differences D_{t+1}/σ_{t+1} should behave like a zero-mean iid sequence with a probability mass of $(1-p)$ at zero. We use the standard t-test to check if the

---

[15] We use 1000 days.

[16] See Section 4.4.3 in McNeil *et al.* (2005).

conditional mean $E[\widehat{D}_{t+1}|\widehat{D}_{t+1} \neq 0]$ of the normalized differences,

$$\widehat{D}_{t+1} = \left(\frac{X_{t+1} - \widehat{\mathrm{AVaR}}_p(X_{t+1}|I_t)}{\hat{\sigma}_{t+1|t}} \right) \times \mathbf{1}_{\{\widehat{\mathrm{VaR}}_p(X_{t+1}|I_t) < X_{t+1}\}},$$

is statistically different from zero.[17]

12.3. Application to Equity Market Returns

We conclude this chapter with an application of the model and methodology described in the previous section to the daily log-returns of 22 developed and 19 emerging equity market indices shown in Table 12.1.[18]

Also shown in Table 12.1 are the corresponding market indices, the starting dates of the samples, and the number of daily observations. The sample period for all indices ends on June 28, 2013.[19]

Backtesting a risk model based on EVT requires availability of relatively large samples which justifies using the full sample for each equity market index in our illustration. However, since data availability are limited for some market indices, comparing results across markets becomes difficult. For this reason, we carry out two calculations, one of them using the full samples in which we test the relevance of the GARCH-EVT model, and another one based on the period from January 1, 2003 to June 28, 2013 which contains rather diverse market conditions and for which data are available for almost all markets.

We report the results of three calculations for our illustration. First, we report the test as to whether the GARCH-EVT model is realistic out-of sample for forecasting VaR and AVaR at 1% tail probability which represents a standard tail probability in risk modeling. Second, for each day in the backtesting time window, we calculate

[17] See also Stoyanov *et al.* (2017) for additional VaR tests.

[18] The data were downloaded from Datastream.

[19] The difference in the number of observations is because of data availability, the number of observations reflects the maximum available daily data for each market index.

Table 12.1. The equity market indices used in the study, the starting date of the samples, and number of daily observations. The sample period for all indices ends on June 28, 2013.

	Stock Market Index	Starting date	No. of observations
Developed Markets			
Austria	ATX	January 1, 1986	7169
Australia	S&P/ASX 200 Index	June 1, 1992	5500
Belgium	BEL 20	January 1, 1990	6129
Canada	S&P/TSX Composite Index	January 1, 1969	11,608
Denmark	OMX Copenhagen Index	January 1, 1996	4565
Finland	OMX Helsinki Index	January 1, 1987	6911
France	CAC 40	July 1, 1987	6777
Germany	DAX 30	January 1, 1965	12,651
Hong Kong	Hang Seng Index	January 1, 1970	11,347
Ireland	Irish SE Overall Index	January 1, 1987	6911
Italy	FTSE MIB Index	January 1, 1998	4042
Japan	NIKKEI 225 Index	January 1, 1950	16,500
New Zealand	NZX 50 Index	January 1, 2001	3260
Netherlands	AEX Index	January 1, 1983	7955
Norway	OSLO Exchange All Share Index	January 1, 1983	7955
Portugal	PSI-20	January 1, 1993	5346
Singapore	Strait Times Index	September 1, 1999	3608
Spain	IBEX 35	January 1, 1987	6910
Sweden	OMX Stockholm Index	January 1, 1987	6910
Switzerland	SMI	July 1, 1988	6521
US	S&P 500 Index	January 1, 1964	12,913
UK	FTSE 100	January 1, 1984	7695
Emerging Markets			
Argentina	Argentina Merval Index	November 1, 1989	6173
Brazil	Brazil Bovespa Index	January 1, 1993	5346
China	Shanghai SE Composite Index	January 1, 1991	5868
Chile	Santiago SE General Index	January 1, 1987	6911
Czech Republic	Prague SE Index	April 1, 1994	5016
Egypt	Egypt Hermes Financial	January 1, 1995	4825
Hungary	Budapest SE Index	January 1, 1991	5868
India	CNX 500 Index	January 1, 1991	5868
Indonesia	IDX Composite Index	April 1, 1983	7890
Malaysia	FTSE Bursa Malaysia KLCI	January 1, 1980	8738
Mexico	IPC Index	January 1, 1988	6650
Peru	Lima SE General Index	January 1, 1991	5868
Philippines	Philippines SE Index	January 1, 1986	7172
Poland	Warsaw General Index	January 1, 1993	5346
Russia	MICEX Index	October 1, 1997	4108
South Korea	KOSPI	January 1, 1975	10,043
Taiwan	Taiwan SE Weighted Index	January 1, 1971	11,084
Thailand	SET	May 1, 1975	9957
Turkey	BIST National 100	January 1, 1998	6650

the 95% confidence bound of the point estimate of the tail index to check if the hypothesis $\xi = 0$ can be rejected. Finally, we backtest the GARCH-EVT model with $\xi = 0$ to test if this model can be rejected out of sample at 1% tail probability.

12.3.1. GARCH-EVT with ξ unconstrained and $\xi = 0$

This section analyzes the performance of the risk model at 1% tail probability when the important ξ parameter is unconstrained in the approximate MLE and when it is constrained to zero. We consider two cases: (i) the samples are used in full and (ii) the period from January 1, 2003 to June 28, 2013. In the following, the subscript L denotes left tail, and R denotes right tail, respectively.

Full samples

The left panel of Table 12.2 provides the results of the unconstrained case. In terms of the VaR-based tests, the performance of the risk model is remarkable for both the left and the right tails; the combined test for the left tail fails for no market while for the right tail it fails only for two countries (Argentina and the Philippines). The out-of-sample t-test for AVaR provided in the left panel of Table 12.3 leads to the same conclusion — the model is overall very successful at this probability level.

The average $\hat{\xi}$ for all countries is provided in the left panel of Table 12.3. The averages are clearly close to zero; the three countries with the highest average $\hat{\xi}_L$ are Hungary (0.1269), Indonesia (0.1269), and Peru (0.1229) — all of them in the group of the emerging markets. The average $\hat{\xi}_R$ of any market is generally lower than the corresponding average $\hat{\xi}_L$. The averages are much lower than estimated tail indices provided in previous studies.[20]

Time series plots of the estimated ξ of the left and the right tails and their 95% confidence bounds for six countries for the period from January 1, 2003 to June 28, 2013 are provided in Figures 12.1 and 12.2.

[20] See for example Poon *et al.* (2004), Jondeau and Rockinger (2003), and LeBaron and Samanta (2005).

Table 12.2. *P*-values of VaR-based statistics of the out-of-sample performance of the

	Unrestricted GARCH-EVT, full sample									
	Left-1%					Right-1%				
	Exc	Ku-test	Ch-test	KC-test	Geo-test	Exc	Ku-test	Ch-test	KC-test	Geo-test
Developed										
Austria	63	0.866	0.678	0.904	0.999	62	0.967	0.262	0.532	0.967
Australia	46	0.880	0.092	0.240	0.909	39	0.358	0.409	0.466	0.305
Belgium	58	0.356	0.690	0.603	0.430	55	0.606	0.623	0.776	0.234
Canada	117	0.294	0.788	0.556	0.428	112	0.566	0.156	0.311	0.488
Denmark	37	0.820	0.402	0.686	0.481	37	0.820	0.402	0.686	0.241
Finland	63	0.614	0.706	0.820	0.073	72	0.103	0.897	0.262	0.289
France	54	0.615	0.313	0.529	0.511	69	0.149	0.850	0.347	0.325
Germany	119	0.817	0.510	0.784	0.839	122	0.611	0.796	0.850	0.617
Hong Kong	110	0.522	0.152	0.292	0.385	98	0.586	0.941	0.860	0.312
Ireland	56	0.683	0.562	0.777	0.867	66	0.376	0.769	0.647	0.808
Italy	38	0.183	0.501	0.328	0.561	30	0.940	0.439	0.740	0.470
Japan	156	0.935	0.740	0.943	0.903	161	0.630	0.570	0.758	0.370
Netherlands	68	0.852	0.247	0.502	0.722	72	0.768	0.777	0.920	0.391
Norway	76	0.443	0.271	0.406	0.714	73	0.679	0.213	0.423	0.206
New Zealand	22	0.900	0.511	0.799	0.435	26	0.481	0.436	0.577	0.113
Portugal	54	0.121	0.183	0.124	0.228	34	0.134	0.271	0.178	0.879
Singapore	23	0.538	0.199	0.362	0.784	20	0.213	0.578	0.394	0.568
Spain	57	0.783	0.582	0.828	0.950	65	0.447	0.748	0.711	0.029
Sweden	65	0.447	0.207	0.337	**0.011**	69	0.207	0.833	0.441	0.075
Switzerland	54	0.871	0.302	0.579	0.204	53	0.765	0.311	0.572	0.380
UK	67	0.994	0.244	0.508	0.664	69	0.801	0.231	0.472	0.505
US	128	0.419	0.614	0.635	0.253	123	0.722	0.546	0.782	0.883
Emerging										
Argentina	59	0.320	0.243	0.309	0.487	50	0.809	**0.013**	**0.044**	0.791
Brazil	37	0.313	0.425	0.437	0.618	35	0.182	0.451	0.309	0.739
China	48	0.923	0.500	0.793	0.245	38	0.110	0.439	0.207	0.375
Chile	60	0.907	0.153	0.358	0.399	44	**0.039**	0.417	0.085	0.133
Czech Republic	50	0.132	0.261	0.172	0.156	38	0.731	0.394	0.656	0.860
Egypt	42	0.548	0.088	0.195	0.468	33	0.383	**0.033**	0.070	0.821
Hungary	54	0.451	0.633	0.671	0.519	51	0.739	0.119	0.280	0.308
Indonesia	64	0.549	0.629	0.743	0.675	79	0.232	0.078	0.103	0.087
India	46	0.698	0.459	0.705	0.926	46	0.698	0.459	0.705	0.305
Mexico	64	0.325	0.757	0.588	0.648	54	0.737	0.307	0.561	0.601
Malaysia	80	0.765	0.269	0.520	0.581	79	0.853	0.258	0.518	0.321
Philippines	67	0.504	0.760	0.764	0.618	47	**0.050**	0.054	**0.023**	0.469
Poland	47	0.593	0.107	0.237	0.139	32	0.067	0.237	0.093	0.500
Peru	52	0.635	0.128	0.281	0.663	52	0.635	0.588	0.771	0.880
Russia	35	0.487	0.372	0.527	0.868	27	0.453	0.491	0.595	0.592
South Korea	81	0.311	0.759	0.571	0.415	81	0.311	0.226	0.288	0.417
Taiwan	107	0.541	0.896	0.822	0.863	106	0.608	0.133	0.284	0.829
Thailand	80	0.301	0.748	0.556	0.740	86	0.704	0.269	0.505	0.584
Turkey	58	0.841	0.147	0.342	0.363	56	0.948	0.128	0.314	0.736

Exc denotes number of exceedances, Ku-test, Ch-test, and KC-test denote Kupiec's, test. Rejection at 5% level is marked in bold. Adapted from Stoyanov *et al.* (2017).

GARCH-EVT model covering the full samples of all markets.

GARCH-EVT with $\xi = 0$, full sample									
Left-1%					Right-1%				
Exc	Ku-test	Ch-test	KC-test	Geo-test	Exc	Ku-test	Ch-test	KC-test	Geo-test
64	0.768	0.698	0.888	0.981	61	0.931	0.270	0.542	0.961
45	0.999	0.085	0.226	0.757	39	0.358	0.409	0.466	0.305
59	0.290	0.712	0.534	0.415	55	0.606	0.623	0.776	0.278
116	0.340	0.802	0.615	0.667	108	0.851	0.130	0.313	0.473
35	0.914	0.357	0.651	0.630	37	0.820	0.402	0.686	0.241
62	0.707	0.685	0.858	0.091	71	0.132	0.875	0.317	0.306
52	0.439	0.331	0.462	0.482	68	0.188	0.828	0.410	0.622
115	0.889	0.459	0.753	0.869	118	0.889	0.853	0.973	0.509
109	0.587	0.145	0.299	0.404	96	0.456	0.909	0.752	0.497
56	0.683	0.562	0.777	0.873	66	0.376	0.769	0.647	0.808
38	0.183	0.501	0.328	0.561	30	0.940	0.439	0.740	0.470
157	0.871	0.752	0.939	0.931	161	0.630	0.570	0.758	0.361
67	0.758	0.254	0.497	0.692	71	0.861	0.758	0.939	0.337
72	0.768	0.223	0.456	0.662	71	0.861	0.226	0.473	0.211
22	0.900	0.511	0.799	0.435	22	0.900	0.511	0.799	0.393
52	0.206	0.159	0.166	0.354	32	0.067	0.237	0.093	0.661
23	0.538	0.522	0.674	0.784	20	0.213	0.578	0.394	0.060
56	0.684	0.301	0.539	0.884	65	0.447	0.748	0.711	**0.029**
64	0.526	0.195	0.353	**0.012**	68	0.255	0.812	0.509	0.159
51	0.565	0.329	0.527	0.106	52	0.662	0.320	0.554	0.168
67	0.994	0.244	0.508	0.664	68	0.897	0.725	0.932	0.375
124	0.655	0.559	0.763	0.737	119	0.991	0.493	0.791	0.772
59	0.320	0.243	0.309	0.487	47	0.503	**0.009**	**0.026**	0.865
36	0.242	0.438	0.373	0.836	35	0.182	0.451	0.309	0.739
47	0.809	0.479	0.756	0.190	39	0.149	0.427	0.258	0.386
62	0.707	**0.031**	0.091	0.253	44	**0.039**	0.417	0.085	0.229
50	0.132	0.261	0.172	0.156	36	0.503	0.420	0.577	0.533
42	0.548	0.088	0.195	0.468	34	0.483	**0.037**	0.089	0.752
57	0.243	0.701	0.470	0.455	46	0.698	0.079	0.199	0.917
62	0.396	0.591	0.604	0.615	78	0.280	0.073	0.112	0.108
46	0.698	0.459	0.705	0.926	46	0.698	0.459	0.705	0.305
67	0.172	0.823	0.384	0.601	51	0.455	0.335	0.476	0.642
78	0.943	0.246	0.509	0.698	77	0.966	0.235	0.494	0.425
67	0.504	0.760	0.764	0.618	47	**0.050**	0.054	**0.023**	0.469
47	0.593	0.107	0.237	0.139	32	0.067	0.237	0.093	0.500
54	0.451	0.148	0.264	0.793	54	0.451	0.633	0.671	0.637
35	0.487	0.372	0.527	0.868	27	0.453	0.491	0.595	0.592
78	0.179	0.707	0.378	0.123	75	0.093	0.263	0.130	0.284
109	0.420	0.865	0.712	0.912	105	0.679	0.137	0.304	0.946
83	0.481	0.237	0.388	0.741	85	0.625	0.258	0.468	0.617
58	0.841	0.147	0.342	0.363	56	0.948	0.128	0.314	0.884

Christoffersen's, and the combined tests, and Geo-test denotes the geometric distribution

Table 12.3. AVaR-based statistics of the out-of-sample performance of the

	Unrestricted GARCH-EVT, full sample									
	Left-1%					Right-1%				
	Avg $\hat{\xi}_L$	Avg $\hat{\beta}_L$	Avg AVaRf	Avg C-Loss	t-test p-value	Avg $\hat{\xi}_R$	Avg $\hat{\beta}_R$	Avg AVaRf	Avg C-Loss	t-test p-value
Developed										
Austria	0.049	0.653	0.045	0.041	0.183	−0.020	0.545	−0.037	−0.035	0.064
Australia	0.016	0.604	0.030	0.030	0.185	−0.150	0.542	−0.024	−0.024	0.234
Belgium	0.070	0.558	0.036	0.030	0.692	−0.002	0.506	−0.031	−0.031	0.492
Canada	0.036	0.639	0.029	0.028	0.178	−0.075	0.565	−0.024	−0.024	0.221
Denmark	−0.078	0.685	0.034	0.032	0.396	−0.046	0.500	−0.029	−0.030	0.147
Finland	0.031	0.600	0.052	0.048	0.076	0.031	0.539	−0.047	−0.044	0.945
France	−0.037	0.612	0.041	0.036	0.256	−0.078	0.517	−0.036	−0.039	0.972
Germany	−0.018	0.574	0.036	0.032	0.097	−0.042	0.497	−0.031	−0.033	0.082
Hong Kong	0.062	0.589	0.054	0.055	0.131	−0.090	0.590	−0.046	−0.046	**0.021**
Ireland	0.097	0.613	0.041	0.037	0.999	−0.040	0.555	−0.032	−0.035	0.509
Italy	−0.043	0.620	0.044	0.036	0.689	−0.076	0.492	−0.036	−0.047	0.381
Japan	0.066	0.624	0.037	0.035	0.165	−0.007	0.516	−0.031	−0.030	0.103
Netherlands	0.051	0.598	0.042	0.038	0.299	−0.064	0.526	−0.033	−0.036	0.283
Norway	0.016	0.634	0.041	0.042	0.149	−0.060	0.537	−0.034	−0.032	0.132
New Zealand	−0.071	0.666	0.021	0.021	0.689	−0.171	0.567	−0.018	−0.018	0.439
Portugal	−0.036	0.662	0.036	0.034	0.909	−0.010	0.575	−0.035	−0.029	0.157
Singapore	−0.090	0.669	0.034	0.030	0.582	−0.119	0.578	−0.030	−0.037	0.855
Spain	0.066	0.572	0.044	0.038	0.309	−0.013	0.500	−0.037	−0.037	0.411
Sweden	0.011	0.613	0.042	0.036	0.607	−0.002	0.504	−0.036	−0.040	0.162
Switzerland	0.003	0.640	0.036	0.032	0.814	0.018	0.472	−0.030	−0.031	0.621
UK	0.002	0.578	0.032	0.030	0.462	−0.038	0.450	−0.027	−0.028	0.440
US	−0.013	0.595	0.031	0.029	0.097	−0.119	0.588	−0.026	−0.029	0.112
Emerging										
Argentina	−0.051	0.737	0.068	0.064	0.444	−0.062	0.627	−0.061	−0.063	0.189
Brazil	−0.008	0.637	0.065	0.065	0.229	−0.017	0.488	−0.055	−0.062	0.079
China	−0.003	0.676	0.060	0.054	0.779	0.150	0.563	−0.070	−0.061	0.364
Chile	0.026	0.593	0.025	0.024	0.411	−0.085	0.601	−0.023	−0.022	0.240
Czech Republic	0.067	0.600	0.046	0.042	0.248	−0.029	0.554	−0.039	−0.039	0.241
Egypt	0.081	0.603	0.055	0.048	0.122	0.137	0.524	−0.055	−0.051	0.156
Hungary	0.127	0.550	0.057	0.058	0.499	−0.018	0.608	−0.052	−0.044	0.552
Indonesia	0.127	0.662	0.052	0.048	0.110	0.116	0.564	−0.050	−0.044	0.074
India	0.025	0.658	0.051	0.050	0.432	0.037	0.516	−0.044	−0.043	0.548
Mexico	0.031	0.638	0.049	0.046	0.978	−0.096	0.624	−0.043	−0.049	0.713
Malaysia	0.093	0.621	0.042	0.043	0.145	−0.013	0.630	−0.038	−0.039	0.984
Philippines	0.016	0.652	0.049	0.048	0.778	0.031	0.590	−0.048	−0.052	0.343
Poland	−0.015	0.619	0.045	0.044	0.221	−0.158	0.628	−0.040	−0.041	**0.025**
Peru	0.123	0.572	0.044	0.046	0.430	−0.061	0.639	−0.040	−0.036	0.468
Russia	0.026	0.675	0.069	0.070	0.505	0.007	0.540	−0.059	−0.067	0.189
South Korea	0.063	0.594	0.051	0.050	0.063	−0.078	0.618	−0.045	−0.043	0.215
Taiwan	0.018	0.647	0.052	0.047	0.981	−0.017	0.576	−0.049	−0.047	0.763
Thailand	0.085	0.588	0.045	0.044	0.258	−0.072	0.673	−0.041	−0.041	**0.024**
Turkey	0.054	0.599	0.082	0.078	0.302	−0.041	0.600	−0.076	−0.076	0.523

Avg denotes average, AVaRf denotes forecasted AVaR, and C-Loss denotes observed the entire sample of estimated values. Rejection at 5% level is marked in bold. Adapted

GARCH-EVT model covering the full samples of all markets.

GARCH-EVT with $\xi = 0$, full sample							
Left-1%				Right-1%			
Avg $\hat{\beta}_L$	Avg AVaRf	Avg C-Loss	t-test p-value	Avg $\hat{\beta}_R$	Avg AVaRf	Avg C-Loss	t-test p-value
0.692	0.043	0.041	0.070	0.536	−0.037	−0.035	0.119
0.612	0.030	0.030	0.161	0.470	−0.025	−0.024	0.055
0.601	0.034	0.030	0.266	0.507	−0.031	−0.031	0.723
0.667	0.028	0.028	0.061	0.531	−0.024	−0.024	0.545
0.630	0.035	0.033	0.607	0.478	−0.030	−0.030	0.274
0.624	0.051	0.048	0.055	0.561	−0.047	−0.044	0.204
0.590	0.042	0.037	0.290	0.480	−0.037	−0.039	0.189
0.573	0.035	0.033	0.116	0.478	−0.032	−0.033	0.204
0.640	0.052	0.055	**0.023**	0.543	−0.048	−0.046	0.622
0.684	0.038	0.037	0.218	0.538	−0.032	−0.035	0.970
0.594	0.045	0.036	0.320	0.457	−0.037	−0.047	0.556
0.673	0.036	0.035	**0.009**	0.516	−0.031	−0.030	**0.045**
0.638	0.040	0.038	0.090	0.495	−0.034	−0.036	0.446
0.646	0.040	0.042	**0.035**	0.510	−0.034	−0.032	0.592
0.623	0.022	0.021	0.838	0.484	−0.019	−0.019	0.374
0.639	0.037	0.034	0.587	0.579	−0.035	−0.030	0.088
0.613	0.036	0.029	0.615	0.514	−0.032	−0.038	**0.039**
0.622	0.042	0.038	0.104	0.495	−0.037	−0.037	0.608
0.628	0.041	0.036	0.732	0.508	−0.036	−0.041	0.089
0.647	0.036	0.032	0.710	0.481	−0.030	−0.032	0.430
0.582	0.032	0.030	0.505	0.438	−0.027	−0.028	0.589
0.592	0.030	0.029	0.056	0.526	−0.028	−0.029	0.444
0.696	0.070	0.064	0.931	0.589	−0.062	−0.064	0.413
0.632	0.064	0.066	0.113	0.479	−0.055	−0.062	0.147
0.673	0.059	0.054	0.375	0.683	−0.061	−0.061	0.113
0.609	0.024	0.024	0.228	0.554	−0.024	−0.022	0.832
0.646	0.044	0.042	0.754	0.539	−0.039	−0.039	0.322
0.667	0.051	0.048	0.058	0.631	−0.050	−0.051	**0.040**
0.637	0.052	0.057	**0.048**	0.601	−0.051	−0.045	0.142
0.760	0.047	0.049	**0.037**	0.662	−0.044	−0.044	**0.016**
0.674	0.050	0.050	0.213	0.543	−0.043	−0.043	0.309
0.657	0.047	0.046	0.781	0.568	−0.045	−0.050	0.356
0.689	0.039	0.042	**0.033**	0.623	−0.038	−0.039	0.911
0.661	0.048	0.048	0.679	0.618	−0.047	−0.052	0.302
0.611	0.045	0.044	0.400	0.543	−0.042	−0.041	0.906
0.651	0.041	0.045	**0.037**	0.602	−0.041	−0.035	0.798
0.693	0.068	0.070	0.265	0.541	−0.058	−0.067	0.155
0.645	0.049	0.051	**0.015**	0.576	−0.047	−0.043	0.895
0.659	0.052	0.047	0.711	0.569	−0.048	−0.047	0.518
0.650	0.043	0.044	0.090	0.629	−0.043	−0.041	0.414
0.633	0.080	0.078	0.059	0.574	−0.077	−0.076	0.932

loss conditioned on the events of VaR exceedances. The averaging is performed over from Stoyanov *et al.* (2017).

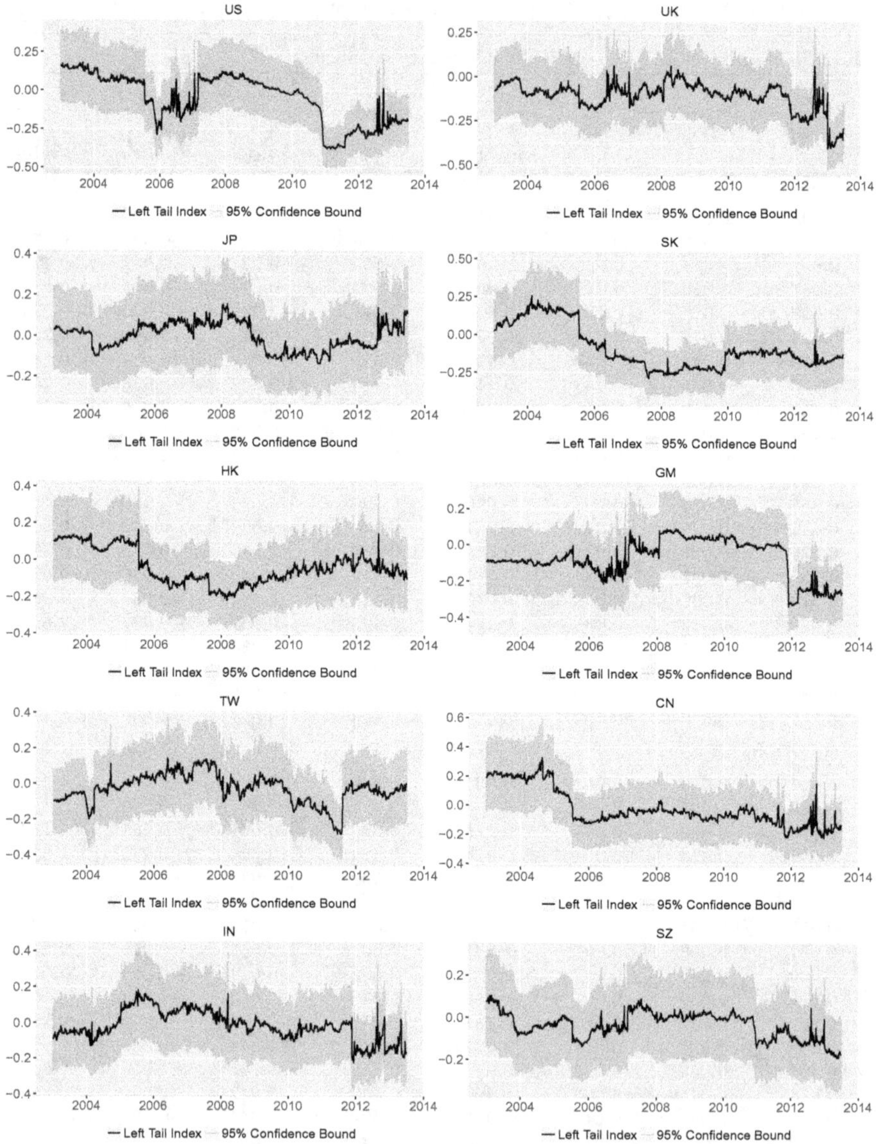

Figure 12.1. The fitted left tail index $\hat{\xi}_L$ for selected market indices. The countries are: US, UK, Japan, South Korea, Hong Kong, Germany, Taiwan, China, India, and Switzerland. Adapted from Stoyanov *et al.* (2017).

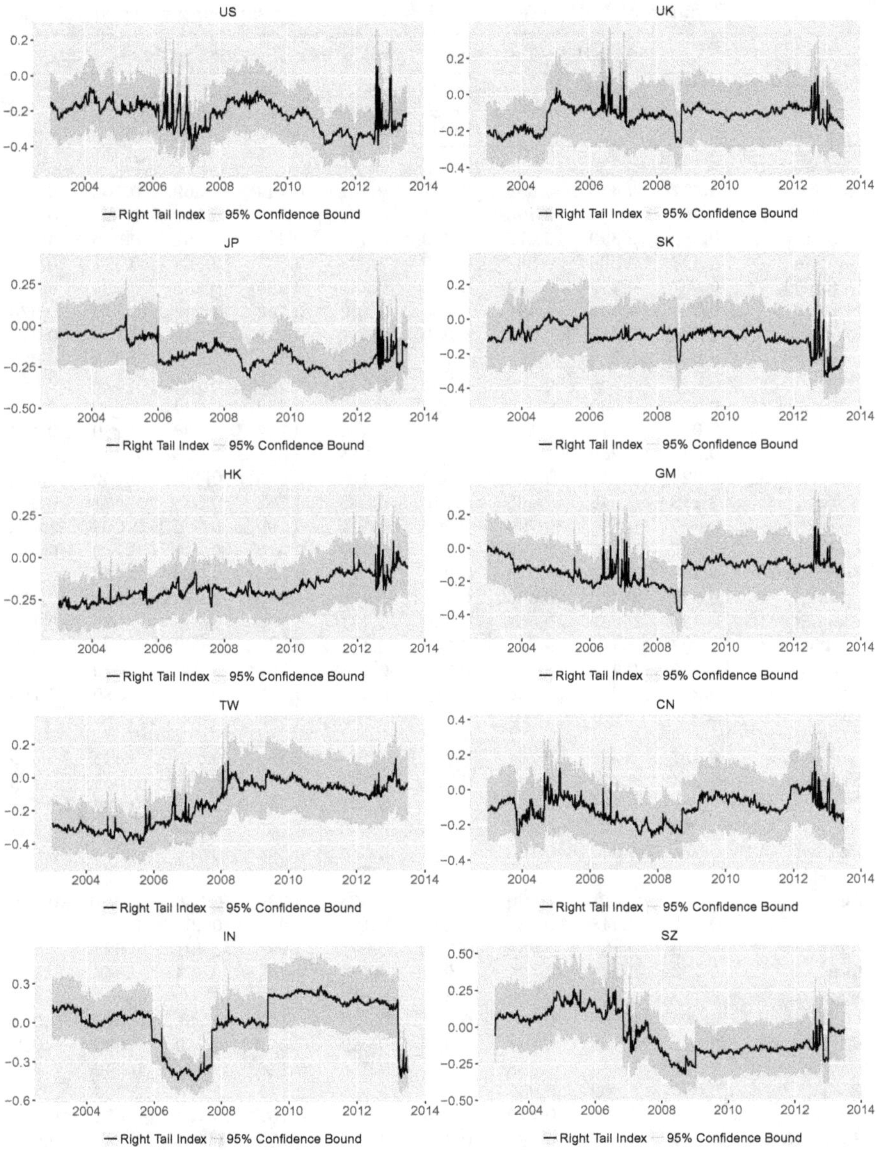

Figure 12.2. The fitted right tail index $\hat{\xi}_R$ for selected market indices. The countries are: US, UK, Japan, South Korea, Hong Kong, Germany, Taiwan, China, India, and Switzerland. Adapted from Stoyanov *et al.* (2017).

Table 12.4. *P*-values of VaR-based statistics of the out-of-sample performance of the

	GARCH-EVT with $\xi = 0$, period from Jan-2003 to June-2013								
	Left-1%			Right-1%			Left-2.5%		
	Ku-test	Ch-test	KC-test	Ku-test	Ch-test	KC-test	Ku-test	Ch-test	KC-test
Developed									
Austria	0.622	0.415	0.635	0.907	0.447	0.744	0.669	0.168	0.353
Australia	0.788	0.248	0.495	0.641	0.497	0.713	0.853	0.395	0.685
Belgium	0.389	0.390	0.476	0.389	0.384	0.472	0.759	0.906	0.947
Canada	0.497	0.399	0.557	0.386	0.533	0.565	0.584	0.471	0.663
Denmark	0.907	0.292	0.570	0.788	0.248	0.495	0.853	0.395	0.685
Finland	0.386	0.533	0.565	0.788	0.480	0.752	0.949	0.372	0.670
France	0.506	0.515	0.648	0.080	0.527	0.177	0.085	0.728	0.214
Germany	0.622	0.339	0.561	0.759	0.431	0.700	0.669	0.465	0.699
Hong Kong	0.221	0.355	0.309	0.386	0.533	0.565	0.761	0.749	0.907
Ireland	0.201	0.569	0.375	0.221	0.355	0.309	0.290	0.569	0.485
Italy	0.080	0.527	0.177	0.788	0.480	0.752	0.108	0.280	0.154
Japan	0.940	0.269	0.542	0.506	0.515	0.648	0.668	0.089	0.214
Netherlands	0.622	0.415	0.635	0.622	0.415	0.635	0.365	0.552	0.556
Norway	0.788	0.248	0.495	0.788	0.480	0.752	0.584	0.471	0.663
New Zealand	0.900	0.511	0.799	0.900	0.511	0.799	0.388	**0.009**	**0.022**
Portugal	0.497	0.364	0.526	0.201	0.569	0.375	0.949	0.130	0.316
Singapore	0.538	0.522	0.674	0.213	0.578	0.394	0.983	0.768	0.957
Spain	0.622	0.415	0.635	0.161	0.470	0.289	0.254	0.346	0.335
Sweden	0.497	0.399	0.557	0.759	0.431	0.700	0.365	0.079	0.142
Switzerland	0.907	0.447	0.744	0.759	0.431	0.700	0.669	0.937	0.910
UK	0.622	0.415	0.635	0.940	0.463	0.762	0.949	0.372	0.670
US	0.297	0.370	0.388	0.641	0.497	0.713	0.108	0.280	0.154
Emerging									
Argentina	0.907	0.447	0.744	0.201	0.153	0.159	0.949	0.843	0.979
Brazil	0.054	0.626	0.140	0.284	0.551	0.471	0.580	0.659	0.779
China	0.907	0.292	0.570	0.907	0.447	0.744	0.949	0.534	0.822
Chile	0.080	0.527	0.177	0.788	0.480	0.752	0.954	**0.033**	0.102
Czech Republic	0.080	0.314	0.130	0.506	0.515	0.648	0.761	0.607	0.837
Egypt	0.641	**0.020**	0.060	0.940	**0.028**	0.089	0.580	**0.000**	0.000
Hungary	0.622	0.415	0.635	0.284	0.170	0.220	0.497	0.658	0.720
Indonesia	0.386	0.188	0.289	**0.032**	0.645	0.090	0.421	0.628	0.643
India	0.788	0.248	0.495	0.940	0.463	0.762	0.954	0.349	0.644
Mexico	0.759	0.431	0.700	0.506	0.515	0.648	0.208	0.846	0.445
Malaysia	0.788	0.248	0.495	0.054	0.626	0.140	0.351	**0.002**	**0.006**
Philippines	0.497	0.364	0.526	**0.018**	0.082	**0.013**	0.290	**0.011**	**0.023**
Poland	0.221	0.443	0.353	0.054	0.626	0.140	0.761	0.749	0.907
Peru	0.080	0.102	0.057	0.054	0.556	0.132	0.584	0.058	0.142
Russia	0.389	0.384	0.472	0.940	0.463	0.762	0.761	0.071	0.187
South Korea	**0.032**	0.645	0.090	0.088	0.606	0.205	0.351	0.741	0.614
Taiwan	0.907	0.447	0.744	0.641	0.497	0.713	0.856	0.327	0.608
Thailand	0.788	0.248	0.495	0.506	0.515	0.648	0.580	0.688	0.791
Turkey	0.907	**0.033**	0.101	0.386	0.533	0.565	0.290	**0.002**	**0.005**

Ku-test, Ch-test, and KC-test denote Kupiec's, Christoffersen's, and the combined tests, bold. Adapted from Stoyanov *et al.* (2017).

GARCH-EVT model in the period 2003–2013.

GARCH-EVT with $\xi = 0$, period from Jan-2003 to June-2013								
Right-2.5%			Left-5%			Right-5%		
Ku-test	Ch-test	KC-test	Ku-test	Ch-test	KC-test	Ku-test	Ch-test	KC-test
0.504	1.000	0.800	0.384	0.267	0.369	0.142	0.537	0.282
0.759	0.488	0.750	0.858	0.269	0.534	0.167	0.069	0.073
0.949	0.059	0.168	0.663	0.299	0.530	**0.040**	0.093	**0.030**
0.584	**0.046**	0.117	0.297	0.306	0.344	0.260	**0.000**	**0.000**
0.759	0.906	0.947	0.663	**0.014**	**0.045**	0.927	0.986	0.996
0.954	0.558	0.841	0.484	**0.029**	0.072	0.249	**0.013**	**0.024**
0.853	0.510	0.791	0.102	0.487	0.206	0.339	**0.034**	0.067
0.208	0.638	0.406	0.384	**0.037**	0.079	0.194	0.178	0.174
0.668	0.075	0.188	0.790	0.745	0.915	0.484	0.126	0.243
0.504	0.497	0.636	0.377	0.641	0.608	0.540	0.594	0.719
0.432	**0.040**	0.089	0.484	0.126	0.243	0.660	0.156	0.331
0.759	0.420	0.689	0.377	0.693	0.627	0.540	0.207	0.374
0.254	**0.032**	0.053	0.339	0.161	0.237	0.484	**0.044**	0.104
0.504	**0.043**	0.102	0.260	0.326	0.327	0.384	**0.037**	0.079
0.736	0.633	0.843	0.108	0.104	0.073	0.293	0.939	0.574
0.580	0.688	0.791	0.934	0.217	0.465	0.540	0.871	0.818
0.190	0.876	0.418	0.678	0.694	0.849	0.628	0.158	0.329
0.759	0.420	0.689	0.598	0.145	0.301	0.260	0.392	0.367
0.761	0.071	0.187	0.598	0.350	0.563	0.864	0.245	0.502
0.759	0.052	0.144	0.260	0.936	0.528	0.484	**0.011**	**0.031**
0.066	**0.021**	**0.013**	0.858	0.392	0.682	0.194	**0.006**	**0.009**
0.504	**0.043**	0.102	0.934	0.756	0.949	0.432	**0.041**	0.091
0.668	0.718	0.855	0.997	0.122	0.303	0.663	0.294	0.524
0.761	0.071	0.187	0.225	0.807	0.465	0.428	**0.049**	0.107
0.668	0.089	0.214	0.858	0.714	0.920	0.663	0.794	0.879
0.853	0.055	0.157	0.599	**0.012**	**0.038**	0.432	0.420	0.531
0.421	0.628	0.643	0.858	**0.002**	**0.008**	0.249	0.867	0.507
0.856	0.108	0.270	0.795	**0.000**	**0.000**	0.129	0.052	**0.048**
0.497	0.658	0.720	0.288	0.431	0.417	0.260	0.760	0.505
0.148	0.120	0.105	0.482	0.057	0.128	0.331	0.362	0.412
0.954	0.065	0.181	0.997	0.726	0.940	0.864	0.101	0.258
0.954	0.558	0.841	0.484	0.279	0.435	0.790	0.640	0.865
0.049	0.372	0.096	0.377	**0.046**	0.093	0.108	**0.006**	**0.006**
0.351	0.058	0.107	0.331	**0.000**	**0.000**	0.377	**0.001**	**0.002**
0.856	0.780	0.946	0.864	**0.003**	**0.012**	0.934	0.771	0.955
0.136	0.698	0.306	0.297	**0.000**	**0.000**	0.121	0.271	0.164
0.497	0.247	0.406	0.728	0.847	0.924	0.428	0.226	0.350
0.290	0.101	0.149	0.384	0.683	0.629	0.288	0.063	0.101
0.421	0.093	0.176	0.599	**0.031**	0.084	0.482	0.358	0.512
0.235	0.799	0.478	0.997	**0.011**	**0.038**	0.858	0.458	0.747
0.088	0.918	0.232	**0.039**	**0.001**	**0.001**	**0.039**	0.714	0.110

and Geo-test denotes the geometric distribution test. Rejection at 5% level is marked in

Table 12.5. AVaR-based statistics of the out-of-sample performance of the GARCH-EVT with $\xi = 0$, period from January 2003 to June 2013

				Left-1%			Right-1%		
	Avg $\hat{\sigma}_t$	Avg $\hat{\beta}_L$	Avg $\hat{\beta}_R$	Avg AVaRf	Avg C-Loss	t-test	Avg AVaRf	Avg C-Loss	t-test
Developed									
Austria	0.224	0.644	0.485	0.046	0.042	0.689	−0.039	−0.038	0.751
Australia	0.157	0.625	0.454	0.032	0.030	0.928	−0.026	−0.026	0.348
Belgium	0.186	0.599	0.473	0.037	0.032	0.184	−0.032	−0.032	0.513
Canada	0.156	0.602	0.430	0.032	0.031	0.735	−0.026	−0.028	0.959
Denmark	0.169	0.625	0.482	0.034	0.032	0.595	−0.030	−0.031	0.187
Finland	0.228	0.627	0.541	0.047	0.044	0.175	−0.042	−0.043	0.232
France	0.213	0.568	0.479	0.042	0.035	0.477	−0.037	−0.042	0.322
Germany	0.213	0.582	0.461	0.043	0.034	0.994	−0.037	−0.047	0.352
Hong Kong	0.221	0.585	0.530	0.044	0.037	0.086	−0.041	−0.044	0.872
Ireland	0.216	0.690	0.474	0.047	0.046	0.369	−0.036	−0.040	**0.018**
Italy	0.216	0.606	0.454	0.045	0.035	0.381	−0.036	−0.048	0.532
Japan	0.225	0.584	0.442	0.045	0.045	0.160	−0.038	−0.035	**0.026**
Netherlands	0.203	0.579	0.478	0.041	0.034	0.265	−0.035	−0.038	0.766
Norway	0.216	0.638	0.448	0.045	0.040	0.366	−0.036	−0.036	0.786
New Zealand	0.108	0.623	0.484	0.022	0.021	0.838	−0.019	−0.019	0.374
Portugal	0.169	0.616	0.554	0.035	0.031	0.234	−0.031	−0.030	0.577
Singapore	0.176	0.613	0.514	0.036	0.029	0.615	−0.032	−0.038	**0.039**
Spain	0.215	0.601	0.474	0.044	0.035	0.180	−0.037	−0.038	0.967
Sweden	0.202	0.630	0.500	0.042	0.035	0.811	−0.036	−0.038	0.355
Switzerland	0.168	0.641	0.478	0.035	0.030	0.928	−0.029	−0.030	0.868
UK	0.174	0.609	0.445	0.036	0.032	0.348	−0.029	−0.029	0.324
US	0.175	0.600	0.508	0.036	0.029	0.751	−0.031	−0.037	**0.049**
Emerging									
Argentina	0.288	0.682	0.627	0.061	0.063	0.498	−0.058	−0.056	0.491
Brazil	0.271	0.620	0.481	0.055	0.056	0.529	−0.048	−0.054	0.417
China	0.251	0.679	0.609	0.054	0.050	0.484	−0.049	−0.047	0.937
Chile	0.116	0.615	0.481	0.023	0.023	0.570	−0.021	−0.021	0.602
Czech Republic	0.211	0.657	0.522	0.044	0.043	0.729	−0.038	−0.042	0.272
Egypt	0.259	0.712	0.563	0.056	0.058	**0.049**	−0.049	−0.053	0.094
Hungary	0.244	0.544	0.554	0.046	0.049	0.336	−0.047	−0.047	0.969
Indonesia	0.217	0.740	0.531	0.048	0.049	0.200	−0.040	−0.046	0.069
India	0.226	0.695	0.450	0.050	0.049	0.640	−0.037	−0.039	0.670
Mexico	0.190	0.663	0.540	0.040	0.038	0.737	−0.035	−0.045	0.313
Malaysia	0.122	0.692	0.625	0.026	0.026	0.165	−0.025	−0.023	0.658
Philippines	0.205	0.678	0.590	0.043	0.045	0.958	−0.040	−0.041	0.492
Poland	0.197	0.602	0.538	0.039	0.038	0.365	−0.037	−0.038	0.759
Peru	0.215	0.619	0.597	0.042	0.046	0.189	−0.043	−0.036	0.770
Russia	0.314	0.711	0.520	0.069	0.071	0.255	−0.057	−0.067	0.155
South Korea	0.225	0.673	0.473	0.048	0.050	0.723	−0.039	−0.033	0.106
Taiwan	0.204	0.672	0.494	0.044	0.039	0.636	−0.036	−0.037	0.154
Thailand	0.203	0.666	0.535	0.042	0.051	0.324	−0.038	−0.043	0.554
Turkey	0.302	0.599	0.594	0.060	0.061	0.226	−0.059	−0.056	0.593

Avg denotes average, AVaRf denotes forecasted AVaR, and C-Loss denotes observed entire sample of estimated values. Rejection at 5% level is marked in bold. Adapted

GARCH-EVT model in the period 2003–2013 and p-values of the t-test.

GARCH-EVT with $\xi = 0$, period from January 2003 to June 2013											
Left-2.5%			Right-2.5%			Left-5%			Right-5%		
Avg AVaRf	Avg C-Loss	t-test	Avg AVaRf	Avg C-Loss	t-test	Avg AVaRf	Avg C-Loss	t-test	Avg AVaRf	Avg C-Loss	t-test
0.038	0.036	0.922	−0.033	−0.033	0.655	0.032	0.031	0.895	−0.028	−0.028	0.488
0.027	0.024	0.582	−0.022	−0.022	0.166	0.023	0.022	0.854	−0.019	−0.019	0.074
0.031	0.028	0.200	−0.027	−0.028	0.343	0.026	0.026	0.184	−0.023	−0.023	0.563
0.027	0.027	0.577	−0.022	−0.021	0.174	0.023	0.023	0.886	−0.019	−0.019	0.188
0.028	0.027	0.795	−0.025	−0.027	0.681	0.024	0.023	0.615	−0.022	−0.022	0.596
0.038	0.034	0.735	−0.035	−0.036	0.566	0.032	0.031	0.893	−0.030	−0.031	0.296
0.035	0.031	0.225	−0.031	−0.036	0.264	0.030	0.028	0.481	−0.027	−0.027	0.991
0.036	0.032	0.878	−0.031	−0.032	0.260	0.030	0.029	0.973	−0.027	−0.027	0.783
0.037	0.034	0.657	−0.034	−0.036	0.436	0.031	0.030	0.943	−0.029	−0.029	0.083
0.038	0.036	0.916	−0.031	−0.033	0.639	0.032	0.031	0.909	−0.027	−0.028	0.918
0.037	0.032	0.872	−0.030	−0.032	0.968	0.032	0.030	0.199	−0.026	−0.028	0.670
0.038	0.038	0.317	−0.032	−0.029	**0.010**	0.032	0.033	0.362	−0.028	−0.026	**0.024**
0.034	0.030	0.795	−0.030	−0.032	0.772	0.029	0.027	0.826	−0.025	−0.027	0.558
0.037	0.034	0.110	−0.031	−0.031	0.783	0.031	0.031	0.149	−0.027	−0.027	0.902
0.018	0.017	0.404	−0.016	−0.015	0.337	0.016	0.016	0.416	−0.014	−0.013	0.228
0.029	0.028	0.828	−0.026	−0.025	0.982	0.024	0.024	0.829	−0.022	−0.021	0.316
0.030	0.028	0.345	−0.027	−0.028	0.140	0.025	0.023	0.333	−0.023	−0.023	**0.029**
0.036	0.032	0.711	−0.032	−0.034	0.322	0.031	0.029	0.328	−0.027	−0.029	0.706
0.035	0.030	0.841	−0.030	−0.033	0.884	0.029	0.027	0.697	−0.026	−0.027	0.944
0.029	0.026	0.575	−0.024	−0.026	0.894	0.024	0.023	0.368	−0.021	−0.022	0.952
0.030	0.028	0.757	−0.025	−0.026	**0.030**	0.025	0.023	0.856	−0.022	−0.022	0.438
0.030	0.027	0.817	−0.026	−0.026	**0.016**	0.025	0.025	0.209	−0.022	−0.023	0.072
0.050	0.049	0.868	−0.048	−0.045	0.170	0.042	0.042	0.795	−0.040	−0.038	0.231
0.046	0.044	0.275	−0.040	−0.044	0.868	0.038	0.036	**0.018**	−0.035	−0.037	0.952
0.044	0.043	0.541	−0.041	−0.038	0.907	0.037	0.036	0.622	−0.034	−0.032	0.993
0.019	0.020	0.055	−0.017	−0.017	0.657	0.016	0.017	0.109	−0.015	−0.014	0.355
0.036	0.035	0.072	−0.032	−0.034	0.546	0.030	0.030	0.537	−0.027	−0.027	0.636
0.045	0.046	0.082	−0.040	−0.040	0.171	0.037	0.038	0.180	−0.034	−0.034	0.090
0.038	0.041	0.165	−0.039	−0.039	0.393	0.032	0.034	0.313	−0.033	−0.033	**0.028**
0.039	0.040	0.392	−0.033	−0.034	0.824	0.032	0.032	0.682	−0.029	−0.029	0.285
0.041	0.039	0.960	−0.032	−0.031	0.698	0.034	0.032	0.833	−0.027	−0.027	0.597
0.033	0.029	0.470	−0.030	−0.033	0.188	0.027	0.026	0.706	−0.025	−0.026	0.233
0.021	0.020	0.273	−0.021	−0.019	0.330	0.017	0.017	0.590	−0.017	−0.016	0.156
0.035	0.038	0.324	−0.033	−0.032	0.201	0.029	0.030	0.552	−0.028	−0.027	0.177
0.032	0.033	0.064	−0.031	−0.029	**0.006**	0.027	0.028	0.277	−0.027	−0.026	**0.024**
0.035	0.037	**0.028**	−0.035	−0.032	0.498	0.029	0.031	0.132	−0.030	−0.028	0.365
0.056	0.061	0.082	−0.048	−0.052	0.251	0.046	0.049	0.227	−0.041	−0.044	0.470
0.040	0.037	**0.028**	−0.033	−0.033	**0.038**	0.033	0.031	**0.002**	−0.028	−0.028	0.070
0.036	0.034	0.534	−0.030	−0.031	0.423	0.030	0.030	0.497	−0.026	−0.026	0.365
0.034	0.038	0.356	−0.031	−0.034	0.702	0.028	0.030	0.423	−0.027	−0.028	0.454
0.049	0.050	0.190	−0.049	−0.047	0.736	0.041	0.042	0.170	−0.041	−0.038	0.365

loss conditioned on the events of VaR exceedances. The averaging is performed over the
from Stoyanov *et al.* (2017)

The corresponding plots for the other markets are very similar. The time variation of $\hat{\xi}$ does not seem to be related in any way to the market conditions in this period. For example, the plot for the US reveals a downward trend while one would expect to see spikes around the financial crisis of 2008 if tail thickness spiked in this period.

The 95% confidence bounds of $\hat{\xi}$ on the plots in Figure 12.1 indicate that $\xi = 0$ is rarely rejected. Regarding differences between the left and the right tail behavior, $\hat{\xi}_L$ appears generally higher than $\hat{\xi}_R$ but because of the wide confidence bounds it is rarely possible to reject $\xi_L = \xi_R$.

The fact that $\xi = 0$ is rarely rejected in-sample justifies considering the GARCH-EVT model with the constraint $\xi_L = \xi_R = 0$ in the statistical estimation. The right panel of Table 12.2 provides the corresponding p-values of the VaR-based tests. The results imply no deterioration of the restricted risk model out-of-sample.

The right panel of Table 12.3 provides the AVaR-based statistics for the restricted model. Although the average forecasted AVaR (Avg AVaRf) and the average loss conditioned on the occurrence of VaR exceedances look similar, the t-test for the left tail indicates a deterioration in out-of-sample performance for eight markets (three developed and five emerging). The same test for the right tail shows results similar to the unrestricted risk model. Although the constraint $\xi = 0$ results in rejections for some markets, it is remarkable that the exponential tail is difficult to reject for the vast majority of them.

The period from January 2003 to June 2013

Since the out-of-sample performance of the restricted risk model does not deteriorate substantially over the full sample periods, we restrict our attention to the period from January 1, 2003 to June 28, 2013 which provides a common timeline for all markets except New Zealand and Singapore. The VaR-based results are provided in Table 12.4 and the AVaR-based statistics are provided in Table 12.5. Both tables are split into three panels corresponding to the three tail probabilities: left (1%), middle (2.5%), and right (5%).

Because EVT provides an asymptotic model for the tail, it is expected to work better for VaR and AVaR at low tail probabilities. Our results are consistent with other empirical papers[21] indicating the GARCH-EVT model works well at the 1% tail probability level. Higher tail probability levels lead to more rejections of the model, which is consistent with the theoretical framework. Nevertheless, it is remarkable that for half of the markets the model is acceptable simultaneously at the three quantile levels and for both tails.

Key Points

- In finance, extreme value theory has been traditionally applied to estimate probabilities of extreme losses or loss thresholds such that losses beyond them occur with a predefined small probability.
- Because the clustering of volatility causes fat tails in the unconditional return distribution, we apply extreme value theory to the residuals of an estimated GARCH model which allows studying the residual tail thickness; that is, after the clustering of volatility effect has been explained away.
- A celebrated result in extreme value theory states that as the threshold increases, the conditional probability of extreme losses beyond the threshold converges to the generalized Pareto distribution.
- The most common approach for testing for extreme values is the peak-over-threshold method.
- In combination with the probabilistic model, two measures of risk in evaluating asset returns can be used to perform out-of-sample tests on the predictive properties of the model at different tail probabilities: value-at-risk and average value-at-risk.

[21] For example, see Danielsson and de Vries (1997), Pownall and Koedij (1999), McNeil and Frey (2000), Bekiros and Georgoutsos (2005), Fernandez (2005), and Tolikas *et al.* (2007). Beyond large-cap equity indices, Fabozzi and Stoyanov (2018) apply a model similar to the one presented in this chapter to U.S. equity factors such as size, value, momentum, investment, and profitability and draw similar conclusions.

References

S.D. Bekiros and D.A. Georgoutsos. Estimation of value-at-risk by extreme value and conventional methods: A comparative evaluation of their predictive performance. *International Financial Markets, Institutions and Money*, 15:209–228, 2005.

V. Chavez-Demoulin and P. Embrechts. Smooth extremal models in finance. *Journal of Risk and Insurance*, 71(2):183–199, 2004.

P.F. Christoffersen. Evaluating interval forecasts. *International Economic Review*, 39(4):841–862, 1998.

J. Danielsson and C.G. de Vries. Beyond the sample: Extreme quantile and probability estimation. Working paper, London School of Economics, 1997.

P. Embrechts, C. Klüppelberg, and T. Mikosch. *Modelling Extremal Events for Insurance and Finance*. Springer, Berlin, 1997.

F.A. Fabozzi and S. Stoyanov. Tail Behavior of Equity Factor Returns. Working Paper, 2018.

E.F. Fama. The behavior of stock market prices. *Journal of Business*, 38(1):34–105, 1965.

V. Fernandez. Risk management under extreme events. *International Review of Financial Analysis*, 14:113–148, 2005.

D. Furió and F.J. Climent. Extreme value theory versus traditional GARCH approaches applied to financial data: A comparative evaluation. *Quantitative Finance*, 13(1):45–63, 2013.

G.D. Gettinby, C.D. Sinclair, D.M. Power, and R.A. Brown. An analysis of the distribution of extreme share returns in the UK from 1975 to 2000. *Journal of Business Finance and Accounting*, 31(5&6):607–645, 2004.

J. Gonzalo and J. Olmo. Which extreme values are really extreme? *Journal of Financial Econometrics*, 2(3):349–369, 2004.

C. Gouriéroux. *ARCH-Models and Financial Applications*. Springer, New York, 1997.

A. Jalal and M. Rockinger. Predicting tail-related risk measures: The consequences of using GARCH filters for non-GARCH data. *Journal of Empirical Finance*, 15:868–877, 2008.

E. Jondeau and M. Rockinger. Testing for differences in the tails of stock-market returns. *Journal of Empirical Finance*, 10:559–581, 2003.

P.H. Kupiec. Techniques for verifying the accuracy of risk measurement models. *Journal of Derivatives*, 3(2):73–84, 1995.

B. LeBaron and R. Samanta. Extreme value theory and fat tails in equity markets. 2005. Available at SSRN: http://ssrn.com/abstract=873656 or http://dx.doi.org/10.2139/ssrn.873656.

F. Longin. The choice of the distribution of asset returns: How extreme value theory can help? *Journal of Banking and Finance*, 29(4):1017–1035, 2005.

B.B. Mandelbrot. The variation of certain speculative prices. *Journal of Business*, 36:394–419, 1963.

A. McNeil, R. Frey, and P. Embrechts. *Quantitative risk management: Concepts, techniques, and tools.* Princeton University Press, Princeton, NJ, 2005.

A.J. McNeil and R. Frey. Estimation of tail-related risk measures for heteroscedastic financial time series: An extreme value approach. *Journal of Empirical Finance*, 7(3–4): 271–300, 2000.

S.H. Poon, M. Rockinger, and J. Tawn. Extreme value dependence in financial markets: Diagnostics, models, and financial implications. *The Review of Financial Studies*, 17(2):581–610, 2004.

R.A.J. Pownall and K.G. Koedij. Capturing downside risk in financial markets: The case of the Asian crisis. *Journal of International Money and Finance*, 18:853–870, 1999.

S.V. Stoyanov, L. Loh, and F.J. Fabozzi. How fat are the tails of equity market indices? *International Journal of Finance and Economics*, 22(3):181–200, 2017.

S.V. Stoyanov, S.T. Rachev, B. Racheva-Iotova, and F.J. Fabozzi. Fat-tailed models for risk estimation. *Journal of Portfolio Management*, 37(2):107–117, 2011.

K. Tolikas and G.D. Gettinby. Modelling the distribution of the extreme share returns in Singapore. *Journal of Empirical Finance*, 16:254–263, 2009.

K. Tolikas, A. Koulakiotis, and R.A. Brown. Extreme risk and value-at-risk in the German stock market. *European Journal of Finance*, 13(4): 373–395, 2007.

How to Run the Matlab Code

There is no perfect programming language available and it is always a challenge when it comes to choosing a particular one. We selected Matlab as the programming language in this book since it is the programming language we have used in our many research projects. R and Python are now a good alternative, but may need a little bit more work (i.e., more code is required) and it is not simple to obtain the speed Matlab has in vectorized operations. However, R and Python are very flexible, free, and many third-party packages/ libraries are available. The Julia programming language seems to be another promising alternative, but it has less libraries (written in Julia itself) available in comparison to R and Python. The language itself is under development (i.e., it is only recently that the version 1.0 has been released).[1] Finally, GNU Octave is an open-source alternative to Matlab with syntax which is largely compatible with it.

We provide here a brief outline on how to use the Matlab code described in this book. The actual implementation of the features we

[1]The Jupyter Notebook is an open-source web application to create and share documents containing code, equations, visualizations, and narrative text. The name itself comes from three languages: Ju(lia) + Py(thon) + (e)R. It is an interactive computing environment which integrates on-the-fly code execution and a display of the results: a very different programming approach compared to what students are taught at many universities until a few years ago. As alternatives, RStudio is a good integrated development environment (IDE) when programming in R and similarly Spyder when programming in Python.

describe is illustrated in the chapters of the book. A useful online documentation is available in the Matlab environment. The reader should take advantage of this whenever a piece of code in the book is unclear.

The Matlab code is available for download at the links

`https://www.worldscientific.com/worldscibooks/10.1142/11118#t=suppl`

`https://www.mathworks.com/matlabcentral/fileexchange/`

and

`https://sites.google.com/site/micheleleonardobianchiwebpage/`.

For each chapter, the corresponding scripts are in the folder with prefix

matlab_code_Ch

and with the number of the corresponding chapter. For simplicity, it is better to add all folders to the path by setting the path by clicking on the *Set path* button in the Matlab graphical interface. This is needed because in some cases the scripts described in a chapter consider the functions implemented in another chapter.

With the exception of Chapters 10 and 12, the reader should be able to replicate the results described in the book. In Chapters 10 and 12 we do not provide detailed instructions to replicate results, though all pieces of code are available.

In Chapter 7 it is possible to use the code without the need for additional data. Although the same is true for Chapter 8, an initial guess for model parameters is needed to evaluate the density to simulate the non-Gaussian multivariate models proposed in the chapter. In Chapters 8, 10, 11, and 12 some empirical applications are provided. These applications consider the log-returns of the main European stock indexes, all German stocks included in the DAX index, all Euro-denominated stocks included in the EuroStoxx 50 index, and 22 developed and 19 emerging equity market indices. We do not provide these data and the estimates based on these data. Stock prices and index values can be downloaded from Bloomberg or Reuters if the user has a license or these can be freely downloaded from Google or Yahoo!Finance. These activities can be performed automatically by using directly the Matlab programming language, or by downloading the *csv* file from the Yahoo!Finance website. For example, to download the time series of stock returns considered in Chapter 11, the following website can be considered:

`https://finance.yahoo.com/quote/DBK.DE/history?p=DBK.DE`.

After having selected the time period, the frequency, and the data type (i.e., historical prices, dividends only or stock splits), by clicking on *Download Data*, it is possible to download the data.

In Chapter 11, stock prices are stored in the file *historicalDataDB.mat*. The code described in the chapter provides all of the necessary data. In this case, the results can be replicated.

License

BSD 3-Clause License

Copyright © 2018 Michele Leonardo Bianchi, Stoyan V. Stoyanov, Gian Luca Tassinari, Frank J. Fabozzi, Sergio M. Focardi

All rights reserved.

Redistribution and use in source and binary forms, with or without modification, are permitted provided that the following conditions are met:

- Redistributions of source code must retain the above copyright notice, this list of conditions and the following disclaimer.
- Redistributions in binary form must reproduce the above copyright notice, this list of conditions and the following disclaimer in the documentation and/or other materials provided with the distribution.
- Neither the name of the copyright holder nor the names of its contributors may be used to endorse or promote products derived from this software without specific prior written permission.

THIS SOFTWARE IS PROVIDED BY THE COPYRIGHT HOLDERS AND CONTRIBUTORS "AS IS" AND ANY EXPRESS OR IMPLIED WARRANTIES, INCLUDING, BUT NOT LIMITED TO, THE IMPLIED WARRANTIES OF MERCHANTABILITY AND FITNESS FOR A PARTICULAR PURPOSE ARE DISCLAIMED. IN NO EVENT SHALL THE COPYRIGHT HOLDER

OR CONTRIBUTORS BE LIABLE FOR ANY DIRECT, INDIRECT, INCIDENTAL, SPECIAL, EXEMPLARY, OR CONSEQUENTIAL DAMAGES (INCLUDING, BUT NOT LIMITED TO, PROCUREMENT OF SUBSTITUTE GOODS OR SERVICES; LOSS OF USE, DATA, OR PROFITS; OR BUSINESS INTERRUPTION) HOWEVER CAUSED AND ON ANY THEORY OF LIABILITY, WHETHER IN CONTRACT, STRICT LIABILITY, OR TORT (INCLUDING NEGLIGENCE OR OTHERWISE) ARISING IN ANY WAY OUT OF THE USE OF THIS SOFTWARE, EVEN IF ADVISED OF THE POSSIBILITY OF SUCH DAMAGE.

Author Index

D

E

F

G

H

Subject Index

D

E

L

M

N

T

U

V

W

Z

www.ingramcontent.com/pod-product-compliance
Lightning Source LLC
LaVergne TN
LVHW010539300926
849162LV00002B/175